Developments in Primatology: Progress and Prospects

Series Editor

Louise Barrett, Lethbridge, Canada

More information about this series at http://www.springer.com/series/5852

Martín M. Kowalewski • Paul A. Garber
Liliana Cortés-Ortiz • Bernardo Urbani
Dionisios Youlatos

Editors

Howler Monkeys

Adaptive Radiation, Systematics, and Morphology

Editors
Martín M. Kowalewski
Estación Biológica Corrientes
Museo Argentino de Ciencias Naturales
Consejo Nacional de Investigaciones
  Científicas y Técnicas (CONICET)
Buenos Aires, Argentina

Liliana Cortés-Ortiz
Museum of Zoology
Department of Ecology and Evolutionary
  Biology
University of Michigan
Ann Arbor, MI, USA

Dionisios Youlatos
Department of Zoology
Aristotle University of Thessalonik
Thessaloniki, Greece

Paul A. Garber
Department of Anthropology
University of Illinois at Urbana-Champaign
Urbana, IL, USA

Bernardo Urbani
Centro de Antropología
Instituto Venezolano de Investigaciones
  Científicas
Caracas, Venezuela

ISBN 978-1-4939-1956-7   ISBN 978-1-4939-1957-4 (eBook)
DOI 10.1007/978-1-4939-1957-4
Springer New York Heidelberg Dordrecht London

Library of Congress Control Number: 2014955676

© Springer Science+Business Media New York 2015
This work is subject to copyright. All rights are reserved by the Publisher, whether the whole or part of the material is concerned, specifically the rights of translation, reprinting, reuse of illustrations, recitation, broadcasting, reproduction on microfilms or in any other physical way, and transmission or information storage and retrieval, electronic adaptation, computer software, or by similar or dissimilar methodology now known or hereafter developed. Exempted from this legal reservation are brief excerpts in connection with reviews or scholarly analysis or material supplied specifically for the purpose of being entered and executed on a computer system, for exclusive use by the purchaser of the work. Duplication of this publication or parts thereof is permitted only under the provisions of the Copyright Law of the Publisher's location, in its current version, and permission for use must always be obtained from Springer. Permissions for use may be obtained through RightsLink at the Copyright Clearance Center. Violations are liable to prosecution under the respective Copyright Law.
The use of general descriptive names, registered names, trademarks, service marks, etc. in this publication does not imply, even in the absence of a specific statement, that such names are exempt from the relevant protective laws and regulations and therefore free for general use.
While the advice and information in this book are believed to be true and accurate at the date of publication, neither the authors nor the editors nor the publisher can accept any legal responsibility for any errors or omissions that may be made. The publisher makes no warranty, express or implied, with respect to the material contained herein.

Printed on acid-free paper

Springer is part of Springer Science+Business Media (www.springer.com)

*M.K.: To Mariana and Bruno,
and to my parents and sisters*

*P.A.G.: To Seymour, Sylvia, Jenni, Sara,
and of course Chrissie*

L.C.O.: To Paloma, Tommy, and Tom

B.U.: To my family, to my mentors, to Padmini

*D.Y.: To Evangelos (Sr. and Jr.), Alexandra,
and Ioanna*

Foreword

It is a privilege to be able to study wild howler monkeys, and an honor to have been invited to write the foreword to this volume of collected papers about them. Thank you to the howlers—everything I know, I learned from you—and thank you to the volume editors for this invitation and to all of the Latin American countries which so generously enabled me carry out research on howler monkeys in their forests.

Alouatta is considered the most successful New World primate genus in terms of ecological dominance as defined by overall biomass. Howler monkeys have a wide geographical distribution, which extends from southern Mexico through Central and South America and into northern Argentina. Their great success as a genus stems in large part from their ability, unusual in a neotropical primate, to use leaves as a primary food source. Fruits and flowers are also popular howler foods but it is their ability to survive for long periods on diets consisting largely of leaves that underlies their great ecological success. This ability has enabled howler monkeys to occupy a tremendous diversity of habitat types throughout the neotropics and to survive in small forest fragments that could not support other primates.

Despite being known for the loud sonorous howling vocalization produced by adult males, howler monkeys are quite subtle, secretive, and quiet monkeys most of the time. They spend a high percentage of their daylight hours throughout the year quietly resting or sleeping to conserve energy—this inactivity is an important feature of their overall foraging strategy. I have studied howler monkeys now for more than 40 years and to me they remain endlessly complex, fascinating, and endearing study subjects. I remember my parents asking me, after a decade or so of howler research and many missed holidays and family celebrations, if I hadn't answered just about all the questions that could possibly be asked about howler monkeys. But as the collection of papers in this volume clearly shows, there is thankfully no end in sight to the array of interesting questions that can be posed about members of the genus *Alouatta*.

Because of howlers' wide distribution and abundance, it's no surprise that over the decades an unusually large number of primatologists have carried out research on wild howler monkeys such that, today, it is considered one of the best studied of all primate genera. Indeed, though perhaps not well appreciated, the first successful systematic naturalistic study of *any* wild primate anywhere in the world was carried out on howler monkeys. In the early 1930s, C. Ray Carpenter travelled from the USA to Panama to begin a field study of mantled howler monkeys (*Alouatta palliata*) on Barro Colorado Island (BCI). Carpenter was motivated to study wild monkeys because he was convinced that a better understanding of primate behavior in the natural environment would provide important insights into key features of human biology and behavior. Though earlier attempts had been made to try and study wild chimpanzees and mountain gorillas, these study subjects proved elusive and little information was compiled. In contrast, Carpenter was able to spend long periods of time both in 1932 and 1933 observing howler monkeys at close range and amassed a wealth of detailed behavioral information. He also collected and identified many important howler food species and censused all individuals in every howler group on BCI two times during his fieldwork—providing invaluable baseline data for future studies of howler monkey population dynamics at this site. Carpenter produced an excellent and meticulously organized scholarly monograph from his field study, one filled with original information about howler monkey behavioral ecology—information as valid and interesting today as it was in 1934 when his original monograph *A Field Study of The Behavior and Social Relations of Howling Monkeys* was published. To say Carpenter was decades ahead of his time does not begin to do him justice.

After Carpenter's pioneering field study, world events intervened, leading to a hiatus in howler monkey research. But in the 1950s and 1960s, a number of young researchers followed in Carpenter's footsteps and travelled to BCI to observe howler monkeys—though generally only for short periods of time. During this period, field studies were also begun on red howlers at Hato Masaguaral in Venezuela and in 1972 on mantled howler monkeys at La Pacifica in Costa Rica. My howler monkey research began in 1974. Barro Colorado Island was an ideal study site because I was interested in dietary questions and by that time, a considerable amount of information was available on features of the BCI forest and the island had an excellent herbarium—essential tools for a dietary study. During my initial fieldwork, there were no other primate researchers on the island. But by 1978, a few short years later, primate field studies had begun to take off and so many graduate students began arriving on BCI to examine one or another attribute of howler monkeys that often we had to take turns collecting data on the more popular study groups near the laboratory buildings.

Though we now know a great deal more about howler monkeys than we did in the Carpenter's day, we still have much to discover about this engaging New World primate. Answering important questions about the ecology and behavior of living systems generally involves a long investment of time and effort and many years of continuous study at particular research sites. It is ironic that as this fact has become more apparent, funding to support long-term field studies has become increasingly difficult to secure. If our knowledge of living systems such as howler monkeys is to progress, researchers must not only ask the interesting questions but also have the time to compile the data needed to answer them. To enable the relevant studies to be carried out, however, our first task is to ensure the successful conservation of howler monkeys and their habitats throughout the neotropics. Without the habitats and monkeys, funding will not be necessary as there will be nothing left for us to study.

Editors Martín Kowalewski, Paul Garber, Liliana Cortés-Ortiz, Bernardo Urbani, and Dionisios Youlatos are to be commended for this timely and informative two-volume series on the genus *Alouatta*. What is particularly special and impressive about this and its companion volume is the international roster of countries represented by the volume's contributors and, in particular, the welcome contribution of so many Latin American scholars. This cohort of productive and dedicated Latin American primatologists represents the single most profound change that has occurred in my 40 years of studying wild howler monkeys. Every country in Latin America but Chile and perhaps Uruguay hosts at least one species of *Alouatta* and some countries are host to two, three, or more howler species. The conservation future of howlers and their habitats depends on the knowledge and expertise of these local Latin American scholars, who are in the best position to validate the importance of howler conservation and who understand the politics and policies of their own countries. Their influence is necessary to develop those policies and implement those decisions about conservation areas which will serve to ensure the survival of all howler monkey species into the indefinite future.

Environmental Science, Policy, & Management   Katharine Milton
University of California
Berkeley, CA, USA

Acknowledgment

We acknowledge and gratefully thank the following scholars for reviewing earlier drafts of chapters in this volume: John Fleagle, Luciana Oklander, Montserrat Ponsà, Dietmar Zinner, Alfred Rosenberger, Karina Keller Marques da Costa Flaibán, Melissa Emery Thompson, G. Rex Gaskins, Sylvia Vitazkova, Gregory Blomquist, Valdir Filgueiras Pessoa, Mario S. Di Bitetti, Eduardo Fernández-Duque, and Charles Snowdon.

Contents

Part I Introduction

1　**Why Is It Important to Continue Studying the Anatomy, Physiology, Sensory Ecology, and Evolution of Howler Monkeys?** 3
   Martín M. Kowalewski, Paul A. Garber, Liliana Cortés-Ortiz, Bernardo Urbani, and Dionisios Youlatos

Part II Taxonomy, Genetics, Morphology and Evolution

2　**Fossil Alouattines and the Origins of *Alouatta*: Craniodental Diversity and Interrelationships** 21
   Alfred L. Rosenberger, Siobhán B. Cooke, Lauren B. Halenar, Marcelo F. Tejedor, Walter C. Hartwig, Nelson M. Novo, and Yaneth Muñoz-Saba

3　**The Taxonomy of Howler Monkeys: Integrating Old and New Knowledge from Morphological and Genetic Studies** 55
   Liliana Cortés-Ortiz, Anthony B. Rylands, and Russell A. Mittermeier

4　**Cytogenetics of Howler Monkeys** 85
   Marta D. Mudry, Mariela Nieves, and Eliana R. Steinberg

5　**Hybridization in Howler Monkeys: Current Understanding and Future Directions** 107
   Liliana Cortés-Ortiz, Ilaria Agostini, Lucas M. Aguiar, Mary Kelaita, Felipe Ennes Silva, and Júlio César Bicca-Marques

6　**Morphology of Howler Monkeys: A Review and Quantitative Analyses** 133
   Dionisios Youlatos, Sébastien Couette, and Lauren B. Halenar

Part III Physiology

7 Hematology and Serum Biochemistry in Wild Howler Monkeys 179
Domingo Canales-Espinosa, María de Jesús Rovirosa-Hernández, Benoit de Thoisy, Mario Caba, and Francisco García-Orduña

8 Endocrinology of Howler Monkeys: Review and Directions for Future Research 203
Sarie Van Belle

9 The Howler Monkey as a Model for Exploring Host-Gut Microbiota Interactions in Primates 229
Katherine R. Amato and Nicoletta Righini

10 Ecological Determinants of Parasitism in Howler Monkeys 259
Rodolfo Martínez-Mota, Martín M. Kowalewski, and Thomas R. Gillespie

Part IV Ontogeny and Sensory Ecology

11 An Ontogenetic Framework for *Alouatta*: Infant Development and Evaluating Models of Life History 289
Melissa Raguet-Schofield and Romina Pavé

12 The Sensory Systems of *Alouatta*: Evolution with an Eye to Ecology 317
Laura T. Hernández Salazar, Nathaniel J. Dominy, and Matthias Laska

13 Production of Loud and Quiet Calls in Howler Monkeys 337
Rogério Grassetto Teixeira da Cunha, Dilmar Alberto Gonçalves de Oliveira, Ingrid Holzmann, and Dawn M. Kitchen

14 Function of Loud Calls in Howler Monkeys 369
Dawn M. Kitchen, Rogério Grassetto Teixeira da Cunha, Ingrid Holzmann, and Dilmar Alberto Gonçalves de Oliveira

Part V Conclusions

15 New Challenges in the Study of Howler Monkey Anatomy, Physiology, Sensory Ecology, and Evolution: Where We Are and Where We Need to Go? 403
Dionisios Youlatos, Martín M. Kowalewski, Paul A. Garber, and Liliana Cortés-Ortiz

Subject Index 415

Taxonomic Index 421

Contributors

Ilaria Agostini Facultad de Ciencias Forestales, Consejo Nacional de Investigaciones Científicas y Técnicas de Argentina (CONICET)—Instituto de Biología Subtropical, Universidad Nacional de Misiones, Puerto Iguazú, Argentina

CeIBA (Centro de Investigaciones del Bosque Atlántico), Puerto Iguazú, Argentina

Lucas M. Aguiar Universidade Federal da Integração Latino-Americana (Unila), Foz do Iguaçu, Brazil

Katherine R. Amato Program in Ecology, Evolution, and Conservation Biology, University of Illinois at Urbana-Champaign, Champaign, IL, USA

Department of Anthropology, University of Colorado Boulder, Boulder, CO, USA

Júlio Cesar Bicca-Marques Faculdade de Biociências, Departamento de Biodiversidade e Ecologia, Pontifícia Universidade Católica do Rio Grande do Sul, Porto Alegre, Rio Grande do Sul, Brazil

Mario Caba Centro de Investigaciones Biomédicas, Universidad Veracruzana, Xalapa, Veracruz, Mexico

Domingo Canales-Espinosa Instituto de Neuroetología, Universidad Veracruzana, Xalapa, Veracruz, México

Siobhán B. Cooke Department of Anthropology, Northeastern Illinois University, Chicago, IL, USA

Liliana Cortés-Ortiz Museum of Zoology, Department of Ecology and Evolutionary Biology, University of Michigan, Ann Arbor, MI, USA

Sébastien Couette Ecole Pratique des Hautes Etudes, Laboratoire Paléobiodiversité et Evolution & UMR uB/CNRS 6282 "Biogéosciences", Dijon, France

Rogério Grassetto Teixeira da Cunha Instituto de Ciências da Natureza- Universidade Federal de Alfenas, Alfenas, Minas Gerais, Brazil

Benoit de Thoisy Laboratoire des Interactions Virus Hôtes, Institut Pasteur de la Guyane, Cayenne, French Guiana

Nathaniel J. Dominy Department of Anthropology, Dartmouth College, Hanover, NH, USA

Paul A. Garber Department of Anthropology, University of Illinois at Urbana-Champaign, Urbana, IL, USA

Francisco García-Orduña Instituto de Neuroetología, Universidad Veracruzana, Xalapa, Veracruz, México

Thomas R. Gillespie Departments of Environmental Sciences and Environmental Health, Emory University, Atlanta, GA, USA

Dilmar Alberto Gonçalves de Oliveira Departamento de Fauna/CBRN, Centro de Manejo de Fauna Silvestre, Secretaria do Meio Ambiente do Estado de São Paulo, São Paulo, Brazil

Lauren B. Halenar Department of Biological Sciences, Bronx Community College, The City University of New York, Bronx, NY, USA

New York Consortium in Evolutionary Primatology (NYCEP), The City University of New York, New York, NY, USA

Walter C. Hartwig Department of Basic Sciences, Touro University College of Osteopathic Medicine Mare Island, Vallejo, CA, USA

Ingrid Holzmann CONICET (Consejo Nacional de Investigaciones Científicas y Técnicas), IBS (Instituto de Biología Subtropical), CeIBA (Centro de Investigaciones del Bosque Atlántico), Buenos Aires, Argentina

Mary Kelaita Department of Anthropology, University of Texas at San Antonio, San Antonio, TX, USA

Dawn M. Kitchen Department of Anthropology, The Ohio State University, Columbus, OH, USA

Department of Anthropology, The Ohio State University-Mansfield, Mansfield, OH, USA

Martín M. Kowalewski Estación Biológica Corrientes, Museo Argentino de Ciencias Naturales, Consejo Nacional de Investigaciones Científicas y Técnicas (CONICET), Buenos Aires, Argentina

Matthias Laska IFM Biology, Section of Zoology, Linköping University, Linköping, Sweden

Rodolfo Martínez-Mota Department of Anthropology, University of Illinois at Urbana-Champaign, Urbana, IL, USA

Russell A. Mittermeier Center for Applied Biodiversity Science, Conservation International, Arlington, Washington, DC, USA

Contributors

Marta D. Mudry Grupo de Investigación en Biología Evolutiva (GIBE), Labs 46/43, 4° piso, Depto. de Ecología, Genética y Evolución, IEGEBA, Facultad de Ciencias Exactas y Naturales, Universidad de Buenos Aires, Ciudad Autónoma de Buenos Aires, Argentina

Consejo de Investigaciones Científicas y Técnicas (CONICET), Ciudad Autónoma de Buenos Aires, Argentina

Yaneth Muñoz-Saba Instituto de Ciencias Naturales, Universidad Nacional de Colombia, Bogotá, Colombia

Stephen Nash Department of Art, State University of New York & Conservation International, Stony Brook, NY, USA

Mariela Nieves Grupo de Investigación en Biología Evolutiva (GIBE), Labs 46/43, 4° piso, Depto. de Ecología, Genética y Evolución, IEGEBA, Facultad de Ciencias Exactas y Naturales, Universidad de Buenos Aires, Ciudad Autónoma de Buenos Aires, Argentina

Consejo de Investigaciones Científicas y Técnicas (CONICET), Ciudad Autónoma de Buenos Aires, Argentina

Nelson M. Novo Centro Nacional Patagónico-CONICET, Puerto Madryn, Provincia de Chubut, Argentina

Romina Pavé Instituto Nacional de Limnología (INALI), Consejo Nacional de Investigaciones Científicas y Técnicas (CONICET), Santa Fe, Argentina

Melissa Raguet-Schofield Department of Anthropology, University of Illinois, Urbana, IL, USA

Nicoletta Righini Department of Anthropology, University of Illinois at Urbana-Champaign, Champaign, IL, USA

Instituto de Ecologia, A.C., Xalapa, Veracruz, Mexico

Alfred L. Rosenberger Department of Anthropology and Archaeology, Brooklyn College, The City University of New York, Brooklyn, NY, USA

The Graduate Center, The City University of New York, New York, NY, USA

New York Consortium in Evolutionary Primatology (NYCEP), New York, NY, USA

Department of Mammalogy, The American Museum of Natural History, New York, NY, USA

María de Jesús Rovirosa-Hernández Instituto de Neuroetología, Universidad Veracruzana, Xalapa, Veracruz, México

Anthony B. Rylands Center for Applied Biodiversity Science, Conservation International, Arlington, Washington, DC, USA

Felipe Ennes Silva Instituto de Desenvolvimento Sustentável Mamirauá, Tefé, Brazil

Laura T. Hernández Salazar Instituto de Neuroetología, Universidad Veracruzana, Xalapa, Veracruz, Mexico

Eliana R. Steinberg Grupo de Investigación en Biología Evolutiva (GIBE), Labs 46/43, 4° piso, Depto. de Ecología, Genética y Evolución, IEGEBA, Facultad de Ciencias Exactas y Naturales, Universidad de Buenos Aires, Ciudad Autónoma de Buenos Aires, Argentina

Consejo de Investigaciones Científicas y Técnicas (CONICET), Ciudad Autónoma de Buenos Aires, Argentina

Marcelo F. Tejedor Centro Nacional Patagónico-CONICET, Puerto Madryn, Provincia de Chubut, Argentina

Bernardo Urbani Centro de Antropología, Instituto Venezolano de Investigaciones Científicas, Caracas, Venezuela

Sarie Van Belle Instituto de Biología, Universidad Nacional Autónoma de México, Coyoacan, Distrito Federal, Mexico

Dionisios Youlatos Department of Zoology, School of Biology, Aristotle University of Thessaloniki, Thessaloniki, Greece

Part I
Introduction

Chapter 1
Why Is It Important to Continue Studying the Anatomy, Physiology, Sensory Ecology, and Evolution of Howler Monkeys?

Martín M. Kowalewski, Paul A. Garber, Liliana Cortés-Ortiz, Bernardo Urbani, and Dionisios Youlatos

1.1 Introduction

The goals of this first chapter to our volume *Howler Monkeys*: *Adaptive Radiation, Systematics, and Morphology* are to highlight the importance of morphological, genetic, and physiological studies for understanding the evolutionary adaptations of this highly successful genus. Many questions continue to exist regarding the systematics, anatomy, and physiology of *Alouatta*. Despite being one of the most commonly studied primate taxa in the Neotropics, the number of howler species is unresolved, and the distribution of many species and subspecies is poorly documented. Several attempts have been made to evaluate howler monkey taxonomic

---

M.M. Kowalewski (✉)
Estación Biológica Corrientes, Museo Argentino de Ciencias Naturales, Consejo Nacional de Investigaciones Científicas y Técnicas (CONICET), Buenos Aires, Argentina
e-mail: martinkow@gmail.com

P.A. Garber
Department of Anthropology, University of Illinois at Urbana-Champaign, Urbana, IL, USA
e-mail: p-garber@illinois.edu

L. Cortés-Ortiz
Museum of Zoology, Department of Ecology and Evolutionary Biology, University of Michigan, Ann Arbor, MI, USA
e-mail: lcortes@umich.edu

B. Urbani
Centro de Antropología, Instituto Venezolano de Investigaciones Científicas, Caracas, Venezuela
e-mail: bernardourbani@yahoo.com

D. Youlatos
Department of Zoology, School of Biology, Aristotle University of Thessaloniki, Thessaloniki, Greece
e-mail: dyoul@bio.auth.gr

diversity based on morphological (e.g., Hill 1962; Gregorin 2006), cytogenetic (e.g., de Oliveira et al. 2002; Steinberg et al. 2014), and molecular (e.g., Bonvicino et al. 2001; Cortés-Ortiz et al. 2003) analyses. More recently, several authors have attempted to integrate available information on genetics, morphology, and biogeography to provide a more comprehensive view of the systematics of this genus (e.g., Groves 2001, 2005; Rylands et al. 2000, 2006; Rylands and Mittermeier 2009; Glander and Pinto 2013; Cortés-Ortiz et al. 2014a, b), identifying two primary radiations: one that originated Amazonian/Atlantic Forest howlers and the other that produced all Central American taxa. Nonetheless, a better understanding of the diversity of howler monkeys will only be possible by continuing integrating different types of data for the same individuals using a thorough sampling across their wide distribution. Howlers (genus *Alouatta*) are distributed from 21°N to 30°S in Central and South America. They occupy the widest range of habitats of any Neotropical genus and are among the most dimorphic of New World primates in body mass and color patterns (Wolfheim 1983; Crockett and Eisenberg 1987; Emmons and Feer 1990; Nowak 1999; Groves 2001; Di Fiore et al. 2010). In all species males are at least 25 % heavier than females and two taxa, *A. caraya* and *A. guariba*, are dichromatic (Crockett and Eisenberg 1987; Neville et al. 1988). There are currently 12 recognized or putatively recognized species in the genus *Alouatta*.

Howlers are found from sea level to ≥3,200 m occupying diverse habitat types from closed canopy wet evergreen forests, including "terra firme" and inundated swamp forests, to open, highly seasonal deciduous and semideciduous woodlands, gallery forests, and llanos habitats containing patches of relatively low trees in open savannah (Crockett and Eisenberg 1987; Camacho and Defler 1985; Wolfheim 1983; Brown and Zunino 1994). Howlers are principally arboreal; however, several species that live in drier areas come to the ground and cross open areas between patches of forest (Crockett 1998; Di Fiore et al. 2010).

Several studies of howler anatomy have been published. These have focused on dental and cranial anatomy, the hyoid apparatus, and the prehensile tail. The study of howler anatomy has been used as a comparative framework for the study of ateline adaptations, studies of parallel evolution of the prehensile tail in atelines and cebines, and as a model to investigate morphological adaptations of Miocene hominoids. Outstanding early examples of anatomical research in howlers include the monograph by W.C.O. Hill (1962), comparison of black howlers (*A. villosa* = *A. pigra*) with other cebid platyrrhines, as part of his comprehensive volume devoted to the group (Hill 1962), and the detailed monograph by Schön (1968) on the muscular anatomy of the ursine howler (*A. arctoidea*). In that same year, M.R. Malinow edited a volume on the biology of *A. caraya*, as part of the series *Biblioteca Primatologica* (Malinow 1968). This volume included chapters on the functional anatomy, skeletal development, ontogeny, hematology, soft tissue anatomy, and general pathology of the black and gold howler monkey.

Following these publications, Schön and colleagues produced a series of papers examining the appendicular, cranial, and hyoid anatomy of the red howlers (Schön 1968; Schön Ybarra 1984, 1998; Schön Ybarra and Schön 1987). The impact of these publications for evaluating the morphological adaptations of howler monkeys is discussed by Youlatos et al. (2014, see also below). During the same period (mid- and late 1970s to early 1980s), Stern and collaborators explored the functional anatomy and positional behavior of howlers and other atelines through detailed comparative anatomy and electromyography. These authors established the framework of using howler monkeys as morphobehavioral analogs for studies of early hominoids (Stern 1971; Stern et al. 1980). This was followed by an increase in publications on howler cranial, hyoid, and appendicular morphology, in which the genus was considered as integral for understanding the adaptive radiation of platyrrhines, and more particularly of the highly apomorphic atelines, with whom they share large body mass, a prehensile tail, and adaptations to a suspensory way of life (e.g., Rosenberger and Strier 1989). Recent studies have used new morphometric analyses (e.g., geometric morphometrics) to better understand howler functional anatomy. New fossil material has firmly identified the ancestral group that gave rise to modern *Alouatta* (see Rosenberger et al. 2014). This research has revealed significant differences in cranial anatomy, the shape of the hyoid, and specific characters of the long bones across howler species, suggesting that the morphology of *Alouatta* is more variable than previously considered. Ongoing and future studies will need to focus on evidence for ages, sex, and populational differences in functional and evolutionary correlates of howler basicranial morphology, degree of airorynchy, skull size, hyoid shape and size, long bone robusticity, shape of proximal and distal humeral, femoral articular facets, and the morphology of the carpals and tarsals to better understand the adaptive radiation of the genus.

A major goal of this volume is to review and evaluate the current data on howler endocrinology (see Van Belle 2014), their gut microbiome (see Amato and Righini 2014), sensory and communication systems (vision, auditory, and vocal) (see Hernández-Salazar et al. 2014; da Cunha et al. 2014, Kitchen et al. 2014), parasitology (Martínez-Mota et al. 2014), and nutritional ecology (see Garber et al. 2014). An understanding of howler monkeys anatomy and physiology provides a framework for examining how social (Kowalewski and Garber 2014) and reproductive strategies (Van Belle 2014; Van Belle and Bicca-Marques 2014), feeding ecology, and foraging decisions (Dias and Rangel-Negrin 2014; Kowalewski and Garber 2014). Although we argue for the recognition of 12 howler species, the majority of field and laboratory studies have focused on 6 species *A. palliata*, *A. pigra*, *A. caraya*, *A. arctoidea*, *A. belzebul*, and *A. guariba*. Ongoing field research has shown that several howler species consume a diet that includes more fruits than leaves (Garber et al. 2014; Behie and Pavelka 2014), engage in nonaggressive forms of intragroup male–male reproductive competition (Kowalewski and Garber 2014), and that females are more sensitive to social and ecological stress than males (Van Belle 2014). These studies also have demonstrated that female mate choice, male and female migration patterns, extragroup copulations, collective action, and kinship are likely to play a critical role in howler male and female reproductive strategies (Garber and Kowalewski 2014). One goal of this volume is to present new frameworks that integrate data on howler behavioral ecology and reproduction with endocrine function, digestive physiology, host-microbe communities, anatomy, and evolution.

Our knowledge of the sensory physiology of *Alouatta* is biased in terms of their unique vocal repertoire. Detailed studies of sound production in *Alouatta* indicate inter- and intraspecific variability such that Central America howlers (*A. palliata* and *A. pigra*) are reported to have differences in call structure compared to species from South America (see below). Recent research suggests that the howler hyoid has undergone important evolutionary changes towards increased pneumatization, with a large, hollow balloon-like basihyal and enlarged laryngeal cartilages (some of these are partly ossified) that serve, along with air sacs, as a resonating chamber. Changes in the hyoid complex in howlers have resulted in significant modifications in cranial and mandibular shape, leading to a large flat face and airorhynchous skull, elongated basicranial shape, a flat and reduced nuchal plane, a vertically positioned *foramen magnum*, and large and deep mandibular ramus (e.g., Rosenberger et al. 2014; Youlatos et al. 2014).

One unexpected aspect of howler physiology that distinguishes them from other New World primates are derived features of their visual system resulting in routine trichromatic color vision in both males and females in *A. seniculus* and *A. caraya* (Jacobs et al. 1996). In other taxa of New World primates, there are sex-linked differences in color vision with males being dichromatic, approximately 60 % of females also are dichromatic, and 40 % of females are trichromatic. The main theories proposed to explain the adaptive advantages of this apomorphy include leaf and fruit selection, visual social signals, increased ability to detect camouflaged predator, and the use of color by males or females to determine health or reproductive condition (Sumner and Mollon 2000, 2002; Dominy and Lucas 2001; Regan et al., 2001; Dominy 2004; Jacobs 2007).

After Malinow's early book on the biology of *A. caraya* (Malinow 1968), and individual publications focusing on howler morphology, systematics, and physiology, Neville et al. published a review of *Alouatta* in the book, *Ecology and Behavior of Neotropical Primates Volume 2* (edited Mittermeier et al. 1988). However, it took another decade before the publication of another volume dedicated to *Alouatta*. This was a special issue of the *International Journal of Primatology* (Vol. 19: issue 3) published in 1998. This issue, edited by M. Clarke, was the result of a symposium entitled *Howlers: Past and Present*, organized by K. Glander at the 1988 Congress of the International Primatological Society held at Brasilia, Brazil. Of the 11 articles in this volume, 2 dealt with anatomical issues: 1 on the forelimb anatomy of *A. arctoidea* (SchönYbarra 1998) and 1 on cranial pathology of *A. palliata* (DeGusta and Milton 1998). The remainder of the articles focused on behavior, ecology, and conservation.

Given significant advances in the tools available to primate researchers coupled with a dramatic increase in the number of howler species and groups studied, we have put together two comprehensive companion volumes, one titled *Howler Monkeys: Adaptive Radiation, Systematics, and Morphology* and a second *Howler Monkeys: Behavior, Ecology and Conservation*. These volumes integrate our current knowledge of the behavioral, ecological, social, and evolutionary processes that have shaped the evolution, biology, physiology, and life history of this taxon. In this first volume we include 15 chapters divided into 5 sections

(1) Introduction; (2) Taxonomy, genetics, morphology, and evolution; (3) Physiology; (4) Ontogeny and sensory ecology; and (5) Conclusions. Each chapter identifies directions for further research on howler monkeys using a comparative framework. In developing this volume, we have relied on the expertise of researchers from habitat countries. Sixty-four percent of the chapters in the 2 volumes are led by a Latin American or non-Latin American that lives permanently in this region, and 89 % of the chapters have at least 1 Latin American coauthor. Thus, we acknowledge the growing number of Latin American scholars that currently study Neotropical primates in situ and emphasize the importance of highlighting this research to ensure the continuity of long-term projects that can increase our understanding of Latin American primates.

1.1.1 The Taxonomy, Genetics, and Evolution of Howler Monkeys

This first part of the volume is focused on the evolutionary history of the genus *Alouatta*. In Chap. 2, Rosenberger and collaborators offer a unique summary of fossil alouattines arguing that fossils such as *Paralouatta* (16.5 Ma), *Stirtonia* (13.5–11.8 Ma), *Solimoea* (6.8–9 Ma), and *Protopithecus* (ca. 20,000 BP) are ancestors to extant howler monkeys. The chapter also includes an examination of some fossil taxa that previously have been only briefly discussed in the literature (e.g., *Protopithecus* and the non-alouattine *Caipora*) as well as *Solimoea*, which was originally considered a stem ateline, but the authors advocate its inclusion in the alouattines. Of these fossil genera, only *Stirtonia* can be considered a committed leaf-eater similar to extant howlers. This is based on detailed functional traits in molar morphology shared with *Alouatta*. In contrast, the molars of *Paralouatta* (and *Solimoea*) are apparently more primitive, while the lesser-known dentition of *Protopithecus* presents a different anatomical pattern, perhaps closer to atelines and thus possibly morphotype-like for alouattines. The authors suggest that these features in *Protopithecus* represent adaptations to howling rather than adaptations to the consumption and mastication of a leaf-based diet. Another important idea from this chapter is that relatively small brain sizes evolved in the alouattines prior to their dental commitment to leaf-eating. Thus, perhaps in alouattine evolution morphological constraints associated with loud howling result in changes in cranial design that limited space available for an expanded brain volume.

In Chap. 3, Cortés-Ortiz and colleagues review the taxonomy of howler monkeys and in comparing morphological and genetic data provide support for nine species: mantled howlers (*A. palliata*), Central American black howlers (*A. pigra*), red howlers (*A. seniculus*), ursine howler (*A. arctoidea*), red-handed howlers (*A. belzebul*), Bolivian red howler monkey (*A. sara*), Guyanan red howler (*A. macconnelli*), brown howlers (*A. guariba*), and black and gold howlers (*A. caraya*). These authors also suggest that three more taxa should be tentatively considered as full species

(*A. nigerrima, A. ululata, A. discolor*). Final confirmation of species status awaits but for which additional genetic and/or morphological studies are required to confirm status. Cortés-Ortiz et al. (2014a, b) also propose five subspecies in *A. palliata* (*A. p. mexicana, A. p. palliata, A. p. coibensis, A. p. trabeata*, and *A. p. aequatorialis*), three subspecies in *A. seniculus* (*A. s. seniculus, A. s. juara*, and *A. s. puruensis*), and two in *A. guariba* (*A. g. guariba* and *A. g. clamitans*). These authors further acknowledge the possibility that *A. pigra* may contain two subspecies (*A. p. pigra* and *A. p. luctuosa*). This chapter constitutes the most complete taxonomic evaluation of howlers to date. Steven Nash has generously provided plates with accurate drawings of each *Alouatta* species and subspecies.

In Chap. 4, Mudry et al. provide a comprehensive review of howler cytogenetic studies, highlighting the differences in chromosome number among the different taxa, some of which are due to the presence of microchromosomes. They review the evidence of multiple sexual systems present in *Alouatta* including the formation of trivalents X1X2Y in males of *A. belzebul* and *A. palliata*; quadrivalents X1X2Y1Y2 in males of *A. seniculus, A. pigra, A. macconnelli, A. sara*, and *A. caraya*; and possible pentavalents X1X2X3Y1Y2 in males of *A. guariba*. Based on cytomolecular analyses they propose an independent origin of the sex chromosome systems in the Mesoamerican and South American lineages.

In Chap. 5, Cortés-Ortiz and colleagues explain the importance of hybridization in the evolutionary history of howler monkeys and examine the morphological, behavioral, and genetic data available from the few known howler monkey hybrid zones: between *A. palliata* and *A. pigra* in Mexico and between *A. guariba* and *A. caraya* in Argentina and Brazil. Morphological data from these hybrid zones indicate the existence of individuals with intermediate phenotypes; however, genetic studies of the *A. palliata* × *A. pigra* hybrid zone show that it is not always possible to distinguish pure forms from admixed individuals. Furthermore, the genetic analyses demonstrated that most individuals in the hybrid zone are multigenerational backcrossed hybrids. The lack of early-generation male hybrids, consistent with Haldane's rule, which states that in hybrid systems if one sex is absent it is the heterogametic sex, provides strong support for the contention that reproductive isolation is already present between these taxa. Further behavioral, cytogenetic, and molecular studies are required to understand the mechanisms promoting reproductive isolation between howler species and the maintenance of species integrity despite hybridization.

1.1.2 The Anatomy and Physiology of Howlers

This section of the volume describes the anatomical and physiological characteristics of howlers. For example, in Chap. 6, Canales-Espinosa and collaborators focus on blood biochemistry and hematology. By doing so, they not only review the published information (*A. caraya*) but also provide novel information from the howlers of Mexico (*A. palliata* and *A. pigra*) and French Guyana (*A. macconnelli*), providing reference values for these species. Among the patterns found include evidence of a higher concentration of white blood cells in females (except in *A. caraya*) than in males and a higher concentration of white blood cells overall in *A. caraya* and *A. palliata* than in other species. Additionally, creatinine levels were found to be higher in males, in relation to body mass differences, and protein levels were found to be lower in Mexican species than in other *Alouatta* species. Although some differences between males and females may follow a sexual dimorphic pattern (i.e., creatinine level), some results may be associated with ontogeny, aging (i.e., mean corpuscular volume and mean corpuscular hemoglobin), or health status (i.e., white blood cells). The variability present among *Alouatta* species may reflect the ability of howler monkeys to live in marginal and highly variable habitats.

In Chap. 7, van Belle reviewed data on hormones and behavior from six species of howlers (*A. palliata*, *A. arctoidea*, *A. caraya*, *A. pigra*, *A. belzebul*, and *A. seniculus*). Although the database is limited, this chapter explores relationships between the concentrations of sexual and stress-related hormones and growth patterns, mating relationships, intra- and extragroup male–male competition, resource scarcity, habitat fragmentation, translocation, and sociality that serve to better understand the physiological response of howlers to changes in the social and ecological environment. Data suggest that in male *A. palliata*, fecal androgens increase at 3 years of age. However these results are equivocal as 3-year-old males that were evicted from their natal groups show lower levels of fecal androgens than males of similar age who remained in their social groups. Data on ovarian cycles are available from three species (*A. arctoidea*, *A. caraya*, and *A. pigra*). Although different techniques have been used to estimate the length of the ovarian cycle, in most species the range falls between 13 and 25 days.

Endocrine function in *Alouatta* also may reflect nutritional status and food availability. In a long-term study on *A. pigra* in Belize (Behie et al. 2010; Behie and Pavelka 2012, 2014), glucocorticoid levels were higher during periods of fruit scarcity compared to periods of fruit abundance in a population recovering from a collapse and habitat destruction caused by a hurricane. In *A. pigra*, glucocorticoid levels were found to increase as a consequence of intragroup competition (Van Belle et al. 2008). In contrast, Rangel-Negrín et al. (2011) reported that in *A. palliata* glucocorticoid levels increased in response to intergroup competition. As it is stated by Van Belle (2014), these differences may reflect differential hormonal responses to variable social situations within and between groups and to changing demographic patterns, provide a framework for understanding behavioral individuals and species-specific differences in male and female mating strategies and social interactions.

In Chap. 8, Amato and Righini evaluate the role of the gut microbiome in howler heath and feeding ecology. Primates and other mammals rely on mutualistic microbial communities in their gut to provide them with energy via the fermentation of otherwise indigestible material such as fiber. Howler monkeys, as are all other primates, are dependent on their gut microbiota for the breakdown of plant structural carbohydrates, and Amato and Righini use recently collected data to describe the gut microbiome of captive and wild black howler monkeys (*A. pigra*) to test two models of host–microbe interactions and bioenergetics. The two models tested focus on (1) general host–microbiota interactions and (2) measures of bioenergetics that include gut microbiota effects. Their results indicate that individual howler monkey microbial community composition differs more across habitats than across seasons, and that these differences are strongly associated with the nutrient composition of the diet. Examining how spatial and temporal fluctuations in resource availability and the plant and animal tissues consumed affect the primate gut microbiome, and in turn, how this influences host nutrition and physiology is critical for examining questions regarding age- and sex-based differences in feeding ecology. In particular, whether adult males, adult female, and juveniles can consume the same diet, but due to differences in their microbiome, differently extract nutrients. This has important implications for examining the role that the gut microbiota plays in primate ecology, health, and conservation. There is only one howler species represented in their dataset (*A. pigra*), and this highlights the need to conduct comparative studies on other howler species.

In Chap. 9, Martínez-Mota and collaborators offer an overview and a meta-analysis of gastrointestinal parasites that are hosted by howler monkeys. They explore how ecological factors affect parasitic infection in this primate genus analyzing eight howler monkey species (*Alouatta palliata, A. pigra, A. macconnelli, A. sara, A. seniculus, A. belzebul, A. guariba, and A. caraya*), at more than 35 sites throughout their distribution. Some factors such as human presence and annual precipitation may influence the prevalence of intestinal parasites. For example, precipitation, latitude, altitude, and human proximity may differentially influence the prevalence of parasite type. For example, nematode prevalence increases with precipitation, trematodes appear to be unaffected by these climatic/anthropogenic variables (no trend was found), cestode presence was higher in remote habitats than in rural habitats, amoebae were found to exhibit higher prevalence at lower latitudes and at sites with high precipitation, *Trypanoxiuris* sp. showed a trend of decreasing prevalence towards higher altitudes, *Giardia* sp. was found to decrease with increasing precipitation, and *Plasmodium* sp. was not found to be strongly associated with any of the variables measured. In addition, the authors found that parasitic infection in howlers appears to be biased towards few individuals within a group. Given that infectious diseases are serious threats for primate survival, this study provides a baseline for evaluating the dynamics of parasite–howler interactions and for comparative studies in other platyrrhines.

In Chap. 10, Youlatos and colleagues provide a comprehensive review of howler morphology. The authors examine howler cranio-mandibular and hyoid shape and form using three-dimensional geometric morphometrics. This methodology offers advantages over more traditional approaches by measuring shape, estimating shape variability, and calculating variance in allometry and form. The authors also review howler dental and postcranial anatomy. The results indicate that howler monkeys possess a skull with a robust prognathic, airorynchous face, small braincase, and posteriorly directed occipital condyles and *foramen magnum* and a hypertrophied hyoid with enlarged laryngeal cartilages. These represent distinctive morphological traits that characterize this genus compared to other atelines and platyrrhines. The results indicate that the unique morphology of *Alouatta*'s cranium and hyoid is strongly associated with a shift to a loud communication lifestyle. Additionally, the arrangement and morphology of the dentition including small incisors and flat elongated crested molars suggest an increased ability to process leaves and possibly seeds, while the appendicular morphology reveals an emphasis on an above-branch quadrupedal positional repertoire and short-distance travel. Limb morphology associated with these positional behaviors includes relatively short forelimb long bones with joints that allow ample movements at the level of the shoulder and elbow and more restricted movement at the wrist. Moreover, the howler hip, knee, and tarsal joints are quite flexible, and both *manus* and *pes* provide stable grasping on arboreal supports, with the help of a comparably short prehensile tail. These major behavioral axes, enhanced sound production functioning in long-distance vocal communication, variable but generally increased ability to dentally process leaves, and above-branch locomotor and posture behavior describe a suite of traits that distinguish *Alouatta* from other atelines.

1.1.3 The Ontogeny and Sensory Ecology of Howlers

The final section of the volume presents information on the ontogeny and sensory systems of howler monkeys. In Chap. 11, Raguet-Schofield and Pavé present data on the ontogeny of *Alouatta* examining the degree to which howler development follows a "fast-slow" continuum and whether individual life history traits are best understood in terms of dissociated development. Although, howlers have traditionally been characterized as having fast life histories compared to other atelines, the authors point out the need for a change of paradigm when interpreting ontogeny. For example, *A. palliata* seem to reach age at first reproduction earlier than *A. caraya* and *A. seniculus*, but has a longer interbirth interval (IBI) and later age at weaning. These patterns do not correspond with the paradigm of a fast vs. slow developmental trajectory. Also, the authors suggest that compared to other atelines, *Alouatta* females shift resources from current to future offspring more rapidly. Thus, howler females reach reproductive age earlier, exhibit a shorter gestation period, shorter IBI, and wean infants earlier than other atelines; however, female growth rates are indistinguishable between *A. caraya* and *Ateles geoffroyi* (Leigh 1994), indicating that the fast-slow evolutionary model misrepresents the pattern and pace of primate development. Sexual dimorphism also is expressed at different phases of development including postnatal growth rate, craniodental maturation, and initiation of solid food intake in *Alouatta*. For example, males exhibit more rapid postnatal growth than females; moreover, male growth does not remain uniformly accelerated and instead alternates between periods of slower growth and periods of faster growth, supporting a pattern of life history dissociability. The authors make a strong argument about the necessity of using a dissociability model to analyze *Alouatta* life history traits, indicating that some traits develop relatively early in ontogeny and others develop late in ontogeny compared to other atelines, and therefore howler development does not conform to the predictions of a fast-slow continuum.

In Chap. 12, Hernández-Salazar and colleagues review data exploring how howler monkeys perceive the world. Although there are several studies on howler vision, there are limited data on other senses. This chapter focuses on a review of the anatomy, physiology, genetics, and behavioral relevance of hearing, tactile communication, taste, vision, and olfactory communication in howler monkeys in comparison to other platyrrhines, and in particular to *Ateles* sp. Some specific differences among howlers and other atelines are (1) howler monkeys exhibit a form of trichromatic color vision that make them more similar to the Old World monkeys, apes, and humans than to other platyrrhines. In this regard, it has been argued that routine trichromatic vision may be linked to a diet where leaves represent a critical component; (2) the ability to use loud calls to increase group cohesion, intergroup communication, and between group male spacing. Unfortunately, we lack specific information of the sense of smell, touch, and taste to better understand its role in food selection and social–sexual interaction within and between groups. Overall, we continue to lack accurate measures of physiological performance for the majority of sensory communication in howlers and other atelines.

In Chaps. 13 and 14, da Cunha, Kitchen, and collaborators provide a thorough review of the diversity of vocal communication in howler monkeys. Chapter 13 focuses on the acoustic structure of the vocalizations, particularly loud calls, and the variation among different species. A striking division between Central and South American howlers is difference in male loud calls parallel genetic differences that separate two identified phylogenetic clades of the genus *Alouatta* (Cortés-Ortiz et al. 2003). Thus, both *A. palliata* and *A. pigra* produce only simple, short-duration roars, their barks are essentially just shorter syllables of their species-typical roars, and both barks and roars usually occur in the same bout of loud calling. However, although their individual vocalizations are shorter than in South American howlers, they are produced during bouts that last much longer than in the southern species, with pauses between calls. The authors present information on the anatomy of the vocal organs, and discuss the limited information available regarding the vocal repertoire of "of more subtle calls" in *Alouatta*. The chapter concludes by offering a set of standardized methodologies to study vocal communication in this genus. Chapter 14 reviews the functional studies conducted to date on loud vocalizations in howlers, highlighting both inter- and intraspecific variation. The authors explore the role of male loud calls in group cohesion, predator avoidance, attraction of females, and competition over resources and address the understudied role of female loud calling. Their main results indicate that calls have a major function in assessment of rivals. The rate and patterns of vocal battles and intergroup encounters and the likelihood that groups will escalate these conflicts after physical aggression vary among species and populations. Although there is strong support that howling evolved at least in part under male intrasexual selective pressures, the importance of resource competition remains unclear.

1.2 Conclusions

A major goal of this volume is to integrate published and unpublished data on howler monkey evolution, systematics, genetics, and anatomy into a framework that can be used to study other primate radiations. Thus, we feel that this book will be of great interest to students and researchers examining a range of issues in evolutionary biology, genetics, anthropology, primatology, physiology, and endocrinology. In addition, encounters with howler monkeys are common in the field, and most primatologists studying in tropical and subtropical America have observed one or more of the currently described taxa. Therefore, we foresee this book as a centerpiece, contributing to the scientific literature on primates, as well as adding to our understanding of Neotropical community ecology. Finally, we want to stress that, although many authors have contributed directly to this volume, there are other scholars who have contributed greatly to our knowledge of howler physiology, anatomy, demography, evolution, and conservation that are not included in this volume. However, their contributions have made this volume possible. Most certainly this includes Clarence Raymond Carpenter, Margaret Clarke, Alejandro Estrada, Kenneth Glander, Robert Horwich, Katharine Milton, Miguel Schön Ybarra, and Gabriel Zunino. Additionally absent are many graduate students currently gathering new and innovative data and whose work will certainly broaden our knowledge in the near future.

So, why is it important to continue studying howlers? As for many other primate species, critical data remain to be collected. We need to promote the development and maintenance of long-term study sites that include populations of the same species living in diverse ecological communities in order to understand the adaptability of the genus *Alouatta*. In addition, we need to collect data to more clearly define the set of conditions that promote phenotypic variability in howlers. Furthermore, long-term data on a broad set of taxa will facilitate comparative analyses needed to explore the underlying mechanisms of behavioral, ecological, morphological, and genetic variability. New available methodologies are critical for addressing twenty-first century questions in primatology. These techniques include molecular genetics, 3D geometric morphometrics, GIS technology, portable high-definition and high-speed video recording, hormone analyses, nutritional analyses of plant foods, and the use of molecular methods for examination of disease, the gut microbiome, and invertebrate and vertebrate DNA present in primate feces. Although these technologies may increase the cost of research, the information they will provide will surely be of significant value in advancing our understanding of howler monkey behavior, ecology, and evolution. These new studies will require the collaboration of multidisciplinary research teams across countries. Many of the chapters in this volume are the result of such collaboration and an irrefutable proof that we, as primatologists, are heading in the right direction.

Acknowledgments M.K. thanks Mariana and Bruno for their support during the edition of these volumes. P.A.G. wishes to acknowledge Chrissie, Sara, and Jenni for their love and support and for allowing him to be himself. While writing this paper L.C.O. was supported by NSF grant BCS-0962807. B.U. thanks his family and Padmini for always being there.

References

Amato KR, Righini N (2014) The howler monkey as a model for exploring host-gut microbiota interactions in primates. In: Kowalewski M, Garber P, Cortés-Ortiz L, Urbani B, Youlatos D (eds) Howler monkeys: adaptive radiation, systematics, and morphology. Springer, New York

Behie AM, Pavelka MSM (2012) Food selection in the black howler monkey following habitat disturbance: implications for the importance of mature leaves. J Trop Ecol 28:153–160

Behie AM, Pavelka SM (2014) Fruit as a key factor in howler monkey population density: conservation implications. In: Kowalewski M, Garber P, Cortés-Ortiz L, Urbani B, Youlatos D (eds) Howler monkeys: behavior, ecology and conservation. Springer, New York

Behie AM, Pavelka MSM, Chapman CA (2010) Ecological sources of variation in fecal cortisol levels in howler monkeys in Belize. Am J Primatol 72:600–606

Bonvicino CR, Lemos B, Seuánez HN (2001) Molecular phylogenetics of howler monkeys (*Alouatta*, Platyrrhini): a comparison with karyotypic data. Chromosoma 110:241–246

Brown AD, Zunino GE (1994) Habitat, density and conservation problems of Argentine primates. Vida Silvestre Neotrop 3(1):30–40

Camacho JH, Defler TR (1985) Some aspects of the conservation of non-human primates in Colombia. Primate Conserv 6:42–50

Canales-Espinosa D, Rovirosa-Hernández M, de Thoisy B, Caba M, García-Orduña F (2014) Hematology and serum biochemistry in wild howler monkeys. In: Kowalewski M, Garber P, Cortés-Ortiz L, Urbani B, Youlatos D (eds) Howler monkeys: adaptive radiation, systematics, and morphology. Springer, New York

Cortés-Ortiz L, Bermingham E, Rico C, Rodríguez-Luna E, Sampaio I, Ruiz-García M (2003) Molecular systematics and biogeography of the Neotropical monkey genus, Alouatta. Mol Phylogenet Evol 26:64–81

Cortés-Ortiz L, Agostini I, Aguiar LM, Kelaita M, Silva FE, Bicca-Marques JC (2014a) Hybridization in howler monkeys: current understanding and future directions. In: Kowalewski M, Garber P, Cortés-Ortiz L, Urbani B, Youlatos D (eds) Howler monkeys: adaptive radiation, systematics, and morphology. Springer, New York

Cortés-Ortiz L, Rylands AB, Mittermeier RA (2014b) The taxonomy of howler monkeys: integrating old and new knowledge from morphological and genetic studies. In: Kowalewski M, Garber P, Cortés-Ortiz L, Urbani B, Youlatos D (eds) Howler monkeys: adaptive radiation, systematics, and morphology. Springer, New York

Crockett CM (1998) Conservation biology of the genus *Alouatta*. Int J Primat 19(3):549–578

Crockett CM, Eisenberg JF (1987) Howlers: variations in group size and demography. In: Smuts BB, Cheney DL, Seyfarth RM, Wangham RW, Struhsaker TT (eds) Primate societies. University of Chicago, Chicago

da Cunha RGT, Oliveira DAG, Holzmann I, Kitchen DM (2014) Production of loud and quiet calls in howler monkeys. In: Kowalewski M, Garber P, Cortés-Ortiz L, Urbani B, Youlatos D (eds) Howler monkeys: adaptive radiation, systematics, and morphology. Springer, New York

de Oliveira EHC, Neusser M, Figueiredo WB, Nagamachi C, Pieczarka JC, Sbalqueiro IJ, Wienberg J, Müller S (2002) The phylogeny of howler monkeys (*Alouatta*, Platyrrhini): reconstruction by multi-color cross-species chromosome painting. Chromosome Res 10:669–683

DeGusta D, Milton K (1998) Skeletal pathologies in a population of *Alouatta palliata*: behavioral, ecological and evolutionary implications. Int J Primatol 19(3):615–650

Di Fiore A, Link A, Campbell CJ (2010) The atelines: Behavioral and socioecological diversity in a New World radiation. In: Campbell CA, Fuentes A, MacKinnon K, Bearder S, Stumpf R (eds) Primates in Perspective, 2nd Edition. Oxford University Press, Oxford

Dias PAD, Rangel-Negrin A (2014) Diets of howler monkeys. In: Kowalewski M, Garber P, Cortés-Ortiz L, Urbani B, Youlatos D (eds) Howler monkeys: behavior, ecology and conservation. Springer, New York

Dominy NJ (2004) Color as an indicator of food quality to anthropoid primates: ecological evidence and an evolutionary scenario. In: Ross CF, Kay RF (eds) Anthropoid origins: new visions. Kluwer Academic, New York

Dominy NJ, Lucas PW (2001) Ecological importance of trichromatic vision to primates. Nature 410:363–366

Emmons LH, Feer F (1990) Neotropical rainforest mammals. A field guide. University of Chicago, Chicago

Garber PA, Kowalewski MM (2014) New challenges in the study of howler monkey behavioral ecology and conservation: where we are and where we need to go? In: Kowalewski M, Garber P, Cortés-Ortiz L, Urbani B, Youlatos D (eds) Howler monkeys: behavior, ecology and conservation. Springer, New York

Garber P, Righini N, Kowalewski M (2014) Evidence of alternative dietary syndromes and nutritional goals in the genus Alouatta. In: Kowalewski M, Garber P, Cortés-Ortiz L, Urbani B, Youlatos D (eds) Howler monkeys: behavior, ecology and conservation. Springer, New York

Glander KE, Pinto LP (2013) Subfamily Alouattinae, *Alouatta* Lacépède, 1799. In: Mittermeier RA, Rylands AB, Wilson DE (eds) Handbook of the mammals of the world. Primates, vol 3. Lynx Edicions, Barcelona

Gregorin R (2006) Taxonomy and geographic variation of species of the genus *Alouatta* Lacépède (Primates, Atelidae) in Brazil. Rev Bras Zool 23:64–144

Groves CP (2001) Primate taxonomy. Smithsonian Institution, Washington, DC

Groves CP (2005) Order primates. In: Wilson DE, Reeder DM (eds) Mammal species of the world: a taxonomic and geographic reference, vol 1, 3rd edn. Johns Hopkins University, Baltimore

Hernández-Salazar LT, Dominy NJ, Laska M (2014) The sensory systems of *Alouatta*: evolution with an eye to ecology. In: Kowalewski M, Garber P, Cortés-Ortiz L, Urbani B, Youlatos D (eds) Howler monkeys: adaptive radiation, systematics, and morphology. Springer, New York

Hill WCO (1962) Primates: comparative anatomy and taxonomy. V. Cebidae, Part B. Edinburgh University, Edinburgh

Jacobs GH (2007) New world monkeys and color. Int J Primatol 28(4):729–759

Jacobs GH, Neitz M, Deegan JF, Neitz J (1996) Trichromatic colour vision in New World monkeys. Nature 382:156–158

Kitchen DM, Teixeira da Cunha RG, Holzmann I, Gonçalves de Oliveira DA (2014) Function of loud calls in howler monkeys. Howler monkeys: adaptive radiation, systematics, and morphology. Springer, New York

Kowalewski M, Garber P (2014) Solving the collective action problem during intergroup encounters: the case of black and gold howler monkeys. In: Kowalewski M, Garber P, Cortés-Ortiz L, Urbani B, Youlatos D (eds) Howler monkeys: behavior, ecology and conservation. Springer, New York

Leigh SR (1994) Ontogenetic correlates of diet in anthropoid primates. Am J Phys Anthropol 94:499–522

Malinow MR (1968) Biology of the howler monkey (*Alouatta caraya*). Bibl primatol 7. Karger AG, Basel

Martínez-Mota R, Kowalewski MM, Gillespie TR (2014) Ecological determinants of parasitism in howler monkeys. In: Kowalewski M, Garber P, Cortés-Ortiz L, Urbani B, Youlatos D (eds) Howler monkeys: adaptive radiation, systematics, and morphology. Springer, New York

Mittermeier RA, Rylands AB, Coimbra-Filho AF (1988) Systematics: species and subspecies – an update. In: Mittermeier RA, Rylands AB, Coimbra-Filho AF, Fonseca GA (eds). Ecology and behavior of Neotropical primates. vol 2. World Wildlife Fund, Washington

Mudry MD, Nieves M, Steinberg ER (2014) Cytogenetics of howler monkeys. In: Kowalewski M, Garber P, Cortés-Ortiz L, Urbani B, Youlatos D (eds) Howler monkeys: adaptive radiation, systematics, and morphology. Springer, New York

Neville MK, Glander KE, Braza F, Rylands AB (1988) The howling monkeys, genus Alouatta. In: Mittermeier RA, Rylands AB, Coimbra-Filho AF, da Fonseca GAB (eds) Ecology and behavior of neotropical primates, vol 2. World Wildlife Fund, Washington, DC

Nowak RM (1999) Walker's primates of the world. Johns Hopkins University, Baltimore

Raguet-Schofield M, Pavé R (2014) An ontogenetic framework for *Alouatta*: infant development and evaluating models of life history. In: Kowalewski M, Garber P, Cortés-Ortiz L, Urbani B, Youlatos D (eds) Howler monkeys: adaptive radiation, systematics, and morphology. Springer, New York

Rangel-Negrín A, Dias PAD, Chavira R, Canales-Espinosa D (2011) Social modulation of testosterone levels in male black howlers (Alouatta pigra). Horm Behav 59:159–166

Regan BC, Juliot C, Simmen B, Viénot F, Charles-Dominique PC, Mollon JD (2001) Fruits, foliage and the evolution of primate colour vision. Philos Trans R Soc Lond B Biol Sci 356:229–283

Rosenberger AL, Strier KB (1989) Adaptive radiation of the ateline primates. J Hum Evol 18:717–750

Rosenberger AL, Cooke SB, Halenar LB, Tejedor MF, Hartwig WC, Novo NM, Muñoz-Saba Y (2014) Fossil alouattines and the origins of Alouatta: craniodental diversity and interrelationships. In: Kowalewski M, Garber P, Cortés-Ortiz L, Urbani B, Youlatos D (eds) Howler monkeys: adaptive radiation, systematics, and morphology. Springer, New York

Rylands AB, Mittermeier RA (2009) The diversity of the New World primates (Platyrrhini). In: Garber PA, Estrada A, Bicca-Marques JC, Heymann EW, Strier KB (eds) South American primates: comparative perspectives in the study of behavior, ecology, and conservation. Springer, New York

Rylands AB, Schneider H, Langguth A, Mittermeier RA, Groves CP, Rodríguez-Luna E (2000) An assessment of the diversity of New World primates. Neotrop Primates 8:61–93

Rylands AB, Groves CP, Mittermeier RA, Cortés-Ortiz L, Hines JJ (2006) Taxonomy and distributions of Mesoamerican primates. In: Estrada A, Garber P, Pavelka M, Luecke L (eds) New perspectives in the study of Mesoamerican primates: distribution, ecology, behavior and conservation. Springer, New York

Schön MA (1968) The muscular system of the red howling monkey. US Natl Mus Bull 273:1–185

Schön Ybarra MA (1984) Locomotion and postures of red howlers in a deciduous forest-savanna interface. Am J Phys Anthropol 63(1):65–76

Schön Ybarra MA (1998) Arboreal quadrupedalism and forelimb articular anatomy of red howlers. Int J Primatol 19(3):599–613

Schön Ybarra MA, Schön MA III (1987) Positional behavior and limb bone adaptations in red howling monkeys (*Alouatta seniculus*). Folia Primatol 49(2):70–89

Steinberg ER, Nieves M, Mudry MD (2014) Multiple sex chromosome systems in howler monkeys (Platyrrhini, *Alouatta*). Comp Cytogen 8:43–69

Stern JT (1971) Functional myology of the hip and thigh of Cebid monkeys and its implications for the evolution of erect posture. Bibl Primatol 14:1–318

Stern JT Jr, Wells JP, Jungers WL, Vangor AK, Fleagle JG (1980) An electromyographic study of serratus anterior in atelines and *Alouatta*: implications for hominoid evolution. Am J Phys Anthropol 52:323–334

Sumner P, Mollon JD (2000) Catarrhine photopigments are optimised for detecting targets against a foliage background. J Exp Biol 203:1963–1986

Sumner P, Mollon JD (2002) Did primate trichromacy evolve for frugivory or folivory? In: Mollon JD et al (eds) Normal and defective colour vision. Oxford University, Oxford

Van Belle S (2014) Endocrinology of howler monkeys: review and directions for future research. In: Kowalewski M, Garber P, Cortés-Ortiz L, Urbani B, Youlatos D (eds) Howler monkeys: adaptive radiation, systematics, and morphology. Springer, New York

Van Belle S, Bicca-Marques JC (2014) Insights into reproductive strategies and sexual selection in howler monkeys. In: Kowalewski M, Garber P, Cortés-Ortiz L, Urbani B, Youlatos D (eds) Howler monkeys: behavior, ecology and conservation. Springer, New York

Van Belle S, Estrada A, Strier KB (2008) Social relationships among male *Alouatta pigra*. Int J Primatol 29:1481–1498

Wolfheim JH (1983) Primates of the World. University of Washington, Seattle

Youlatos D, Couette S, Halenar LB (2014) Morphology of howler monkeys: a review and quantitative analyses. In: Kowalewski M, Garber P, Cortés-Ortiz L, Urbani B, Youlatos D (eds) Howler monkeys: adaptive radiation, systematics, and morphology. Springer, New York

Part II
Taxonomy, Genetics, Morphology and Evolution

Chapter 2
Fossil Alouattines and the Origins of *Alouatta*: Craniodental Diversity and Interrelationships

Alfred L. Rosenberger, Siobhán B. Cooke, Lauren B. Halenar, Marcelo F. Tejedor, Walter C. Hartwig, Nelson M. Novo, and Yaneth Muñoz-Saba

Abstract The howler monkey clade includes species of *Alouatta* and four extinct genera, *Stirtonia*, *Paralouatta*, *Protopithecus*, and probably *Solimoea* as well. Contrary to expectations, this radiation may have originated as a largely frugivorous group; advanced, *Alouatta*-like leaf-eating is a novelty well-developed in the *Alouatta-Stirtonia* sublineage only. Revised body mass estimates place *Stirtonia* and *Paralouatta* within the size range exhibited by the living forms and confirm the

---

A.L. Rosenberger
Department of Anthropology and Archaeology, Brooklyn College, The City University of New York, 2900 Bedford Avenue, Brooklyn, NY 11210, USA

The Graduate Center, The City University of New York, 365 Fifth Avenue, New York, NY 10016, USA

New York Consortium in Evolutionary Primatology (NYCEP), The City University of New York, New York, NY, USA

Department of Mammalogy, The American Museum of Natural History, Central Park West at 79th St., New York, NY 10024, USA

S.B. Cooke
Department of Anthropology, Northeastern Illinois University,
5500 N. St. Louis Avenue, Chicago, IL 60625, USA

L.B. Halenar
Department of Biological Sciences, Bronx Community College, The City University of New York, 2155 University Avenue, Bronx, NY 10453, USA

New York Consortium in Evolutionary Primatology (NYCEP), The City University of New York, New York, NY, USA

M.F. Tejedor (✉) • N.M. Novo
Centro Nacional Patagónico-CONICET,
Boulevard Brown 2915, (9120), Puerto Madryn, Provincia de Chubut, Argentina
e-mail: tejedor@cenpat.edu.ar

W.C. Hartwig
Department of Basic Sciences, Touro University College of Osteopathic Medicine, Mare Island, Vallejo, CA 94592, USA

Y. Muñoz-Saba
Instituto de Ciencias Naturales, Universidad Nacional de Colombia, Bogotá, DC, Colombia

place of *Protopithecus* in a larger, baboon-like size range. While their dentitions are more primitive than the *Alouatta-Stirtonia* pattern, the cranial anatomy of *Protopithecus* and *Paralouatta* is distinctly similar to living howler monkeys in highly derived features relating to enlargement of the subbasal space in the neck and in head carriage, suggesting that ancestral alouattines may have had an enlarged hyolaryngeal apparatus. All alouattines also have relatively small brains, including *Protopithecus*, a genus that was probably quite frugivorous. The successful origins of the alouattine clade may owe more to key adaptations involving communication and energetics than dental or locomotor breakthroughs. While the fossil record confirms aspects of previous character-analysis reconstructions based on the living forms, alouattines experienced a complexity of adaptive shifts whose history cannot be recoverable without a more complete fossil record.*

Resumen El clado de los monos aulladores incluye las especies de *Alouatta* y cuatro géneros extintos, *Stirtonia*, *Paralouatta*, *Protopithecus* y probablemente *Solimoea*. Contrario a las expectativas, esta radiación pudo haberse originado a partir de hábitos frugívoros. La avanzada folivoría de *Alouatta* es una novedad desarrollada solamente en el sublinaje de *Alouatta-Stirtonia*. Las estimaciones de masa corporal ubican a *Stirtonia* y *Paralouatta* dentro del rango que exhiben las formas vivientes y confirman la posición de *Protopithecus* en un rango de tamaño mayor, similar al de los babuinos africanos. Considerando que la dentición es más primitiva que el patrón observado en *Alouatta-Stirtonia*, la anatomía craneana de *Protopithecus* y *Paralouatta* es similar a la de los aulladores vivientes debido a los rasgos altamente especializados relacionados al agrandamiento del espacio sub-basal en el cuello, así como en la posición de la cabeza, sugiriendo que los alouatinos ancestrales pudieron haber tenido un gran aparato hiolaríngeo. Todos los alouatinos también presentan un cerebro pequeño, incluyendo *Protopithecus*, género que probablemente haya sido frugívoro. El origen exitoso del clado de los alouatinos pudo deberse más a adaptaciones de comunicación y energéticas que a cambios dentarios o locomotores. Mientras que el registro fósil confirma ciertos aspectos de análisis de caracteres previos basados en formas vivientes, los alouatinos experimentaron una complejidad de adaptaciones cuya historia no podría reconstruirse sin el registro fósil.*

Keywords Fossil primates • Howler monkeys • Craniodental morphology • Adaptation • Phylogeny

---

*Since this chapter was written, additional study by Halenar and Rosenberger (2013) of the material discussed here as *Protopithecus* led to the conclusion that the two samples actually represent two different genera. The essentially complete Bahian skeleton, which forms the basis of the present discussion, is being assigned to a new genus and species, *Cartelles coimbrafilhoi*, within subfamily Alouattinae. The original Lund material from Minas Gerais bears the original name *Protopithecus*, but its affinities are more likely to be found among atelines than alouattines.

Abbreviations

%	Percent
CT	Computed Tomography
e.g.	For example
Fig.	Figure
Figs.	Figures
i.e.	In other words
kg	Kilograms
m1	First lower molar
m3	Third lower molar
M1	First upper molar
MA	Millions of years
mm	Millimeters
NWM	New World monkeys
P3	Third upper premolar
p4	Fourth lower premolar
P4	Fourth upper premolar

2.1 Introduction

Fossils discovered in recent years have added important information to our knowledge of the diversity and evolution of platyrrhines closely related to one of the most anatomically divergent members of the radiation, the living howler monkeys, *Alouatta*. While the record is still scant, these additions mean the alouattine-plus-ateline clade, i.e., the fully prehensile-tailed New World monkeys (NWMs), is becoming one of the better-known lineages among the platyrrhines. Only pitheciines are better represented taxonomically among Tertiary and Quaternary remains (Rosenberger 2002).

The first historical narratives of the evolution of howler monkeys are of recent vintage, and they relied extensively on character analysis of the morphology and behavioral ecology of living atelids rather than paleontology (e.g., Rosenberger and Strier 1989; Strier 1992). Out of necessity, these studies focused on the contrasts between the living members of the two sister clades, alouattines (*Alouatta*) and atelines (*Lagothrix, Ateles, Brachyteles*). Few relevant, informative fossils were known prior to the 1980s. The one exception was *Stirtonia tatacoensis* from the middle Miocene La Venta beds of Colombia, 13.5–11.8 MA (Flynn et al. 1997). It was first found as dental remains in the late 1940s (Stirton 1951) and the species has been widely recognized as being both similar and related closely to *Alouatta* (e.g., Szalay and Delson 1979; Setoguchi et al. 1981; Delson and Rosenberger 1984; Rosenberger 1992; Hartwig and Meldrum 2002; but see Hershkovitz 1970). In the late 1980s, a second species, *S. victoriae*, was discovered at La Venta (Kay et al. 1987), and an isolated *Stirtonia* molar from the younger, late middle Miocene

Solimões Formation in western Brazil, about 8 MA, also came to light recently (Kay and Frailey 1993). *Stirtonia* reinforced the notion that leaf-eating was an enduring and essential aspect of the howler monkey's ecophylogenetic biology. The type specimen of another species related to *Alouatta*, *Protopithecus brasiliensis* from the Quaternary of Brazil, had been known since 1838 (Lund 1838), but the fossil was based on a partial humerus and femur and could not be properly interpreted for another 150 years (Hartwig and Cartelle 1996; see footnote above).

Finds in Brazil and Cuba add another dimension of complexity to the *Stirtonia-Alouatta* story, introducing an unexpected anatomical diversity. This panorama of diversity highlights the unusual nature of living *Alouatta* as a genus and suggests a need to reevaluate the Rosenberger and Strier (1989)/Strier (1992) model of alouattine evolution. Besides *Protopithecus brasiliensis*, the *Alouatta* clade also includes *Paralouatta varonai* from the Quaternary of Cuba (Rivero and Arredondo 1991) and perhaps *Paralouatta marianae* from the Miocene of Cuba (see MacPhee et al. 2003). If the latter species, known only from a single astragalus, is indeed an alouattine, these congeners represent a lengthy span of geological time. More problematic is *Solimoea acrensis*, described from a small set of isolated dental elements, two specimens including three teeth, discovered in Brazil's Solimões Formation (Kay and Cozzuol 2006). The best evidence of its affinities consists of a single lower molar, which has distinctive crown morphology. The species was originally interpreted as a stem ateline, but we present reasons why it is probably an alouattine. Finally, also from a late Pleistocene cave of Bahia, Brazil, is a little known extinct species of howler monkey, *Alouatta mauroi* (Tejedor et al. 2008), which we mention only for the sake of completeness.

Our purpose here is to establish the taxonomic composition and interrelationships of living and extinct alouattines, present new information pertaining to their craniodental diversity, and explore several aspects of alouattine evolutionary history as an adaptive array. The phylogenetics and differentiation of this group has not been discussed previously. Part of the reason for this is that the composition of the subfamily Alouattinae is a matter of debate. In addition to the question of *Solimoea*, raised here for the first time, there are different views about the affinities of *Paralouatta* (e.g., Rivero and Arredondo 1991; MacPhee and Horovitz 2002; Rosenberger 2002), which MacPhee and colleagues (MacPhee et al. 1995; Horovitz and MacPhee 1999) maintain is monophyletically related to the other extinct Caribbean primates and, among the extant forms, to mainland *Callicebus*, a pitheciid. The present study emphasizes why, from a functional-morphological perspective, an affinity with alouattines is the more parsimonious hypothesis, as Rivero and Arredondo (1991) originally proposed.

2.2 Methods

Craniodental measurements of the modern samples used in this study are largely from collections in the American Museum of Natural History, the United States National Museum, the Field Museum of Natural History, the Natural History

Museum (London), the Museu Nacional de Rio de Janeiro, and the Zoologisk Museum, Statens Naturhistoriske Museum (Copenhagen). Species identifications and sample sizes are given where appropriate. Standard linear craniodental measurements were taken to the 0.10 mm with digital calipers. Some teeth were measured using high-resolution laser scans of epoxy casts, using Landmark Editor (Wiley et al. 2005). Endocranial volumes were taken by pouring small plastic beads or other filler into the cavity then transferring the mass to a graduated glass beaker, except in the case of *Paralouatta varonai*. It was CT scanned in Havana, Cuba, using a medical scanner and a slice thickness of 0.8 mm. Using ImageJ (http://rsb.info.nih.gov/ij/), the endocranial cavity was then outlined as individual slices, composited, and measured. Some measurement error was unavoidable due to difficulty in separating bone from the matrix-filled cavity, but our figures here are consistent with other measurements used in the context of our assessment of relative brain size (see below).

Genealogical interrelationships were inferred using conventional, non-algorithmic procedures of character analysis and cladistic reconstruction. Our methodology is based on the functional-adaptational approach (see Szalay and Bock 1991). Reviews of the methodology as applied to atelids can be found in Rosenberger and Strier (1989), Rosenberger et al. (1990), and Rosenberger (1992), where additional references to the literature on cladistic phylogeny reconstruction can be found. Our intent has been to produce a character analysis that elucidates the homologies and polarities of functionally relevant anatomical features. We thus use functional-adaptive inference as well as taxonomic distributional information. The latter relies on commonality and out-group comparisons in order to develop hypotheses about the directionality of change in traits, but functional-adaptive information is necessary to hypothesize *why* such changes may have taken place. Although we do not specifically present distributional information on non-atelids, we draw on the morphology of the other platyrrhines, living and extinct, as a collective out-group in working out polarities.

We focus on large-scale morphological features that are demonstrably important in distinguishing *Alouatta* from other atelids at the genus level and are also relevant functionally to the evolution of howler monkey craniodental adaptations, since we are interested in establishing how unit characters evolved within functional complexes as a part of the phylogenetic history of alouattines. Our rationale presumes that the *Alouatta* cranium and dentition, which is radically different from most primates in many ways, is composed of an assortment of features that are derived relative to other atelids and platyrrhines. We hypothesize that the evolution of many craniodental features has been driven specifically by a novel adaptive complex relating to howling and folivory.

Solimoea, which we limit to a single molar tooth as discussed below, is referenced only sparingly in the character analysis, which emphasizes cranial anatomy. The basis for our interpretation of the fossil's affinities is presented in the body of the text following the same functional-adaptive lines employed to assess the cranium.

One feature we address but do not examine through a structured character analysis is body mass. While it has always been evident that body size would figure prominently in narrative explanations of platyrrhine evolution (see Hershkovitz 1972; Rosenberger 1980), its importance for atelids has become stunningly reaffirmed by discovering the large subfossils *Protopithecus* and *Caipora*. The initial body weight estimates for these genera (Cartelle and Hartwig 1996; Hartwig and Cartelle 1996) placed both well outside the range of modern forms. However, they were made using regression equations based on a catarrhine reference sample, a phylogenetically less desirable methodology (Hartwig 1995). Statistically robust equations based on platyrrhine postcranial elements which have been shown to be closely linked with body size (e.g., Ruff 2003) have recently been published and confirm the original estimates (Halenar 2011a, b). They have also been used to confirm an estimate of approximately 7–9.5 kg for *Paralouatta* (Cooke and Halenar 2012). We have taken a less formalistic approach in order to factor in this new information on size and integrate it with the broader analysis. The taxonomic terminology we use divides the monophyletic family Atelidae into alouattines (subfamily Alouattinae: extant *Alouatta*; extinct *Stirtonia, Paralouatta, Protopithecus,* and *Solimoea*) and atelines (subfamily Atelinae: extant *Ateles, Brachyteles,* and *Lagothrix*; extinct *Caipora*).

2.3 Results

2.3.1 Craniodental Morphology and Paleontological Synopsis

Two of the three fossil alouattine genera are represented by very good crania (Table 2.1). The third, *Stirtonia*, is known by excellent dental remains (e.g., Hershkovitz 1970; Szalay and Delson 1979; Setoguchi et al. 1981; Kay et al. 1987; Fleagle et al. 1997; Fleagle 1999; Hartwig and Meldrum 2002). The latter preserves both upper and lower cheek teeth that are unmistakably similar to *Alouatta* (see Figs. 2.1 and 2.2), so much so that Delson and Rosenberger (1984) suggested that generic separation obscures the possibility that *Stirtonia* and *Alouatta* may share an ancestor–descendant relationship and that classifying them as congeners ought to be considered. However, more work needs to be done to more accurately determine the relationships between *Stirtonia* and *Alouatta*.

Like *Alouatta*, the upper molars of *Stirtonia* (Fig. 2.1) are relatively square, with an elevated, lobe-like hypocone; high-relief buccal cusps carrying a long ectoloph; deeply notched centrocrista; and a well-developed stylar shelf area. Lower molars have a small, elevated trigonid with protoconid and metaconid set at an oblique angle and a long talonid with a sharply angled, elongate cristid obliqua. This pattern of features, including elements that have been assessed quantitatively in *Alouatta* (e.g., Kay 1975; Rosenberger and Kinzey 1976; Kay and Hylander 1978; Kay et al. 1987), is universally interpreted as shearing, leaf-eating characteristics. The upper and lower premolars of *Stirtonia* are also consistent with an *Alouatta*-like morphology, as are the tooth proportions. Incisors are not known for *Stirtonia*, but the intercanine span in the type mandible appears to be relatively narrow; *Alouatta* incisors

2 Evolution of Alouattines

Table 2.1 Summary of known fossil alouattines

Genus and species	Locality	Age	Attributed material	Body size (kg)	Major references
Stirtonia tatacoensis	La Venta, Colombia	Middle Miocene	Mandible, isolated teeth	6	Stirton (1951), Hershkovitz (1970), Setoguchi (1980), Setoguchi et al. (1981), Setoguchi and Rosenberger (1985)
Stirtonia victoriae	La Venta, Colombia	Middle Miocene	Gnathic-dental (including deciduous teeth)	10	Kay et al. (1987)
Stirtonia sp.	Rio Acre, Brazil	Late Miocene	Isolated lower molar		Kay and Frailey (1993)
Paralouatta varonai	Cueva de Mono Fosil, Cuba	Pleistocene	Craniodental and postcranial	9.5	Rivero and Arredondo (1991), Horovitz and MacPhee (1999), MacPhee and Meldrum (2006)
Paralouatta marianae	Domo de Zaza, Cuba	Early Miocene	Astragalus		MacPhee and Iturralde-Vinent (1995), MacPhee et al. (2003)
Protopithecus brasiliensis	Lagoa Santa, Brazil; Toca da Boa Vista, Brazil	Late Pleistocene	Proximal femur, distal humerus; nearly complete skeleton	25	Lund (1838), Hartwig (1995), Hartwig and Cartelle (1996), Hartwig (2005); (see footnote above)
Solimoea acrensis	Rio Acre, Brazil	Late Miocene	Isolated lower molar (provisionally m1)	5.4–6	Kay and Cozzuol (2006)
Alouatta mauroi	Gruta dos Brejoes, Brazil	Late Pleistocene	Craniodental	NA	Tejedor et al. (2008)

Fig. 2.1 Laser scan generated occlusal views of atelid left maxillary molars [digitized at 25 μm point intervals (here and below) from epoxy casts]. Teeth at left are first molars, in most cases brought to about the same mesiodistal lengths. (1) *Ateles geoffroyi*, (2) *Caipora bambuiorum*, (3) *Lagothrix lagotricha*, (4) *Brachyteles arachnoides*, (5) *Alouatta seniculus*, (6) *Stirtonia tatacoensis*, (7) *Paralouatta varonai*, (8) *Protopithecus brasiliensis*

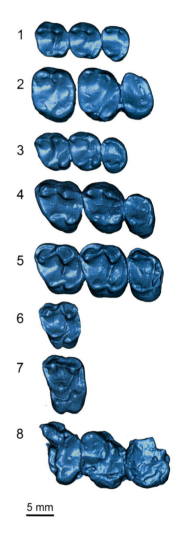

are relatively small (see below). There is no information on the posterior part of the mandible of *Stirtonia*, the extreme expansion of which is diagnostic of *Alouatta*.

Protopithecus brasiliensis is now known from a nearly complete skeleton with a very well-preserved skull (Figs. 2.3, 2.4, and 2.5) that includes the anterior teeth, premolars, and a partial upper molar, as well as a mandible with anterior teeth and premolars. It presents an interesting mosaic of craniodental and postcranial traits not found in any other NWM (Hartwig and Cartelle 1996). It shares several cranial features exhibited only in *Alouatta* among the living platyrrhines, including a relatively extended basicranium and a compound temporo-nuchal crest, which led Hartwig and Cartelle to recognize its alouattine affinities. The teeth of *Protopithecus* are still incompletely analyzed. They are nonetheless highly informative for the present purpose (see below).

2 Evolution of Alouattines

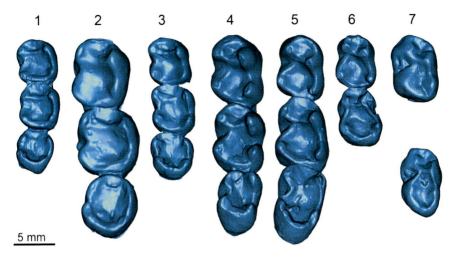

Fig. 2.2 Laser scan generated occlusal views of atelid left mandibular molars (protocols as above). (1) *Ateles geoffroyi*, (2) *Caipora bambuiorum*, (3) *Lagothrix lagotricha*, (4) *Brachyteles arachnoides*, (5) *Alouatta seniculus*, (6) *Stirtonia tatacoensis*, (7) *Paralouatta varonai* (m1, m3)

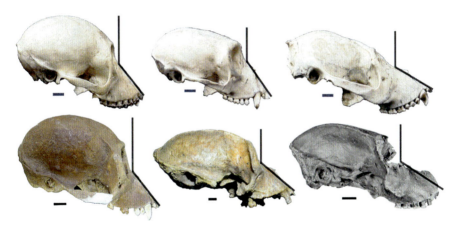

Fig. 2.3 Crania of extant and extinct members of the alouattine and ateline radiations (*lateral view*). *Left to right, top row*: *Brachyteles*, *Lagothrix*, *Alouatta*. *Bottom row*: *Caipora*, *Protopithecus*, *Paralouatta*. Scale bars represent 1 cm. Note the similarities linking *Alouatta*, *Protopithecus*, and *Paralouatta* to the exclusion of the other three genera, especially size and shape of the neurocranium and the airorynchous facial skeleton. The latter trait is indicated by the more acute angle superimposed upon those three skulls between the nasal bridge and the tip of the incisors

Paralouatta has been classified as two species, *P. varonai* and *P. marianae* (Rivero and Arredondo 1991; MacPhee et al. 2003). The latter is known only by an astragalus. The former is represented by a fairly well-preserved but broken skull with worn teeth, a mandible, various isolated teeth (Figs. 2.1, 2.2, 2.3, 2.4, and 2.5), and postcranial material (Rivero and Arredondo 1991; Horovitz and MacPhee 1999; MacPhee and Meldrum 2006). The phylogenetic connection to *Alouatta* that we

Fig. 2.4 Basal view of (*left to right*) *Lagothrix*, *Alouatta*, *Protopithecus*, and *Paralouatta*. Scale bars represent 1 cm. Note the anterior-posterior elongation of the alouattine cranial base, as well as the more marked postorbital constriction. The orientation of the foramen magnum and nuchal region of the fossils is intermediate between the ateline condition of *Lagothrix* and the alouattine condition of extant howler monkeys

Fig. 2.5 Posterior view of (*left* to *right*) *Caipora*, *Protopithecus*, and *Paralouatta*, brought approximately to same cranial width. Contrast the relatively small, low braincase; cylindrical brain shape; and prominence of both the temporal (*red arrow*) and nuchal (*blue arrow*) crests of *Protopithecus* and *Paralouatta* with the rounded, globular braincase; lack of marked temporal lines; and a much less rugose nuchal plane of *Caipora*

advocate is a matter of controversy. The first specimen, the skull, was found prior to the recovery of the new Brazilian *Protopithecus* material which, as we explain below, supports the case for the alouattine affinities of *Paralouatta*. When initially described, its overall morphology convinced Rivero and Arredondo (1991) that *Paralouatta* is closely related to its namesake *Alouatta*. However, MacPhee and colleagues argued that *Paralouatta* belongs to a newly recognized clade of Greater Antillean primates (MacPhee et al. 1995; Horovitz and MacPhee 1999; MacPhee and Horovitz 2004) most closely related as a group to *Callicebus*. This was based on the finding by Horovitz and MacPhee (1999) of three alleged unambiguous, observable craniodental characters that support the clade including *Antillothrix bernensis*, *Xenothrix mcgregori*, and *Paralouatta varonai*: nasal fossa wider than palate at level of M1, lower canine alveolus buccolingually smaller than p4, and m1 protoconid with bulging buccal surface. While this is an intriguing result given the isolation of these taxa from the mainland, it is far from definitive. Thus, Rosenberger (2002) held that Rivero and Arredondo (1991) were correct, as we further elaborate below. An added dimension to the paleobiology of *Paralouatta* was recently introduced by study of the postcranium. It led MacPhee and Meldrum (2006) to suggest *Paralouatta* may have been semiterrestrial.

The fourth fossil species we present as alouattine is *Solimoea acrensis* (Kay and Cozzuol 2006). The type specimen is an isolated lower molar with good crown morphology, identified as an m1. The general description given above for *Alouatta* and *Stirtonia* lower molars, which as we stated appears to be universally regarded as howler monkey-like and largely unique to NWMs, compares favorably with the pattern of *Solimoea*. All are relatively long teeth, with a compact, small elevated trigonid, obliquely oriented trigonid wall (postvallid), elongate talonid, and a long and deeply inflected cristid obliqua.

Caipora bambuiorum, from the same cavern that produced *Protopithecus* (Cartelle and Hartwig 1996), is in our view the only known extinct ateline (but see footnote above). It is included here for its comparative value in assessing the morphocline polarity of traits among the atelids.

2.3.2 Body Size

Body size deserves special mention here and we consider it separately from the rest of the character analysis for reasons given above. We provide a series of alternative weight estimates for the fossils based on regressions using different taxonomic samples of anthropoids and different independent variables, both dental and cranial (Conroy 1987; Kay et al. 1998; Sears et al. 2008) (Fig. 2.6). We caution, however, that difficulties remain and, as indicated above, the postcranial skeleton may be more suitable for estimating body size in *Protopithecus* and *Caipora* (Halenar 2011a, b). Some equations using skulls have relatively low R^2 values so they cannot be considered highly reliable for projections. While the equations for molars have R^2 values of 0.9 or greater, lower molars are missing from *Protopithecus*. *Caipora*, which is probably a frugivore, may also have relatively small teeth, which may bias a molar-based weight estimation. Nevertheless, in our analysis *Stirtonia* and *Paralouatta* fall within the range of modern howler monkeys in body mass, as does *Solimoea*. As noted, new body mass estimates for *Protopithecus* and *Caipora* were deemed necessary as the original estimates were calculated from regression equations based on a catarrhine reference sample (Hartwig 1995; Cartelle and Hartwig 1996; Hartwig and Cartelle 1996). Alternative regression equations to estimate size were calculated using a sample of primates encompassing a wide range of body sizes and locomotor patterns (for sample composition see Halenar 2011a, b). For this exercise, the centroid sizes of the epiphyses of various long bones were employed as the skeletal estimator and equations were generated based on the entire

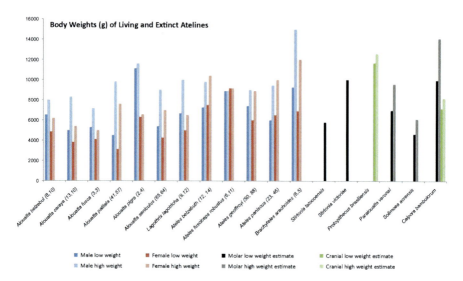

Fig. 2.6 Male and female body weights as reported in the literature for the living atelids (DiFiore and Campbell 2007) and their fossil relatives, the latter based on tooth and/or skull measurements. Weights for *Stirtonia* and *Paralouatta* are from Fleagle (1999) and MacPhee and Meldrum (2006), respectively. For other fossil species, including *Paralouatta* for which additional estimates are included, weights were calculated using the monkey, anthropoid, all primate, and female anthropoid regression equations of Conroy (1987) and the female platyrrhine equation of Kay et al. (1998). Body size estimates based on skull length and bizygomatic width were derived from Sears et al. (2008) equations. The highs and lows are shown instead of averages to demonstrate the wide and overlapping range of body sizes seen in the living atelids, making body mass a difficult character to code and interpret via character analysis. Estimates based on cranial measures are deemed less reliable because of low coefficients of determination (R^2) in the original regressions. Body mass estimates for *Protopithecus* and *Caipora* using craniodental measures are substantially below previous reports, but the original estimates of 20–25 kg are confirmed based on postcranial regression equations (Halenar 2011a)

sample, the platyrrhines only, and the atelids only. Three aspects of "body size" were predicted for the fossil: body weight (kg), total length (TOTL; mm) which includes the length of the tail (TAIL), and trunk length, head, and body (TrL; mm) which includes the length of the skull and trunk (TOTL = TAIL + TrL; Ford and Corruccini 1985). A relatively wide range of body size estimates was thus recovered for *Protopithecus*: 12–35 kg, 1,479–1,887 mm TOTL, and 613–831 mm TrL. This range reflects the use of different skeletal elements, reference samples, and regression models. The equation with the combined highest R^2 (=0.98), lowest %SEE (=11.0), MPE (=14.7), and QMLE (=1.005) is that which uses the distal humerus with a platyrrhine-only reference sample; this gives an estimate of 28 kg for the more recently discovered specimen from Toca da Boa Vista and 24 kg for the original specimen discovered by Lund in Lagoa Santa. Condensing all of the estimates into an average, disregarding the obvious extreme outliers in estimate and confidence statistics, gives a body weight of approximately 23 kg, 1,675 mm TOTL, and 710 mm TrL. As an alternative to compiling an average value, a histogram of all body weight predictions shows 19, 21, and 25 kg as the most frequent estimates with a reasonable range from 17 to 29 kg. This puts *Protopithecus* in the size range of a large male baboon or proboscis monkey and confirms its place in a large-bodied category that no longer exists among extant platyrrhines.

For simplicity, and taking into account the considerations discussed above, within atelids we code the range of character states (Table 2.2) describing body mass at the generic level as medium and large, choosing these terms in part as a semantic device to distinguish them from other platyrrhines often regarded as being middling in size for the radiation, e.g., pitheciids and *Cebus* (e.g., Hershkovitz 1977). We recognize this grossly underrepresents intrageneric diversity (and likely selection for body mass at the species level) and especially the nature and complex distribution of sexual dimorphism among atelids. But it is a useful, operational approximation considering the foci of this study, fossils and genus-level systematics.

2.3.3 Character Analysis

Table 2.2 also summarizes the taxonomic distribution of the ten features we assess in detail. As mentioned, the major reasons for selecting these are that they tend to diagnose *Alouatta* as a genus, defining it morphologically, phylogenetically, and adaptively relative to other living NWM, and they are well represented in the cranial remains of three fossil genera. The fourth, *Solimoea*, is obviously an exception.

2.3.3.1 Facial Proportions

Rosenberger (1992) and Rosenberger and Strier (1989) suggested that the *Lagothrix*-like condition of the facial skeleton, here termed "moderately large", is ancestral overall for atelids (Figs. 2.3 and 2.4). This was based, in part, on the interpretation that there are two other extremes in the atelid morphocline, exemplified by the *Ateles* and the *Alouatta* poles, each one highly likely to be derived since they are associated functionally with novel adaptations. In *Ateles*, ripe fruit frugivory is linked with reduction of the cheek teeth, well-developed anterior teeth (e.g., Rosenberger 1992; Anthony and Kay 1993), and a small face. This pattern occurs in *Caipora* as well. In *Alouatta*, massive changes in the placement and orientation of the large facial skeleton are associated with specializations of the cranial base related to extreme enlargement of the hyoid and the production of stentorian vocalizations (see Biegert 1963). Cheek teeth are also relatively large and anterior teeth are proportionately small (e.g., Rosenberger 1992; Anthony and Kay 1993). Regarding the fossils, we interpret the face of *Protopithecus* as moderately large, hence similar to the condition seen in *Lagothrix*, although more work needs to be done on the allometry of this region in the large-bodied fossil. *Paralouatta*, however, clearly does have a relatively large, long face resembling *Alouatta* in its proportions. Of the fossils under consideration, it is most comparable to *Alouatta* with a markedly prognathic snout, but similar prognathism is also evident in *Protopithecus*.

Table 2.2 Character-analysis distributions

	Atelid morphotype	Ateline morphotype	Caipora	Alouattine morphotype	Protopithecus	Paralouatta	Solimoea	Stirtonia	Alouatta
Body size	Medium	Medium	Large	Medium	Large	Medium	Medium	Medium	Medium
1. Facial proportions	Moderately large	Moderately large	Small	Large	Moderately large	Large	NA	NA	Large
2. Craniofacial haft	Non-airorhynchous	Non-airorhynchous	Non-airorhynchous	Airorhynchous	Airorhynchous	Airorhynchous	NA	NA	Airorhynchous
3. Postorbital constriction	Moderate	Moderate	Moderate	Marked	Marked	Marked	NA	NA	Marked
4. Cranial crests	Moderate lines	Moderate lines	Reduced lines	Marked lines	Marked lines, compound crests	Marked lines, compound crests	NA	NA	Marked lines, compound crests
5. Nuchal plane	Flat, unreduced, subvertical	Flat, unreduced, subvertical	Rounded, unreduced	Flat, unreduced, subvertical	Flat, enlarged, subvertical	Flat, reduced, subvertical	NA	NA	Flat, reduced, vertical
6. Foramen magnum	Unreduced, posterior	Unreduced, posterior	Unreduced, posterior	Reduced, far posterior	Reduced, far posterior	Reduced, far posterior	NA	NA	Reduced, far posterior
7. Brain size and shape	Unreduced, non-cylindrical	Unreduced, non-cylindrical	? Enlarged, globular	Reduced, non-cylindrical	Reduced, cylindrical	Reduced, cylindrical	NA	NA	Reduced, cylindrical
8. Basicranial shape	Not elongate	Not elongate	Short	Elongate	Elongate	Elongate	NA	NA	Elongate
9. Incisor proportions	Intermediate	? Intermediate	Enlarged	Intermediate	Enlarged	? Intermediate	NA	? Reduced	Reduced
10. Molar relief and crown shape	Intermediate	Intermediate	Reduced relief, short	Cristodont, elongate	Cristodont	Moderately cristodont, elongate	? Cristodont, elongate	Cristodont, elongate	Cristodont, elongate

The atelid morphotype conditions generally reflect the states evident in non-atelid platyrrhines, which we consulted as a collective out-group. No transformations are inferred for the ateline morphotype. See text for explanations regarding the questionable character states for incisor proportions and molar relief in the fossil taxa

2.3.3.2 Craniofacial Haft

A feature correlated with facial size and prognathism is the orientation of the face relative to the braincase. *Alouatta*, is unusual and highly derived among platyrrhines in having an uptilted rostrum, a condition known as airorhynchy (Figs. 2.3 and 2.4). This design contributes to the expansion of space in the neck for the permanently inflated air sacs inside the hollowed-out hyoid bone and its associated cartilages. Airorhynchy is also linked functionally with elongation of the cranial base (see below). *Paralouatta* closely resembles *Alouatta* in this respect, although the dorsal tilt of the face seems to be less exaggerated. Even though the tip of the fossil's snout is broken near the level of the canines, it is evident that the toothrow is nearly as arched in lateral view, forming an exaggerated curve of Spee. *Protopithecus* has a modestly uptilted face as well. The rostra of other platyrrhines are constructed differently and are generally non-airorhynchous, as in *Caipora*. The lateral profile of the *Brachyteles* dental arcade, with large postcanine teeth and a moderately deep but non-prognathic face, is slightly curved upward anteriorly.

2.3.3.3 Postorbital Constriction

The degree of postorbital constriction is influenced by braincase size and shape, craniofacial proportions, and the anteroposterior alignment of the face at the craniofacial junction (Fig. 2.3). The modern alouattines and atelines present contrasting character states. The constriction is moderate in atelines, including *Caipora*, but it is marked (i.e., narrow or waisted) in *Alouatta*. In atelines such as *Ateles* and *Brachyteles*, with retracted, subcerebral (below the horizontal axis of the brain) faces and large, relatively globular braincases, width at the craniofacial junction is not constricted. But even in *Lagothrix*, where the braincase is not globular, the constriction is unimpressive, as it tends to be in other platyrrhines, suggesting that this state is ancestral in atelids. In *Alouatta*, in contrast, the combination of a precerebral, uptilted face, massive width of the posterior face, and narrow braincase produces the markedly constricted effect. In ventral view (Fig. 2.4), *Paralouatta* resembles howler monkeys in these factors. The same is evident in *Protopithecus*, but it manifests differently because the braincase is quite wide posteriorly, owing largely to well-developed exocranial superstructures.

2.3.3.4 Cranial Crests

The development of exocranial temporal lines and nuchal crests may be strongly influenced by size, age, gender, and sexual dimorphism, indicating caution in making comparisons without population samples of fossil atelids (Figs. 2.3 and 2.5). Of the fossil specimens considered here, *Caipora* is a young adult; *Protopithecus* is an adult but with relatively unworn teeth; *Paralouatta* is an adult with advanced postcanine tooth wear. Judging by canine prominence, anterior premolar size, and the known level of sexual dimorphism in the living species, *Caipora* and

Protopithecus appear to be males. The canine crowns of *Paralouatta* are broken away, but the expression of cranial crests suggests the skull may also be male.

Among modern atelids, moderate to prominent temporal lines, evenly developed anteriorly and posteriorly, are present in *Lagothrix*, *Brachyteles*, and *Alouatta*. Strong nuchal lines or crests tend to occur in the robust *Alouatta* males and are quite common interspecifically. Neither temporal lines nor nuchal crests are well-developed in *Ateles*, or in *Caipora*, which corresponds with their reduced cheek teeth and rounded, large braincases, among other factors. We surmise this is a correlate of the soft/ripe-frugivory feeding complex seen in *Ateles*. It is also related to what may be termed a semi-orthograde head carriage, i.e., the head is not strongly cantilevered off the vertebral column but tends to rest atop the cervical vertebrae in compliance with tail-assisted climbing and other semi-orthograde positional behaviors.

With a small braincase and large temporal and nuchal muscles, a compound temporo-nuchal crest is well-developed in *Alouatta*, although its distribution among the modern species has not been mapped out. Nevertheless, in the larger and more robust males, laterally away from the midline, the temporal enthuses fuse with the nuchal line to form a raised lateral margin of the nuchal region. By comparison with other atelids, these features are extremely well-developed in *Protopithecus*, probably as an elaboration of an *Alouatta*-like pattern exaggerated by the allometrics of a very large body size. The compound temporo-nuchal crest is present also in *Paralouatta* but exhibited less dramatically, comparing more favorably with the variations seen in *Alouatta*.

2.3.3.5 Nuchal Plane

Alouatta is unusual among platyrrhines in having a nuchal plane that is flat, often rugose in texture, reduced in size, vertically oriented (Fig. 2.5), and exhibiting a semicircular dorsal perimeter when viewed from behind—all features corresponding with the cylindrical shape of the braincase and pronounced set of muscle attachments on the occiput. Sex differences exist, but this overall *gestalt* is fixed in howler monkeys. It relates to head carriage and craniofacial mass. The foramen magnum and occipital condyles are directed posteriorly rather than ventrally as in other NWM, meaning that the large, heavy head of *Alouatta*, which is eccentrically loaded up front due to its snouty prognathic design, tends to be extended dog- or lemur-like out from the shoulders and neck, in typical pronograde fashion. The flat, vertical nuchal plane presumably gives the trapezius and other neck muscles apt mechanical advantage in supporting the horizontally disposed skull. Following previous arguments, we regard the *Lagothrix*-like condition, a relatively flat, subvertical, and unreduced nuchal plane as ancestral in atelids. The contrasting rounded and unreduced morphology of *Ateles* and *Caipora* is considered derived for atelines. *Paralouatta* resembles *Alouatta* generally, but the plane of the nuchal region appears to be more primitive, slanted in a manner that compares with *Lagothrix*. Similarly, *Protopithecus* retains an inclined nuchal plane but it is also greatly expanded laterally, owing to the hypertrophic compound temporo-nuchal crests. We hypothesize that this is at least partly an allometric contingency but it may also reflect differences in the proportions of the jaw adductor muscles. The apparent lack of gonial expansion in comparison to *Alouatta* suggests that *Protopithecus* had a less elaborate masseter complex, while the enlarged temporo-nuchal crests suggest the posterior part of the temporalis muscle was exaggerated instead.

2.3.3.6 Foramen Magnum

Both the position (see Schultz 1955) and relative size of the foramen magnum differs among atelids. These features are related to head posture and endocranial volume. As indicated, it is extremely posteriorly positioned in *Alouatta*, *Paralouatta*, and *Protopithecus*, especially so in howler monkeys (Fig. 2.4), and the particulars conform to the degree of nuchal plane modifications in these genera. *Alouatta* exhibits the most derived pattern. The more anterior location of the foramen magnum in atelines is consistent with the more common location documented by Schultz, which is ancestral for NWMs and atelines. For convenience we code it as posterior, offsetting it from the condition in *Saimiri* and *Cebus*. They have foramina magna that are distinctly more "centrally" located within the long axis of the skull.

The foramen magnum also varies in proportions, with atelines and alouattines clearly having different scaling patterns (Fig. 2.7a). Relative to basicranial length, foramen magnum area (length × breadth) is small in *Alouatta*, *Paralouatta*, and *Protopithecus*, falling well below the scatter of points and the regression line representing modern atelines and *Caipora*. The size of the foramen is also closely correlated with endocranial volume across primates (e.g., Jerison 1973; Martin 1990). Brain size is relatively larger in atelines than alouattines (Fig. 2.7b), which helps explain why the foramen magnum is proportionately smaller in the latter. Again, the alouattine condition is very likely the derived pattern among atelids, given the rarity of de-encephalization, which is often associated in mammals with herbivory or folivory (see section below for an expanded explanation). But it is also possible that to some degree, relatively small brain size in this group reflects primitive platyrrhine proportions. The status of atelines also requires further examination. While *Ateles* and *Brachyteles* have been singled out as having derived, elevated relative brain sizes (Cole 1995, in Hartwig 2005), it appears from this assessment that all the atelines, including *Lagothrix* and *Caipora*, jointly share this pattern. Even *Brachyteles*, a genus that might be expected to have experienced selection for a reduced relative brain size as a correlate to its more leafy diet, follows the ateline pattern and is relatively larger-brained than any alouattine (Rosenberger et al. 2011).

2.3.3.7 Brain Size and Shape

As indicated, among modern platyrrhines, it is well established that howler monkeys have an unusually small brain size relative to body mass (e.g., Stephan and Andy 1964; Hershkovitz 1970; Stephan 1972; Clutton-Brock and Harvey 1980; Eisenberg 1981; Martin 1984, 1990; Hartwig 1996), and this likely represents, at least in part, an

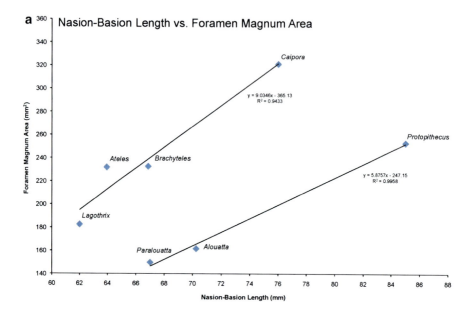

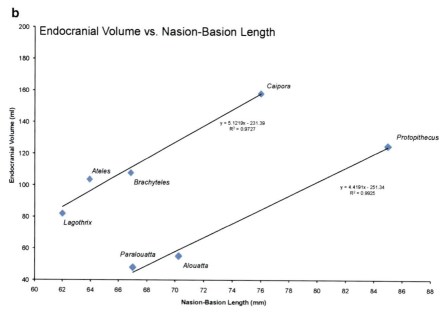

Fig. 2.7 Bivariate plots of (**a**) foramen magnum area and (**b**) endocranial volume relative to nasion-basion length in atelids. Note the separate distributions of the atelines, including *Caipora*, toward the *top* of the graph and the alouattines, including *Protopithecus* and *Paralouatta*, toward the *bottom*. Data points for the living genera are sex-pooled means from the following samples: *Brachyteles arachnoides*, 3; *Ateles belzebuth*, 16; *Lagothrix lagotricha*, 15; *Alouatta belzebuth*, 16. Alouattines have relatively smaller brains, even the frugivorous *Protopithecus*, while the leaf-eating ateline *Brachyteles* does not have a reduced brain size

adaptation to folivory (e.g., Clutton-Brock and Harvey 1980; Eisenberg 1981; Martin 1984, 1990; Harvey and Clutton-Brock 1985; Rosenberger et al. 2011). Since folivory is clearly a derived habit among NWM, the correlative, relatively small *Alouatta* brain may have evolved via de-encephalization. This does not, however, mean there is no component of primitiveness in this character state, for early platyrrhines probably had smaller brain sizes than modern members (see Tejedor et al. 2006; Sears et al. 2008), parallel increases in relative brain size occurred, and basal lineages of the major clades may logically be expected to retain the primitive platyrrhine condition (see Hartwig et al. 2011).

The conjunction of a relatively small brain in howler monkeys with a posteriorly positioned foramen magnum, small nuchal plane, extended basicranial platform, and precerebral, airorhynchous face makes it likely that the cylindrical shape of the *Alouatta* braincase is a derived by-product of a spatial packaging phenomenon (i.e., Biegert 1963; Gould 1977; Ross and Ravosa 1993). The *Protopithecus* skull closely resembles *Alouatta* in this respect although its braincase differs in shape for it is wider posteriorly than anteriorly, a pattern not seen elsewhere among platyrrhines. However, some of this is an exocranial effect of the very wide nuchal plane, with well-developed lateral nuchal crests and a massive set of temporal roots supporting the zygomatic arches. The finding of Krupp et al. (2012) that the *Protopithecus* brain resembles *Alouatta* in overall shape helps explain why *Protopithecus* cannot share the globular braincase shape of *Ateles, Brachyteles,* and *Caipora,* all at the opposite end of the spectrum. Roughly speaking, the *Protopithecus* braincase may more closely resemble *Lagothrix,* whose morphology may be described as non-cylindrical for convenience. This would suggest it shares the ancestral condition for atelids.

2.3.3.8 Basicranial Shape

Alouatta is unusual among platyrrhines and other primates in having an elongate basicranium (Fig. 2.4), presumably as another derived correlate of subbasal spatial packaging, i.e., making room for the enlarged hyoid complex (Biegert 1963). However, it should be noted that within *Alouatta*, there is considerable interspecific variation in cranial base shape, with *A. palliata* showing a shorter, more rounded condition (Halenar 2008). *A. seniculus* males appear to be the most exaggerated, perhaps because the foramen magnum is shifted posteriorly to such an extreme degree. We designate the contrasting character states of *Ateles* and *Caipora* as short, but their modified, encephalized skulls suggest this may not be the ancestral atelid or ateline condition. We hypothesize that the deeper morphotype condition is more moderate and designate the primitive condition as "not elongate." Hartwig and Cartelle (1996) pointed out that the *Alouatta*-like elongate pattern is evident in *Protopithecus*, and it is exhibited in *Paralouatta* as well (Rivero and Arredondo 1991; Halenar 2012). We consider the *Protopithecus* morphology less derived than in *Alouatta* and *Paralouatta*, largely because the nuchal plane continues to extend behind it. In agreement with many of the qualitative statements made above regarding facial proportions and airorynchy, 3D geometric morphometric analysis of the

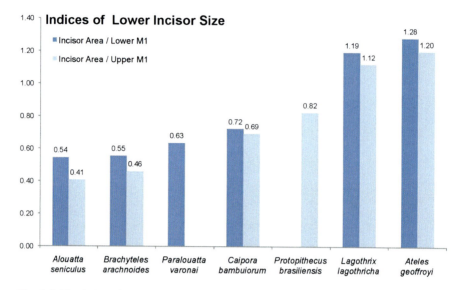

Fig. 2.8 The index of lower incisor size was calculated by dividing the incisal crown cross-sectional area (length×breadth of i1) by first molar area (length×breadth). This exercise was repeated for maxillary and mandibular first molars so as to be able to include both *Paralouatta* and *Protopithecus*. Lower values, which indicate relatively small incisors, correspond with a more folivorous diet, as in *Alouatta*. The position of *Protopithecus* toward the higher end of the scale, with an index proportionately twice the size of *Alouatta*, suggests it was considerably frugivorous

Protopithecus cranial base suggests that it exhibits an intermediate morphology between the extremely derived *Alouatta* and *Ateles* conditions; principal components analysis aligns the fossil with extant *Lagothrix* in terms of its degree of basicranial elongation and flexion (Halenar 2012).

2.3.3.9 Incisor Proportions

Morphologically, the incisors of atelids appear to show an acute sensitivity to selection reflecting critical dietary preferences (Fig. 2.8). Thus, *Alouatta* and *Brachyteles*, the most folivorous platyrrhines, have evolved relatively small-crowned incisors, probably independently (Eaglen 1984; Rosenberger 1992; Anthony and Kay 1993), whereas the other atelids have relatively larger incisors with the lower incisors being distinctly spatulate in shape. Reduced crowns like those of *Alouatta* and *Brachyteles* are not prevalent among other platyrrhines, making it likely that the unreduced condition is ancestral for atelids. The much enlarged incisors of *Ateles* and *Lagothrix* may be another specialization related to intensive fruit harvesting behaviors. This makes it difficult to specify the morphotype ateline condition. By default, we regard it as being intermediate. Importantly, the proportions of *Paralouatta* more closely resemble the condition seen in *Alouatta* and *Brachyteles* than *Protopithecus* or any modern atelines. *Protopithecus* incisors are quite large proportionately, although not to the extent seen in *Ateles* and *Lagothrix*. *Stirtonia* specimens lack incisors, but the well-preserved-type mandible of *S. tatacoensis* has canines positioned relatively close together, suggesting these teeth were not especially enlarged.

2.3.3.10 Molar Relief and Crown Shape

Taking a very abstract approach in order to characterize the morphology of upper and lower molars simultaneously, we define two crown patterns as character states: "low relief", with relatively low cusps and shallow, broad basins, and, "cristodont", having more relief and an emphasis on relatively elevated cusps and lengthy crests, which thus restricts lower molar basins and lengthens the crown (Figs. 2.1 and 2.2). *Alouatta* is the archetypical example of the cristodont pattern with upper molars also exhibiting a set of strongly developed buccal ectoloph crests (especially the centrocrista between paracone and metacone) as well as a stylar region with a robust buccal cingulum, which is associated with localized crest development. *Brachyteles* (see Rosenberger 1992) shares several features of the cristodont pattern with *Alouatta* but appears to have evolved aspects of it by a different, convergent pathway emphasizing lingual, as opposed to buccal, shear. Hence, the massively developed metaconids and entoconids seen in *Brachyteles* molars (Fig. 2.2).

Among platyrrhines, cristodont molars like those of *Alouatta* and *Brachyteles* do not occur outside of the atelids, so it is reasonable to regard this state as derived (in parallel). The low-relief pattern of *Ateles* and *Lagothrix* is also part of an unusual, large-basin occlusal morphology among NWMs, functionally related to masticating soft, ripe fruit (Kay 1975; Rosenberger 1992; Anthony and Kay 1993). Hence, we interpret both patterns as derived from a still hypothetical architecture we term "intermediate" for convenience. Among the fossils, *Paralouatta* upper molars (Fig. 2.1) clearly share with *Alouatta* well-developed buccal and stylar cristodont features, but the crown is more primitive lingually, retaining the well-differentiated hypocone, for example, that is broadly similar to many living NWM and middle Miocene fossils. The *Paralouatta* cusps and crests also tend to be more blunted than sharp. The morphology of *Protopithecus* is poorly known since the specimen lacks lower molars and the single M1 is broken; however, it evidently does not display the cristodont pattern. The occlusal surface of the upper molar appears to be relatively flat and the premolar cusps are bulbous. Both species of *Stirtonia* have very *Alouatta*-like, cristodont upper molars. *Caipora* exhibits an ateline-like, low-relief pattern.

The cross-sectional crown shape of lower molars also tends to distinguish most atelines from alouattines (Fig. 2.9). All alouattines have relatively long first lower molars. Length exceeds breadth by approximately 25 % or more. Here, again, *Brachyteles* converges on *Alouatta, Stirtonia, Paralouatta,* and *Solimoea*, while *Caipora* is an outlier among atelines. Other modern NWMs tend to have proportions similar to living atelines. First molars of species of *Aotus, Callicebus, Pithecia,* and *Cebus*, for example, have length/breadth ratios of 1.0–1.1 (Fig. 2.9). The overall functional continuity of this aspect with others that are part of the cristodont molar pattern indicates that elongation is a homologously derived element of crown design in alouattines, probably related to maximizing the linear length of shearing blades.

2.4 Discussion

2.4.1 Implications: Taxonomic Composition of the Fossil Alouattines and the Problem of Solimoea

The status of two of the three fossils at the focus of our character analysis has not been challenged. *Stirtonia* and *Protopithecus* present a robust, persuasive series of craniodental features tying them to *Alouatta*. There are also several postcranial features of the hip and thigh that may link *Protopithecus* and *Alouatta* (Halenar 2012). The affinities of *Paralouatta* have been debated (Rivero and Arredondo 1991; Horovitz and MacPhee 1999; Rosenberger 2002). As evident above, we have proceeded with the working hypothesis that the Cuban genus is an alouattine and refer readers elsewhere (Rosenberger 2002; Rosenberger et al. 2008) for arguments countering the notion that *Paralouatta* is a member of a monophyletic Caribbean group most closely affiliated with *Callicebus*. In nine of ten cranial features assessed here, *Paralouatta* shares the same derived state with *Alouatta* (Table 2.2). In two characters *Paralouatta* is "one step" less derived. In no cases are there any phenetic discrepancies to challenge the notion that these individual, intercorrelated elements are not homologous or functionally contrastive. We thus conclude that *Paralouatta* is a well-established alouattine.

The other species requiring attention is *Solimoea acrensis*. Kay and Cozzuol (2006) maintain that *Solimoea acrensis* is a stem ateline. The claim is based on a cladistic analysis using PAUP (Swofford 2002) of the two specimens they allocate to the taxon, an isolated lower molar inferred to be m1, the type specimen, and a referred maxillary fragment with P3–4, which is in poor condition. It is important to note that the Kay and Cozzuol (2006) analysis is not an independent assessment of morphological evolution among atelids because it is based on the "molecular scaffold" approach. In other words, the results of a molecular study were first used to arrange the topology of the tree. Then PAUP mapped characters onto the tree to produce the most economical distribution of states among the taxa.

We do not find the arguments compelling and suggest, alternatively, that *Solimoea* is an alouattine. There are major concerns that raise questions and warrant discussion: (1) the existence of distinctive phenetic similarities as well as a unique constellation of derived features shared by the type of *Solimoea* and alouattines, exclusively, and (2) Kay and Cozzuol's reliance on characters from the maxillary specimen which may, in fact, not belong to the same taxon as the type.

The small-basin crown morphology of the type lower molar is far more similar to an alouattine than any of the wide-basined, extant atelines (Fig. 2.2). While *Solimoea* exhibits a crown pattern that appears to be less modified than the highly

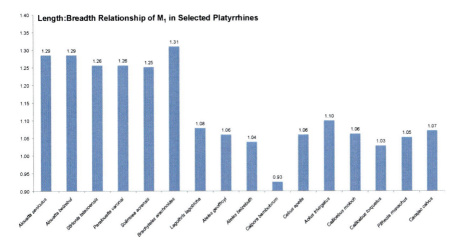

Fig. 2.9 Length breadth ratio (length/breadth) of m1 in selected platyrrhine species. Higher values indicate relatively longer and narrower teeth, a correlate of lengthened shearing blades, and are a derived feature associated with the alouattine clade and, independently, the semi-folivorous species *Brachyteles*. The crown morphology and length: breadth ratio of *Solimoea* aligns it with alouattines

distinctive *Stirtonia* and *Alouatta*, it conforms to expectations of the alouattine morphotype. This configuration appears to be derived for atelids, based on character analysis and, especially, the expanded sense of alouattine diversity that is informed by taking *Paralouatta* into account. *Solimoea* shares with *Alouatta* and *Stirtonia* a morphological combination not seen elsewhere among NWM: (a) abbreviated, mesially narrowing, elevated trigonid and low, elongate, basined talonid; (b) obliquely oriented postvallid; (c) sharply angled cristid obliqua, forming a prominent ectoflexid; and (d) relatively long and narrow crown shape (Fig. 2.9). Buccally, the *Solimoea* lower molar also exhibits a resemblance to *Paralouatta*, whose maxillary molars demonstrate a primitive version of *Alouatta*-like ectoloph features as noted above. This is consistent with the *Solimoea* lower molar simply being more primitive, i.e., less of a "shearing folivore," than the highly committed leaf-eaters *Alouatta* and *Stirtonia*. This functional and dietary inference is a conclusion also reached by Kay and Cozzuol (2006) based on quantification of shearing potential. Concerning resemblances between *Solimoea* and the ateline *Brachyteles*, some are evident in the angularity of the buccal aspect of the crown. However, this is probably partly due to primitiveness as well as a joint emphasis on shearing features.

The allocation of the maxillary specimen to the taxon is not convincing, for it is by no means evident that it is associated with the type lower molar. While there is a general conformity in the sizes of the lower molar and the upper premolars and they were recovered from the same locality, it would not be unusual for there to be several sympatric primate species and genera of similar size at an Amazonian locale (in this regard we note with interest that the gigantic *Protopithecus* and *Caipora* were found almost side by side in a cave, but their taphonomic histories remain a mystery). Kay and Cozzuol (2006) justify this allocation quantitatively, referencing the proportions of the crown areas (length × breadth) of the two specimens. They present a bivariate plot of m1 area vs. P4 area (Fig. 5, p. 677) based on a series of 13 *Lagothrix lagotricha* specimens and note that the plot point for the paired set of Acre fossils falls within the minimum convex polygon that bounds the distribution. We replicated and extended this exercise (Fig. 2.10) but arrive at a different conclusion. In our larger sample population of *L. lagotricha*, when jointly plotted the Acre specimens (termed *Solimoea acrensis* in the figure) do not lie within the polygon. It is also evident there is considerable overlap in the size relationships of m1 and P4 among species and genera of platyrrhines across a broad spectrum of body sizes at the 95 % confidence limits of populations, which undermines the taxonomic usefulness of this criterion (Fig. 2.10). The ellipses show that if the corresponding upper and lower teeth of most *Brachyteles* and *Alouatta*, or of most *Ateles*, *Lagothrix*, and *Cebus*, for example, were potted interchangeably by permutation, there would be no way of distinguishing or sorting confidently any individual tooth or tooth set to a species. Furthermore, our sample of howler monkeys uses *A. seniculus* only. If a smaller species was included in this case study, incidental sampling bias may even confound the metric associations of as many as six genera, *Alouatta*, *Brachyteles*, *Ateles*, *Lagothrix*, *Cebus*, and *Solimoea*.

In essence, the preserved morphology of the upper premolars is insufficient to properly test for an occlusal match with the lower molar, and compatibility in size is of little consequence. The premolars appear to be bunodont, rectangular, and of low relief, with large lingual occlusal surfaces, which is inconsistent with the non-bunodont, moderately high-relief morphology of the lower molar or with the latter's abbreviated, oblique trigonid. It appears to us that these teeth may be mismatched taxonomically and, if so, this negates their utility in the generic diagnosis and cladistic analysis.

2.4.2 Interrelationships, Craniodental Morphology, and Adaptations of Fossil Alouattines

The cladistic interrelationships derived from our character analysis are summarized in Fig. 2.11. Our overall results continue to support prior arguments that *Stirtonia* is the fossil most closely related to *Alouatta*. For example, the upper molar morphology of *Paralouatta* tends to reinforce the *Alouatta-Stirtonia* linkage by default because the Cuban form's crowns are blunter, but its upper molars present a W-shaped ectoloph and moderately well-developed stylar elements, structural features that eventually became trenchant shearing surfaces in *Alouatta* and *Stirtonia*. The lingual aspect of *Paralouatta* upper molars also had not yet developed the sharp, lobe-like hypocone, which is prominent in *Alouatta* and *Stirtonia*.

With a cylindrical braincase and constricted nuchal region, synapomorphies shared with *Alouatta* but combinatorially absent in *Protopithecus*, *Paralouatta* is not only more derived than *Protopithecus* in the direction of *Alouatta*. It also bears

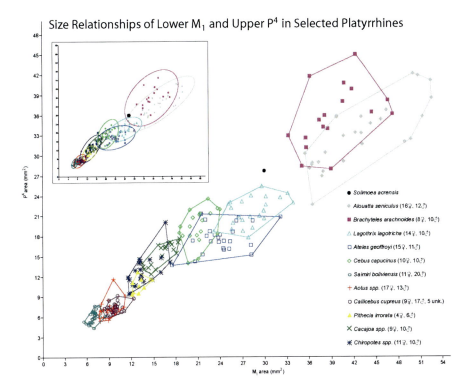

Fig. 2.10 Bivariate plot of m1 area (length × breadth) vs. P4 area (length × breadth) in selected platyrrhine species. A mixed species sample was used for *Aotus*, *Cacajoa*, and *Chiropotes*, including: *A. vociferans*, *A. lemurinus*, *A. infulatus*, *A. nigriceps*, *A. trivirgatus*, and *A. brumbacki*; *Chiropotes albinasus* and *C. satans*; and *Cacajao calvus* and *C. melanocephalus*. Mixed species samples were included to increase sample size and were only marginally more variable than single-species samples for this measure. Minimal convex polygons are shown in the main body of the figure and the *inset* shows 95 % confidence intervals. *Solimoea* is identified by an enlarged *black dot* in the *inset*. The multiple taxonomically overlapping proportions of these teeth across the range of sizes exhibited by platyrrhine species and genera, some of which can occur sympatrically, means that such size associations are not reliable as taxonomic identifiers in fossil assemblages like the Rio Acre sample involving *Solimoea*

a distinctly closer phenetic resemblance to howler monkeys. This offsets *Protopithecus* as a basal member of the alouattine clade given what we know currently of their diversity. In the basicranium as well, *Protopithecus* appears to exhibit the ancestral end of the alouattine morphocline while *Alouatta* and *Paralouatta* occupy the opposite pole. Equally important, *Protopithecus* has a very different dental *gestalt*, lacking both the reduced incisors of *Alouatta* and *Paralouatta* and a cristodont molar pattern as seen in *Alouatta*. Among all the fossils, *Protopithecus* retains the largest combination of dental and cranial features consistent with the alouattine morphotype. In a general way, the morphological pattern is concordant with Rosenberger and Strier's (1989) proposal of a *Lagothrix*-like craniodental morphology being ancestral for alouattines and atelines.

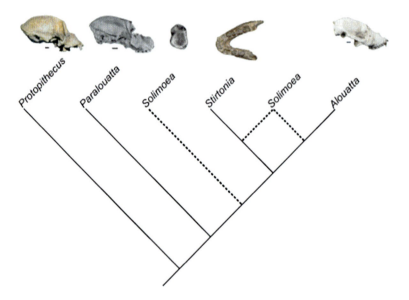

Fig. 2.11 Cladogram showing the proposed interrelationships of alouattine platyrrhines. Dotted lines indicate all possible positions for *Solimoea*. Images are for illustrative purposes and are not shown to scale

Based on one lower molar, the position of *Solimoea* is still difficult to fix. The shape of this tooth does not conform to the apparently open-basin crown morphology of the damaged *Protopithecus* upper molar (or with the bunodont pattern of existing lower premolars). Its small-basin design is not consistent with the advanced atelines, *Lagothrix* and *Ateles*. Resemblances to *Brachyteles* are confined to the more primitive buccal aspect of the crowns, and this generalized angularity of shape is also shared with alouattines. *Solimoea* most closely resembles *Alouatta* and *Stirtonia* overall. The buccally flaring, elevated protoconid; strongly angled pre- and post-hypoconid cristae; and strong trigonid-talonid height differential appear to be a derived combination that would facilitate folivory through selection for additional or more efficient shearing.

Cranially, both *Protopithecus* and *Paralouatta* present a mixture of features resembling the unique patterns exhibited by *Alouatta* in areas of the basicranium and nuchal region, overall braincase shape, and facial structure. The more detailed resemblances shared by *Paralouatta* and *Alouatta* imply an important functional overlap that appears to be related to a novel organization of the skull. These probably relate to fixation of an enlarged hyolaryngeal mechanism in the neck between the rami of the mandible and to a large, cantilevered head. The general organization of *Protopithecus* clearly indicates a shift from a more general, *Lagothrix*-like pattern in the *Paralouatta-Alouatta* direction.

An unexpected outcome of this study involves two related indicators of relative brain size, endocranial volume and foramen magnum area. The same reduced scaling conditions were observed among all alouattines known from cranial material but not in any of the atelines, not even the one genus with a pervasive tendency to eat leaves, *Brachyteles*. As noted previously, the relatively small brain size in *Alouatta* has been regarded as part of its folivorous feeding adaptation, so its occurrence in *Protopithecus*, with a decidedly frugivorous dentition, is thus counterintuitive. Noteworthy is a recent study (Krupp et al. 2012) that independently confirms the relatively small size of the *Protopithecus* brain and demonstrates, via an endocranial cast, its *Alouatta*-like shape and surface morphology. Given that the *Protopithecus* dentition is not howler monkey-like, these findings are consistent with the hypothesis that the shift toward folivory in *Alouatta* may have been predicated on an earlier reliance on seeds (Rosenberger et al. 2011). This initial dependence on seeds could have benefitted from similar energy-saving features like small brains (Rosenberger et al. 2011), if selection for large brains (possibly driven by features such as functional association with a large, complex social organization) was absent. If correct, the comparable small brains of *Protopithecus*, *Paralouatta*, and *Alouatta* more likely reflect a derived de-encephalization than a primitively atelid small-brain pattern, though parceling out these historical factors remains a difficult proposition as we noted above.

Of the three fossil genera, only *Stirtonia*, the howler monkey's closest relative known thus far, can be considered a comparably committed leaf-eater by the detailed functional similarities shared with *Alouatta* in molar morphology and by its inferred body mass. The cheek teeth of *Paralouatta* are more bunodont and given to wearing more flatly, contrasting with *Alouatta*, which exposes lines of dentine along the crown perimeter and stems from a thin-enamel occlusal design that emphasizes and maintains shearing. Thus is it likely that *Paralouatta* molars are not designed optimally for shredding leaves.

Upper molars of *Protopithecus* lack crested shearing perimeter lines, and the large-basined crowns do not seem to resemble either the *Alouatta-Stirtonia* pattern or the morphology of *Paralouatta*. The lower premolars are also notably bunodont. Its incisor teeth are wide, spatulate, and relatively large in cross section, which is consistent with a generalized frugivory as opposed to pitheciine-like, sclerocarpic harvesting (Kinzey 1992; Rosenberger 1992). The summed cross-sectional area of *Protopithecus* incisors is 82 % of the area of the upper molar. Comparable ratios for leaf-eating folivores are 41 % in *Alouatta seniculus* and 46 % for *Brachyteles arachnoides*. Thus, one of the three alouattine fossils shows consistently strong indications of a non-leafy diet in aspects involving both anterior and posterior teeth. *Paralouatta* is also suggestive of the same.

2.5 Conclusion

With the addition of fossils, the subfamily Alouattinae now consists of four extinct genera in addition to *Alouatta*: *Stirtonia, Paralouatta, Protopithecus,* and *Solimoea.* It understates the case to say that the only extant alouattine is a platyrrhine outlier in its morphology, trophic adaptations, and clamorous mode of communicating. But as the diversity of this group is filled in by the discovery of related fossils, it becomes apparent that living howler monkeys are also something of an outlier among alouattines as well, for there is more than one "kind" of alouattine. *Stirtonia* is currently the howler monkey's closest relative and its teeth are barely distinguishable from *Alouatta*, which suggests a fundamentally similar diet. The other fossils help strip back the feeding specializations of the *Alouatta-Stirtonia* group to disclose more primitive anatomies and different dietary adaptations and thus help close the trophic gap between alouattines and atelines. The other crucial aspect regarding evolutionary history revealed by the fossils comes from the cranium, which helps trace another signature adaptation of *Alouatta*, howling. We arrive at these conclusions regarding phylogeny, diversity, divergence, and evolutionary adaptation as outcomes reflecting the particular approach used in our analysis.

Confidence that *Solimoea* is an alouattine is elevated by the observation that its morphology falls *within or along* a morphocline that circumscribes the anatomical patterns of genera whose monophyletic affinities with *Alouatta* are corroborated independently by cranial morphology. It is also affirmed by the observation that what limited functional morphology can be drawn from the *Solimoea* tooth, i.e., its inferred mechanical shearing potential, is consistent with the notion that alouattines, living and extinct, exhibit a range of dental features relating to frugivory-folivory but that neither of the two most basal genera are projected to be as highly committed to folivory as are *Alouatta* and *Stirtonia*. In other words, the functional morphology of *Solimoea* is consistent with models of alouattine evolution, which predict what is self-evident in the broader context of NWM evolution—alouattines more primitive than *Alouatta* would have existed, and they would have exhibited a lesser emphasis on shearing features. On the other hand, there is nothing in any of the models that would predict primitive atelines would also resemble alouattines, only that they are not likely to be comparable to either *Ateles* or *Brachyteles* dietarily and morphologically. A *Lagothrix*-like dentition may still serve adequately as the default model of a morphotypic ateline dentition (Rosenberger and Strier 1989).

Of the two extinct genera known by cranial remains, the alouattine affinities of the Brazilian *Protopithecus* seems well established although there has been debate about the Cuban *Paralouatta* (e.g., MacPhee and Horovitz 2002; Rosenberger 2002). Here, too, functional morphology and a morphoclinal perspective weigh heavily in favor of *Paralouatta* being related cladistically to howler monkeys. With *Protopithecus* and *Alouatta* representing the range of alouattine extremes in terms of cranial character states and patterns, *Paralouatta* seems comfortably nested near the middle anatomically but decidedly closer to *Alouatta* at the more derived end of the spectrum. The functional explanation we propose as the underlying engine behind this transformation series relates more to howling adaptations than to mastication and diet. *Protopithecus*, with its relatively extended basicranium and uptilted face, exhibits the beginnings of a trend toward a greatly enlarged subbasal space, and this may represent a primitive version of the architecture supporting an enlarged hyolaryngeal apparatus. *Paralouatta* is clearly even more *Alouatta*-like in this regard (see also Halenar 2012).

The literature's essentially unanimous endorsement of the hypothesis that *Stirtonia* is an alouattine quite closely related to modern howler monkeys is reinforced by finding that comparable aspects of the molars of *Paralouatta* (and *Solimoea*) are apparently more primitive, while the lesser known dentition of *Protopithecus* presents a different anatomical pattern, perhaps closer to atelines and thus possibly morphotype-like for alouattines. The evidence points strongly to the assessment that this most basal genus of the clade was decidedly frugivorous.

It is difficult to say how long the alouattine clade has been evolving. Recent molecular studies vary in their estimates. For example, Opazo et al. (2006) posit the origins of the clade at 16.75 million years. Schrago (2007) estimates a mean divergence date for *Alouatta* at 12.4 MA but this involves a broad range of ages, 9.1–18.6 MA. In a narrower study, Hodgson et al. (2009) estimate the divergence of *Ateles* relative to non-atelid platyrrhines, thus *Alouatta* by implication, at 18.0 MA, with a range of 15.7–21.6 MA. The fossil record offers an indication of a Miocene differentiation as well. There is one report (Tejedor 2002) of possible alouattines existing in Patagonia in the late-early Miocene, but it is based on meager evidence, isolated canine teeth showing certain resemblances to *Alouatta*. This would be a pre-La Venta occurrence, about four million years prior to *Stirtonia victoriae* and the younger *S. tatacoensis*. The isolated astragalus from Cuba allocated to *Paralouatta marianae* is dated stratigraphically to ~17 MA (MacPhee et al. 2003), also antedating La Venta, but the affinities of this bone must be considered tenuous.

What can be said with some confidence is that by La Venta times, ~11–13 MA, and at the younger Acre site, ~ 8 MA, the modernized members of the *Alouatta* branch of the radiation existed, probably as committed howler monkey-like leaf-eaters living in the formative Amazon basin as this ecological community was being assembled (Rosenberger et al. 2009). The fossils outside this zone, *Paralouatta varonai* in Cuba and *Protopithecus brasiliensis* in eastern Brazil, shed light in a different direction, toward the remote origins of *Alouatta*. Despite the recent geological ages of these two species, they retain a variety of primitive morphologies and occupy positions on the alouattine cladogram basal to the differentiation of the *Stirtonia-Alouatta* lineage. This means that alouattines branching off before the La Venta horizon and outside of Amazonia may have been less committed adaptively to masticating leaves, and may thus come closer to approximating the original adaptive *gestalt* of the group.

Dentally, the fossils, all relatively large-bodied platyrrhines and all expected to have used, as atelids, fully prehensile tails, comprise an adaptive radiation of mixed feeders within the frugivore-folivore spectrum. At least two "stages" in the morphological evolution of the skull that relates to howling can be discerned. *Paralouatta* is sufficiently similar to living howler monkeys in the cranial base and occipital region to suggest the same set of novel specializations were present in terms of the biological roles of the hyoid complex and occipital region as they relate to vocalization and head carriage. *Protopithecus* is less advanced in that direction. But it signifies that at the basal branch of the radiation, the alouattine clade had already shifted toward some semblance of the loud-calling lifestyle of *Alouatta* before the clade produced the specialized capacity to harvest and masticate leaves. Long call adaptations also seem to have preceded the evolution of the howler monkey's postcranial skeletal adaptations (see Hartwig and Cartelle 1996; Jones 2008; Halenar 2011a, b), also emerging in an alouattine that was more frugivorous, as exemplified by *Protopithecus*. As morphologists, we emphasize here that a shift in social behavior, possibly imprinted on the ancestral cranial morphology of the lineage, may have been instrumental in the successful differentiation of alouattines prior to the evolution of the modified dental and locomotor adaptations that one might have expected as essential niche characteristics of this radiation.

We have elsewhere suggested (Rosenberger et al. 2009) that the natural history and biogeography of living *Alouatta* species, potential pioneers due to their dietary flexibility, suggests the possibility that the genus arose not in the greater Amazon basin but elsewhere on the continent in less lush habitats. This idea appears to be consistent with the interpretations we present here, since two genera more basal to the *Alouatta-Stirtonia* clade occur outside Amazonia.

Another important insight is that relatively small brain sizes evolved among alouattines before their intense dental commitment to leaf-eating. *Protopithecus* appears to be a rare example of a small-brained, frugivorous anthropoid. This raises several interesting questions. Is there an evolutionary link between ostentatious howling, which may well have been part of the *Protopithecus* repertoire, and relatively modest brains, perhaps as a morphological constraint on cranial design? Have we overemphasized the physiological and adaptive connections between small brains and leaf-eating? Can facultative leaf-eating in a mixed feeder, perhaps enabled by large body size and concomitantly large guts—*Protopithecus* may be such an example—form a trophic substrate that would motivate selection for de-encephalization? Is relative brain size more sensitive to selection supporting folivorous or semi-folivorous diets (Rosenberger et al. 2011), or facultative leaf-eating, than dentition? Could de-encephalization have evolved as a seed-eating adaptation in the absence of selection for brain size increase? We can only speculate that howling, small brains, and leaf-eating are interconnected as low-energy balancing factors of potential adaptive value: long-distance advertisement that requires little movement or exposure, a brain that can be metabolically maintained relatively cheaply, and a food source that requires little exercise to acquire and produces energy slowly and at low dosages. These characteristics aptly describe facets central to the howler monkey lifestyle, but they offer little in the way of explaining how and why *Alouatta* came to be. The first batch of diverse alouattine fossils suggests some answers lay buried.

Acknowledgements We owe much to many: to the editors of this volume for inviting us to contribute; to the museums mentioned above that make our research possible, especially our home institution, the American Museum of Natural History; to Leandro Salles and Castor Cartelle and their museums, MNRJ and PUC MINAS, for making this project possible; to Marilyn Norconk for sharing her insights on platyrrhines; to Andi Jones and Mike Rose for discussions on the Brazilian fossils and use of their photographs; to the agencies at Brooklyn College (Tow) and the City University of New York (PSC CUNY) for financial assistance to ALR; to the Wenner-Gren Foundation and CUNY NYCEP for a postdoctoral fellowship awarded to MFT which helped support our collaboration; and to NSF DDIG awards to LBH (0925704) and SBC (0726134) and an Alumnae Association of Barnard College Graduate Fellowship to help support SBC in her research on Caribbean primates. We thank the reviewers and editors for many helpful suggestions. The software package PAST was employed for several computations and charts (http://folk.uio.no/ohammer/past/).

References

Anthony MRL, Kay RF (1993) Tooth form and diet in ateline and alouattine primates: reflections on the comparative method. Am J Sci 293:356–382

Biegert J (1963) The evaluation of characteristics of the skull, hands and feet for primate taxonomy. In: Washburn SL (ed) Classification and human evolution. Aldine, Chicago

Cartelle C, Hartwig WC (1996) A new extinct primate among the Pleistocene megafauna of Bahia, Brazil. Proc Natl Acad Sci U S A 93:6405–6409

Clutton-Brock TH, Harvey PH (1980) Primates, brains and ecology. J Zool 207:151–169

Cole TM (1995) Comparative craniometry of the Atelinae (Platyrrhini, Primates): function, development, and evolution. PhD Dissertation. State University of New York at Stony Brook, New York

Conroy GC (1987) Problems of body-weight estimation in fossil primates. Int J Primatol 8:115–137

Cooke SB, Halenar LB (2012) The evolution of body size in the Caribbean primates (Abstract). International Primatological Society meetings, 13 August 2012, Cancun

Delson E, Rosenberger AL (1984) Are there any anthropoid primate living fossils? In: Eldredge N, Stanley SM (eds) Living fossils. Springer, New York

DiFiore A, Campbell C (2007) The Atelines: Variation in ecology, behavior, and social organization. In: Campbell C, Fuentes A, MacKinnon K, Panger M, Bearder S (eds) Primates in Perspective. New York, Oxford University Press

Eaglen RH (1984) Incisor size and diet revisited: the view from a platyrrhine perspective. Am J Phys Anthropol 64:263–275

Eisenberg JF (1981) The mammalian radiations. University of Chicago, Chicago

Fleagle JG (1999) Primate adaptation and evolution. Academic, New York

Fleagle JG, Kay RF, Anthony MRL (1997) Fossil New World monkeys. In: Kay RF, Madden RH, Cifelli RL, Flynn JJ (eds) The history of a Neotropical Fauna: vertebrate paleobiology of the Miocene of tropical South America. Smithsonian Institution, Washington, DC

Flynn JJ, Guerrero J, Swisher CC (1997) Geochronology of the Honda group, Columbia. In: Kay RF, Madden RH, Cifelli RL, Flynn JJ (eds) The history of a Neotropical fauna: vertebrate paleobiology of the Miocene of tropical South America. Smithsonian Institution, Washington, DC

Ford SM, Corruccini RS (1985) Intraspecific, interspecific, metabolic and phylogenetic scaling in platyrrhine primates. In: Jungers WL (ed) Size and scaling in primate biology. Plenum, New York

Gould SJ (1977) Ontogeny and phylogeny. Harvard University, Cambridge

Halenar LB (2008) Agreement between interspecific variation in vocalization patterns and cranial base morphology in *Alouatta*: preliminary results and future directions (Abstract). Am J Phys Anthropol (Suppl. 46):111

Halenar LB (2011a) Reconstructing the locomotor repertoire of *Protopithecus brasiliensis*. I. Body size. Anat Rec 294:2024–2047

Halenar LB (2011b) Reconstructing the locomotor repertoire of *Protopithecus brasiliensis*. II. Forelimb morphology. Anat Rec 294:2048–2063

Halenar LB (2012) Paleobiology of *Protopithecus brasiliensis*, a plus-sized Pleistocene Platyrrhine from Brazil. PhD Dissertation, CUNY Graduate Center, New York

Halenar, LB, Rosenberger, AL (2013) A closer look at the "*Protopithecus*" fossil assemblages: new genus and species from Bahia, Brazil. J Hum Evol 65:374–390

Hartwig WC (1995) A giant New World monkey from the Pleistocene of Brazil. J Hum Evol 28:189–196

Hartwig WC (1996) Perinatal life history traits in New World monkeys. Am J Primatol 40:99–130

Hartwig WC (2005) Implications of molecular and morphological data for understanding atelid phylogeny. Int J Primatol 26:999–1015

Hartwig WC, Cartelle C (1996) A complete skeleton of the giant South American primate *Protopithecus*. Nature 381:307–311

Hartwig W, Meldrum DJ (2002) Miocene platyrrhines of the northern Neotropics. In: Hartwig WC (ed) The primate fossil record. Cambridge University, Cambridge

Hartwig WG, Rosenberger AL, Norconk M, Young Owl M (2011) Relative brain size, gut size and evolution in New World monkeys. Anat Rec 294:2207–2221

Harvey PH, Clutton-Brock TH (1985) Life history variation in primates. Evolution 39:559–581

Hershkovitz P (1970) Notes on Tertiary platyrrhine monkeys and description of a new genus from the late Miocene of Colombia. Folia Primatol 12:1–37

Hershkovitz P (1972) The recent mammals of the Neotropical Region: a zoogeographic and ecological review. In: Keast A, Erk FC, Glass B (eds) Evolution, mammals, and southern continents. State University of New York, Albany

Hershkovitz P (1977) Living New World monkeys (Platyrrhini), with an introduction to the primates, vol 1. University of Chicago, Chicago

Hodgson JA, Sterner KN, Matthews LJ, Burrell AS, Jania RA, Raaum RL, Stewart C-B, Disotell TR (2009) Successive radiations, not stasis, in the South American primate fauna. Proc Natl Acad Sci U S A 106:5534–5539

Horovitz I, MacPhee RDE (1999) The quaternary Cuban platyrrhine *Paralouatta varonai* and the origin of the Antillean monkeys. J Hum Evol 36:33–68

Jerison HJ (1973) Evolution of the brain and intelligence. Academic, New York

Jones AL (2008) The evolution of brachiation in ateline primates, ancestral character states and history. Am J Phys Anthropol 137:123–144

Kay RF (1975) The functional adaptations of primate molar teeth. Am J Phys Anthropol 43:195–216

Kay RF, Cozzuol MA (2006) New platyrrhine monkeys from the Solimoes Formation (late Miocene, Acre State, Brazil). J Hum Evol 50:673–686

Kay RF, Frailey CD (1993) Large fossil platyrrhines from the Rio Acre local fauna, late Miocene, western Amazonia. J Hum Evol 25:319–327

Kay RF, Hylander WI (1978) The dental structure of mammalian folivores with special reference to Primates and Phalangeroidea (Marsupialia). In: Montgomery GG (ed) The ecology of arboreal folivores. Smithsonian Institution, Washington, DC

Kay RF, Madden R, Plavcan JM, Cifelli RL, Diaz JG (1987) *Stirtonia victoriae*, a new species of Miocene Colombian primate. J Hum Evol 16:173–196

Kay RF, Johnson D, Meldrum DJ (1998) A new pitheciin primate from the middle Miocene of Argentina. Am J Primatol 45:317–336

Kinzey WG (1992) Dietary adaptations in the Pitheciinae. Am J Phys Anthropol 88:499–514

Krupp A, Cartelle C, Fleagle JG (2012). Size and external morphology of the brains of the large fossil platyrrhines *Protopithecus* and *Caipora*. (Abstract) Am J Phys Anthropol (Suppl. 54):186

Lund PW (1838) Blik paa Brasiliens dyreverden for sidste jordomvaeltning. Det Kongelige Danske Videnskabernes Selskabs Naturvidenskabelige og Mathematiske Afhandlinger 8:61–144

MacPhee RDE, Horovitz I (2002) Extinct quaternary platyrrhines of the Greater Antilles and Brazil. In: Hartwig WC (ed) The primate fossil record. Cambridge University, Cambridge

MacPhee RDE, Horovitz I (2004) New craniodental remains of the Quaternary Jamaican monkey *Xenothrix mcgregori* (Xenotrichini, Callicebinae, Pitheciidae), with a reconsideration of the *Aotus* hypothesis. Am Mus Novitates 3434:1–51

MacPhee RDE, Iturralde-Vinent MA (1995) Origin of the Greater Antillean land mammal fauna, 1: new Tertiary fossils from Cuba and Puerto Rico. Am Mus Novitates 3141:1–32

MacPhee RDE, Meldrum J (2006) Postcranial remains of the extinct monkeys of the Greater Antilles (Platyrrhini, Callicebinae, Xenotrichini), with a consideration of semiterrestriality in *Paralouatta*. Am Mus Novitates 3516:1–65

MacPhee RDE, Horovitz I, Arredondo O, Vasquez OJ (1995) A new genus for the extinct Hispaniolan monkey *Saimiri bernensis* Rimoli, 1977, with notes on its systematic position. Am Mus Novitates 3134:1–21

MacPhee RDE, Iturralde-Vinent M, Gaffney ES (2003) Domo de Zaza, an early Miocene vertebrate lin South-Central Cuba, with notes on the tectonic evolution of Puerto Rico and the Mona Passage. Am Mus Novitates 3394:1–42

Martin RD (1984) Body size, brain size and feeding strategies. In: Chivers D, Wood B, Bilsborough A (eds) Food acquisition and processing in primates. Plenum, New York

Martin RD (1990) Primate origins and evolution: a phylogenetic reconstruction. Chapman and Hall, London

Opazo JC, Wildman DE, Prychitko T, Johnson RM, Goodman M (2006) Phylogenetic relationships and divergence times among New World monkeys (Platyrrhini, Primates). Mol Phylo Evol 40:274–280

Rivero M, Arredondo O (1991) *Paralouatta varonai*, a new quaternary platyrrhine from Cuba. J Hum Evol 21:1–11

Rosenberger AL (1980) Gradistic views and adaptive radiation of platyrrhine primates. Z Morphol Anthropol 71:157–163

Rosenberger AL (1992) Evolution of feeding niches in new world monkeys. Am J Phys Anthropol 88:525–562

Rosenberger AL (2002) Platyrrhine paleontology and systematics: the paradigm shifts. In: Hartwig WC (ed) The primate fossil record. Cambridge University, Cambridge

Rosenberger AL, Kinzey WG (1976) Functional patterns of molar occlusion in platyrrhine primates. Am J Phys Anthropol 45:281–298

Rosenberger AL, Strier KB (1989) Adaptive radiation of the ateline primates. J Hum Evol 18:717–750

Rosenberger AL, Setoguchi T, Shigehara N (1990) The fossil record of callitrichine primates. J Hum Evol 19:209–236

Rosenberger AL, Halenar LB, Cooke SB, Hartwig WC (2008) Morphology and evolution of the spider monkey, genus *Ateles*. In: Campbell C (ed) Spider monkeys: behavior, ecology and evolution of the genus *Ateles*. Cambridge University, New York

Rosenberger AL, Tejedor MF, Cooke SB, Pekkar S (2009) Platyrrhine ecophylogenetics, past and present. In: Garber P, Estrada A, Bicca-Marques JC, Heymann EW, Strier KB (eds) South American primates: comparative perspectives in the study of behavior, ecology and conservation. Springer, New York

Rosenberger AL, Halenar LB, Cooke SB (2011) The making of platyrrhine semi-folivores: models for the evolution of folivory in primates. Anat Rec 294:2112–2130

Ross C, Ravosa M (1993) Basicranial flexion, relative brain size, and facial kyphosis in nonhuman primates. Am J Phys Anthropol 91:305–324

Ruff CB (2003) Long bone articular and diaphyseal structure in Old World monkeys and apes II: estimation of body mass. Am J Phys Anthropol 120:16–37

Schrago CG (2007) On the time scale of the New World primate diversification. Am J Phys Anthropol 132:344–354

Schultz AH (1955) The position of the occipital condyles and of the face relative to the skull base in primates. Am J Phys Anthropol 13:97–120

Sears KE, Finarelli JA, Flynn JJ, Wyss AR (2008) Estimating body mass in New World "monkeys" (Platyrrhini, Primates) from craniodental measurements, with a consideration of the Miocene platyrrhine, *Chilecebus carrascoensis*. Am Mus Novitates 3617:1–29

Setoguchi T (1980) Discovery of a fossil primate from the Miocene of Colombia. Monkey 24:64–69

Setoguchi T, Rosenberger AL (1985) Miocene marmosets: first evidence. Intl J Primatol 6:615–625

Setoguchi T, Watanabe T, Mouri T (1981) The upper dentition of *Stirtonia* (Ceboidea, Primates) from the Miocene of Colombia, South America and the origin of the postero-internal cusp of upper molars of howler monkeys (*Alouatta*). Kyoto University Overseas Research Reports of New World Monkeys 3:51–60

Stephan H (1972) Evolution of primate brains: a comparative anatomical investigation. In: Tuttle RH (ed) The functional and evolutionary biology of primates. Aldine, Chicago

Stephan H, Andy OJ (1964) Quantitative comparison of brain structures from insectivores to primates. Am Zool 4:59–74

Stirton RA (1951) Ceboid monkeys from the Miocene of Colombia. Univ Calif Pub Geol Sci 28:315–356

Strier KB (1992) Ateline adaptations: behavioral strategies and ecological constraints. Am J Phys Anthropol 88:515–524
Swofford DL (2002) PAUP. Phylogenetic analysis using parsimony. Version 4.0b10 (Altivec). Sinauer, Sunderland
Szalay FS, Bock WJ (1991) Evolutionary theory and systematics: relationships between process and patterns. J Zool Syst Evol Res 29:1–39
Szalay FS, Delson E (1979) Evolutionary history of the primates. Academic, New York
Tejedor MF (2002) Primate canines from the early Miocene Pinturas Formation, southern Argentina. J Hum Evol 43:127–141
Tejedor MF, Tauber AA, Rosenberger AL, Swisher CC, Palacios ME (2006) New primate genus from the Miocene of Argentina. Proc Natl Acad Sci U S A 103:5437–5441
Tejedor MF, Rosenberger AL, Cartelle C (2008) Nueva especie de *Alouatta* (Primates, atelineae) del Pleistoceno tardío de Bahía, Brasil. Ameghin 45:247–251
Wiley DF, Amenta N, Alcantara DA, Ghosh D, Kil YK, Delson E, Harcourt-Smith W, Rohlf FJ, St. John K, Hamann B (2005) Evolutionary morphing (extended abstract and associated video presentation). Proceedings of IEEE visualization conference 2005, pp. 1–8

Chapter 3
The Taxonomy of Howler Monkeys: Integrating Old and New Knowledge from Morphological and Genetic Studies

Liliana Cortés-Ortiz, Anthony B. Rylands, and Russell A. Mittermeier

Abstract The taxonomic history of the howler monkeys, genus *Alouatta*, has been long, complex, and filled with omissions and mistakes. This has created confusion over the validity of different taxa. Here we review the taxonomic history of the genus and evaluate the validity of the different taxa based on current knowledge generated through morphological and genetic studies. We recognize nine species of howlers (*A. palliata, A. pigra, A. seniculus, A. arctoidea, A. sara, A. macconnelli, A. guariba, A. belzebul, A. caraya*) and three more taxa that we tentatively consider full species (*A. nigerrima, A. ululata, A. discolor*), but for which genetic and/or morphological studies are required to confirm this status. We recognize five subspecies in *A. palliata* (*A. p. mexicana, A. p. palliata, A. p. coibensis, A. p. trabeata*, and *A. p. aequatorialis*), three in *A. seniculus*, (*A. s. seniculus, A. s. juara*, and *A. s. puruensis*), two in *A. guariba* (*A. g. guariba* and *A. g. clamitans*), and acknowledge the possibility that *A. pigra* may have two subspecies (*A. p. pigra* and *A. p. luctuosa*). Most species and subspecies require field studies to determine their actual distribution ranges. Furthermore, a combination of morphological and genetic analyses is needed to confirm the validity of several taxa. Given the broad presence of howler monkeys in the Neotropics, these studies would require the collaboration of a multidisciplinary network of researchers across the range of distribution of the genus.

Resumen La historia taxonómica de los monos aulladores, género *Alouatta*, ha sido larga, compleja y llena de omisiones y errores. Esto ha creado confusión respecto la validez de los distintos taxa. En este capítulo revisamos la historia taxonómica del género y evaluamos la validez de los distintos taxa con base en el conocimiento actual generado a través de estudios morfológicos y genéticos.

L. Cortés-Ortiz (✉)
Museum of Zoology, Department of Ecology and Evolutionary Biology,
University of Michigan, Ann Arbor, MI, USA
e-mail: lcortes@umich.edu

A.B. Rylands • R.A. Mittermeier
Conservation International, 2011 Crystal Drive, Arlington, Washington, DC, USA
e-mail: a.rylands@conservation.org; r.mittermeier@conservation.org

Reconocemos nueve especies de monos aulladores (*A. palliata, A. pigra, A. seniculus, A. arctoidea, A. sara, A. macconnelli, A. guariba, A. belzebul, A. caraya*) y tres taxa que tentativamente consideramos como especies verdaderas (*A. nigerrima, A. ululata, A. discolor*), pero que requieren de estudios morfológicos y/o genéticos para confirmar su estatus específico. Reconocemos cinco subespecies en *A. palliata* (*A. p. mexicana, A. p. palliata, A. p. coibensis, A. p. trabeata* y *A. p. aequatorialis*), tres en *A. seniculus*, (*A. s. seniculus, A. s. juara* y *A. s. puruensis*), dos en *A. guariba* (*A. g. guariba* y *A. g. clamitans*) y consideramos la posibilidad de que *A. pigra* pueda tener dos subspecies (*A. p. pigra* y *A. p. luctuosa*). La mayoría de las especies y subespecies requieren de trabajos de campo que permitan delimitar sus rangos de distribución. Además, se precisa de análisis de datos genéticos y morfológicos para confirmar la validez de varios de estos taxa. Dada la amplia presencia de los monos aulladores en el Neotrópico, estos estudios requerirán de la colaboración de una red multidisciplinaria de científicos a través del rango de distribución del género.

Keywords *Alouatta* • Morphology • Genetics • Systematics

3.1 Introduction

Howler monkeys (*Alouatta* Lacépède, 1799) are the most widespread primate genus in the Neotropics. They range from southern Veracruz State in Mexico to northern Argentina (Fig. 3.1), and can be found in numerous forest types across the region (Neville et al. 1988; Glander and Pinto 2013). They are among the largest of the platyrrhines (Hill 1962; Peres 1994) along with the muriquis (*Brachyteles*), the spider monkeys (*Ateles*), and woolly monkeys (*Lagothrix*).

Until the 1980s, *Alouatta* was classified in the Cebidae (with all of the non-clawed platyrrhines) in the subfamily Alouattinae Elliot, 1904 (Hill 1962; Napier and Napier 1967; Napier 1976; Hershkovitz 1977). The revision of Rosenberger (1980, 1981, 2011; see also Schneider and Rosenberger 1996) placed *Alouatta* in the family Atelidae—the large, prehensile-tailed platyrrhines that also include *Ateles, Lagothrix*, and *Brachyteles*. Initially (1981), Rosenberger's revolutionary rearrangement had what he termed the "large suspensory frugivore-folivores" in a subfamily, Atelinae, as part of the family Pitheciidae, and, as a consequence, moved the howlers one step down to a tribe, Alouattini in the Atelinae; the spider monkeys, woolly monkeys, and muriquis being placed in the tribe Atelini. Subsequent compelling molecular phylogenetic evidence has placed the large prehensile-tailed frugivore-folivores as a distinct family, the Atelidae, with two subfamilies Alouattinae and Atelinae (for a review, see Schneider and Sampaio 2013), an arrangement accepted by Rosenberger (2011; see also Perelman et al. 2011; Halenar and Rosenberger 2013). The tribe Alouattini is composed of *Alouatta* and four extinct fossil genera *Stirtonia, Paralouatta, Solimoea*, and *Cartelles* Halenar and Rosenberger, 2013 (Rosenberger et al. 2015). The Atelini contains *Ateles, Brachyteles*, and *Lagothrix*, and two extinct forms, *Caipora* and *Protopithecus*.

3 Howler Monkey Taxonomy

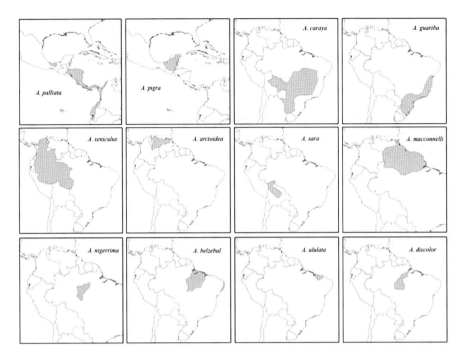

Fig. 3.1 Geographic distribution of currently recognized species of howler monkeys, genus *Alouatta* (distributions modified from IUCN 2013. IUCN Red List of Threatened Species. Version 2013.1. http://www.iucnredlist.org. Downloaded on September 24th, 2013)

Elliot (1913) provided the first comprehensive twentieth-century review of the howler monkey species. He listed twelve, *A. beelzebul* [sic] (Linnaeus, 1766), *A. seniculus* (Linnaeus, 1766), *A. caraya* (Humboldt, 1812), *A. ursina* (Humboldt, 1805), *A. villosus* (Gray, 1845), *A. palliata* (Gray, 1849) (with subspecies *mexicana* Merriam 1902, and *coibensis* Thomas 1902), and *A. aequatorialis* Festa, 1903, and five that he himself had described and named, *A. macconnelli* Elliot, 1910, *A. insulanus* Elliot, 1910, *A. juara* Elliot, 1910, *A. sara* Elliot, 1910, and *A. ululata* Elliot, 1912 (see Tables 3.1 and 3.2).

Ihering (1914) reviewed the Brazilian species of howlers and clarified the confusion created by Humboldt (1805) when he described the Venezuelan howler (*A. ursina*) but provided an illustration of the brown howler of the Brazilian Atlantic forest. Elliot (1913) did not catch this mistake, and Ihering (1914) named the brown howler *A. fusca* (Geoffroy Saint-Hilaire, 1812). Despite Humboldt's confusion, Cabrera (1957), Hill (1962), and Groves (2001, 2005) attributed the name *guariba* Humboldt, 1812 to the brown howler (see Rylands and Brandon-Jones 1998), although Gregorin (2006) affirmed that *A. fusca* is the correct name, as was also argued by Hershkovitz (1963).

Table 3.1 Names given to different forms of howler monkeys over time and the correspondence to the taxa recognized in this publication (derived or misspelled names are not included)

Name	Publication	Taxon as considered in this publication	Reference	Type locality
seniculus	Linnaeus (1766)	A. seniculus	*Syst. Nat.*, 12th Ed. 1:37	Cartagena, Colombia
belzebul	Linnaeus (1766)	A. belzebul	*Syst. Nat.*, 12th Ed. 1:37	Brazil, Pará, Rio Capim
ursina	Humboldt and Bonpland (1805)	A. arctoidea	*Rec. Obs. Zool. Anat. Comp.* 1:8	Caracas, Venezuela
caraya	Humboldt (1812)	A. caraya	In: Humboldt & Bonpland 1811 (1812). *Rec. Obs. Zool. Anat. Comp.* 1:355	Paraguay
guariba	Humboldt (1812)	A. guariba	In: Humboldt & Bonpland 1811 (1812). *Rec. Obs. Zool. Anat. Comp.* 1: pl. 30	Brazil, restricted by Cabrera (1957) to the Rio Paraguassú, Bahia
niger	Geoffroy Saint-Hilaire (1812)	A. caraya	*Ann. Mus. d'Hist. Nat. Paris* 19:108	"Bresil"
stramineus	Geoffroy Saint-Hilaire (1812)	A. caraya	*Ann. Mus. d'Hist. Nat., Paris* 19:108	"Gran Pará"
fusca	Geoffroy Saint-Hilaire (1812)	A. guariba	*Ann. Mus. d'Hist. Nat. Paris* 19:108	Brazil, restricted by Cabrera (1957) to the Rio Paraguassú, Bahia
rufimanus	Kühl (1820)	A. belzebul	*Beitr. Zool. Vergl. Anat.* 1:31	Unknown
barbatus	von Spix (1823)	A. caraya	*Sim. Vespert. Brasil.* p.48, pl. XXXII and XXXIII	Between the rios Negro and Solimões
discolor	von Spix (1823)	A. discolor	*Sim. Vespert. Brasil.* p.48, pl. XXXIV	Forte Gurupá, state of Pará
chrysurus	I. Geoffroy Saint-Hilaire (1829)	A. macconnelli or A. seniculus?	1828 (1829). *Mem. Mus. d'Hist. Nat. Paris*, 17:166, 171	Spanish Guiana (Venezuela) or Colombia

3 Howler Monkey Taxonomy

laniger	Gray (1845)	*A. seniculus*	*Ann. Mag. Nat. Hist.* 16:219	Colombia
bicolor	Gray (1845)	*A. guariba*	*Ann. Mag. Nat. Hist.* 16:219.	Brazils
villosus	Gray (1845)	*A. pigra*	*Ann. Mag. Nat. Hist.* 16:220	Vera Paz, Guatemala
auratus	Gray (1845)	*A. macconnelli*	*Ann. Mag. Nat. Hist.*, 16:220	Brazils (Orinoco on label)
palliata	Gray (1849)	*A. palliata*	*Proc. Zool. Soc. Lond.*, 1849:138, pl. VI	Shores of Lake Nicaragua
nigra	Slack (1863)	*A. caraya*	*Proc. Acad. Nat. Sci. Philadelphia* 1863 (1862):518	Brazil, Paraguay and Bolivia
mexicana	Merriam (1902)	*A. palliata mexicana*	*Proc. Biol. Soc. Washington* 15:67	Minatitlan, state of Veracruz, Mexico
coibensis	Thomas (1902)	*A. palliata coibensis*	*Novit. Zool.* 9:135	Coiba Island, off the Pacific coast of Panama
aequatorialis	Festa (1903)	*A. palliata aequatorialis*	*Boll. Mus. Zool. ed Anat. Comp. Univ. Torino* 18:3	Vinces, Guayas Province, Ecuador
caucensis	J. A. Allen (1904)	*A. seniculus*	*Bull. Am. Mus. Nat. Hist.* 20:462, figs. 2, 4, Nov. 28, 1904	Charingo, upper Río Cauca, Colombia
rubicunda	J. A. Allen (1904)	*A. seniculus*	*Bull. Am. Mus. Nat. Hist.* 20:458, figs. 1, 3	Bonda, Santa Marta, Colombia
metagalpa	J. A. Allen (1908)	*A. palliata palliata*	*Bull. Am. Mus. Nat. Hist.* 24:670	Lavala, Nicaragua
mexianae	Hagmann (1908)	*A. belzebul*	*Arch. Rass. Ges. Biol.* p.6	Isle of Mexiana, Amazon Delta
insulanus	Elliot (1910)		*Ann. Mag. Nat. Hist.* 8th Series, 5:79	Island of Trinidad
macconnelli	Elliot (1910)	*A. macconnelli*	*Ann. Mag. Nat. Hist.* 8th Series, 5:80	Coast of Demerara

(continued)

Table 3.1 (continued)

Name	Publication	Taxon as considered in this publication	Reference	Type locality
juara	Elliot (1910)	A. juara	Ann. Mag. Nat. Hist. 8th Series, 5:80	Río Juara, Peruvian Amazon
sara	Elliot (1910)	A. sara	Ann. Mag. Nat. Hist. 8th Series, 5:81	Province of Sara, Bolivia
ululata	Elliot (1912)	A. ululata	Bull. Am. Mus. Nat. Hist. 31:32	Miritiba, Maranhão, northeastern Brazil
inclamax	Thomas (1913)	A. palliata aequatorialis	Ann. Mag. Nat. Hist., 8th Series, 12:567	Intac, about 50 miles north of Quito, Ecuador
quichua	Thomas (1913)	A. palliata aequatorialis	Ann. Mag. Nat. Hist., 8th Series, 12:567	Río Blanco 20 miles west of Mindo, about 75°10′W on Equator, Alt. 2500 ft., northwest Ecuador
inconsonans	Goldman (1913)	A. palliata aequatorialis	Smithson. Misc. Coll., 60 (22):17	Cerro Azul, Panama
bogotensis	J. A. Allen (1914)	A. seniculus	Bull. Am. Mus. Nat. Hist. 33:648	Subia, Cundinamarca near Bogotá, Colombia
caquetensis	J. A. Allen (1914)	A. seniculus	Bull. Am. Mus. Nat. Hist. 33:650	La Murelia = Muralla, Caquetá, Colombia
trabeata	Lawrence (1933)	A. palliata trabeata	Bull. Mus. Comp. Zool. 75:328	Capina, Herrera Province, Panama
pigra	Lawrence (1933)	A. pigra	Bull. Mus. Comp. Zool. 75:333	Uaxactun, Petén, Guatemala
luctuosa	Lawrence (1933)	A. pigra	Bull. Mus. Comp. Zool. 75:337	Mountain Cow, Cayo District, British Honduras (Belize)

3 Howler Monkey Taxonomy

arctoidea	Cabrera (1940)	*A. arctoidea*	*Ciencia, Méx.* 1:404	Caracas, Venezuela (same as that of *ursina* Humboldt 1812)
clamitans	Cabrera (1940)	*A. guariba clamitans*	*Ciencia Méx.* 1:404	State of São Paulo
amazonica	Lönnberg (1941)	*A. seniculus*	*Ark. Zool.* 33A (10):16	Codajáz, north of Rio Solimões, Amazonas, Brazil
puruensis	Lönnberg (1941)	*A. puruensis*	*Ark. Zool.* 33A (10):16	Rio Purus, Amazonas, Brazil
juruana	Lönnberg (1941)	*A. juara*	*Ark. Zool.* 33A (10):18	Upper Rio Juruá
iheringi	Lönnberg (1941)	*A. guariba clamitans*	*Ark. Zool..* 33A (10):23	Serra de Cantareira, São Paulo, Brazil
beniensis	Lönnberg (1941)	*A. sara?*	*Ark. Zool.* 33A (10):24	Puerto Salinas, Río Beni, Bolivia
tapajozensis	Lönnberg (1941)	*A. discolor*	*Ark. Zool.* 33A (10):27	Brazil, restricted by Bonvicino et al. (1989) to Aveiros, Amazonas
nigerrima	Lönnberg (1941)	*A. nigerrima*	*Ark. Zool.* 33A (10):33	West of Rio Tapajóz and south of the Amazon, Brazil (Patinga, Amazonas, Brazil)

Table 3.2 The taxonomy of the howlers (*Alouatta*) according to Elliot (1911), Cruz Lima (1945), Cabrera (1957), Hill (1962), Rylands et al. (2000), Groves (2001), Gregorin (2006), and Glander and Pinto (2013)

Elliot (1912)	Cruz Lima (1945) Brazilian Amazonian species	Cabrera (1957) South American species	Hill (1962)
A. belzebul [sic] (Linnaeus, 1766)	*A. b. belzebul*	*A. b. belzebul*	*A. b. belzebul*
A. ululata Elliot, 1912	*A. b. ululata*	*A. b. ululata*	*A. b. ululata*
discolor (von Spix, 1823) is a synonym of *A. belzebul*	*A. b. discolor* (Spix, 1823)	*A. b. discolor*	*A. b. discolor*
A. b. mexianae (Hagmann, 1908) was not considered	Not considered	*A. b. mexianae*	*A. b. mexianae*
–	*A. nigerrima* Lönnberg, 1941	*A. b. nigerrima*	*A. b. nigerrima*
A. seniculus (Linnaeus, 1766)	*A. s. seniculus*	*A. s. seniculus*	*A. s. seniculus*
A. s. stramineus[a] is a synonym of *A. seniculus*	*A. s. straminea*[a]	*A. s. straminea*[a]	*A. s. stramineus*
A. macconnelli Elliot, 1910	Synonym of *A. s. straminea*	Synonym of *A. s. straminea*	*A. s. macconnelli*
A. insulanus Elliot, 1910	–	Not considered	*A. s. insulanus*
A. juara Elliot, 1910	*A. s. juara*	Synonym of *A. s. seniculus*	*A. s. juara*
–	*A. s. puruensis* Lönnberg, 1941	Synonym of *A. s. seniculus*	*A. s. puruensis* Lönnberg, 1941
–	*A. s. amazonica* Lönnberg, 1941 is a synonym of *A. s. straminea*	Synonym of *A. s. straminea*	*A. s. amazonica*

3 Howler Monkey Taxonomy

		A. s. arctoidea Cabrera, 1940	A. s. arctoidea
A. sara Elliot, 1910	–	A. s. sara	A. s. sara
A. caraya (Humboldt, 1812)	–	A. caraya	A. caraya
A. ursina (Humboldt, 1812)	Synonym of A. s. seniculus	Synonym of A. s. arctoidea	Synonym of A. s. arctoidea
A. palliata palliata (Gray, 1849)	–	A. p. palliata	A. p. palliata
A. p. mexicana Merriam, 1902	–	–	A. p. mexicana
A. aequatorialis Festa, 1903	–	A. p. aequatorialis	A. p. aequatorialis
A. palliata coibensis Thomas, 1902	–	–	A. p. coibensis
–	–	–	A. p. trabeata Lawrence, 1933
A. villosus (Gray, 1845)	–	–	A. villosa
			A. p. pigra Lawrence, 1933
			A. p. luctuosa Lawrence, 1933
A. guariba considered to be a synonym of A. ursinus	–	A. guariba guariba (Humboldt, 1812)	A. g. guariba
–	–	A. guariba clamitans Cabrera, 1940	A. g. clamitans
–	–	A. guariba beniensis Lönnberg, 1941	A. g. beniensis

(continued)

Table 3.2 (continued)

Rylands et al. (2000)	Groves (2001, 2005)	Gregorin (2006) Brazilian species	Glander and Pinto (2013)
A. belzebul belzebul	A. belzebul	A. belzebul	A. belzebul
A. b. ululata	Synonym of A. belzebul	A. ululata	A. ululata
A. b. discolor	Synonym of A. belzebul	A. discolor	A. discolor
A. b. mexianae not considered	Synonym of A. belzebul	Synonym of A. discolor	Synonym of A. discolor
A. nigerrima	A. nigerrima	A. nigerrima	A. nigerrima
A. s. seniculus	A. s. seniculus	–	A. s. seniculus
Type of A. s. straminea is a female A. caraya	Type of A. s. straminea is a female A. caraya	Synonym of A. macconnelli	Type of A. s. straminea is a female A. caraya
A. seniculus ssp.	A. macconnelli	A. macconnelli	A. macconnelli
A. s. insulanus	Synonym of A. macconnelli	–	Synonym of A. macconnelli
A. s. juara	A. s. juara	A. juara	A. s. juara
A. s. puruensis	Synonym of A. s. juara	A. puruensis	A. s. puruensis
A. s. amazonica	Synonym of A. s. juara	Synonym of A. s. juara	Synonym of A. s. juara
A. s. arctoidea	A. s. arctoidea	–	A. arctoidea
A. sara	A. sara	–	A. sara
A. caraya	A. caraya	A. caraya	A. caraya
A. ursinus not considered	Synonym of A. s. arctoidea	–	Not considered
A. palliata palliata	A. palliata	–	A. palliata palliata
A. p. mexicana	Synonym of A. palliata	–	A. p. mexicana
A. p. aequatorialis	Synonym of A. palliata	–	A. p. aequatorialis

A. c. coibensis	A. coibensis	–	A. p. coibensis
A. c. trabeata	Synonym of A. coibensis	–	A. p. trabeata
A. pigra	A. pigra	–	A. pigra
A. villosa and A. p. luctuosa are synonyms of A. pigra[b]	A. villosa and A. p. luctuosa are synonyms of A. pigra		
A. g. guariba	A. g. guariba	A. fusca	A. g. guariba
A. g. clamitans	A. g. clamitans	A. clamitans	A. g. clamitans
A. g. beniensis not considered	Synonym of A. guariba	–	Not considered

[a]Elliot (1912), Cruz Lima (1945), Cabrera (1957) and Hill (1962) attribute authorship to Humboldt (1812), we attribute the name to É. Geoffroy Saint-Hilaire (1812) (see text)
[b]Following the revision by Smith (1970)

A number of reviews of the howler monkey taxa from particular geographic regions (e.g., Lawrence 1933; Tate 1939; da Cruz Lima 1945; Hershkovitz 1949; Cabrera 1957) continued to modify the taxonomy of the genus for the next 30 years (Table 3.2). Osman Hill published another complete review of the genus in 1962. The taxonomic arrangement that he proposed was accepted with minor modifications (e.g., Smith 1970) until the late 1980s, when cytogenetic and molecular genetic analyses began to contribute new evidence that challenged the taxonomy based primarily on morphological characters.

In this chapter, we integrate the knowledge provided by studies on morphological and genetic characters and propose a taxonomic arrangement for the genus that is congruent with both. We briefly summarize the main issues regarding the taxonomy of each recognized taxa and provide our reasoning on why they are considered in their respective taxonomic levels. We also attempt to guide the reader on the important gaps of information that permeate the taxonomy of the genus in an attempt to stimulate new integrated genetic and morphological studies, based on individuals of known geographic origin.

3.2 Mesoamerican Howler Monkeys

The current taxonomy of Mesoamerican howler monkeys is based on morphological and genetic studies (Lawrence 1933; Smith 1970; Horwich 1983; Cortés-Ortiz et al. 2003; Steinberg et al. 2008). Lawrence (1933) conducted a comprehensive study of Mesoamerican howlers, reviewing the taxonomy of all taxa described for the region through an analysis of both cranial features and pelage coloration of a large number of specimens, including type specimens. She considered all Mesoamerican howlers to be subspecies of *A. palliata*, seven in all: *A. p. mexicana*, *A. p. palliata*, *A. p. aequatorialis*, *A. p. pigra*, *A. p. luctuosa*, *A. p. coibensis*, and *A. p. trabeata*. *Alouatta villosa* (Gray, 1845) she argued was indeterminable given the lack of a skin of the holotype (only a damaged skull remained) and confusion regarding the actual locality of the type. She thus assigned the name *A. palliata pigra* for the Guatemalan howler and *A. p. luctuosa* (Lawrence 1933) as the form in Belize. Hall and Kelson (1959), Hill (1962) and Napier (1976) continued to use the name *A. villosa*. Brandon-Jones (2006) provided a detailed analysis of the evidence that the *A. villosa*-type specimen is recognizably distinct from *A. palliata* (and other black howlers such as *A. nigerrima* and *A. caraya*) and that the type locality was recorded as Mexico. He argued, as such, that *A. villosa* is correctly a senior synonym of *A. pigra*. However, as the type specimen is reduced to a cranium (the skin was lost, Napier 1976) and without a proper study of this specimen, we conservatively continue listing this species using the name *A. pigra*, as did Smith (1970), Hall (1981), Groves (2001), and Rylands et al. (2006).

3.2.1 Alouatta palliata *(Gray, 1849)*

Type: Syntypes, adult female, skin and skull, No. 1848.10.26.1, and adult male, skin, 1848,10.26.2, British Museum (Natural History) (Napier 1976).
Type locality: Nicaragua (shores of Lake Nicaragua).
Common name: Mantled howler.

This species ranges from southern Veracruz State in Mexico, south through Central America and the Pacific slopes of the Andes in Colombia and Ecuador reaching the Tumbes region in northern Peru (Fig. 3.1).

Only five of the seven subspecies recognized by Lawrence (1933) for *A. palliata* are currently accepted: *A. p. mexicana*, *A. p. palliata*, *A. p. aequatorialis*, *A. p. coibensis*, and *A. p. trabeata*. Lawrence (1933) commented on the difficulty in distinguishing the subspecies of *A. palliata*, with the exception of the Guatemalan (*A. p. pigra*) and the Coiba Island (*A. p. coibensis*) forms. Smith (1970), likewise, analyzed cranial and dental patterns, and pelage coloration of *A. palliata*, and concluded that the forms *palliata*, *mexicana*, *coibensis*, and *trabeata* were only weakly definable as subspecies. Froehlich and Froehlich (1986, 1987), however, analyzed patterns of dermal ridges of howler monkeys from Nicaragua, Costa Rica, and Panama, and found that individuals from Coiba Island and the Azuero Peninsula were distinct from those of Nicaragua, Costa Rica, and the rest of Panama. They considered that differences in the dermal ridge patterns of the hands and feet were congruent with genetic distances, and that the difference found between *coibensis/trabeata* and *palliata/aequatorialis* groups was comparable to that of either of them and the South American howler monkey species. They, therefore, suggested that the forms *coibensis* and *trabeata* be treated as subspecies of a distinct species, *A. coibensis*. Glander and Pinto (2013) adhered to this view, Groves (2001, 2005) considered *trabeata* to be junior synonym of *A. coibensis*, and Rylands et al. (1995, 2000) maintained them as subspecies of *A. palliata* (as per Lawrence 1933). The molecular genetic studies (mtDNA) of Cortés-Ortiz et al. (2003) produced no evidence to back this distinction. *Alouatta palliata coibensis* individuals were found to share mitochondrial haplotypes with *A. palliata* from Costa Rica and Mexico, and *A. p. trabeata* individuals shared haplotypes with *A. p. aequatorialis* from central Panama. These two clades represented northern and southern mitochondrial lineages separated around the Sona Peninsula in western Panama (Cortés-Ortiz et al. 2003). Interestingly, recent phylogeographic analyses based on nuclear markers (microsatellites) support the phylogeographic brake between *A. p. palliata* and *A. p. aequatorialis*, but show closer proximity of *A. p. coibensis* and *A. p. trabeata* to *A. p. aequatorialis* (Cortés-Ortiz et al. unpublished). This discrepancy between mitochondrial and nuclear markers in the Coiba population may be the result of a secondary intergradation between formerly distinct lineages when Coiba Island was connected to the mainland (see Ford 2006). Studies including a larger number of individuals from Coiba Island, the Azuero Peninsula, and other regions in western and central Panama are necessary to corroborate the subspecific status of *A. p. coibensis* and *A. p. trabeata*, as well as to identify the limits of the ranges of *A. p. palliata* and *A. p. aequatorialis*.

3.2.2 Alouatta pigra *Lawrence, 1933*

Type: Adult male, collected by A. Murie, 4 May 1931, Museum of Zoology, University of Michigan.
Type locality: Uaxactun, Petén, Guatemala.
Common name: Central American black howler.

Alouatta pigra is distributed across the Peninsula of Yucatán in Mexico, Belize, and north and eastern Guatemala (Fig. 3.1). Most authors followed Lawrence's taxonomic arrangement of the Mesoamerican howlers, until 1970, when James D. Smith analyzed two sympatric populations of howler monkeys from the state of Tabasco, Mexico. He examined the skulls, teeth, and/or pelage of 238 specimens from across the range of *A. palliata* (*sensu* Lawrence 1933) and maintained that the two (partially sympatric) forms from his sample constituted two well-differentiated species, *A. palliata* and *A. pigra*. *Alouatta palliata luctuosa* Lawrence, 1933, he considered a junior synonym of *A. pigra*, given that the only extant specimen (from Mountain Cow, Cayo District, Belize) fell within the range of variation of *A. pigra*. Horwich (1983) supported Smith's recognition of two distinct species and commented on differences in group size between *A. palliata* and *A. pigra* (the latter having consistently smaller group sizes than the former), as well as differences in male genitalia; *A. pigra* males have fully descended testes from infancy, the testes of *A. palliata* males do not descend until they are subadults. Cortés-Ortiz et al. (2003) using mitochondrial DNA sequence data corroborated that *A. palliata* and *A. pigra* fall into two reciprocally monophyletic groups that separated approximately 3 MA, supporting their status as distinct species. Steinberg et al. (2008) also supported this distinction based on chromosome number (*A. palliata* 2n=54 and *A. pigra* 2n=58) and on the male sex determination system (X_1X_2Y in *A. palliata* and $X_1X_2Y_1Y_2$ in *A. pigra*). All samples of *A. pigra* analyzed by Cortés-Ortiz et al. (2003) and Steinberg et al. (2008) came from Mexico, and the validity of *A. pigra luctuosa* from the Cayo District in Belize has yet to be evaluated with genetic data.

Regarding the proper name for the Central American black howler monkey, as already mentioned, Brandon-Jones (2006) provided evidence that the *A. villosa* type (skull of a young adult female in the British Museum [Natural History]) is recognizably distinct from *A. palliata* and that the type locality can be identified as "Mexico." However, further studies exploring the similarities of the *A. villosa*-type specimen to a large sample of crania from *A. pigra* from Mexico, Guatemala, and Belize, as well as the differences with other black howlers (i.e., *A. nigerrima* and *A. caraya*) of known geographical provenance would provide compelling evidence for the suggested status of *A. villosa* as senior synonym of *A. pigra*. Given the widespread use of *A. pigra* and the benefits of taxonomic stability, we continue to call this taxon *A. pigra* until further studies are completed.

3.3 South American Howler Monkeys

The taxonomy of South American howler monkeys has been subject to multiple revisions. Linnaeus (1766) was the first to describe species of South American howler monkeys under his genus *Simia*. He recognized two: *S. belzebul* and *S. seniculus*. To date there have been over 48 names used to refer to the different forms of South American howlers (Table 3.1), which has caused considerable confusion in the taxonomy of the genus. Many of these names have become synonyms, and some forms are still recognized as distinct despite the need for taxonomic revision. Below we summarize the currently recognized species and subspecies following Groves (2001), Gregorin (2006), Rylands et al. (2000, 2012), Rylands and Mittermeier (2009), and Glander and Pinto (2013). It is reasonable to assume, however, that some of these species still require specific studies to determine their validity, and so this taxonomic arrangement should be considered a working hypothesis.

3.3.1 Alouatta caraya *(Humboldt, 1812)*

Type: No type preserved.
Type locality: Paraguay.
Common name: Black and gold howler, Paraguayan howler.

This species has a broad distribution including Brazil (Pantanal, and parts of the Cerrado and Caatinga), northern Argentina, Paraguay, and Bolivia (Fig. 3.1). Most authors have recognized this taxon as a full species given the presence of a distinct sexual dichromatism: males are mostly black and females have a yellowish or brindled tawny color (Groves 2001). Morphological analyses of the hyoid bone in males distinguish this species from other howlers, particularly by the lack of a tentorium (see Hershkovitz 1949; Gregorin 2006). *Alouatta caraya* has a diploid number of $2n=52$, which does not vary in specimens from distinct localities (de Oliveira et al. 2002). This species presents a sex chromosome system of $X_1X_1X_2X_2/X_1X_2Y_1Y_2$ (Mudry et al. 2001; de Oliveira et al. 2002), also observed in taxa of the *A. seniculus* group (*A. macconnelli, A. sara, A. arctoidea*) (Seuánez et al. 2005) and in *A. pigra* (Steinberg et al. 2008). This quadrivalent sex chromosome system differs from those of *A. guariba, A. belzebul*, and *A. palliata* (see chapter by Mudry et al. 2015). Chromosome painting with human chromosome probes suggests that the rearrangements that gave rise to this system in *A. caraya* and the *A. seniculus* group may have a single origin (Mudry et al. 2001), whereas the origin of the *A. pigra* quadrivalent sex chromosome system seems to be separate (see chapter by Mudry et al. 2015). The possible single origin of the sex chromosome system between *A. caraya* and the *A. seniculus* group suggests that *A. caraya* may have a close phylogenetic affinity with the Amazonian red howler taxa. Indeed, *A. caraya* shares one chromosome painting association pattern with the red howlers (*A. seniculus*, *A. sara*, *A. macconnelli*) and *A. guariba*, but not with *A. belzebul* (see Stanyon et al. 2011). A phylogenetic reconstruction by de Oliveira et al. (2002) based on parsimony analysis of chromosomal changes in different species of *Alouatta* places *A. caraya* as a sister group of *A. belzebul*, but phylogenetic studies based on mitochondrial and/or nuclear sequence data include *A. caraya* in a clade with species from the *A. seniculus* group (Cortés-Ortiz et al. 2003; Nascimento et al. 2005; Perelman et al. 2011), or show this species as basal to all howlers (e.g., Bonvicino et al. 2001). The actual phylogenetic position of *A. caraya* in the genus may require additional multilocus analyses that contain representatives of all the main lineages of howler monkeys, and include multiple samples from distinct geographic localities for each taxon.

Nascimento et al. (2005, 2007) and Ascunce et al. (2007) analyzed interspecific variation in *A. caraya* using mitochondrial DNA markers (the control region and the cytochrome *b* gene). In both cases they found evidence of divergent mitochondrial haplotypes consistent with individuals from different geographical regions (Mato Grosso, Brazil vs. Santa Cruz, Bolivia vs. Goiás, Brazil in Nascimento et al. 2005, and Mato Grosso, Brazil vs. northern Argentina/Paraguay vs. Goiás, Brazil in Ascunce et al. 2007). However, some of these divergent haplotypes can be found in both Argentina and Brazil and therefore they may be the result of ancestral polymorphism or the expansion and secondary contact of formerly allopatric populations (Ascunce et al. 2007). The structuring of mitochondrial haplotypes in these localities suggests that these populations could represent different subspecies. Whether these populations constitute different taxa, however, remains to be explored with further studies that include samples of individuals from a wider range of locations.

3.3.2 Alouatta guariba *(Humboldt, 1812)*

Type: No original type available. Hill (1962) notes for *Mycetes bicolor* (Gray, 1845) (in the British Museum, adult male, skin and skull, 1844. 5.14.16) that the label indicates it was collected by Parzudaki on the Orinoco. Napier (1976, p. 76) noted that "it appears to be more closely related to *A. belzebul*," and she catalogued it under *Alouatta* sp. (p. 88).
Type locality: Brazil. Restricted by Cabrera (1957) to Rio Paraguassú, Bahia (Hill 1962; Napier 1976).
Common name: Brown howler.

This species inhabits the Atlantic Forest of Brazil, south of the Rio São Francisco (Rylands et al. 1996) (Fig. 3.1). The taxonomy of this taxon is quite complex and there is some confusion about its correct name. Rylands and Brandon-Jones (1998) argued that É. Geoffroy Saint-Hilaire (1806) mentioned the name "guariba" in reference to *A. belzebul* not as a binomial but as a common name. "Guariba" is the name for howlers in the Brazilian Amazon and É. Geoffroy Saint-Hilaire used it to distinguish *A. belzebul* from the spider monkey *Ateles belzebuth*. The name *guariba*, used for the brown howler by Humboldt (1812), is not as such a junior homonym as argued by Hershkovitz (1963) and is available for this taxon. (Humboldt [1812] predated the *Stentor fuscus* of É. Geoffroy Saint-Hilaire [1812] by 2 months [Thomas 1913]). Gregorin (2006) disagrees with this, however, and supports the view of Hershkovitz (1963) that *guariba* is an objective junior homonym and that the correct name to use for this taxon is *A. fusca* (Geoffroy Saint-Hilaire, 1812). Here we follow the interpretation of Rylands and Brandon-Jones (1998) in using the name *A. guariba* for this species.

Two different forms have been identified based on pelage coloration: a northern brown howler *A. g. guariba* and a southern form in which males are darker than females, the southern brown howler *A. g. clamitans* Cabrera, 1940. This distinction is supported by cytogenetic (de Oliveira 1995; 1996; de Oliveira et al. 2002) and molecular (Harris et al. 2005; de Mello Martins et al. 2011) data. Based on morphological analyses of cranial and hyoid bone features, Gregorin (2006) considers that these two forms should be considered full species. Analyses of mitochondrial DNA variation among the populations of *A. guariba* in three localities (in the states of Rio de Janeiro, São Paulo, and Santa Catarina) show the existence of two mitochondrial lineages (de Mello Martins et al. 2011). These monophyletic lineages are apparently consistent with the northern and southern forms, which are in contact in the state of São Paulo. Unfortunately, de Mello Martins et al. (2011) do not provide information on the genetic distance between these lineages, but according to their Fig. 2, it is similar to that observed among different haplotypes of *A. belzebul*. Although this study supports some distinction between *A. g. guariba* and *A. g. clamitans*, the taxonomic recognition as full species will require further genetic analyses that include a larger number of samples across the range of this species, representatives of other recognized howler species, as well as the use of multiple loci.

3.3.3 Alouatta belzebul *Group*

3.3.3.1 *Alouatta belzebul* (Linnaeus, 1766)

Type: Inexistent or unknown.
Type locality: Brazil. Thomas (1911) restricted the type locality to the state of Pernambuco, based on the materials collected by Marcgrave (1648), on which Linnaeus based his description (Gregorin 2006).
Common name: Red-and-black howler, red-handed howler.

Alouatta belzebul has been recognized as a full species ever since Linnaeus (1766) first described it in his genus *Simia*. Hill (1962) recognized five subspecies: *A. b. belzebul*, *A. b. discolor*, *A. b. mexianae*, *A. b. ululata*, and *A. b. nigerrima*. Morphological, cytogenetic, and molecular studies have provided evidence that *A. nigerrima* is more closely related to *A. seniculus* than to *A. belzebul*, and here we consider it tentatively as a full species (see below). Groves (2001, 2005) considered *mexianae*, *discolor*, and *ululata* as synonyms of *A. belzebul*; however, here we cautiously follow Gregorin (2006) who placed *mexianae* as a junior synonym of *A. discolor* and recognized *belzebul*, *discolor*, and *ululata* as full species based on morphological analyses of the cranium and hyoid apparatus, and pelage color pattern. Below we comment on the need to validate some of these species with genetic data.

Alouatta belzebul has a disjunct distribution with populations in the lower Amazon region in Brazil, to the south of the Rio Amazonas, as well as in isolated populations in the coastal forests of northeastern Brazil (Langguth et al. 1987; Coimbra-Filho et al. 1995) (Fig. 3.1). Genetic analyses have shown little divergence in haplotypes of populations of *A. belzebul* in north-east Brazil and the Amazon basin (Bonvicino et al. 2001; Cortés-Ortiz et al. 2003), and a phylogeographic study based on the mitochondrial cytochrome *b* gene fails to distinguish populations from Paraíba from those of the left margin of the Rio Tocantins in the state of Pará (Nascimento et al. 2008).

3.3.3.2 *Alouatta discolor* (von Spix, 1823)

Type: Juvenile male, Munich Museum.
Type locality: Forte Curupá (=Gurupá), south margin of Rio Amazonas, between Rio Tapajós and Rio Xingu, state of Pará, Brazil. [Not the Island of Gurupá].
Common name: Spix's red-handed howler monkey.

This species is distributed from the right bank of the rios Tapajós and Juruena to the rios Xingú and Irirí (Gregorin 2006; Glander and Pinto 2013) (Fig. 3.1). Elliot (1913) regarded this taxon as a synonym of *A. belzebul* given that he considered that the type (a juvenile with a darker pelage) presented pelage coloration within the range of this species. Lönnberg (1941) described a new subspecies of *A. belzebul* to the east of the Rio Tapajós, naming it *A. b. tapajozensis*, but Cruz Lima (1945) argued that Lönnberg's *tapajozensis* was a synonym of Spix's *discolor*. He included this form as a subspecies of *A. belzebul* (*A. b. discolor*). Later authors (e.g., Cabrera 1957; Hill 1962) followed Cruz Lima in recognizing the form *discolor* as a subspecies of *A. belzebul*. Groves (2001, 2005) considered *discolor* as a synonym of a monotypic *A. belzebul*, but Gregorin (2006), after a detailed analysis of morphometric data and pelage coloration patterns of over 70 individuals from Pará, concluded that the diagnostic characters of the hyoid bone and pelage coloration clearly distinguish *A. discolor* from the other two phylogenetically close taxa (*A. belzebul* and *A. ululata*) and are sufficiently trustworthy to validate its specific status. To date, there are no genetic data from individuals of this region, so we only tentatively consider *A. discolor* as a full species following Gregorin (2006).

3.3.3.3 *Alouatta ululata* Elliot, 1912

Type: Adult male, skin and skull, No. 1911.10.16.10, British Museum (Natural History) (Napier 1976).
Type locality: Miritiba, northern Maranhão State, Brazil.
Common name: Maranhão red-and-black howler.

Alouatta ululata is distributed in the north-east of Brazil, in the north of the states of Maranhão, Piauí, and Ceará (Gregorin 2006) (Fig. 3.1). The typical pelage coloration pattern of this form was first described by Dollman (1910) from individuals collected in Miritiba (Maranhão), but he believed that the specimens that he was analyzing belonged to *A. discolor*, given that their pelage coloration pattern was similar to the description given by Spix (1823). Later, Elliot (1912) recognized the specimens as part of a distinct species, given that he considered that the type specimen and figure from Spix (1823) portrayed a young *A. belzebul* individual with the typical darker coloration, and did not accurately match Spix's description of *discolor*. Furthermore, Elliot considered that Spix's description could not be applied to any species of howler known at that time (Elliot 1913). However, Elliot found consistent differences between the specimens from Miritiba and the specimens from the west of Pará, which he considered to be *A. belzebul* (here *A. discolor*) and he therefore described it as a distinct species, *A. ululata*. Ihering (1914), in his review of the genus *Alouatta*, analyzed a number of specimens of howler monkeys from Brazil and Venezuela with the aim of resolving the problematic positioning of the forms *discolor*, *ululata*, and *belzebul*. Based on his analyses of cranial measurements and pelage coloration, he concluded that the three forms belonged to a single species, *A. belzebul*. Since then, most authors either considered *ululata* as a synonym (e.g., Lönnberg 1941; Groves 2001) or as subspecies of *A. belzebul* (e.g., da Cruz Lima 1945; Hill 1962). Gregorin (2006) considered that the specimens he analyzed from Ceará and Maranhão presented a conspicuous coloration (sexually dichromatic) not described for any individual of *A. belzebul* or *A. discolor*, and therefore agreed with the recognition of *A. ululata* as a distinct species. As mentioned earlier, there are no genetic studies that include specimens of this taxon, and therefore the question of their genetic distinctiveness from *A. belzebul* and *A. discolor* remains an open question.

3.3.4 **Alouatta seniculus** *Group*

3.3.4.1 *Alouatta seniculus* (Linnaeus, 1766)

Type: Inexistent or unknown.
Type locality: Cartagena, department of Bolivar, Colombia.
Common name: Colombian red howler.

This is one of the two species of howler monkeys originally described by Linnaeus (1766) in his genus *Simia*. It was originally believed that it was broadly distributed in South America to the north of the Rio Amazonas, but morphological and genetic analyses have yielded enough evidence to conclude that the howlers in this area belong to a species complex (see below) rather than to a single species. Hill (1962) recognized nine subspecies of *Alouatta seniculus* (*seniculus*, *arctoidea*, *stramineus*, *insulanus*, *amazonica*, *macconnelli*, *juara*, *puruensis*, and *sara*), most of which are considered valid taxa today; a couple are considered synonyms, and others have even been reclassified as full species (see below). The only taxon not recognized as valid is *stramineus*, the holotype of which is a female *A. caraya* (see Rylands and Brandon-Jones 1998). Here we recognize three possible subspecies of *A. seniculus*: *A. s. seniculus*, *A. s. juara*, and *A. s. puruensis*. Three more taxa formerly included as part of the *seniculus* group (*A. arctoidea*, *A. macconnelli*, and *A. sara*) are considered to be distinct species, based primarily on genetic information (see below). Unfortunately, genetic information is not available for the entire species group. Gregorin (2006) elevated *A. s. juara* and *A. s. puruensis* to the species level; however, we consider that more studies (particularly on their geographic distribution and genetics) are needed before categorizing these taxa as distinct species. Nonetheless, we present each subspecies separately to allow a better understanding of the variation already observed by Gregorin.

3.3.4.2 *Alouatta seniculus seniculus* (Linnaeus, 1766)

The distribution of *A. s. seniculus* is restricted to Colombia (east of the Andes), northwestern Venezuela (around Maracaibo Lake), Brazilian Amazon to the north of the Rio Solimões and south of the Rio Negro, eastern Ecuador, and eastern Peru (east of the Río Huallaga, to the upper Marañon, and rios Napo and Putumayo).

This subspecies represents the typical *seniculus* described by Linnaeus (1766). A number of genetic studies have included individuals sampled from within the distribution range of this subspecies (e.g., Yunis et al. 1976; Cortés-Ortiz et al. 2003) showing clear genetic differences with other species. However, phylogeographic studies that include representatives of the different subspecies here considered are still lacking, preventing a better understanding of the distribution of genetic variation within the species and limiting our abilities to correctly classify these forms.

3.3.4.3 *Alouatta seniculus juara* Elliot, 1910

The Juruá red howler monkey is distributed through the western portion of the Brazilian Amazon, in the states of Acre and Amazonas, south of the Rio Solimões and west of the Rio Purus (Fig. 3.1). Its range extends to Peru, but the range limits are not known. The taxonomic position of *juara* has been debated, sometimes considering it a junior synonym of *A. seniculus* (e.g., Cabrera 1957), other times including it as one of its subspecies (e.g., Groves 2001) or as a distinct species (e.g., Elliot 1910; Gregorin 2006). These discrepancies are mainly due to the lack of comprehensive studies. Gregorin (2006) examined 31 specimens from Brazil that occur within the supposed range of this taxon. He found statistical differences on morphometric variables between *juara* and *A. macconnelli* and *A. nigerrima*, but he did not present data for these types of analyses comparing *juara* to *A. seniculus*. Nonetheless, he found differences in pelage coloration between *juara* and *A. seniculus* from northern Colombia. Regarding hyoid morphology, Gregorin (2006) found a high similarity between *juara* and *A. seniculus* and concluded that they must be phylogenetically proximate.

Lima and Seuánez (1991) reported on the karyotype of one individual sampled in Tefé, in the range of *juara*, and mentioned that the karyotype of this individual was "basically the same as that of the Colombian specimens" analyzed by Yunis et al. (1976)

and that "this finding indicates that the geographic range of this subspecies might extend further south to the Solimões River." Gregorin (2006) used this information as evidence supporting the distinction of *juara* from *A. macconnelli* and *A. nigerrima*, but it remains unclear whether this taxon is distinct from *A. seniculus*.

3.3.4.4 *Alouatta seniculus puruensis* Lönnberg, 1941

This subspecies is distributed along both margins of the Rio Purus, to the lower Rio Madeira and the middle Rio Aripuanã, extending eastward to the Rio Teles Pires. To the south, it is restricted to the northern margin of the Rio Abunã on the border of Bolivia (Glander and Pinto 2013) (Fig. 3.1). Gregorin (2006) reported the southernmost locality for this species to be Placido de Castro in Acre, Brazil.

The main diagnostic character for this subspecies is the sexual dichromatism, with males being dark rufous (back mahogany red and flanks and limbs maroon [Hill 1962]) and females golden orange with the distal portions of the limbs, tails, and beard dark rufous (pale yellowish, with the flank fringe partially orange and the limbs and tail also showing some orange [Hill 1962]). Gregorin (2006) found morphological differences in the shape of the hyoid bone, and proposed that it be considered a full species. We continue to place this taxon as a subspecies of *A. seniculus* until genetic studies reveal the degree of divergence (if any) with the other members of this species.

3.3.4.5 *Alouatta arctoidea* Cabrera, 1940

Type: Adult male, Paris Museum.
Type locality: Caracas, Venezuela, fixed by J. A. Allen (1916) on the basis of Humboldt's vernacular name (Hill 1962), further restricted by Cabrera (1957) to the valley of Aragua (Groves 2001).
Common name: Venezuelan red howler, Ursine red howler.

Alouatta arctoidea occurs on the island of Trinidad and in northern Venezuela, from the coastal region of Falcón to the state of Miranda (north of the Orinoco) (Fig. 3.1). Bodini and Pérez-Hernández (1987) reported a possibly distinct form in the Venezuelan llanos; however, no further studies have reported the existence of this howler and here we consider it as part of *A. arctoidea*. This species was initially described by Humboldt and Bonpland (1805) as *Simia ursina* but he created confusion due to the disagreement between the description and the figure given by the author (which depicted *A. guariba*). The name *ursina* (Humboldt) is now considered a synonym of *guariba*. J. A. Allen (1916) designated this form as a subspecies of *A. seniculus* (*A. s. ursina*), and Cabrera (1940) proposed to call it *A. s. arctoidea*, as the name *ursina* was inadmissible for this form given that it is a homonym to *Simia hamadryas ursinus* Kerr, 1792 and *Simia ursina* (Bechstein 1800) used to refer to the "ursine baboon" (Cabrera 1940).

Based on cytogenetic differences recognized by Stanyon et al. (1995), this taxon is currently accepted as a full species. The 14 chromosomal rearrangements found by these authors between *A. arctoidea* (four individuals from Hato Masaguaral, Venezuela) and *A. sara* (one individual captured in Bolivia, held at the San Diego Zoo) are more typical of differences between species, and are on the same order of magnitude as those found between *A. sara* and *A. seniculus* by Minezawa et al. (1985). However, until now no *A. arctoidea* specimens have been analyzed using molecular techniques and no monkeys from northern Venezuela have been cytogenetically characterized. Therefore, further validation of this species is necessary and the extent of its distribution range remains to be studied.

3.3.4.6 *Alouatta macconnelli* Elliot, 1910

Type: Adult male, skin and skull, No. 1908.3.7.3, British Museum (Natural History) (Napier 1976).
Type locality: Coast of Demerara, Guiana.
Common name: Guianan red howler, Golden howler.

Alouatta macconnelli is distributed throughout the Guiana Shield, including French Guiana, Suriname and Guyana, southern Venezuela (south of the Río Orinoco), and northern Brazil (from the coast of the state of Amapá to the eastern margins of the rios Negro and Branco, including Gurupá island in the Amazon delta) (Fig. 3.1).

There has been confusion about the name that should be used for the howler monkeys that inhabit the Guiana Shield, as well as whether one or two taxa should be recognized. The confusion started with É. Geoffroy Saint-Hilaire's (1812) description of *Stentor stramineus*,[1] which was based on a specimen of a female *A. caraya* from Central Brazil (Elliot 1913; Rylands and Brandon-Jones 1998; Gregorin 2006). Later, Elliot (1910) described *A. macconnelli* from one specimen from French Guiana, and synonymized *stramineus* with *A. seniculus* (Elliot 1913), which he deemed as clearly different from *A. macconnelli*. Tate (1939) considered both *macconnelli* and *stramineus* as subspecies of *A. seniculus*, and da Cruz Lima (1945) considered *A. macconnelli* (Elliot 1910) a synonym of *A. seniculus straminea* (Geoffroy Saint-Hilaire 1812). Hill (1962) still included both *straminea* and *macconnelli* as subspecies of *A. seniculus*, but wrote "There is every indication that both *macconnelli* and *amazonica* fall within the range of variation of *stramineus* and they should accordingly be treated as synonyms thereof, an action already taken by Cabrera (1957)."

Bonvicino et al. (1995), based on morphological analyses of individuals in the range of the two putative subspecies (*A. s. stramineus* and *A. s. macconnelli*), concluded that the howlers from the northern bank of the Amazon can be divided into two species separated by the Rio Trombetas, and argued that this is supported

---

[1] Although some authors, for example, Hill (1962), attribute the name *stramineus* to Humboldt (1812) (*Simia straminea*) published two months before É. Geoffroy Saint Hilaire's *Stentor stramineus*, Humboldt (1812) gave specific credit to Geoffroy Saint-Hilaire for the name; he merely placed it in Linnaeus' genus *Simia* and changed the gender accordingly. (See Article 50.1.1 of the *International Code of Zoological Nomenclature* <http://iczn.org/code>.)

by the biochemical and karyological analyses of Sampaio et al. (1991) and Lima et al. (1990) and Lima and Seuánez (1991), respectively. However, as discussed by Rylands and Brandon-Jones (1998, p. 887), the main "confusion regarding the identity of the Guianan howler is compounded by the inadequate, or imprecise information and uncorroborated conjectures about its geographic distribution." Rylands and Brandon-Jones (1998) found that the same specimens analyzed by different authors as part of one taxon were regarded as the alternative taxon by others. This is true throughout the taxonomic history of these two taxa, but was particularly true in the case of the analyses done by Bonvicino et al. (1995).

Molecular analyses by Sampaio et al. (1996) using biochemical data concluded that the genetic distance between populations from opposite sides of the Rio Trombetas, and those to the east of the Rio Jari are too small to justify their separation even at the subspecific level. Figueiredo et al. (1998) analyzed mitochondrial DNA sequence data of individuals from the same locations as those of Sampaio et al. (1996) and found that genetic distances among them were small and similar to those found among populations, and that individuals from different localities were sometimes less divergent than individuals from the same population. They concluded that both biochemical and mitochondrial DNA data strongly suggested gene flow among the three studied populations and the existence of a single species within the studied area.

Gregorin (2006) studied pelage coloration and morphometric data from specimens sampled in Brazil, and compared them to specimens from the rest of the Guiana Shield (French Guiana, Suriname, and southern Venezuela). He found that the Brazilian and "Guianan" specimens presented great variation in pelage coloration, which was not distinguished geographically. The same was true when analyzing the morphology of the hyoid bone. Principal Component Analyses of morphometric data also did not support the recognition of two taxa for the Guiana Shield (Gregorin 2006). In all, Gregorin's study supported Figueiredo et al.'s (1998) conclusion that there is a single taxon of howler in the Guiana Shield, which is easily distinguishable from the other recognized species of howlers. The description of Elliot's *A. macconnelli* type falls within the spectrum of variation found in the specimens from the Guiana Shield by Gregorin (2006), and he concluded that this taxon should be conservatively named *A. macconnelli*, but acknowledged the observation by Rylands and Brandon-Jones (1998) that given that the type locality of *Mycetes auratus* Gray is in the west of the range of *A. macconnelli*, the correct name for this species may well be *Alouatta auratus* (Gray, 1845).

3.3.4.7 *Alouatta sara* Elliot, 1910

Type: Adult female, skin and skill, No. 1907.8.2.1 British Museum (Natural History) (Napier 1976).
Type locality: Province of Sara, Bolivia.
Common name: Bolivian red howler.

Distributed in Bolivia, this species occurs from the department of Pando south along the Andean Cordillera, and east into Bolivia, including the Río Beni basin and east as far as the Mamoré-Guaporé interfluvium (Anderson 1997; Büntge and Pyritz 2007) (Fig. 3.1).

Alouatta sara was initially considered a distinct species by Elliot (1910) and later regarded as a subspecies of *A. seniculus* by Cabrera (1957) and Hill (1962). Minezawa et al. (1985) conducted cytogenetic analyses based on a sample of 33 red howler monkeys from the region of Santa Cruz de la Sierra, Bolivia, and concluded that their karyotype differed considerably from those of Colombian red howlers (*A. s. seniculus*), and therefore should be considered a different species. Similarly, comparative cytogenetic analyses by Stanyon et al. (1995) demonstrated karyotype differences between *A. sara* and *A. arctoidea* congruent with typical differences observed between species belonging to different genera. Groves (2001) considered *sara* to be a full species. Cortés-Ortiz et al. (2003) and Perelman et al. (2011) included mitochondrial and nuclear sequence data of *A. sara* individuals in their phylogenetic analyses, but only the former authors also included samples from *A. seniculus* from Colombia. Phylogenetic analyses based on mitochondrial DNA data show that *A. sara* and *A. seniculus* are sister taxa that diverged approximately 2.4 MA (Cortés-Ortiz et al. 2003). Based on cytogenetic and molecular data, the recognition of *A. sara* as a full species is strongly supported. Nonetheless, further studies are needed to understand the actual borders of its geographic range, its relationships with other howler species, and the levels of genetic variation across the distribution of this taxon.

3.3.4.8 *Alouatta nigerrima* Lönnberg, 1941

Type: Originally in the Stockholm Museum (none of the seven specimens was designated as holotype in the original description [da Cruz Lima 1945]). Lectotype in the Swedish Museum of Natural History NRM A63 3316 (indicated by Cabrera [1957], and officially designated by Gregorin [2006]).
Type locality: The left margin of the Rio Tapajós (restricted by Cabrera [1957] to Patinga, state of Pará, Brazil).
Common name: Amazonian black howler.

Alouatta nigerrima is endemic to Brazil, with a geographic range that extends between the rios Madeira and Tapajós, north to the Rio Amazonas (Fig. 3.1). A few specimens have also been collected in the northern margin of the Rio Amazonas, in the regions of Oriximiná and Obidos, in the state of Pará (Gregorin 2006).

Lönnberg (1941) described *A. nigerrima* as a full species, distinct from its neighbor *A. belzebul tapajoensis* (here *A. discolor*, see above), and considered that these two forms were phylogenetically close. One of the most important distinctive characters described by Lönnberg (1941) as diagnostic of *A. nigerrima* is related to the morphology of the hyoid bone, which clearly distinguishes it from *A. discolor* (Gregorin 2006). Cruz Lima (1945) also considered *A. nigerrima* to be a distinct species, but Hershkovitz (1949) placed it as a subspecies of *A. belzebul* (although recognizing that the morphology of the hyoid bone could place it closer to *A. seniculus* than to *A. belzebul*). Cabrera (1957) and Hill (1962) followed Hershkovitz in listing *nigerrima* as a subspecies of *A. belzebul*.

Armada et al. (1987) studied the karyotypes of 10 *A. belzebul* individuals captured on the left margin of the Rio Tocantins, state of Pará, Brazil, and of one captive individual of unknown origin tentatively identified as *A. belzebul nigerrima*, based on pelage coloration. They found considerable differences in the karyotypes of these two forms, and suggested a taxonomic reevaluation of these taxa. Oliveira (1996) noted that the g-banding pattern of *A. b. nigerrima* presented by Armada et al. (1987) was more similar to the one observed in *A. seniculus* than in *A. b. belzebul*. Bonvicino et al. (2001) analyzed mitochondrial sequence data of seven Brazilian *Alouatta* species, including a sample from the same individual identified as *A. belzebul nigerrima* by Armada et al. (1987), and found that it was phylogenetically closer to *A. macconnelli* and *A. seniculus* than to *A. belzebul*.

Gregorin (2006) analyzed the pelage coloration and cranial morphology of 98 specimens of *nigerrima*, and found that the distinctive black pelage was invariable across most of the specimens he analyzed. He also noticed that the morphology of the hyoid bone was distinct from any other Brazilian howler (confirming the observations previously made by Cruz Lima [1945]), but that it is more similar to the hyoid of *A. macconnelli* than to that of *A. belzebul*.

These morphological and genetic studies have supported the early recognition of *A. nigerrima* as a distinct species from *A. belzebul* and phylogenetically closer to the *seniculus* group. The genetic analyses have, however, been based on a single individual of unknown origin. Given its proximity to *A. macconnelli* and the fact that this latter taxon presents great variation in pelage coloration that includes individuals that are completely black (see Gregorin 2006), the question arises as to whether this specimen really originates from the range of *A. nigerrima* or is in fact a dark phase individual of *A. macconnelli*. Further genetic studies from individuals of known origin are needed to discard this possibility and validate the status of *A. nigerrima*.

3.4 Concluding Remarks

This review of the taxonomic history of the howler monkeys allows us to recognize nine distinct species: *A. palliata*, *A. pigra*, *A. seniculus*, *A. arctoidea*, *A. belzebul*, *A. caraya*, *A. guariba*, *A. sara*, and *A. macconnelli*. Three other taxa are tentatively considered here as species: *A. discolor*, *A. ululata*, and *A. nigerrima*, but their full validation as species still requires thorough genetic and/or morphological studies. Furthermore, we include *puruensis* and *juara* as subspecies of *A. seniculus*, as there is no data available to strongly support their taxonomic position as distinct species. The possibility remains that *A. guariba* is composed of two or more subspecies or even species. We recognize *A. pigra* and *A. palliata* as the only howler species present in Mesoamerican, with *A. palliata* including *coibensis* and *trabeata* as subspecies (besides *palliata*, *mexicana*, and *aequatorialis*) given that molecular data does not support the separation of these two taxa as distinct species from *A. palliata*.

Throughout this review, it should be evident that we still have a long way to go to fully resolve the taxonomy of *Alouatta*. This is not surprising given the wide distribution of the genus—the largest of any Neotropical genus—and the long history of diversification of the living taxa, which started about 7 MA. Particular efforts should be made to understand the distributions of the different forms and the genetic and morphological variation within taxa. Ideally, studies should use specimens from known geographical locations, and include individuals from as many localities as possible within the known range of the taxa, ensuring the inclusion of individuals from or near the type locality and from locations near the known or presumed boundaries. They also should include representatives of closely related taxa, not only those from geographic neighbors, but also those that are considered phylogenetically closest. This endeavor requires collaboration among researchers from different countries or geographical regions, and across different disciplines.

Our ability to properly identify and classify different taxa that represent distinct evolutionary units requires integrated taxonomic, biogeographic, and evolutionary studies (i.e., addressing the Linnean, Wallacean, and Darwinian shortfalls in biodiversity conservation; Diniz-Filho et al. 2013), and has immediate implications for the long-term survival of these taxa. Most species of howlers and other Neotropical primates face great and varied threats to their survival, and our ability as scientists to properly portray the extent of diversity within this group will contribute to the public understanding of the importance of maintaining this diversity. As a first step in this direction, we need to be able to clearly distinguish the different forms based on strong biological (genetic and morphological) evidence. As such, the taxonomy that prevails in a particular moment is a scientific hypothesis based on current knowledge, and changes are to be expected as more knowledge is acquired.

Taxonomic instability due to rapid changes in taxonomic arrangements may produce confusion at the moment of identifying proper units for conservation. However, these changes help to identify meaningful evolutionary lineages that require immediate conservation attention. In this respect, what is worrisome is not that the taxonomy of a group changes over time, but whether or not the changes made (or proposed) are supported by thorough and solid scientific evidence. When changes in taxonomic arrangements arise due to a better understanding of the evolutionary relationships within a group, they must be accepted as part of the development of science and should not be considered fickle and inconvenient to conservation efforts. Rather, they should be seen as elements to consider that strengthen our efforts to conserve biological diversity and meaningful evolutionary units.

As studies of howler monkeys in new geographical areas and using new genetic and morphological technique increase, it is likely that the taxonomic arrangement proposed here will change. Our responsibility lies in ensuring that the information used to make the decisions to do these changes is solid and comprehensive before these changes are accepted.

Acknowledgments We thank Bernardo Urbani, Dionisios Youlatos, and an anonymous reviewer on the first draft of this manuscript for their helpful comments. L.C.O. was supported by an NSF grant (BCS-0962807) while writing this chapter.

References

Allen JA (1916) List of mammals collected in Colombia by the American Museum of Natural History expeditions, 1910–1915. Bull Am Mus Nat Hist 35:191–238

Anderson S (1997) Mammals of Bolivia, taxonomy and distribution. Bull Am Mus Nat Hist 231:1–652

Armada JLA, Barroso CML, Lima MMC, Muniz JAPC, Seuánez HN (1987) Chromosome studies in *Alouatta belzebul*. Am J Primatol 13:283–296

Ascunce MS, Hasson E, Mulligan CJ, Mudry MD (2007) Mitochondrial sequence diversity of the southernmost extant New World monkey, *Alouatta caraya*. Mol Phylogenet Evol 43:202–215

Bodini R, Pérez-Hernández R (1987) Distribution of the species and subspecies of cebids in Venezuela. Fieldiana Zool 39:231–244

Bonvicino CR, Fernandes MEB, Seuánez HN (1995) Morphological analysis of *Alouatta seniculus* species group (Primates, Cebidae). A comparison with biochemical and karyological data. Hum Evol 10:169–176

Bonvicino CR, Lemos B, Seuánez HN (2001) Molecular phylogenetics of howler monkeys (*Alouatta*, Platyrrhini); a comparison with karyotypic data. Chromosoma 110:241–246

Brandon-Jones D (2006) Apparent confirmation that *Alouatta villosa* (Gray, 1845) is a senior synonym of *A. pigra* Lawrence, 1933 as the species-group name for the black howler monkey of Belize, Guatemala and Mexico. Primate Conserv (21):41–43

Büntge ABS, Pyritz LW (2007) Sympatric occurrence of *Alouatta caraya* and *Alouatta sara* at the Río Yacuma in the Beni Department, northern Bolivia. Neotrop Primates 14:82–83

Cabrera A (1940) Los nombres científicos de algunos monos americanos. Cienc Rev Hisp Am Cienc Puras y Aplic 1:402–405

Cabrera A (1957) Catálogo de los mamíferos de América del Sur. Rev Mus Argentino de Cienc Nat "Bernardino Rivadavia" 4:1–307

Coimbra-Filho AF, Câmara I de G, Rylands AB (1995) On the geographic distribution of the red-handed howling monkey, *Alouatta belzebul*, in north-east Brazil. Neotrop Primates 3:176–179

Cortés-Ortiz L, Bermingham E, Rico C, Rodríguez-Luna E, Sampaio I, Ruiz-García M (2003) Molecular systematics and biogeography of the Neotropical monkey genus, *Alouatta*. Mol Phylogenet Evol 26:64–81

da Cruz Lima E (1945) Mammals of Amazônia, Vol. 1. General introduction and Primates. Contribuições do Museu Paraense Emílio Goeldi de História Natural e Etnografia, Belém do Pará – Rio de Janeiro

de Mello Martins F, Gifalli-Iughetti C, Koiffman CP, Harris EE (2011) Coalescent analysis of mtDNA indicates Pleistocene divergence among three species of howler monkey (*Alouatta* spp.) and population subdivision within the Atlantic Coastal Forest species, *A. guariba*. Primates 52:77–87

de Oliveira EHC (1995) Chromosomal variation in *Alouatta*. Neotrop Primates 3:181–182

de Oliveira EHC (1996) Cytogenetic and phylogenetic studies of *Alouatta* from Brazil and Argentina. Neotrop Primates 4:156–157

de Oliveira EHC, Neusser M, Figueiredo WB, Nagamachi C, Pieczarka JC, Sbalqueiro IJ, Wienberg J, Müller S (2002) The phylogeny of howler monkeys (*Alouatta*, Platyrrhini): reconstruction by multi-color cross-species chromosome painting. Chromosome Res 10:669–683

Diniz-Filho JAF, Loyola RD, Raia P, Mooers AO, Bibi, LM (2013) Darwinian shortfalls in biodiversity conservation. Trends Ecol Evol 28:689–695

Dollman JG (1910) A note on *Alouatta discolor* of Spix. Ann Mag Nat Hist 8th Series 6:422–424

Elliot DG (1910) Descriptions of new species of monkeys of the genera *Seniocebus*, *Alouatta*, and *Aotus*. Ann Mag Nat Hist 8th Series 5:77–83

Elliot DG (1912) New species of monkeys of the genera *Galago*, *Cebus*, *Alouatta*, and *Cercopithecus*. Bull Am Mus Nat Hist 31:31–33

Elliot DG (1913) A review of the Primates. Monograph series, Vol. 1. American Museum of Natural History, New York [Published on 15 June 1913, not 1912 see correction slip.]

Figueiredo WB, Carvalho-Filho NM, Schneider H, Sampaio I (1998) Mitochondrial DNA sequences and the taxonomic status of *Alouatta seniculus* populations in northeastern Amazonia. Neotrop Primates 6:73–77

Ford SM (2006) The biogeographic history of Mesoamerican primates. In: Estrada A, Garber PA, Pavelka MSM, Luecke L (eds) New Perspectives in the study of Mesoamerican primates: distribution, ecology, behavior and conservation. Kluwer/Springer, New York

Froehlich JW, Froehlich PH (1986) Dermatoglyphics and subspecific systematics of mantled howler monkeys (*Alouatta palliata*). In: Taub DM, King FA (eds) Current perspectives in primate biology. Van Nostrand Reinhold, New York

Froehlich JW, Froehlich PH (1987) The status of Panama's endemic howling monkeys. Primate Conserv 8:58–62

Geoffroy Saint-Hilaire É (1806) Memoire sur les singes a main imparfaite ou les ateles. Ann Mus Hist Nat Paris 7:260–273

Geoffroy Saint-Hilaire É (1812) Tableau des quadrumanes ou des animaux composant le premier ordre de la classe des mammifères. Ann Mus Hist Nat Paris 19:85–122

Glander KE, Pinto LP (2013) Subfamily Alouattinae, *Alouatta* Lacépède, 1799. In: Mittermeier RA, Rylands AB, Wilson DE (eds) Handbook of the mammals of the world, vol 3, Primates. Lynx Edicions, Barcelona

Gregorin R (2006) Taxonomy and geographic variation of species of the genus *Alouatta* Lacépède (Primates, Atelidae) in Brazil. Rev Bras Zool 23:64–144

Groves CP (2001) Primate taxonomy. Smithsonian Institution Press, Washington, DC

Groves CP (2005) Order primates. In: Wilson DE, Reeder DM (eds) Mammal species of the world: a taxonomic and geographic reference, vol 1, 3rd edn. Johns Hopkins University Press, Baltimore

Halenar LB, Rosenberger AL (2013) A closer look at the '*Protopithecus*' fossil assemblages: new genus and species from the Pleistocene of Minas Gerais, Brazil. J Hum Evol 65:374–390

Hall ER (1981) The mammals of North America, vol 1. Wiley, New York

Hall ER, Kelson KR (1959) The mammals of North America, vol 1. The Ronald Press Company, New York

Harris EE, Gifalli-Inghetti C, Braga ZH, Koiffman CP (2005) Cytochrome b sequences show subdivision between populations of the brown howler monkey *Alouatta guariba* from Rio de Janeiro and Santa Catarina, Brazil. Neotrop Primates 13:16–17

Hershkovitz P (1949) Mammals of northern Colombia. Preliminary report No. 4: monkeys (Primates), with taxonomic revisions of some forms. Proc US Natl Mus 98:323–427

Hershkovitz P (1963) Primates. Comparative anatomy and taxonomy, [volume] V, *Cebidae*, part B., *A Monograph*, by W. C. Osman Hill. Edinburgh University Press. 1962, xxix 537 pp., 31 pls., 94 figs., 3 maps. $32.00. A critical review with a summary of the volumes on New World Primates. Am J Phys Anthropol 21:391–398

Hershkovitz P (1977) Living New World monkeys (Platyrrhini) with an introduction to Primates, vol 1. Chicago University Press, Chicago

Hill WCO (1962) Primates comparative anatomy and taxonomy V. Cebidae Part B. Edinburgh University Press, Edinburgh

Horwich RH (1983) Species status of the black howler monkey, *Alouatta pigra*, of Belize. Primates 24:288–289

Humboldt A (1812) Tableau synoptique des singes de l'Amérique. In: Humboldt A, Bonpland A. Recueil d'observations de zoologie et d'anatomie comparée, faites dans l'océan Atlantique, dans l'intérieur du nouveau continent et dans la mer du sud pendant les années 1799, 1800, 1801, 1802 et 1803, vol 1. Deuxième partie. Observations de zoologie et d'anatomie comparée. Schoell and Dufour and Co, Paris

Humboldt A, Bonpland A (1805) Recueil d'observations de zoologie et d'anatomie comparée, faites dans l'océan Atlantique, dans l'intérieur du nouveau continent et dans la Mer du Sud pendent les années 1799, 1800, 1801, 1802 et 1803. Book 1. Paris, Levrault, Schoell et Comp

Ihering HV (1914) Os bugios do gênero *Alouatta*. Rev Mus Paulista 9:231–256
Kerr R (1792) The animal kingdom. J Murray & R Faulder, London
Langguth A, Teixeira DM, Mittermeier RA, Bonvicino C (1987) The red-handed howler monkey in northeastern Brazil. Primate Conserv 8:36–39
Lawrence B (1933) Howler monkeys of the *palliata* group. Bull Mus Comp Zool 75:313–354
Lima MMC, Seuánez HN (1991) Chromosome studies in the red howler monkey, *Alouatta seniculus stramineus* (Platyrrhini, Primates): description of an $X_1X_2Y_1Y_2/X_1X_1X_2X_2$ sex-chromosome system and karyological comparison with other subspecies. Cytogenet Cell Genet 57:151–156
Lima MMC, Sampaio MIC, Schneider MPC, Scheffrahn W, Schneider H, Salzano FM (1990) Chromosome and protein variation in red howler monkeys. Braz J Genet 13:789–802
Linnaeus C (1766) Systema naturae per regna tria naturae secundum classes, ordines, genera, species, cum characteribus, differentiis, synonymis, locis, 12 edn. Laurentius Salvius, Stockholm
Lönnberg E (1941) Notes on members of the genera *Alouatta* and *Aotus*. Ark Zool 33A:1–44
Marcgrave G (1648) Historiae rerum naturalium Brasilae, Book 6. De Quadrupedibus et Serpentibus, Amsterdam
Minezawa M, Harada M, Jordan OC, Valdivia Borda CJ (1985) Cytogenetics of the Bolivian endemic red howler monkeys (*Alouatta seniculus sara*): accessory chromosomes and Y-autosome translocation related numerical variations. Kyoto Univ Overseas Res Rep New World Monkeys 5:7–16
Mudry MD, Rahn MI, Solari AJ (2001) Meiosis and chromosome painting of sex chromosome systems in Ceboidea. Am J Primatol 54:65–78
Mudry MD, Nieves M, Steinberg ER (2015) Cytogenetics of howler monkeys. In: Kowalewski M, Garber PA, Cortés-Ortiz L, Urbani B, Youlatos D (eds) Howler monkeys: adaptive radiation, systematics, and morphology. Springer, New York
Napier PH (1976) Catalogue of primates in the British Museum (natural history). Part 1: Families Callitrichidae and Cebidae. British Museum (Natural History), London
Napier JR, Napier PH (1967) A handbook of living primates. Academic, London
Nascimento FF, Bonvicino CR, da Silva FCD, Schneider MPC, Seuánez HN (2005) Cytochrome b polymorphisms and population structure of two species of *Alouatta* (Primates). Cytogenet Genome Res 108:106–111
Nascimento FF, Bonvicino CR, Seuánez HN (2007) Population genetic studies of *Alouatta caraya* (Alouattinae, Primates): inferences on geographic distribution and ecology. Am J Primatol 69:1093–1104
Nascimento FF, Bonvicino CR, de Oliveira MM, Schneider MPC, Seuánez HN (2008) Population genetic studies of *Alouatta belzebul* from the Amazonian and Atlantic Forests. Am J Primatol 70:423–431
Neville MK, Glander KE, Braza F, Rylands AB (1988) The howling monkeys, genus *Alouatta*. In: Mittermeier RA, Rylands AB, Coimbra-Filho AF, da Fonseca GAB (eds) Ecology and behavior of Neotropical primates, vol 2. World Wildlife Fund, Washington, DC
Perelman P, Johnson WE, Roos C, Seuánez HN, Horvath JE, Moreira MA, Kessing B, Pontius J, Roelke M, Rumpler Y, Schneider MPC, Silva A, O'Brien SJ, Pecon-Slattery J (2011) A molecular phylogeny of living primates. PLoS Genet 7:e1001342. doi:10.1371/journal.pgen.1001342
Peres CA (1994) Which are the largest New World monkeys. J Hum Evol 26:245–249
Rosenberger AL (1980) Gradistic views and adaptive radiation of platyrrhine primates. Z Morphol Anthropol 71:157–163
Rosenberger AL (1981) Systematics: the higher taxa. In: Coimbra-Filho AF, Mittermeier RA (eds) Ecology and behavior of neotropical primates, vol 1. Academia Brasileira de Ciências, Rio de Janeiro
Rosenberger AL (2011) Evolutionary morphology, platyrrhine evolution and systematics. Anat Rec 294:1955–1974
Rosenberger AL, Cooke SB, Halenar L, Tejedor MF, Hartwig WC, Novo NM, Munoz-Saba Y (2015) Fossil Alouattines and the origin of *Alouatta*: craniodental diversity and interrelationships. In: Kowalewski M, Garber PA, Cortés-Ortiz L, Urbani B, Youlatos D (eds) Howler monkeys: adaptive radiation, systematics, and morphology. Springer, New York

Rylands AB, Brandon-Jones D (1998) Scientific nomenclature of the red howlers from the northeastern Amazon in Brazil, Venezuela, and the Guianas. Int J Primatol 19:879–905

Rylands AB, Mittermeier RA (2009) The diversity of the New World primates (Platyrrhini). In: Garber PA, Estrada A, Bicca-Marques JC, Heymann EW, Strier KB (eds) South American primates: comparative perspectives in the study of behavior, ecology, and conservation. Springer, New York

Rylands AB, Mittermeier RA, Rodríguez-Luna E (1995) A species list for the New World primates (Platyrrhini): distribution by country, endemism, and conservation status according to the Mace-Lande system. Neotrop Primates 3:113–160

Rylands AB, da Fonseca GAB, Leite YLR, Mittermeier RA (1996) Primates of the Atlantic forest: origin, endemism, distributions and communities. In: Norconk MA, Rosenberger AL, Garber PA (eds) Adaptive radiations of the neotropical primates. Plenum Press, New York

Rylands AB, Schneider H, Langguth A, Mittermeier RA, Groves CP, Rodríguez-Luna E (2000) An assessment of the diversity of New World primates. Neotrop Primates 8:61–93

Rylands AB, Groves CP, Mittermeier RA, Cortés-Ortiz HJJ (2006) Taxonomy and distributions of Mesoamerican primates. In: Estrada A, Garber P, Pavelka M, Luecke L (eds) New perspectives in the study of Mesoamerican primates: distribution, ecology, behavior and conservation. Springer, New York

Rylands AB, Mittermeier RA, Silva JS Jr (2012) Neotropical primates: taxonomy and recently described species and subspecies. Int Zoo Yb 46:11–24

Sampaio MIC, Schneider MPC, Barroso CML, Silva BTF, Schneider H, Encarnación F, Montoya E, Salzano FM (1991) Carbonic anhydrase II in New World monkeys. Int J Primatol 12: 389–402

Sampaio MIC, Schneider MPC, Schneider H (1996) Taxonomy of the *Alouatta seniculus* group: biochemical and chromosome data. Primates 37:67–73

Schneider H, Rosenberger AL (1996) Molecules, morphology, and platyrrhine systematics. In: Norconk MA, Rosenberger AL, Garber PA (eds) Adaptive radiations of neotropical primates. Plenum, New York

Schneider H, Sampaio I (2013) The systematics and evolution of New World primates—a review. Mol Phylogenet Evol (in press). doi:10.1016/j.ympev.2013.10.017

Seuánez HN, Bonvicino CR, Moreira MAM (2005) The primates of the Neotropics: genomes and chromosomes. Cytogenet Genome Res 108:38–46

Smith JD (1970) The systematic status of the black howler monkey, *Alouatta pigra* Lawrence. J Mammal 51:358–369

Stanyon R, Tofanelli S, Morescalchi MA, Agoramoorthy G, Ryder OA, Wienberg J (1995) Cytogenetic analysis shows extensive genomic rearrangements between red howler (*Alouatta seniculus* Linnaeus) subspecies. Am J Primatol 35:171–183

Stanyon R, Garofalo F, Steinberg ER, Capozzi O, Di Marco S, Nieves M, Archidiacono N, Mudry MD (2011) Chromosome painting in two genera of South American monkeys: species identification, conservation, and management. Cytogenet Genome Res 134:40–50

Steinberg ER, Cortés-Ortiz L, Nieves M, Bolzán AD, García-Orduña F, Hermida-Lagunes J, Canales-Espinosa D, Mudry MD (2008) The karyotype of *Alouatta pigra* (Primates: Platyrrhini): mitotic and meiotic analyses. Cytogenet Genome Res 122:103–109

Tate GHH (1939) The mammals of the Guiana region. Bull Am Mus Nat Hist 76:151–229

Thomas O (1911) The mammals of the tenth edition of Linnaeus; an attempt to fix the types of the genera and the exact bases and localities of the species. Proc Zool Soc Lond 1911:120–145

Thomas O (1913) New mammals from South America. Ann Mag Nat Hist 8th Series 12:567–574

von Spix JB (1823) Simiarum et vespertilionum Brasiliensium species novae, ou, histoire naturelle des espèces nouvelles de singes et de chauves-souris, observés et recueillies pendant le voyage dans l'interieur du Brésil. F.S. Hubschmann, Munich

Yunis EJ, Torres de Caballero OM, Ramírez C, Ramírez ZE (1976) Chromosomal variation in the primate *Alouatta seniculus seniculus*. Folia Primatol 25:215–224

Chapter 4
Cytogenetics of Howler Monkeys

Marta D. Mudry, Mariela Nieves, and Eliana R. Steinberg

Abstract Cytogenetic studies of howler monkeys show diploid numbers ranging from 2N=43 in *Alouatta seniculus* to 2N=58 in *A. pigra* with several interspecific chromosomal rearrangements such as translocations and inversions. Other remarkable genetic features are the multiple sex chromosome systems and the presence of microchromosomes. Multiple sexual systems are originated by Y-autosome translocations, resulting in the formation of trivalents X_1X_2Y in males of *A. belzebul* and *A. palliata* and quadrivalents $X_1X_2Y_1Y_2$ in males of *A. seniculus*, *A. pigra*, *A. macconnelli*, and *A. caraya*. Fluorescence *in situ* hybridization (FISH) analyses in the South American species have revealed that segments with homeology to human chromosomes #3 and #15 (synteny 3/15) are involved in these sexual systems. Different authors agreed with the assumption that these diverse sex chromosome systems share the same autosomal pair and the rearrangement may have occurred once. Recent cytogenetic characterization of *A. pigra* and *A. palliata* has shown that the autosomes involved in the translocation that formed the sex chromosome systems in the Mesoamerican and South American species are different. Two independent events of Y-autosome translocations might have led to different sexual systems. Together with the multiple autosomal rearrangements found in the genus, the howler monkey's sex chromosome systems constitute an illustrative example of the possible chromosomal evolutionary mechanisms in Platyrrhini.

Resumen Los estudios citogenéticos realizados en monos aulladores muestran números diploides que van de 2N=43 para *Alouatta seniculus* a 2N=58 para *A. pigra*, con reordenamientos cromosómicos interespecíficos de tipo translocaciones e inversiones. Otras características genéticas notables son los sistemas sexuales múltiples y la presencia de microcromosomas. Los sistemas sexuales múltiples son resultado de translocaciones Y-autosoma, originando trivalentes X_1X_2Y en los machos de *A. belzebul* y *A. palliata* y cuadrivalentes $X_1X_2Y_1Y_2$ en los machos de

---

M.D. Mudry (✉) • M. Nieves • E.R. Steinberg
Grupo de Investigación en Biología Evolutiva (GIBE), Labs 46/43, 4° piso, Depto. de Ecología, Genética y Evolución, IEGEBA, Facultad de Ciencias Exactas y Naturales, Universidad de Buenos Aires, Pab II. Ciudad Universitaria, Ciudad Autónoma de Buenos Aires, Argentina

Consejo Nacional de Investigaciones Científicas y Técnicas (CONICET),
Ciudad Autónoma de Buenos Aires, Argentina
e-mail: martamudry@yahoo.com.ar; mnieves@ege.fcen.uba.ar; steinberg@ege.fcen.uba.ar

A. seniculus, *A. pigra*, *A. macconnelli* y *A. caraya*. La Hibridación in situ Fluorescente (FISH) ha revelado que en la formación de los sistemas sexuales de las especies sudamericanas están involucrados ciertos segmentos homeólogos de los cromosomas humanos #3 y #15 (sintenia 3/15). Diferentes autores están de acuerdo con la hipótesis de que estos sistemas de cromosomas sexuales comparten un mismo par autosómico y que el reordenamiento que les dio lugar habría ocurrido una única vez. Una caracterización citogenética reciente de *A. pigra* y *A. palliata* puso en evidencia que diferentes cromosomas están involucrados en los sistemas sexuales de las especies mesoamericanas y sudamericanas. Dos eventos independientes de translocaciones Y-autosoma parecerían haber originado los diferentes sistemas sexuales. Junto con los múltiples reordenamientos autosómicos que se observan en el género, los sistemas de cromosomas sexuales presentes en aulladores son un ejemplo ilustrativo de los posibles mecanismos de evolución cromosómica en Platyrrhini.

Keywords Karyosystematics • Multiple sex chromosome systems • Cytogenetics • Chromosomal syntenies

Abbreviations

FISH	Fluorescence in situ hybridization
G-banding	Giemsa banding
NWP	New World Primates
PAR	Pseudoautosomal region
R-banding	Reverse banding
REs banding	Restriction enzymes banding
SC	Synaptonemal complex

4.1 Introduction

Karyosystematics, or the study of the natural relationships of species using the information provided by chromosomes, can provide valuable information for taxonomic classifications and evolutionary analyses. In primates, during the last three decades, researchers have proposed chromosomal speciation as a probable evolutionary mechanism to interpret diversity of living primates (De Grouchy et al. 1972; Dutrillaux et al. 1975, 1980; Seuánez 1979; Dutrillaux and Couturier 1981; De Grouchy 1987; Clemente et al. 1990). More recently, chromosomal data began to be used as phylogenetic markers, since they are inherited as Mendelian characters and are conserved within species (Sankoff 2003; Dobigny et al. 2004; Stanyon et al. 2008). Following the maximum parsimony criterion, karyological comparisons allow the identification of chromosomal forms shared by common ancestrality (Dobigny et al. 2004).

In order to obtain better taxonomic inferences, a wide battery of variables including genetic, morphological, and ecological should be employed into the theoretical framework referred to as "Total Evidence" (Kluge 1989). The systematics of New World Primates (NWP) is still under discussion (Rylands 2000; Groves 2001, 2005; Schneider et al. 2001; Rylands and Mittermeier 2009; Perelman et al. 2011; Rylands et al. 2012), and the genus *Alouatta* is not the exception (Cabrera 1957; Hill 1962; Groves 2001; Gregorin 2006; Rylands and Mittermeier 2009; Rylands et al. 2012; Cortés-Ortiz et al. 2014). In this context, cytogenetic studies have become an important tool complementing molecular and morphological data traditionally used in systematic studies.

4.2 Karyological Features of the Genus *Alouatta*

4.2.1 Classical Cytogenetic Analysis

Since the first cytogenetic characterization of the red howler monkey, *Alouatta seniculus* (Bender and Chu 1963), the karyological features of the genus have attracted the scientists' attention, due to the multiple interspecific autosomal rearrangements and, particularly, for the multiple sex chromosome systems observed. The first cytogenetic studies employed standard staining techniques (Fig. 4.1), established the diploid numbers, and allowed to arrange the chromosomes by size and morphology (*A. seniculus*: Bender and Chu 1963; *A. palliata*: Hsu and Benirschke 1970; *A. caraya*: Egozcue and de Egozcue 1965, 1966). The advent of the chromosome banding techniques in the 1970s (Giemsa banding or G-banding, reverse banding or R-banding, and restriction enzyme banding or REs banding) made it possible to reveal a characteristic pattern of specific dark and light bands along each chromosome (Fig. 4.2).

Using these differential banding techniques, homologous chromosomes, chromosome segments, and chromosomal rearrangements could be identified with precision in each karyotype, thus allowing the first interspecific comparisons (Koiffmann and Saldanha 1974; Yunis et al. 1976; Mudry et al. 1981, 1984, 1994; Minezawa et al. 1985; Armada et al. 1987; Lima and Seuánez 1991; Stanyon et al. 1995; Rahn et al. 1996; Vassart et al. 1996; Steinberg et al. 2008). Cytogenetic analyses using these staining techniques in somatic cells showed high chromosomal variability with drastic differences in the chromosome number among species (Table 4.1). Howler monkeys exhibit diploid numbers (2N) ranging from 2N = 43 in *A. seniculus* to 2N = 58 in *A. pigra*. Several interspecific chromosomal rearrangements such as translocations and inversions have been described (De Boer 1974; Mudry et al. 1994; Consigliere et al. 1996, 1998; de Oliveira et al. 2002). This interspecific chromosomal variation coincides with the one observed in the night monkey, genus *Aotus* (Ma 1981; Torres et al. 1998; Ruiz Herrera et al. 2005; Defler and Bueno 2007), but contrasts with other platyrrhine genera such as the squirrel monkeys, genus *Saimiri* (Hershkovitz 1984; Moore et al. 1990), or the capuchin monkeys, genus *Cebus* (Matayoshi et al. 1987; Mudry et al. 1987; Ponsà et al. 1995), in which interspecific variation in 2N is not observed (species show constant diploid numbers).

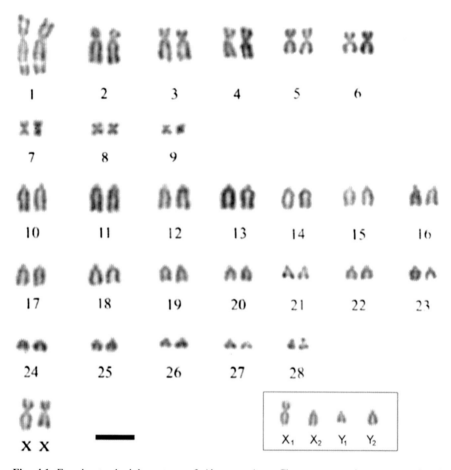

Fig. 4.1 Female standard karyotype of *Alouatta pigra*. Chromosome pairs were numbered consecutively following decreasing size within each type of morphology. *Inset*: sex chromosome system observed in the male *Alouatta pigra*. Bar = 10 μm

Alouatta chromosomal variability is not restricted to differences among species. A few intraspecific polymorphisms have been described in *A. caraya*, but other species, such as *A. guariba*, show multiple intraspecific rearrangements involving differences in diploid number and sexual systems (see Table 4.1). This high intraspecific variability suggests that these may be species complexes rather than single species (Stanyon et al. 1995; Consigliere et al. 1996, 1998; de Oliveira et al. 2000). Within *Alouatta*, karyological studies have contributed to the taxonomic reassessment of several taxa previously considered as subspecies, elevating them to the species level: *A. nigerrima*, previously *A. seniculus nigerrima*; *A. macconnelli*, previously *A. seniculus macconnelli*; *A. stramineus*, previously *A. seniculus stramineus* (here considered within *A. macconnelli*); and *A. belzebul*, previously *A. seniculus belzebul* (Lima and Seuánez 1991; de Oliveira 1996).

4 Cytogenetics of Howler Monkeys

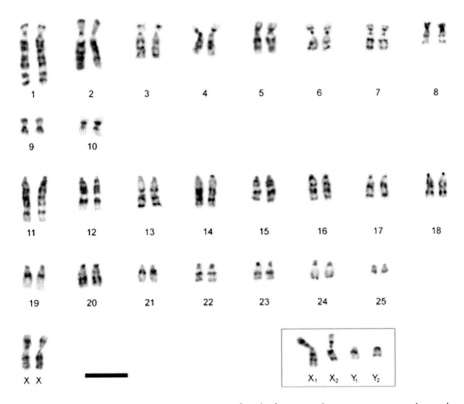

Fig. 4.2 G-banded karyotype of *Alouatta caraya* female. *Inset*: sex chromosome system observed *Alouatta caraya* males. Bar = 10 μm

4.2.1.1 Heterochromatin

Traditionally, the term heterochromatin was used to denote chromosomal regions of the karyotype showing increased condensation (Heitz 1928), revealed as regions of intense staining (denoted as C+ bands) (Fig. 4.3) by C-banding (Sumner 1971). Later, cytomolecular techniques showed that these regions are constituted by distinct medium and highly repetitive DNA sequences (Copenhaver et al. 1999; Fransz et al. 2000; Avramova 2002). In Platyrrhini, heterochromatic regions are highly variable in quantity, quality, and location. C-banding techniques and restriction enzymes digestion have demonstrated that there are different kinds of heterochromatin, and particularly among platyrrhine species, a great variability was described, not only in sequence types but also in their location (centromeric, interstitial, and telomeric blocks) (Matayoshi et al. 1987; Mudry de Pargament and Labal de Vinuesa 1988; Pieczarka et al. 2001; García et al. 2003). Compared with other platyrrhine genera, like *Saimiri*, *Cebus*, and *Aotus*, that show big C+ blocks of extracentromeric heterochromatin,

Table 4.1 Cytogenetic data published for howler monkeys

Species	2N ♀	2N ♂	#NA	#A	Sexual system (female/male)	References
A. caraya	52	52	20	30	XX/XY	Egozcue and de Egozcue (1965) (specimens from Argentina)
A. caraya	–	52	20	30	XX/XY	Egozcue and de Egozcue (1966) (specimens from Argentina)
A. caraya	52	52	20	30	XX/XY	Mudry et al. (1981, 1990), Mudry de Pargament et al. (1984) (specimens from Argentina)
A. caraya	52	52	18	30	$X_1X_1X_2X_2/X_1X_2Y_1Y_2$	Rahn et al. (1996), Mudry et al. (1998, 2001) (specimens from Argentina)
A. guariba clamitans	50	50	20	28	XX/XY	Koiffmann and Saldanha (1974) (São Paulo, Brazil)
A. guariba clamitans	50	49	20	27	$X_1X_1X_2X_2/X_1X_2Y$	Koiffmann and Saldanha (1981) (São Paulo, Brazil)
A. guariba clamitans	46	45	22	20	$X_1X_1X_2X_2/X_1X_2Y$	de Oliveira et al. (1995) (specimens from Parana and Santa Catarina, south of Brazil)
A. guariba clamitans	–	52	22	28	XY	de Oliveira et al. (1995, 2000) (one male from Espirito Santo, Brazil)
A. guariba guariba	–	49	14	32	X_1X_2Y	de Oliveira et al. (1998) (Río de Janeiro, Brazil)
A. guariba clamitans	–	45	22	18	$X_1X_2X_3Y_1Y_2$	de Oliveira et al. (2002) (in captivity in Parana, Brazil)
A. guariba guariba	50	49	20	24	$X_1X_1X_2X_2X_3X_3/X_1X_2X_3Y_1Y_2$	de Oliveira et al. (2002) (in captivity in Minas Gerais, Brazil)
A. nigerrima	50	–	18	30	XX	Armada et al. (1987) (geographic origin unknown)
A. villosa (=A. palliata)	54	–	22	30	XX	Hsu and Benirschke (1970) (one female from Guatemala)
A. palliata	54	53	20	30	$X_1X_1X_2X_2/X_1X_2Y$	Ma et al. (1975) (specimens from Panama)
A. palliata	56	56	20	34	XX/XY	Torres and Ramírez (2003) (specimens from Colombia)
A. palliata	–	53	22	28	X_1X_2Y	Solari and Rahn (2005) (specimens from Mexico)

A. pigra	58	58	18	36	$X_1X_1X_2X_2/$ $X_1X_2Y_1Y_2$	Steinberg et al. (2008) (specimens from Mexico)
A. seniculus	44	44	12	30	XX/XY	Bender and Chu (1963) (specimens from Colombia)
A. seniculus (previously considered A. s. seniculus by the author)	43–45[a]	43–45[a]	12	26	XX/XY	Yunis et al. (1976) (specimens from Colombia)
A. macconnelli (previously considered A. seniculus by the author)	47–49[a]	47–49[a]	18	24	$X_1X_1X_2X_2/$ $X_1X_2Y_1Y_2$	Vassart et al. (1996) (specimens from French Guiana)
A. seniculus	43–45[a]	43–45[a]	12	26	XX/XY	Torres and Leibovici (2001) (specimens from Colombia)
A. macconnelli (previously considered A. s. stramineus by the author)	47–49[a]	47–49[a]	20	22	$X_1X_1X_2X_2/$ $X_1X_2Y_1Y_2$	Lima and Seuánez (1991) (specimens from Brazil)
A. arctoidea (previously considered A. s. arctoidea by the author)	44	45	10	30	$X_1X_1X_2X_2/$ $X_1X_2Y_1Y_2$	Stanyon et al. (1995) (specimens from Venezuela; variation in diploid number between sexes is due to microchromosome number)
A. sara (previously considered A. s. sara by the author)	48–51[a]	48–51[a]	16	24	$X_1X_1X_2X_2/X_1X_2Y$	Minezawa et al. (1985) (specimens from Bolivia)
A. sara (previously considered A. s. sara by the author)	–	46+4m	14	28	$X_1X_2Y_1Y_2$	Stanyon et al. (1995) (specimens from Bolivia)
A. macconnelli (previously considered A. seniculus by the author)	47–49[a]	47–49[a]	20	22	$X_1X_1X_2X_2/$ $X_1X_2Y_1Y_2$	Lima et al. (1990) (specimens from Brazil)
A. belzebul	50	49	22	24	$X_1X_1X_2X_2/X_1X_2Y$	Armada et al. (1987) (specimens from Brazil)

NA non-acrocentric chromosomes, A acrocentric chromosomes, m microchromosomes

[a]Differences are due to the presence of microchromosomes

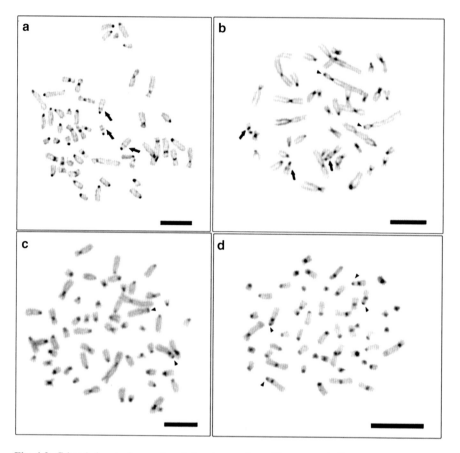

Fig. 4.3 C-banded metaphases of male howler monkeys. The *arrows* indicate the interstitial C+ bands. *Single arrowheads* indicate telomeric C+ bands. Bar = 10 μm. (**a**) *Alouatta caraya* 2N = 52, $X_1X_2Y_1Y_2$, (**b**) *A. guariba* 2N = 45, $X_1X_2X_3Y_1Y_2$, (**c**) *A. pigra* 2N = 58, $X_1X_2Y_1Y_2$, and (**d**) *A. palliata* 2N = 53, X_1X_2Y

Alouatta is the genus with the lowest heterochromatin proportion (Ma et al. 1976; García et al. 1979; 1983; Matayoshi et al. 1987; Mudry de Pargament and Slavutsky 1987; Mudry de Pargament et al. 1984; Mudry 1990; Mudry et al. 1990; 1991; Ponsà et al. 1995; Rahn et al. 1996; Nieves et al. 2005a; Nieves 2007). Within howler monkeys, *A. seniculus* group shows the lowest C+ heterochromatin proportion, with centromeric location only (Lima et al. 1990); meanwhile *A. guariba clamitans* (de Oliveira et al. 1998) and *A. palliata* (Ma et al. 1975) show the highest heterochromatin content within the genus, with centromeric, interstitial, and telomeric C+ blocks (Table 4.2).

4 Cytogenetics of Howler Monkeys

Table 4.2 Heterochromatin location in somatic cells of howler monkeys revealed by C-banding technique

Species	C-bands location	References
A. caraya	C+ Cent in all chromosomal pairs	Mudry et al. (1984, 1994)
	C+ Int in pairs #16 and #21	
A. guariba clamitans	C+ Cent in all pairs; C+ Int in pair #17	de Oliveira et al. (1998)
	C+ Tel in pair #2; 5p and 6p heterochromatic	
A. guariba clamitans	C+ Cent in all pairs; C+ Int in pairs #16, #17	Steinberg (2011)
	C+ Tel in pair #2	
A. nigerrima	C+ Cent in all pairs	Armada et al. (1987)
A. pigra	C+ Cent in all pairs	Steinberg et al. (2008)
	C+ Tel in pair #2	
A. palliata	C+ Cent in all pairs	Steinberg (2011)
	C+ Tel in pairs #3 and #5	
A. palliata	C+ Cent in all pairs; C+ Int in pairs A1, A5, B19, B25 and C+ Tel in pairs A3, A6	Ma et al. (1975)
A. macconnelli (previously considered *A. s. stramineus* by the author)	C+ Cent in all pairs Microchromosomes C-negative	Yunis et al. (1976)
A. macconnelli (previously considered *A. s. stramineus* by the author)	C+ Cent in all pairs, except in 2 biarmed ones; C+ band polymorphism in #19; Microchromosomes C+ positive	Lima and Seuánez (1991)
A. macconnelli (previously considered *A. seniculus* by the author)	C+ Cent in all pairs, except in #4 and #8 9q heterochromatic Microchromosomes C+ positive	Vassart et al. (1996)
A. sara (previously considered *A. s. sara* by the author)	C+ Cent in all pairs Microchromosomes C+ positive and C-negative	Minezawa et al. (1985)
A. seniculus	C+ Cent in all pairs Microchromosomes C+ positive and C-negative	Torres and Leibovici (2001)
A. macconnelli (previously considered *A. seniculus* by the author)	C+ Cent in all pairs, except in #4 and #8 Microchromosomes C+ positive	Lima et al. (1990)
A. belzebul belzebul	C+ Cent in all pairs; C+ Int in 2 pairs; C+ Tel in 2 pairs	Armada et al. (1987)

C+ Cent = C+ Centromeric bands, C+ Tel = C+ Telomeric bands, C+ Int = C+ Interstitial bands

4.2.1.2 Microchromosomes

Microchromosomes are supernumerary chromosomes and their presence is unusual among Primates. Some authors denominate these chromosomes as "B chromosomes" in contrast to the chromosome complement, generally denominated as "A chromosomes". Their name is derived from the fact that many of these supernumerary chromosomes are smaller than the smallest A chromosomes (reviewed in Vujoševič and Blagojevič 2004). In Platyrrhini, *A. seniculus* (Yunis et al. 1976; Torres and Leibovici 2001), *A. sara* (Minezawa et al. 1985), and *A. macconnelli* (Lima et al. 1990; Vassart et al. 1996) exhibit this kind of accessory chromosomes. The structure of microchromosomes has mostly been characterized by C-banding technique (Arrighi and Hsu 1971; Sumner 1971). These minute chromosomes have been reported as small segments of heterochromatin (Patton 1977). However, in howler monkeys, there are reports of microchromosomes being either C-band negative (C−) such as in *A. seniculus* (Yunis et al. 1976) or C-band positive (C+) like in *A. macconnelli* (previously *A. stramineus*, Lima and Seuánez 1991; Lima et al. 1990; Vassart et al. 1996). In *A. sara*, some microchromosomes are C-positive and others are C-negative, and there is no formal hypothesis about this particularity (Minezawa et al. 1985; Torres and Leibovici 2001).

In some howler monkey species, there is variation in microchromosome number between sexes and between individuals (Yunis et al. 1976; Minezawa et al. 1985; Lima et al. 1990; Lima and Seuánez 1991; Vassart et al. 1996). Battaglia (1964) suggested that variation in microchromosome number in mammals could affect the frequency of chiasmata as well as growth, viability, and fertility. Little is known about the meiotic behavior of these microchromosomes, since few analyses in germ cells have been performed. More studies need to be developed in order to elucidate the nature and transmission of these supernumerary chromosomes in howler monkeys.

4.2.1.3 Sex Chromosomes

In mammals, and particularly in Primates, the most frequently described sexual system is the XX/XY (reviewed in Solari 1993). As variants of the ancestral male XY sexual system, Y-autosome translocations that generate multiple sex chromosome systems in males (Fig. 4.4) have been described in Platyrrhini (Hsu and Hampton 1970; Ma et al. 1975; Dutrillaux et al. 1981; Armada et al. 1987; Lima and Seuánez 1991; Rahn et al. 1996; Mudry et al. 1998, 2001; Solari and Rahn 2005; Steinberg et al. 2008). Only the karyological study of germ cells (meiotic analysis) allows the identification and confirmation of these sexual systems; however, meiotic studies are remarkably scarce.

Meiotic karyotypes of platyrrhines have only been described for a small number of species, confirming the sex determination XX/XY in *Cebus libidinosus* (formerly *C. apella paraguayanus*; Seuánez et al. 1983; Mudry et al. 2001), *Ateles geoffroyi* and *Ateles paniscus* (Mudry et al. 2001; Nieves et al. 2005b), and *Saimiri boliviensis boliviensis* (Egozcue 1969; Steinberg et al. 2007) and multiple sex chromosome systems in *Aotus azarae* (Ma et al. 1976), *Callimico* sp. (Hsu and Hampton 1970), *Cacajao* sp. (Dutrillaux et al. 1981), and five species of *Alouatta* (Armada et al. 1987; Lima and Seuánez 1991; Rahn et al. 1996; Mudry et al. 1998, 2001; Solari and Rahn 2005; Steinberg et al. 2008). All meiotic studies of howler monkeys were performed using testes biopsies and confirmed two types of multiple sex chromosome systems (Table 4.3): (1) $X_1X_2Y_1Y_2$ (which forms a chain of four elements or quadrivalent at Metaphase I) in *A. macconnelli* (Lima and Seuánez 1991), *A. caraya*

4 Cytogenetics of Howler Monkeys

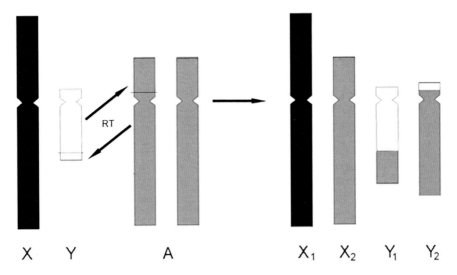

Fig. 4.4 Hypothesis on the origin of the sexual system $X_1X_2Y_1Y_2$ in mammalian males. The ancestral X is shown in *black*, the ancestral Y in *white*, and the autosomal pair (A) involved in the translocation in *gray*. From an XY sexual system, two simultaneous breaks on the proximal region of the short arm (Ap$_{prox}$) and the terminal region of the long arm (Yq$_{ter}$), followed by a reciprocal translocation (RT), give origin to the chromosomes Y_1 and Y_2. The homologous chromosome of the autosomal pair not involved in the translocation became known as X_2, and the ancestral X is now denominated X_1

Table 4.3 Confirmation of howler monkey sexual systems through meiotic studies in males

Species	2N (♂)	#NA	#A	Confirmed sexual system	References
A. belzebul	49	22	24	X_1X_2Y	Armada et al. (1987)
A. macconnelli (previously considered A. s. stramineus by the author)	47–49[a]	20	22	$X_1X_2Y_1Y_2$	Lima and Seuánez (1991)
A. caraya	52	18	30	$X_1X_2Y_1Y_2$	Rahn et al. (1996), Mudry et al. (1998, 2001)
A. palliata	53	22	28	X_1X_2Y	Solari and Rahn (2005)
A. pigra	58	18	36	$X_1X_2Y_1Y_2$	Steinberg et al. (2008)

NA number of non-acrocentric chromosomes, *A* number of acrocentric chromosomes
[a]These differences are due to presence of microchromosomes

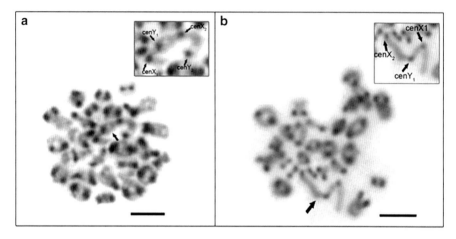

Fig. 4.5 Howler monkey C-banded spermatocytes in Metaphase I showing the location of the C+ heterochromatic regions. This staining technique allows the identification of C+ centromeres, thus revealing the structure of the multivalents. Bar=5 μm. (**a**) *A. caraya* spermatocyte. The *arrow* indicates the sexual $X_1X_2Y_1Y_2$ (four centromeres C+). *Inset*: detail of the sexual quadrivalent. The *arrows* indicate the centromeres of each of the four chromosomal components. (**b**) *A. palliata* spermatocyte. The *arrow* indicates the sexual $X_1X_2Y_1$ (three centromeres C+). *Inset*: detail of the sexual trivalent. The *arrows* indicate the centromeres of each of the three chromosomal components

(Fig. 4.5a) (Mudry et al. 1998, 2001), and *A. pigra* (Steinberg et al. 2008) and (2) X_1X_2Y (which forms a chain of three elements or trivalent at Metaphase I) in *A. belzebul* (Armada et al. 1987) and *A. palliata* (Fig. 4.5b) (Solari and Rahn 2005). The multivalent configurations observed in Metaphase I of howler monkey spermatocytes (Fig. 4.5) allow an alternate segregation in Anaphase I, ensuring a balanced gamete production and maintaining the fertility of the individual carriers.

Some mitotic studies of *Alouatta* described the presence of a typical XY male sexual system in *A. seniculus* (Yunis et al. 1976), *A. guariba clamitans* (Koiffmann and Saldanha 1974; de Oliveira et al. 1995, 2000), and *A. palliata* (Torres and Ramírez 2003), and other studies suggested the presence of a $X_1X_2X_3Y_1Y_2$ sexual system (which would form a chain of five elements or pentavalent in Metaphase I) in *A. guariba guariba* and *A. g. clamitans* (de Oliveira et al. 2002), a X_1X_2Y system in *A. guariba guariba* (de Oliveira et al. 1998) and *A. sara* (Minezawa et al. 1985), and a $X_1X_2Y_1Y_2$ system in *A. sara* (Consigliere et al. 1998). However, the occurrence of all these sex chromosome systems in *Alouatta* still awaits confirmation by meiotic analysis.

The meiotic behavior in early meiotic stages of the howler monkey sexual multivalents has been studied in just a few cases (Rahn et al. 1996; Mudry et al. 1998, 2001; Solari and Rahn 2005). The analysis of the sexual quadrivalent of *A. caraya* at pachytene showed that the maximum extent of synapsis in Y_2 is 51 % of the length of X_2, whereas the maximum extent of synapsis of Y_1 with X_2 is 42.9 %. The synaptonemal complex (SC) between the X_1 axis and Y_1 is the smallest pairing segment in the whole quadrivalent (Mudry et al. 1998). In the sexual trivalent of *A. palliata*, however, the long arms of X_2 and Y_1 are paired in almost all their length, and the short arm of Y_1 forms the pseudoautosomal region (PAR) with X_1p (Solari and Rahn 2005). The end-to-end joining between the X_1 and Y_1 chromosomes is similar in both multivalents, although the Y_1 is a much longer acrocentric in *A. palliata*. In both *A. caraya* and *A. palliata*, the X_1 axis has the typical characteristics (branchings, tangling) that are common in spermatocytes at pachytene stages described for other mammalian X axis (Solari 1993), including other Neotropical primates with XX/XY sexual systems (Mudry et al. 2001).

To understand the mechanisms underlying chromosomal evolution and speciation in mammalian species, experimental descriptions of recombination maps are needed. Few studies using "in situ" immunolocalization of recombination proteins have been applied in nonhuman primates (Garcia-Cruz et al. 2009, 2011; Hassold et al. 2009). Only one of such studies has been carried out in howler monkeys (on spermatocytes of *A. caraya* by Garcia-Cruz et al. 2011), analyzing MLH1 *foci*, which correspond to recombination spots and are equivalent to the chiasmata observed in Metaphase I. The mean MLH1 *foci* number per autosomal set was 40.6 ± 4.3 (standard deviation), with a range of 31–50 MHL1 *foci* per cell. This value is lower than the one observed for human males (49.8 ± 4.3, Sun et al. 2004) but similar to those observed in *Cebus libidinosus* (41.3 ± 4.8), *C. nigritus* (39.2 ± 3.3) (Garcia-Cruz et al. 2011), and *Macaca mulatta* (39.0 ± 3.0) (Hassold et al. 2009). The sexual quadrivalent formed a convoluted sex body, which folded back onto itself, not allowing for a correct visualization of the MLH1 *foci*. More studies are needed in order to understand the meiotic process in howler monkeys.

Classical cytogenetic analysis showed that the chromosomal pair involved in the sexual systems in Mesoamerican howler monkeys, *A. pigra* (API) and *A. palliata* (APA), share no homeology (see Sect. 4.2.2) with the pair involved in the South American species (Steinberg et al. 2008). Chromosomal pair API17 (denominated $APIX_2$ in males) is involved in *A. pigra*'s multiple sexual system, and APA19 (APA19 in females, $APAX_2$ in males) is involved in the multiple sexual system of *A. palliata*. These chromosomal pairs share homeology with *A. caraya* autosome 14 (ACA14) (Fig. 4.6). Therefore, the autosomal pair that is involved in the formation of the sexual systems in *A. pigra* and *A. palliata* is not homeologous with the one involved in the sexual system in *A. caraya* (known as ACA7 in females and $ACAX_2$ in males). ACA7 shares homeology with the autosomal pairs involved in the sexual systems of all South American howler monkey species studied so far (Rahn et al. 1996; Mudry et al. 1998, 2001): *A. seniculus* (Lima and Seuánez 1991), *A. belzebul* (Armada et al. 1987), *A. sara* (Minezawa et al. 1985), *A. guariba* (de Oliveira et al. 2002), and *A. macconnelli* (Lima et al. 1990). This suggests that the multiple sexual systems originated independently in South American and Mesoamerican howler monkeys.

Fig. 4.6 Chromosome homeologies among *A. caraya* (ACA), *A. pigra* (API), *A. palliata* (APA), and *A. guariba* (AGU). (**a**) ACA7 (X_2) shares homeology with API26, API19, APA23, APA18, and AGU7 (X_2). (**b**) ACA14 shares homeology with API17 (X_2), APA19 (X_2), and AGU6

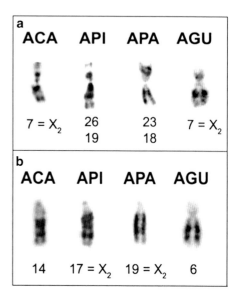

4.2.2 Cytomolecular Analysis

Homeologies at chromosomal level refer to the recognition of chromosome pairs carrying the same information among different organisms. Homologous chromosomes are defined as chromosome pairs of approximately the same length, centromere position, and staining pattern, with genes for the same characteristics at corresponding *loci*. One homologous chromosome is inherited from the organism's mother and the other from the organism's father. When we consider genetic information in different karyotypes of different organisms, we apply the term "homeolog" (Andersson et al. 1996), which becomes useful for phylogenetic analysis. The homeologies identified by G-banding pattern and employed for karyological comparisons are often not informative enough. This is the case when the homeologies involve complex chromosomal rearrangements, small translocated chromosomal fragments or highly rearranged karyotypes, such as *Alouatta*'s (Dobigny et al. 2004; Stanyon et al. 2008).

In the past 20 years, the Fluorescence in situ Hybridization (FISH) technique has proven to be a fast and reliable method to establish chromosomal homeologies between taxa. In this technique, labeled DNA probes specific for entire chromosomes or chromosome regions of a given species are used to hybridize to entire chromosome or chromosome segments in target metaphases (John et al. 1969; Pardue and Gall 1969; Pinkel 1986; Wienberg and Stanyon 1997). Several authors applied this technique to characterize genome conservation in primates (Wienberg et al. 1990; Morescalchi et al. 1997; Consigliere et al. 1998; Stanyon et al. 2004, 2011; Dumas et al. 2007; Amaral et al. 2008). FISH technique provides an unequivocal confirmation of the homeologies previously described by G-banding, giving a higher definition at the cytomolecular level (Wienberg et al. 1990; Wienberg 2005;

Müller 2006; Stanyon et al. 2008). Conserved chromosomal syntenies (regions of chromosomes that can be located together side by side on the same chromosome arm) may be used as markers to investigate possible common evolutionary origin. Cytogeneticists refer to a broken chromosomal synteny when a region in a single chromosome of one taxon is found located in different chromosomes of another taxon. Therefore, it is said that the synteny is broken in the later taxon. Chromosomal syntenies can be broken by fissions or translocations. The analysis of these syntenies throughout the phylogeny of a group, such as Primates, allows the identification of the chromosomal rearrangements that might be involved in the speciation process. In the last two decades, cross-hybridization using probes of human chromosomes and/or other primate species (Wienberg and Stanyon 1997; Wienberg 2005; Stanyon et al. 2008) has been especially helpful to analyze genomic conservation.

Only five species of *Alouatta* have been analyzed using FISH: *A. caraya* (Mudry et al. 2001; Stanyon et al. 2011), *A. guariba* (both *A. g. clamitans* and *A. g. guariba*, de Oliveira et al. 2002; Stanyon et al. 2011), *A. sara* and *A. arctoidea* (Consigliere et al. 1996), and *A. belzebul* (Consigliere et al. 1998). FISH analyses in *Alouatta* were concordant with the G-banding studies, showing high levels of interspecific chromosomal variability, with interchromosomal rearrangements (Consigliere et al. 1996, 1998; Mudry et al. 2001; de Oliveira et al. 2002; Stanyon et al. 2011). Two syntenic associations (4/15 and 10/16) and the loss of the ancestral association 2/16 were proposed as synapomorphies of *Alouatta*. All howler monkeys share the syntenies 14/15 from the ancestral mammalian karyotype and 8/18 from the ancestral Platyrrhini (Consigliere et al. 1996, 1998; de Oliveira et al. 2002; Stanyon et al. 2011).

The synteny 3/15 was found in all South American *Alouatta* species with the exception of *A. belzebul* (Fig. 4.7), and it is involved in their multiple sexual systems (Consigliere et al. 1996, 1998; Mudry et al. 2001; de Oliveira et al. 2002; Stanyon et al. 2011) (see Sect. 4.2.1.3). The 3/15 synteny was also observed in other atelids, such as *Ateles geoffroyi* and *A. belzebuth hybridus*, but is not involved in their XY "human-like" sexual systems (Morescalchi et al. 1997; García Haro 2001; de Oliveira et al. 2005). This synteny was not found in other platyrrhine genera

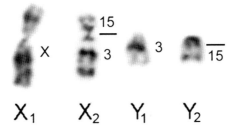

Fig. 4.7 Fluorescence in situ Hybridization with human chromosome probes X, #3, and #15 performed in South American species of *Alouatta* with male sexual system $X_1X_2Y_1Y_2$. The numbers to the right of the chromosome indicate the hybridization signal of each human chromosome on its corresponding *Alouatta* homeolog chromosome

such as *Cebus libidinosus* or *Saimiri boliviensis boliviensis* (Mudry et al. 2001). It was then proposed that this synteny could be ancestral for the Atelidae, and an association with multiple sex chromosomes would have only occurred in South American howler monkeys (de Oliveira et al. 2012). However, this 3/15 synteny is not involved in the sexual systems of *A. pigra* and *A. palliata* (see Fig. 4.6), and it is not conserved in the autosomes of these species (Steinberg et al. 2014). Considering that in other species of atelids, such as *Lagothrix* and *Brachyteles*, the 3/15 synteny has also not been found (Stanyon et al. 2001; de Oliveira et al. 2005), the hypothesis of the ancestrality of this association is not supported. The 3/15 association may had arisen independently in both *Ateles* and the South American howler monkeys. More cytogenetic studies in both genera are needed in order to confirm this last hypothesis.

4.3 Concluding Remarks

Alouatta is a genus with high chromosomal variability, showing multiple interspecific chromosomal rearrangements. C+ heterochromatin is scarce, suggesting that it might not play a prominent role in *Alouatta*'s chromosomal speciation, which contrasts with observations in other platyrrhines (such as *Cebus* sp.). Instead, structural rearrangements might be the main factor promoting the karyological evolution of the genus.

The high inter- and intraspecific karyological variability in the genus needs to be considered when assessing the taxonomy of *Alouatta*. Several species still lack a cytogenetic characterization (either requiring mitotic or meiotic studies, or both). Considering that karyology contributed to the reassessment of several taxa in the past, it seems plausible that the number of species and subspecies could be underestimated (or overestimated) if genetic data are not considered. The characterization of meiotic behavior in *A. sara*, *A. guariba*, and *A. seniculus*, as well as studies in somatic and germ cells in *A. p. coibensis* and *A. nigerrima*, would contribute to *Alouatta* taxonomy and allow testing hypotheses on the chromosomal evolution in the genus.

As stated in Sect. 4.2.1.3, *Alouatta* is one of the NWP genera that present multiple sex chromosome systems, together with *Aotus*, *Callimico*, and *Cacajao*. In Old World Primates, multiple sexual systems have only been suggested by mitotic studies for one species, the silvered leaf monkey *Presbytis cristata* (Bigoni et al. 1997). The involvement of the NWP Y-chromosome in multiple sex chromosome systems, together with the absence of homeology with the human Y-chromosome observed by FISH analysis (Consigliere et al. 1996, 1998; Mudry et al. 2001; de Oliveira et al. 2002), highlights the highly different genomic composition and behavior of Y-chromosomes in platyrrhines compared to those of catarrhines.

Finally, the complex multiple sex chromosome systems observed in *Alouatta* constitute an interesting case study to understand the evolution of sex chromosomes, not only for the diversity of sexual systems but also because it is the only reported case of an independent origin of multiple sex chromosome systems in Primates.

References

Amaral P, Finotelo L, de Oliveira E, Pissinatti A, Nagamachi C, Pieczarka J (2008) Phylogenetic studies of the genus *Cebus* (Cebidae-Primates) using chromosome painting and G-banding. BMC Evol Biol 8:169

Andersson L, Archibald A, Ashburner M et al (1996) Comparative genome organization of vertebrates. Mamm Genome 7:717–734

Armada JLA, Barroso CML, Lima MMC, Muniz JAPC, Seuánez HN (1987) Chromosome studies in *Alouatta belzebul*. Am J Primatol 13:283–296

Arrighi FE, Hsu TC (1971) Localization of heterochromatin in human chromosomes. Cytogenet Genome Res 10:81–86

Avramova ZV (2002) Heterochromatin in animals and plants. Similarities and differences. Plant Physiol 129:40–49

Battaglia E (1964) Cytogenetics of B chromosomes. Caryologia 17:245–299

Bender MA, Chu EHY (1963) The chromosomes of primates. In: Buettner-Janusch J (ed) Evolutionary and genetic biology of primates. Academic Press Inc., New York

Bigoni F, Koehler U, Stanyon R, Ishida T, Wienberg J (1997) Fluorescence in situ hybridization establishes homology between human and silvered leaf monkey chromosomes, reveals reciprocal translocations between chromosomes homologous to human Y/5, 1/9, and 6/16, and delineates an $X_1X_2Y_1Y_2/X_1X_1X_2X_2$ sex-chromosome system. Am J Phys Anthropol 102:315–327

Cabrera A (1957) Catálogo de los mamíferos de América del Sur. Revista del Museo Argentino de Ciencias Naturales IV. MACN, Buenos Aires

Clemente IC, Ponsà M, García M, Egozcue J (1990) Evolution of the simiiformes and the phylogeny of human chromosomes. Hum Genet 84:493–506

Consigliere S, Stanyon R, Koehler U, Agoramoorthy G, Wienberg J (1996) Chromosome painting defines genomic rearrangements between red howler monkey subspecies. Chromosome Res 4:264–270

Consigliere S, Stanyon R, Koehler U, Arnold N, Wienberg J (1998) In situ hybridization (FISH) maps chromosomal homologies between *Alouatta belzebul* (Platyrrhini, Cebidae) and other primates and reveals extensive interchromosomal rearrangements between howler monkey genomes. Am J Primatol 46:119–133

Copenhaver GP, Nickel K, Kuromori T, Benito M-I, Kaul S, Lin X, Bevan M, Murphy G, Harris B, Parnell LD, McCombie WR, Martienssen RA, Marra M, Preuss D (1999) Genetic definition and sequence analysis of *Arabidopsis* centromeres. Science 286:2468–2474

Cortés-Ortiz L, Rylands AB, Mittermeier R (2014) The taxonomy of howler monkeys: integrating old and new knowledge from morphological and genetic studies. In: Kowalewski M, Garber P, Cortés-Ortiz L, Urbani B, Youlatos D (eds) Howler monkeys: adaptive radiation, systematics, and morphology. Springer, New York

De Boer LEM (1974) Cytotaxonomy of the Platyrrhini. Genen Phaenen 17:1–115

De Grouchy J (1987) Chromosome phylogenies of man, great apes, and old world monkeys. Genetica 73:37–52

De Grouchy J, Turleau C, Roubin M, Klein M (1972) Karyotypic evolution in man and chimpanzees. A comparative study of band topographies after controlled denaturation. Ann Genet 15:79–84

de Oliveira EHC (1996) Estudos citogenéticos e evolutivos nas espécies brasileiras e argentinas do genero *Alouatta* (Atelidae, Platyrrhini). Masters Thesis, Universidade Federal do Paraná, Brazil

de Oliveira EHC, Lima MMC, Sbalqueiro IJ (1995) Chromosomal variation in *Alouatta fusca*. Neotrop Primat 3:181–183

de Oliveira EHC, de Lima MMC, Sbalqueiro S, Pissinatti A (1998) The karyotype of *Alouatta fusca clamitans* from Rio de Janeiro, Brazil: evidence for a y-autosome translocation. Genet Mol Biol. 21(3):361–364. doi: http://dx.doi.org/10.1590/S1415-47571998000300012

de Oliveira EHC, Suemitsu E, Da Silva FDC, Sbalqueiro IJ (2000) Geographical variation of chromosomal number in *Alouatta fusca clamitans* (Primates, Atelidae). Caryologia 53:163–168 de Oliveira EHC, Neusser M, Figueiredo WB, Nagamachi C, Pieczarka JC, Sbalqueiro IJ, Wienberg J, Müller S (2002) The phylogeny of howler monkeys (*Alouatta*, Platyrrhini): reconstruction by multicolor cross-species chromosome painting. Chromosome Res 10:669–683

de Oliveira EHC, Neusser M, Pieczarka JC, Nagamachi C, Sbalqueiro IJ, Müller S (2005) Phylogenetic inferences of Atelinae (Platyrrhini) based on multi-directional chromosome painting in *Brachyteles arachnoides*, *Ateles paniscus paniscus* and *Ateles b. marginatus*. Cytogenet Genome Res 108:183–190

de Oliveira EHC, Neusser M, Müller S (2012) Chromosome evolution in New World monkeys (Platyrrhini). Cytogenet Genome Res 137:259–272

Defler TR, Bueno ML (2007) *Aotus* diversity and the species problem. Primate Conserv 22:55–70

Dobigny G, Ducroz J-F, Robinson TJ, Volobouev V (2004) Cytogenetics and cladistics. Syst Biol 53:470–484

Dumas F, Stanyon R, Sineo L, Stone G, Bigoni F (2007) Phylogenomics of species from four genera of New World monkeys by flow sorting and reciprocal chromosome painting. BMC Evol Biol 7:S11

Dutrillaux B, Couturier J (1981) The ancestral karyotype of platyrrhine monkeys. Cytogenet Cell Genet 30:232–242

Dutrillaux B, Rethoré MO, Lejeune J (1975) Analyse du caryotype de *Pan paniscus*. Comparaison avec les autres Pongidae et l'Homme. Hum Genet 28:113–119

Dutrillaux B, Couturier J, Fosse M (1980) The use of high resolution banding in comparative cytogenetics: comparison between man and *Lagothrix lagotricha* (Cebidae). Cytogenet Cell Genet 27:45–51

Dutrillaux B, Descailleaux J, Viegas-Pequignot E, Couturier J (1981) Y-autosome translocation in *Cacajao c. rubicundus* (Platyrrhini). Ann Genet 24:197–201

Egozcue J (1969) Aneuploidía cromosómica: nuevas observaciones sobre la separación precoz de bivalentes en meiosis I. Sangre 14:442–452

Egozcue J, de Egozcue V (1965) The chromosomes of the howler monkey (*Alouatta caraya*, HUMBOLDT 1812). Mamm Chrom Newslett 17:84

Egozcue J, de Egozcue V (1966) The chromosome complement of the howler monkey (*Alouatta caraya*, HUMBOLDT 1812). Cytogenet Genome Res 5:20–27

Fransz PF, Armstrong S, De Jong JH, Parnell LD, van Drunen C, Dean C, Zabel P, Bisseling T, Jones GH (2000) Integrated cytogenetic map of chromosome arm 4S of *A. thaliana*: structural organization of heterochromatic knob and centromere region. Cell 100:367–376

García Haro F (2001) Evolución cromosómica en simiiformes. Homologías, reorganizaciones y heterocromatina. PhD dissertation, Universidad Autónoma de Barcelona, España

García M, Mirò R, Ponsà M, Egozcue J (1979) Chromosomal polymorphism and somatic segregation in *Saimiri sciureus*. Folia Primatol 31:312–323

García M, Miró R, Estop A, Ponsà M, Egozcue J (1983) Constitutive heterochromatin polymorphism in *Lagothrix lagothricha cana*, *Cebus apella*, and *Cebus capucinus*. Am J Primatol 4:117–126

García F, García M, Mora L, Alarcón L, Egozcue J, Ponsà M (2003) Qualitative analysis of constitutive heterochromatin and primate evolution. Biol J Linn Soc 80:107–124

García-Cruz R, Robles P, Steinberg ER, Camats N, Brieño MA, García-Caldés M, Mudry MD (2009) Pairing and recombination features during meiosis in *Cebus paraguayanus* (Primates: Platyrrhini). BMC Genet 10:25

García-Cruz R, Pacheco S, Brieño MA, Steinberg ER, Mudry MD, Ruiz-Herrera A, García-Caldés M (2011) A comparative study of the recombination pattern in three species of Platyrrhini monkeys (Primates). Chromosoma 120:521–530

Gregorin R (2006) Taxonomia e variação geográfica das espécies do gênero *Alouatta* Lacépède (Primates, Atelidae) no Brasil. Rev Bras Zool 23:64–144

Groves CP (2001) Primate taxonomy. Smithsonian Institution Press, Washington, DC

Groves CP (2005) Order Primates. In: Wilson DE, Reeder DM (eds) Mammal species of the world: a taxonomic and geographic reference. John Hopkins University Press, Maryland

Hassold T, Hansen T, Hunt P, VandeVoort C (2009) Cytological studies of recombination in rhesus males. Cytogenet Genome Res 124:132–138

Heitz E (1928) Das heterochromatin der moose. Jahrb Wiss Bot 69:762–818
Hershkovitz P (1984) Taxonomy of squirrel monkeys genus *Saimiri* (Cebidae, Platyrrhini): a preliminary report with description of a hitherto unnamed form. Am J Primatol 6:257–312
Hill W (1962) Primates: comparative anatomy and taxonomy. Edinburgh University Press, Edinburgh
Hsu TC, Benirschke K (1970) An atlas of mammalian chromosomes, vol 4. Springer, Berlin
Hsu TC, Hampton SH (1970) Chromosomes of Callithricidae with special reference to an XX/XO sex chromosome system in Goeldi marmoset (*Callimico goeldii* Thomas 1904). Folia Primatol 13:183–195
John HA, Birnstiel ML, Jones KW (1969) RNA-DNA hybrids at the cytological level. Nature 223:582–587
Kluge AG (1989) A concern for evidence and a phylogenetic hypothesis of relationships among *Epicrates* (Boidae, Serpentes). Syst Zool 38:7–25
Koiffmann CP, Saldanha PH (1974) Cytogenetics of Brazilian monkeys. J Hum Evol 3:275–282
Koiffmann CP, Saldanha PH (1981) Chromosome variability in the family Cebidae (Platyrrhini). Brazil J Genetics IV (4):667–677
Lima MMC, Seuánez H (1991) Chromosome studies in the red howler monkey, *Alouatta seniculus stramineus* (Platyrrhini, Primates): description of an $X_1X_2Y_1Y_2/X_1X_1X_2X_2$ sex chromosome system and karyological comparisons with other subspecies. Cytogenet Cell Genet 57:151–156
Lima MMC, Sampaio MIC, Schneider MPC, Scheffrahn W, Schneider H, Salzano FM (1990) Chromosome and protein variation in red howler monkeys. Rev Bras Genet 13:789–802
Ma NSF (1981) Chromosome evolution in the owl monkey, *Aotus*. Am J Phys Anthropol 54:293–303
Ma NSF, Jones TC, Thorignton RW, Miller A, Morgan L (1975) Y-autosome translocation in the howler monkey, *Alouatta palliata*. J Med Primatol 4:299–307
Ma NSF, Elliott MW, Morgan L, Miller A, Jones TC (1976) Translocation of Y chromosome to an autosome in the Bolivian owl monkey, *Aotus*. Am J Phys Anthropol 45:191–201
Matayoshi T, Seuánez HN, Nasazzi N (1987) Heterochromatic variation in *Cebus apella* (Cebidae, Platyrrhini) of different geographic regions. Cytogenet Cell Genet 44:158–162
Minezawa M, Harada M, Jordan OC, Borda CJV (1985) Cytogenetics of Bolivian endemic red howler monkeys (*Alouatta seniculus sara*): accessory chromosomes and Y-autosome translocation-related numerical variations. Res Rep New World Monkeys 5:7–16
Moore CM, Harris CP, Abee CR (1990) Distribution of chromosomal polymorphisms in three subspecies of squirrel monkeys (genus *Saimiri*). Cytogenet Cell Genet 53:118–122
Morescalchi M, Schempp W, Consigliere S, Bigoni F, Wienberg J, Stanyon R (1997) Mapping chromosomal homology between humans and the black-handed spider monkey by fluorescence in situ hybridization. Chromosome Res 5:527–536
Mudry MD (1990) Cytogenetic variability within and across populations of *Cebus apella* in Argentina. Folia Primatol 54:206–216
Mudry de Pargament MD, Labal de Vinuesa ML (1988) Variabilidad en bandas C de dos poblaciones de *Cebus apella*. Mendeliana 8:79–86
Mudry de Pargament MD, Slavutsky I (1987) Banding patterns of the chromosomes of *Cebus apella*: comparative studies between specimens from Paraguay and Argentina. Primates 28:111–117
Mudry de Pargament MD, Labal de Vinuesa ML, Colillas OJ, Brieux de Salum S (1984) Banding patterns of *Alouatta caraya*. Rev Bras Genet 7:373–379
Mudry MD, Brieux S, Colillas O (1981) Estudio citogenético del mono aullador *Alouatta caraya* de la República Argentina. Physis 40:63–70
Mudry MD, Brown AD, Zunino GE (1987) Algunas consideraciones citotaxonómicas sobre *Cebus apella* de Argentina. Bol Primatol Arg 5:1–2
Mudry MD, Slavutsky I, Labal de Vinuesa M (1990) Chromosome comparison among five species of platyrrhini (*Alouatta caraya*, *Aotus azarae*, *Callithrix jacchus*, *Cebus apella*, and *Saimiri sciureus*). Primates 31:415–420
Mudry MD, Slavutsky I, Zunino GE, Delprat A, Brown AD (1991) A new karyotype of *Cebus apella* from Argentina. Rev Bras Genet 14:729–738

Mudry MD, Ponsà M, Borrell A, Egozcue J, García M (1994) Prometaphase chromosomes of the howler monkey (*Alouatta caraya*): G, C, NOR, and restriction enzyme (RES) banding. Am J Primatol 33:121–132

Mudry MD, Rahn M, Gorostiaga M, Hick A, Merani MS, Solari AJ (1998) Revised Karyotype of *Alouatta caraya* based on synaptonemal complex and banding analyses. Hereditas 128:9–16

Mudry MD, Rahn MI, Solari AJ (2001) Meiosis and chromosome painting of sex chromosome systems in Ceboidea. Am J Primatol 54:65–78

Müller S (2006) Primate chromosome evolution. In: Lupski JR, Stankiewicz P (eds) Genomic disorders: the genomic basis of disease. Humana Press, Totowa

Nieves M (2007) Heterocromatina y evolución cromosómica en primates Neotropicales. PhD dissertation, Universidad de Buenos Aires, Argentina

Nieves M, Mühlmann M, Mudry MD (2005a) Heterochromatin and chromosome evolution: a FISH probe of *Cebus apella paraguayanus* (Primate: Platyrrhini) developed by chromosome microdissection. Genet Mol Res 4:675–683

Nieves M, Ascunce MS, Rahn MI, Mudry MD (2005b) Phylogenetic relationships among some *Ateles* species: the use of chromosomic and molecular characters. Primates 46:155–164

Pardue ML, Gall JG (1969) Molecular hybridization of radioactive DNA to the DNA of cytological preparations. Proc Natl Acad Sci U S A 64:600–604

Patton JL (1977) B-chromosome systems in the pocket mouse, *Perognathus baileyi*: meiosis and C-band studies. Chromosoma 60:1–14

Perelman P, Johnson WE, Roos C, Seuánez HN, Horvath JE, Moreira MAM, Kessing B, Pontius J, Roelke M, Rumpler Y, Schneider MPC, Silva A, O'Brien SJ, Pecon-Slattery J (2011) A molecular phylogeny of living primates. PLoS Genet 7:e1001342

Pieczarka JC, Nagamachi CY, Muniz JA, Barros RM, Mattevi MS (2001) Restriction enzyme and fluorochrome banding analysis of the constitutive heterochromatin of *Saguinus* species (Callitrichidae, Primates). Cytobios 105:13–26

Pinkel D (1986) Cytogenetic analysis using quantitative, high-sensitivity, fluorescent hybridization. Proc Natl Acad Sci U S A 83:2934–2938

Ponsà M, García M, Borell A, García F, Egozcue J, Gorostiaga MA, Delprat A, Mudry MD (1995) Heterochromatin and cytogenetic polymorphisms in *Cebus apella* (Cebidae, Platyrrhini). Am J Primatol 37:325–331

Rahn MI, Mudry MD, Merani MS, Solari AJ (1996) Meiotic behavior of the $X_1X_2Y_1Y_2$ quadrivalent of the primate *Alouatta caraya*. Chromosome Res 4:1–7

Ruiz Herrera A, García F, Aguilera M, García M, Ponsà Fontanals M (2005) comparative chromosome painting in *Aotus* reveals a highly derived evolution. Am J Primatol 65:73–85

Rylands AB (2000) An assessment of the diversity of New World primates. Neotrop Primates 8:61–93

Rylands AB, Mittermeier RA (2009) The diversity of the New World primates (Platyrrhini): an annotated taxonomy. In: Garber PA, Estrada A, Bicca-Marques JC, Heymann EW, Strier KB (eds) South American primates: comparatives perspectives in the study of behavior ecology, and conservation. Springer, New York

Rylands AB, Mittermeier RA, Silva JS Jr (2012) Neotropical primates: taxonomy and recently described species and subspecies. Int Zoo Yearb 46(1):11–24

Sankoff D (2003) Rearrangements and chromosomal evolution. Curr Opin Genet Dev 13:583–587

Schneider H, Canavez FC, Sampaio I, Moreira MAM, Tagliaro CH, Seuánez HN (2001) Can molecular data place each neotropical monkey in its own branch? Chromosoma 109:515–523

Seuánez HN (1979) The phylogeny of human chromosomes. Springer, Berlin

Seuánez HN, Armada JL, Barroso C, Rezende C, da Silva VF (1983) The meiotic chromosomes of *Cebus apella* (Cebidae, Platyrrhini). Cytogenet Genome Res 36:517–524

Solari AJ (1993) Sex chromosomes and sex determination in vertebrates. CRC, Boca Ratón

Solari AJ, Rahn MI (2005) Fine structure and meiotic behaviour of the male multiple sex chromosomes in the genus *Alouatta*. Cytogenet Genome Res 108:262–267

Stanyon R, Tofanelli S, Morescalchi MA, Agoramoorthy G, Ryder OA, Wienberg J (1995) Cytogenetic analysis shows extensive genomic rearrangements between red howler (*Alouatta seniculus*, Linnaeus) subspecies. Am J Primatol 35:171–183

Stanyon R, Consigliere S, Bigoni F, Ferguson-Smith M, O'Brien PCM, Wienberg J (2001) Reciprocal chromosome painting between a New World primate, the woolly monkey, and humans. Chromosome Res 9:97–106

Stanyon R, Bigoni F, Slaby T, Muller S, Stone G, Bonvicino CR, Neusser M, Seuánez HN (2004) Multi-directional chromosome painting maps homologies between species belonging to three genera of New World monkeys and humans. Chromosoma 113:305–315

Stanyon R, Rocchi M, Capozzi O, Roberto R, Misceo D, Ventura M, Cardone MF, Bigoni F, Archidiacono N (2008) Primate chromosome evolution: ancestral karyotypes, marker order and neocentromeres. Chromosome Res 16:17–39

Stanyon R, Garofalo F, Steinberg ER, Capozzi O, Di Marco S, Nieves M, Archidiacono N, Mudry MD (2011) chromosome painting in two genera of South American monkeys: species identification, conservation, and management. Cytogenet Genome Res 134:40–50

Steinberg ER (2011) Determinación sexual en primates Neotropicales: el caso de los monos aulladores. PhD dissertation, Universidad de Buenos Aires, Argentina

Steinberg ER, Nieves M, Mudry MD (2007) Meiotic characterization and sex determination system of Neotropical primates: Bolivian squirrel monkey *Saimiri boliviensis* (Primates: Cebidae). Am J Primatol 69:1236–1241

Steinberg ER, Cortés-Ortiz L, Nieves M, Bolzán AD, García-Orduña F, Hermida-Lagunes J, Canales-Espinosa D, Mudry MD (2008) The karyotype of *Alouatta pigra* (Primates: Platyrrhini): mitotic and meiotic analyses. Cytogenet Genome Res 122:103–109

Steinberg ER, Nieves M, Mudry MD (2014) Multiple sex chromosome systems in howler monkeys (*Alouatta*: Platyrrhini). Comp Cytogenet 8(1):43–69.

Sumner AT (1971) A simple technique for demonstrating centromeric heterochromatin. Exp Cell Res 75:304–306

Sun F, Oliver-Bonet M, Liehr T, Starke H, Ko E, Rademaker A, Navarro J, Benet J, Martin RH (2004) Human male recombination maps for individual chromosomes. Am J Hum Genet 74:521–531

Torres OM, Leibovici M (2001) Caracterización del mono aullador colorado *Alouatta seniculus* que habita en Colombia. Caldasia 23:537–548

Torres OM, Ramírez C (2003) Estudios Citogenético de *Alouatta palliata* (Cebidae). Caldasia 25:193–198

Torres OM, Enciso S, Ruiz F, Silva E, Yunis I (1998) Chromosome diversity of the genus *Aotus* from Colombia. Am J Primatol 44:255–275

Vassart M, Guédant A, Vlé JC, Kéravec J, Séguéla A, Volobouev VT (1996) Chromosomes of *Alouatta seniculus* (Platyrrhini, Primates) from French Guiana. J Hered 87:331–334

Vujošević M, Blagojević J (2004) B chromosomes in populations of mammals. Cytogenet Genome Res 106:247–256

Wienberg J (2005) Fluorescence in situ hybridization to chromosomes as a tool to understand human and primate genome evolution. Cytogenet Genome Res 108:139–160

Wienberg J, Stanyon R (1997) Comparative painting of mammalian chromosomes. Curr Opin Genet Dev 7:784–791

Wienberg J, Jauch A, Stanyon R, Cremer T (1990) Molecular cytotaxonomy of primates by chromosomal in situ suppression hybridization. Genomics 8:347–350

Yunis EJ, De Caballero OMT, Ramírez C, Efraín Ramírez E (1976) Chromosomal variations in the primate *Alouatta seniculus seniculus*. Folia Primatol 25:215–224

Chapter 5
Hybridization in Howler Monkeys: Current Understanding and Future Directions

Liliana Cortés-Ortiz, Ilaria Agostini, Lucas M. Aguiar, Mary Kelaita, Felipe Ennes Silva, and Júlio César Bicca-Marques

Abstract Hybridization, or the process by which individuals from genetically distinct populations (e.g., species, subspecies) mate and produce at least some offspring, is of great relevance to understanding the basis of reproductive isolation and, in some cases, the origins of biodiversity. Natural hybridization among primates has been well documented for a few taxa, but just recently the genetic confirmation of hybridization for a number of taxa has produced new awareness of the prevalence of this phenomenon within the order and its importance in primate evolution. The study of hybridization of *Alouatta pigra* and *A. palliata* in Mexico was among the first to

L. Cortés-Ortiz (✉)
Museum of Zoology, Department of Ecology and Evolutionary Biology,
University of Michigan, 1109 Geddes Avenue, Ann Arbor, MI 48109, USA
e-mail: lcortes@umich.edu

I. Agostini
Facultad de Ciencias Forestales, Consejo Nacional de Investigaciones Científicas y Técnicas de Argentina (CONICET)—Instituto de Biología Subtropical, Universidad Nacional de Misiones, Puerto Iguazú, Argentina

CeIBA (Centro de Investigaciones del Bosque Atlántico), Puerto Iguazú, Argentina
e-mail: agostini.ilaria@gmail.com

L.M. Aguiar
Universidade Federal da Integração Latino-Americana. (UNILA), Foz do Iguaçu, Brazil
e-mail: lucas.aguiar@unila.edu.br

M. Kelaita
Department of Anthropology, University of Texas at San Antonio, San Antonio, TX, USA
e-mail: mary.kelaita@utsa.edu

F.E. Silva
Instituto de Desenvolvimento Sustentável Mamirauá, Tefé, Brazil

J.C. Bicca-Marques
Faculdade de Biociências, Departamento de Biodiversidade e Ecologia, Pontifícia Universidade Católica do Rio Grande do Sul, Porto Alegre, Rio Grande do Sul, Brazil
e-mail: jcbicca@pucrs.br

genetically confirm the current occurrence of hybridization in primates. Following this study, other reports of hybridization have shown that this phenomenon is more widespread among primates than previously anticipated. Within the genus *Alouatta*, there have been reports on the presence of hybridization between *A. caraya* and *A. guariba* in a number of contact zones in Brazil and Argentina, and various studies are currently ongoing in some of these sites to understand the extent and patterns of hybridization between these species. In this chapter, we evaluate the extent of hybridization in the genus *Alouatta*, revise the current knowledge of the genetic and morphological aspects of these hybrid systems, and identify future directions in the study of hybridization within this genus, to understand the possible implications of the hybridization process in the evolutionary history of howler monkeys.

Resumen Hibridación, o el proceso mediante el cual individuos de poblaciones genéticamente distintas (especies o subespecies) se aparean y producen descendencia, tiene gran relevancia en la comprensión de las bases para el aislamiento reproductivo entre distintos taxa y, en algunos casos, para entender el origen de la biodiversidad. La hibridación natural entre primates ha sido bien conocida para unas cuantas especies, pero sólo recientemente la confirmación genética de hibridación entre numerosos taxa de primates ha sido posible y ha conducido a una nueva percepción de la prevalencia de este fenómeno entre los primates y su importancia en la evolución de este grupo. El estudio de la hibridación entre *Alouatta pigra* and *A. palliata* en México fue uno de los primeros que confirmó con evidencia genética la ocurrencia de hibridación en primates. Después de este estudio, otros reportes de hibridación en distintos taxa de primates han puesto de manifiesto que este fenómeno es más común en el orden Primates de lo que inicialmente se pensaba. Dentro del género *Alouatta*, también han habido reportes de hibridación entre *A. caraya* y *A. guariba* en distintas zonas de contacto en Brasil y Argentina, y varios estudios actualmente están en curso en algunas de estas áreas para entender la magnitud de este fenómeno y los patrones de hibridación entre estas especies. En este capítulo evaluamos la presencia de hibridación en el género *Alouatta*, revisamos lo que se conoce sobre los aspectos genéticos y morfológicos en estos sistemas híbridos y planteamos direcciones futuras en el estudio de la hibridación en este género, para entender las implicaciones del proceso de hibridación en la historia evolutiva de los monos aulladores.

Keywords Evolution • Morphology • Genetic admixture • Hybrid zone • Sympatry

5.1 Introduction

Hybridization is the crossing of genetically distinct taxa that produces some viable offspring (Arnold 1997; Mallet 2005). Crosses of pure individuals from different genetic lineages result in first-generation hybrids (F1s), but hybrid individuals can backcross with pure individuals of one of the parental species or crossbreed with other hybrid individuals, producing offspring with variable levels of genetic admixture. Although hybridization was initially considered a process mainly occurring among plants, and with limited representation in animals, a variety of genetic studies in the past few decades have shown that this phenomenon is rather common among sexually reproducing animals, especially between closely related species (Dowling and Secor 1997; Mallet 2005).

In primates, hybridization has been reported in captivity for a number of taxa (e.g., Chiarelli 1973; Tenaza 1984; Coimbra-Filho et al. 1984; Jolly et al. 1997); however, only few cases of natural hybridization in primates were known and studied before the twenty-first century, and most of these involved cercopithecine monkeys (Bernstein 1966; Struhsaker 1970; Nagel 1973; Dunbar and Dunbar 1974; Samuels and Altmann 1986). Identification of hybrids in these studies primarily relied on behavioral and morphological features of individuals that showed mixed characteristics typical of each parental taxon.

The widespread use of molecular techniques to address different aspects of primate systematics, behavior, and ecology during the last two decades has allowed the detection of an increased number of cases of hybridization in different primate taxa (e.g., Merker et al. (2009) in tarsiers; Cortés-Ortiz et al. (2007) in howler monkeys; Wyner et al. (2002) in lemurs; da Silva et al. (1992) in squirrel monkeys), including those in our own lineage (Green et al. 2010). However, there are still large gaps in our understanding of the genetic and morphological outcomes of hybridization at the individual and population levels, as well as their implications for the evolutionary trajectories of primate lineages.

In this chapter we review our current understanding of the prevalence of hybridization among howler monkeys. *Alouatta* is among of the first Neotropical primate genera for which genetic confirmation of hybridization is available (Cortés-Ortiz et al. 2007). We summarize demographic, morphological, behavioral, and genetic studies currently available, and make recommendations on future directions in the study of *Alouatta* hybrid zones and the implications of hybridization in primate evolution.

5.2 Distribution of Howler Monkey Contact Zones

As illustrated throughout this volume, howler monkeys are distributed across the Neotropics and have the broadest distribution of any Neotropical primate genus (Fig. 3.1). Phylogenetic studies have identified between 10 and 14 species and 22 taxa (species and subspecies), but there are a number of poorly known forms that still remain to be studied to allow an adequate evaluation of their taxonomic status (e.g., Peruvian species/subspecies) (see Cortés-Ortiz et al. 2014).

Although howler monkey species maintain allopatric/parapatric distributions in most of their range, small areas of overlap have been reported for some species (Fig. 5.1), including contact between *A. palliata* and *A. pigra* in Mexico (Smith 1970; Horwich and Johnson 1986; Cortés-Ortiz et al. 2007), *A. palliata* and

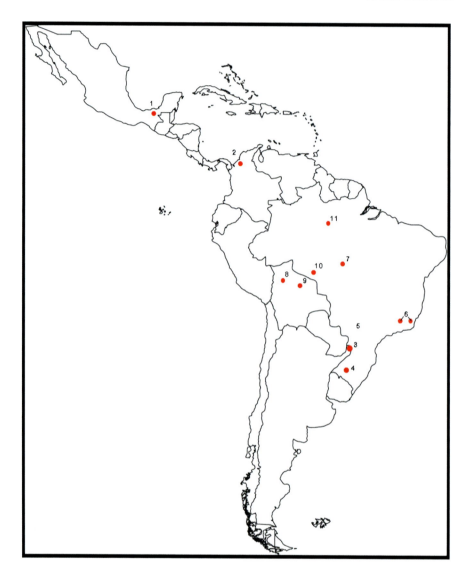

Fig. 5.1 Approximate location of the reported areas of contact between howler monkey species. (1) *A. palliata* and *A. pigra* in Mexico (Horwich and Johnson 1986; Smith 1970), (2) *A. palliata* and *A. seniculus* in northwestern Colombia (Defler 1994; Hernández-Camacho and Cooper 1976), (3) *A. caraya* and *A. guariba clamitans* in northern Argentina (Agostini et al. 2008; Di Bitetti 2005), (4) *A. caraya* and *A. g. clamitans* in southern Brazil (Bicca-Marques et al. 2008), (5) *A. caraya* and *A. g. clamitans* in southern Brazil (Aguiar et al. 2007; 2014; Gregorin 2006), (6) *A. g. guariba* and *A. g. clamitans* in Brazil (Kinzey 1982), (7) *A. discolor* and *A. s. puruensis* in Brazil (Pinto and Setz 2000), (8) *A. caraya* and *A. sara* in Bolivia (Büntge and Pyritz 2007), (9) *A. caraya* and *A. sara* in Bolivia (Wallace et al. 2000), (10) *A. caraya* and *A. sara* in Brazil (Iwanaga and Ferrari 2002), (11) *A. macconnelli* and *A. nigerrima* in Brazil (Napier 1976 and Cruz Lima 1945, cited in Gregorin 2006)

A. seniculus in northwestern Colombia (Hernández-Camacho and Cooper 1976; Defler 1994), *A. caraya* and *A. guariba* in northern Argentina (Di Bitetti 2005; Agostini et al. 2008), *A. caraya* and *A. guariba* in southern Brazil (Gregorin 2006; Aguiar et al. 2007, 2008, 2014; Bicca-Marques et al. 2008), *A. g. guariba* and *A. g. clamitans* in Brazil (Kinzey 1982), *A. discolor* and *A. s. puruensis* in Brazil (Pinto and Setz 2000), *A. caraya* and *A. sara* in Bolivia (Wallace et al. 2000; Büntge and Pyritz 2007) and Brazil (Iwanaga and Ferrari 2002), and *A. macconnelli* and *A. nigerrima* in Brazil (Napier 1976 and Cruz Lima 1945, cited in Gregorin 2006). It is likely that these areas of sympatry are due to secondary contact as a consequence of range expansions after periods of isolation (Cortés-Ortiz et al. 2003; Ford 2006; Gregorin 2006), and therefore, many other areas of contact among different *Alouatta* species may also exist. However, few surveys have been conducted in areas of potential contact within the limits of the distribution of parapatric howler monkey species, and those that exist show that sympatry is rare, but more common than previously anticipated. In some of the areas of sympatry among howler monkeys, individuals with intermediate or mosaic features have been observed (Cortés-Ortiz et al. 2003, 2007; Gregorin 2006; Aguiar et al. 2007; Agostini et al. 2008; Bicca-Marques et al. 2008; Silva 2010), suggesting at least some degree of crossbreeding between taxa and the formation of hybrid zones.

5.3 Studies of Hybridization in Howler Monkeys: Mixed Groups and Demographic Features of Syntopic Hybridizing Species

Evidence of hybridization has been reported for only two pairs of species of howler monkeys: *A. palliata* × *A. pigra* and *A. caraya* × *A. guariba*. These species are distinguishable on the basis of both morphological (Hill 1962; Groves 2001; Gregorin 2006) and genetic (de Oliveira et al. 2002; Cortés-Ortiz et al. 2003; Steinberg et al. 2008) features. The hybridizing species of each of these pairs diverged at approximately 3 and 5 MA, respectively (Cortés-Ortiz et al. 2003). Reports of possible hybridization were initially based on morphological and behavioral observations of individuals living in proximity or in mixed species groups. Later, demographic, behavioral, and genetic studies confirmed or strongly suggested the presence of hybrid offspring in the wild (Cortés-Ortiz et al. 2007; Agostini et al. 2008; Aguiar et al. 2008; Bicca-Marques et al. 2008) and in captivity (de Jesus et al. 2010).

5.3.1 A. palliata × A. pigra *Hybrid Zone in Tabasco, Mexico*

Smith (1970) first reported a possible area of sympatry of *A. palliata* and *A. pigra* in Tabasco, Mexico, based on museum specimens collected ~8 km SE of Macuspana (17°45′40″N, 92°35′35″W). More than a decade later, Horwich and Johnson (1986) surveyed the area where the specimens studied by Smith were collected as well as other nearby areas, but failed in finding direct evidence of the presence of howler monkeys. Nonetheless, through interviewing of local people, they identified a possible area of sympatry in the vicinity of Teapa (17°33′25″N, 92°56′50″W), about 40 km SE of Macuspana. In the early 1990s, Francisco García Orduña, Domingo Canales Espinosa, and Ernesto Rodríguez Luna from the Universidad Veracruzana (UV) in Mexico surveyed several areas across the state of Tabasco and found groups of *A. palliata* and *A. pigra* living in close proximity, as well as mixed groups composed of individuals of both species, and groups with individuals that emitted distinct vocalizations that sounded "intermediate" between the calls of either species (García-Orduña et al. unpubl. data; see also Kitchen et al. 2014). Later excursions to the area with the aim of collecting biological samples for genetic studies revealed that a number of individuals possessed mixed morphological features distinctive of each species (mainly subtle facial features, as well as pelage coloration) (Cortés-Ortiz unpubl. data; see Fig. 5.2 for an example of differences in facial features). Cortés-Ortiz and collaborators sampled 44 groups within this contact zone between 1998 and 2010 (Table 5.1). Most groups (*N*=28) were phenotypically monospecific

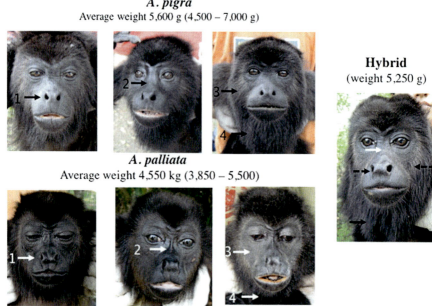

Fig. 5.2 Example of facial differences between *A. pigra* and *A. palliata* females and mixed features in a hybrid female. All pictures are from adult females: (1) nostrils more frontal in *A. pigra* and nasal alar walls more prominent in *A. palliata*, (2) prominent ridge of the nasal bone in *A. palliata* and not apparent in *A. pigra*, (3) hair covering a larger area of the cheeks in *A. pigra* than in *A. palliata*, and (4) longer beard in *A. pigra* than in *A. palliata*. Black arrows denote *A. pigra* features, *white arrows* denote *A. palliata* features, and *dashed arrows* denote intermediate features in the hybrid. Weight averages from Kelaita et al. (2011)

Table 5.1 Groups of howler monkeys surveyed in the areas of contact of known hybridizing species. Individuals are assigned to the different categories based on phenotype and genotype in the Mexican hybrid zone, and only on phenotype in the Brazilian and Argentinian hybrid zones

	Phenotype	Genotype
A. pigra × *A. palliata* (Cortés-Ortiz et al. unpubl.)		
Apa	17	5
Api	11	3
ApaH	0	7
ApiH	7	2
Mix	3	4
Apa-like Hyb	6	15
Api-like Hyb	0	1
Total groups	*44*	*37*
A. guariba × *A. caraya* (Aguiar et al. 2007, 2008)		
Aca	8	–
Agu	5	–
AcaH	0	–
AguH	2	–
Mix	5	–
Hyb	0	–
Total groups	*20*	
A. guariba × *A. caraya* (Bicca-Marques et al. 2008; Silva 2010)		
Aca	11	–
Agu	10	–
AcaH	5	–
AguH	8	–
Mix	5	–
Hyb	4	–
Total groups	*43*	
A. guariba × *A. caraya* (Agostini et al. 2008)		
Aca	3	–
Agu	5	–
AcaH	0	–
AguH	0	–
Mix	1	–
Total groups	*9*	

Apa = *A. palliata*, Api = *A. pigra*, ApaH = group of *A. palliata* with some hybrids, ApiH = group of *A. pigra* with some hybrids, Mix = mixed groups of *A. palliata* and *A. pigra* individuals or *A. caraya* and *A. guariba*, Api-like Hyb = all group members are hybrids resembling *A. pigra*, Apa-like Hyb = all group members are hybrids resembling *A. palliata*. Aca = *A. caraya*, Agu = *A. guariba*, AcaH = group of *A. caraya* with some hybrids, AguH = group of *A. guariba* with some hybrids, Hyb = group entirely composed of hybrid individuals

(17 *A. pigra* and 11 *A. palliata*), but three groups were mixed with individuals phenotypically resembling either species living together, and the remaining 13 groups included individuals with intermediate/mosaic features (detected via either morphology or vocalizations; see Figs. 14.1 and 14.2 in da Cunha et al. 2014 for differences in vocalizations) suggestive of a hybrid origin (but see Sect. 5.4 for a better understanding of the complex relationship between morphology and genetics in this system). Based on these surveys and data, we now know that the *A. palliata* × *A. pigra* hybrid zone in Tabasco is about 20 km wide and covers at least 67 km², with a patchwork of pure, mixed, and hybrid groups (Cortés-Ortiz et al. 2003, 2007) (see Table 5.1 for details on group composition).

5.3.2 A. caraya × A. guariba *Hybrid Zones in Brazil*

Records of mixed groups formed by *A. caraya* and *A. guariba* can be traced back to the beginning of the nineteenth century in the State of Rio Grande do Sul in Brazil (Isabelle 1983). However, Lorini and Persson (1990) were the first to report possible hybridization between these species in Brazil based on morphological studies of museum specimens collected in the 1940s by A. Meyer in the region of the Upper Parana River in the northwestern extreme of the State of Paraná. These specimens had a mosaic pelage coloration pattern representing a mixture of the typical patterns of the two parental species. In his comprehensive review of Brazilian howler monkeys, Gregorin (2006) analyzed the same specimens and also concluded that they represented hybrid individuals. Aguiar et al. (2007, 2008) surveyed a nearby area in the surroundings of the Ilha Grande National Park (23°24′S, 53°49′W) and found both monospecific groups of *A. caraya* and *A. guariba* living in sympatry and groups containing individuals with mosaic coloration patterns (see Fig. 2 in Aguiar et al. 2007 and Fig. 1 in Aguiar et al. 2008). They reported a total of 11 groups living within the boundaries of a 150 ha forest fragment (two monospecific groups of each species, two groups with *A. guariba* + putative hybrids, and five polyspecific groups of *A. caraya* + *A. guariba* + putative hybrids), as well as five *A. caraya* and two *A. guariba* groups living along a 17 km stretch of riverine forest and two monospecific groups (one of each species) living in sympatry in a near forest fragment ("Paredão das Araras," 23°21′10.1″S, 53°44′08.5″W). They found *A. guariba* as the most abundant species in the area, perhaps as a consequence of the prevalence of Atlantic Forest in the area, which is a type of habitat usually inhabited by this species rather than by *A. caraya*. The proportion of putative hybrids was similar to the proportion of *A. caraya* individuals in the area.

Another area of sympatry and hybridization between these taxa in Brazil occurs in the region of São Francisco de Assis, State of Rio Grande do Sul (Bicca-Marques et al. 2008; Silva 2010). Between 2006 and 2009 the team of primatologists and students headed by Bicca-Marques surveyed six localities within an area of approximately 600 km² in this region (29°33′50″–29°35′10″S, 54°58′40″–54°59′50″W), finding a total of 43 groups, 22 of which included at least one potential hybrid individual (i.e., with a mosaic phenotype) (Silva 2010). Interestingly, the distribution of phenotypically *A. guariba* groups decreased westwards and the opposite trend was observed for *A. caraya* groups. The westernmost locality surveyed contained only *A. caraya* groups, and a high percentage of hybrid individuals (42 %) was still present in the easternmost surveyed locality, suggesting that the area of contact and hybridization between these taxa may extend beyond the approximately 20 km wide strip surveyed.

5.3.3 A. caraya × A. guariba *Hybrid Zones in Argentina*

In Argentina, *A. guariba* and *A. caraya* have overlapping distributions in a small region in the province of Misiones, where syntopic populations have been detected in the strictly protected area of El Piñalito Provincial Park (Agostini et al. 2008). In a survey of approximately 800 ha, Agostini et al. (2008) detected three groups of *A. caraya*, five of *A. guariba*, and one mixed group composed of one adult *A. guariba* male, two *A. guariba* females, and one *A. caraya* female. The latter female was observed copulating with *A. guariba* males and giving birth twice to individuals with mosaic phenotypes, similar to those reported in Brazil (see Sect. 5.3.2). The extent of hybridization in this area is still unknown, but the absence of adults with mosaic pelage coloration patterns suggests that hybridization may be less common in this site than in the Brazilian contact zones. More recent surveys in the State of Misiones (one by I. Holzmann during and immediately after a yellow fever outbreak in 2008 [Holzmann 2012] and one by Agostini in 2010 [Agostini unpubl. data]) found no morphological or demographic evidence of hybridization. However, without extensive surveys in other localities within this contact zone, any statement about the lack of hybridization in this region would be premature.

5.4 Morphological Signals of Hybridization

The finding of individuals with intermediate phenotypes (i.e., diagnostic traits of each parental species co-occurring in the same individual) is often seen as evidence of hybridization. However, our understanding of the effects of hybridization on the morphological development of an individual is rather poor. On the one hand, we lack a clear understanding of the extent of phenotypic variation in hybrid individuals (Ackermann 2010), and on the other, many studies have only been able to detect hybridization when genetic markers are used (i.e., when hybridization is cryptic; e.g., Jasinska et al. (2010) in plants; Neaves et al. (2010) in marsupials; Gaubert et al. (2005) in carnivores). The slowly increasing number of studies incorporating genetic and morphological data in the study of the hybridization process suggests that morphologically intermediate and cryptic hybrids are the extremes of a continuum in the morphological expression of hybridization (e.g., Ackermann et al. 2006; Ackermann and Bishop 2010; Kelaita and Cortés-Ortiz 2013).

Much of what it is known about primate hybrid morphology comes from studies of Old World monkeys such as baboons (e.g., Jolly 2001; Ackermann et al. 2006), macaques (e.g., Bynum 2002; Schillaci et al. 2005), and some cercopithecine species (Detwiler 2002). Only a handful of studies addressing the morphology of hybrid New World monkeys have been carried out (e.g., Cheverud et al. 1993 and Kohn et al. 2001 for captive tamarins; Peres et al. 1996 for wild saddled back tamarins). In this section we discuss patterns of morphological variation observed in both presumed (based on phenotype) and genetically confirmed howler monkey hybrids, and discuss the reliability of using morphological cues to identify hybrid individuals.

Howler monkey species differ in numerous phenotypic attributes. Among the most conspicuous are the pelage color patterns that distinguish parapatric species. This is particularly true for the four species that are known to hybridize: *A. caraya*, *A. guariba*, *A. palliata*, and *A. pigra*. In *A. caraya*, adult males are completely black and adult females are pale yellowish-brown, whereas males of *A. guariba* are red and females are dark brown (Gregorin 2006). Coat coloration of *A. palliata* adults is black with light golden hairs on the flanks, whereas the pelage coloration of *A. pigra* is completely black and hairs have a softer texture than in *A. palliata* (Smith 1970). Intermediate pelage coloration between *A. caraya* and *A. guariba* was the trait used by Lorini and Persson (1990) to recognize some of the museum specimens of their study as putative hybrids. This identification generated expectations of hybrid morphotypes represented by mosaic combinations of coat color polymorphisms (Gregorin 2006), which were later used to classify putative *A. caraya*×*A. guariba* hybrids in the wild (Aguiar et al. 2007, 2008; Agostini et al. 2008; Bicca-Marques et al. 2008; Silva 2010). The distinctive pelage coloration of adult males and females of the sexually dichromatic *A. guariba* and *A. caraya* presumably results in easily distinguishable mosaic and/or intermediate features in the hybrid individuals, with up to 20 morphotypes identified in the wild (Aguiar et al. 2008; Silva 2010).

While the detection of *A. caraya*×*A. guariba* hybrids may be possible based on pelage coloration (at least to a certain extent), the recognition of the more similarly colored *A. palliata*×*A. pigra* hybrids using the same methods is not always possible. *Alouatta palliata* and *A. pigra* display some cranial and facial shape differences that can be used to distinguish individuals of each species in the field (see the example in Fig. 5.2). However, these traits show considerable intraspecific variation, and the intermixing of these features produces a broad range of hybrid morphotypes that compromised attempts to generate a clear criterion to accurately distinguish genetically confirmed hybrid and non-hybrid individuals (Kelaita and Cortés-Ortiz 2013).

Morphometric data, in contrast, have shown several quantifiable size differences between *A. palliata* and *A. pigra* for several variables (Kelaita et al. 2011), but analyses of morphological variation based on quantitative (metric) measurements of body size also showed a great variation in the hybrid phenotypes in Mexico (Kelaita and Cortés-Ortiz 2013). Kelaita and Cortés-Ortiz (2013) confirmed the hybrid status of individuals using diagnostic genetic markers (see Sect. 5.5 for details). The genetic data revealed that only 12 % of 128 identified hybrids had similar portions of their genome coming from each parental species. Although none of these individuals were F1 individuals, they were classified as "intermediate" and likely represent early-generation hybrids. The majority of identified hybrids were multigenerational backcrosses, probably resulting from the crossing of first-generation hybrids and their descendants with purebred individuals of one or the other parental species, or from the continued mating among hybrids during multiple generations. Depending on the number of diagnostic alleles of each species present in hybrid individuals, they were classified as *A. palliata*-like or *A. pigra*-like multigenerational backcrossed hybrids. A comparison of 14 morphometric variables among purebred and hybrid adult individuals showed that genetically intermediate hybrids exhibited great variation in morphometric characters. Both male and female intermediates ran the gamut of potential states for each variable, in some cases resembling *A. palliata*, while in others resembling *A. pigra*, or exhibiting values intermediate between or overlapping with the two parental species. On the other hand, multigenerational backcrossed hybrids only resembled the parental species with which they shared most of their alleles (Kelaita and Cortés-Ortiz 2013), compromising their accurate identification as hybrids.

These results indicate that instances of hybridization between well-established taxonomic groups can be underestimated if only a morphological criterion is utilized to identify hybrids. In the case of *A. palliata* × *A. pigra* hybrids, the majority of hybrid individuals are morphologically indistinguishable from parental species. The *A. guariba* × *A. caraya* hybrid studies revealed that hybrid individuals, identified based on pelage coloration patterns, comprised between 14 % (Aguiar et al. 2008) and 25 % (Silva 2010) of all individuals sampled from the respective hybrid zones. Considering that in the howler monkey hybrid zone in Mexico genetically intermediate hybrids comprise 12 % of all sampled individuals, it is likely that the purported *A. guariba* × *A. caraya* hybrids may also represent genetically intermediate individuals. The incorporation of molecular methods will help to test this prediction in the *A. guariba* × *A. caraya* hybrid zones.

5.5 Genetic Studies in the Howler Monkey Hybrid Zones

Genetic confirmation of hybridization in howler monkeys only exists for the *A. palliata* × *A. pigra* hybrid system. An initial study by Cortés-Ortiz et al. (2007) in Tabasco, Mexico determined the hybrid status of 13 individuals based on mitochondrial (mtDNA) and Y-chromosome (SRY gene) sequence data that, respectively, track the maternal and paternal lineages of hybrids, as well as on eight bi-paternally inherited microsatellite loci (three of which had diagnostic alleles for the parental species). Individuals were considered "hybrids" whenever discordance between mtDNA, SRY, and/or microsatellites occurred or when microsatellite loci in the same individual contained a combination of alleles diagnostic of each species. This study suggested unidirectional hybridization in this population, in which the cross of *A. palliata* males and *A. pigra* females only produced F1 fertile females, but the cross of *A. pigra* males and *A. palliata* females appeared to fail in producing fertile offspring. This result is consistent with the prediction of Haldane's rule, which establishes that it is more likely for the heterogametic sex (i.e., males for mammals) to be inviable or sterile (Haldane 1922). Nonetheless, the genetic variability at the uni- and bi-parentally inherited loci found among hybrids showed that backcrossing was occurring and that the production of fertile multigenerational backcrossed males was possible (Cortés-Ortiz et al. 2007). Preliminary results of an ongoing study based on a larger sample size of individuals ($N=178$) from the same hybrid zone and using 15 diagnostic microsatellite loci (which have a higher power to detect mixed ancestry) give support to the directional bias in hybridization and subsequent backcrossing. These new results also show novel genetic combinations (see Fig. 5.3) and a much higher percentage of hybrid individuals in the area of contact than initially recognized (Cortés-Ortiz unpubl. data). Most hybrids in the area are multigenerational, and only a handful of individuals are likely the product of crosses between purebreds and recent generation hybrids. Figure 5.4 summarizes

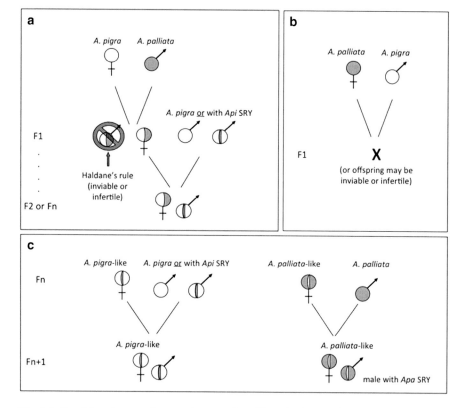

Fig. 5.3 Possible outcomes of crosses between *A. palliata*, *A. pigra* and hybrid individuals based on genotypic data of individuals from the Mexican hybrid zone. (**a**) Crosses between *A. pigra* females and *A. palliata* males only produce fertile females. These F1 females may mate with either *A. pigra* males or backcrossed males with *Api* SRY type and produce female offspring. It is unknown whether males with *Api* SRY type may be produced in this or only in later generations of backcrossing. (**b**) Crosses between *A. palliata* females and *A. pigra* males either do not occur, do not produce offspring, or rarely occur and produce unfertile offspring. (**c**) Further generation hybrids may continue to backcross with either purebred or backcrossed individuals and eventually produce males with *Apa* SRY type (Modified from Fig. 3 of Cortés-Ortiz et al. 2007)

5 Hybridization in Howler Monkeys

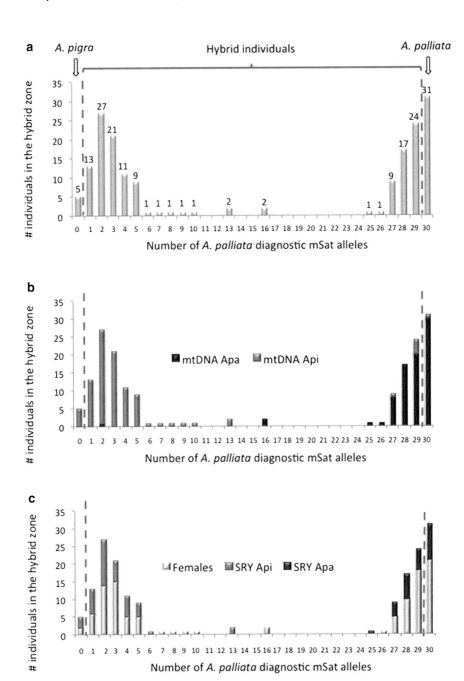

Fig. 5.4 Genetic composition of individuals from the Mexican hybrid zone. The X-axis represents the number of *A. palliata* diagnostic alleles. Individuals with 0 *A. palliata* diagnostic alleles represent pure *A. pigra* individuals whereas those with 30 *A. palliata* diagnostic alleles represent pure *A. palliata* individuals. (**a**) Variation based on 15 diagnostic microsatellite loci, (**b**) composition of mitochondrial DNA (mtDNA) haplotypes, and (**c**) composition of sex determination gene (SRY) haplotypes

the individual genetic variation found in this contact zone. Interestingly, when analyzing the genetic composition of hybrids, it is apparent that mtDNA haplotypes from *A. pigra* are more likely to be present in individuals with most of their nuclear genome (represented by the microsatellite alleles) of the *A. palliata* type, but only one hybrid with mostly *A. pigra* nuclear background has an *A. palliata* mtDNA haplotype. It is also remarkable that all male hybrids have the SRY gene type (reflecting paternal lineage) coincident with the majority of their nuclear background. These observations also support the predictions of Haldane's rule in the *A. palliata*×*A. pigra* hybrid system (Cortés-Ortiz et al. 2007), in which only females are produced in the first generation of crossing, and viable or fertile males appear in the population only after extensive backcrossing among multigenerational hybrids or between hybrids and purebred individuals (see Fig. 5.3C). The patterns of genetic variation observed among hybrid/backcrossed individuals suggest that the directionality in hybridization may be due to chromosomal, cytonuclear, or genomic incompatibilities. Steinberg et al. (2008) studied the chromosomal arrangements of Mesoamerican howler monkeys (see also Mudry et al. 2015) and found that *A. pigra* and *A. palliata* have different modal chromosome numbers ($2n = 58$ for *A. pigra* and $2n = 53$ and 54 for *A. palliata* males and females, respectively), and males have different sex determination systems ($X_1X_2Y_1Y_2$ quadrivalent in *A. pigra* and X_1X_2Y trivalent in *A. palliata*). Whether the apparent lack of early-generation male hybrids is a consequence of chromosomal incompatibilities due to these chromosomal differences is still an open question.

Although molecular data for the *A. caraya*×*A. guariba* hybrid zones are not yet available, the demographic and morphological patterns observed in their contact zones allow some inferences based on the knowledge generated from the *A. palliata*×*A. pigra* genetic studies. First, the presence of mosaic coat color features in putative hybrid males (one subadult male in Aguiar et al. (2007), one infant male in Agostini et al. (2008), four in Bicca-Marques et al. (2008), one in Jesus et al. (2010), and eight adult males in Silva (2010)) suggests that at least some male hybrids are viable. This inference is supported by a case of hybridization in captivity between putatively purebred individuals (Jesus et al. 2010). Second, if the mosaic individuals represent early-generation hybrids, as found in the *A. palliata*×*A. pigra* hybrid zone (Kelaita and Cortés-Ortiz 2013), it is possible that Haldane's rule is not operating in the *A. caraya*×*A. guariba* system. However, the absence of information on the longevity of the morphological signal of hybridization and the lack of molecular data makes it impossible to come to strong conclusions on this respect. Third, *A. caraya* and *A. guariba* also have different modal chromosome numbers ($2n = 52$ for *A. caraya* and $2n = 45–52$ for *A. guariba*; de Oliveira et al. 2002) and males have different sex determination systems ($X_1X_2Y_1Y_2$ quadrivalent in *A. caraya* and $X_1X_2X_3Y_1Y_2$ pentavalent in *A. guariba clamitans*; de Oliveira et al. 2002); therefore, the production of one viable F1 male hybrid in captivity (Jesus et al. 2010) is at least unexpected. Comparative genetic studies in the hybrid zones will provide an outstanding opportunity to explore whether molecular and/or cytogenetic mechanisms (or both) are responsible for the observed levels of reproductive isolation and the maintenance of species integrity despite hybridization.

5.6 Future Directions in the Study of Hybridization of Howler Monkeys

It has been recently recognized that hybridization is a powerful force that has shaped the evolutionary trajectory of a wide range of animal taxa (Dowling and Secor 1997; Arnold 1997; Grant et al. 2004; Mallet 2005). When hybridization occurs, genetic material of one lineage may enter the genetic pool of another, introducing genetic novelty to the latter (a process known as genetic introgression) (Rheindt and Edwards 2011). If this introduction of genetic novelty is advantageous to the recipient individuals, it may influence the evolutionary trajectory of the hybrid population or of one or both of the parental lineages (e.g., Grant and Grant 2010). Therefore, instances of hybridization may contribute to the adaptive radiation and diversification of species.

Several historical, demographic, behavioral, and ecological processes are involved in the origin and maintenance of hybrid zones, and a number of different mechanisms may operate together to maintain the hybridization process. Most of our future research is directed towards understanding the mechanisms that influence the hybridization process in howler monkeys, as well as the effect of hybridization in the ecology and behavior of the interacting taxa.

5.6.1 Endogenous and Exogenous Selective Forces in Hybridization

In general, hybridization may influence evolution in a variety of ways, and it mostly depends on endogenous and exogenous selective forces operating on each hybrid system (Barton 2001). When there is an intrinsic loss of fitness in hybrids, due, for example, to genetic incompatibilities between the two parental genomes (endogenous selection), it is likely that the hybrid zone will constitute a barrier preventing gene flow between the parental taxa. On the other hand, it has been argued that hybrid zones may be maintained by adaptation to different environments (exogenous selection), in which hybrid individuals may be more adapted to fluctuating or intermediate environments (e.g., Cruzan and Arnold 1993). In this case, individuals within the hybrid zone may exhibit a greater variance in fitness. Hybrids with higher fitness will contribute to adaptation either by introgression of alleles to parental taxa or by the establishment of recombinant genotypes (Barton 2001). However, these two selective forces (endogenous and exogenous) are not mutually exclusive and can operate together in the same system: whereas hybrid zones can be maintained by the selection against hybrids and represent barriers to gene flow, the divergence between interacting populations may be generated by adaptation to fluctuating environments (Barton 2001).

Studies have only recently been directed to understanding the effects of hybridization and gene introgression in the evolutionary history of primates (e. g., Arnold and Meyer 2006; Arnold 2009; Ackermann 2010; Green et al. 2010; Zinner et al. 2011).

In addition, only a few examples have actually provided some insight into the patterns of hybridization among primate taxa using genetic data (Cortés-Ortiz et al. 2007; Tung et al. 2008; Zinner et al. 2009; Merker et al. 2009; Ackermann and Bishop 2010).

In the case of the hybridization of howler monkeys in Mexico, there is some support for the operation of endogenous selection (e.g., Haldane's Rule effect), and there are no current environmental differences between the habitats of *A. palliata* and *A. pigra* throughout their distribution range that suggest strong influence of exogenous selection in this hybrid system. The responsible mechanisms for the partial reproductive isolation between the two species remain unknown, but genetic analyses suggest that some of these mechanisms could be attributed to chromosomal differences or to incompatibilities between nuclear and mitochondrial genomes (see Sect. 5.5). Cytogenetic and molecular studies comparing chromosomal and genomic regions associated with hybrid incompatibility should be a next step in our attempts to understand the endogenous mechanisms driving distinct levels of reproductive isolation in howler monkey hybrid zones.

On the other hand, exogenous selection may be strongly influencing the *A. caraya*×*A. guariba* hybrid zones. The currently known hybrid zones between these species in Brazil are located within regions of contact between two biomes (the Atlantic Forest and the Pampas, Bicca-Marques et al. 2008; and the Atlantic Forest, the Pantanal, and the Cerrado, Aguiar et al. 2007, 2008), with forests that are typically inhabited by each species (the Atlantic Forest by *A. guariba* and the Pantanal, the Pampas and the Cerrado by *A. caraya*). In Argentina, the hybrid zone lies within the Atlantic Forest ecoregion, for which *A. guariba* is endemic, but it is not a typical habitat for *A. caraya*. However, both species have very similar trophic niches (Bicca-Marques et al. 2008; Agostini et al. 2010) and are quite tolerant to habitat disturbance (Zunino et al. 2007; Bicca-Marques et al. 2008), which has been recently occurring in this area (Agostini et al. 2008). Therefore, the presence of both species in the area is likely the result of relatively recent secondary contact, with *A. caraya* individuals spreading into areas typically inhabited by *A. guariba*, as a consequence of forest disturbance. These incursions may occur infrequently generating an asymmetrical proportion of individuals of both species. The demography (i.e., abundance, sex ratios, rates of dispersal, etc.) and behavior of hybridizing taxa can affect levels and patterns of gene introgression in hybrid zones (Barton and Hewitt 1989; Wirtz 1999; Rohwer et al. 2001; Field et al. 2011; Gompert et al. 2012), generating different outcomes in the distribution of genetic backgrounds among hybrid zones with different ecological conditions. The availability of multiple contact zones between *A. caraya* and *A. guariba* with important ecological differences among them, as well as differences in the demographic composition of the two hybridizing species, offers a rare opportunity for testing the role that these factors may play on the occurrence and maintenance of the hybrid zones and the patterns of gene introgression. Comparative ecological studies within and outside these three hybrid zones between *A. caraya* and *A. guariba* would provide the grounds to understand the effect of exogenous selection in the fitness of hybrid individuals with distinct genetic architectures, and the differential effects of exogenous versus endogenous selection in the hybridization of howler monkeys.

5.6.2 Habitat Fragmentation and Its Effect on the Hybridization of Howler Monkeys

All howler monkey hybrid zones currently known are located in or surrounded by highly fragmented environments. It has been suggested that human-induced activities may play an important role in promoting hybridization in primates (Detwiler et al. 2005). Based on paleoecological data from the São Francisco de Assis region, Bicca-Marques et al. (2008) suggested that the contact between *A. caraya* and *A. guariba* is a recent consequence of the expansion of the two forests biomes in the past 2,000 years. Similarly, the current contact zone of *A. palliata* and *A. pigra* in Mexico seems to be the result of a secondary contact due to a two-wave colonization process (Cortés-Ortiz et al. 2003; Ford 2006) with a recent northward expansion of *A. palliata* (Cortés-Ortiz 2003). Therefore, it is likely that the origins of these howler monkey hybrid zones are due to paleoecological processes and not to habitat fragmentation. However, howler monkeys are strictly arboreal primates that only descend to the ground to cross canopy gaps or to disperse between fragments (Bicca-Marques and Calegaro-Marques 1995; Pozo-Montuy and Serio-Silva 2007), a task strongly compromised when inter-patch distances are longer than 200 m (Mandujano and Estrada 2005). Therefore, it is possible that habitat disturbance and fragmentation may influence the hybridization process in howler monkeys either by isolating their populations and reducing contact between hybridizing species, or by confining individuals of different species within particular fragments and promoting interbreeding. Dias et al. (2013) analyzed habitat configuration in fragmented landscapes both within the hybrid zone in Tabasco Mexico and in nearby areas where only purebred individuals occur. They concluded that hybridization between Mexican howler monkeys is facilitated in fragmented landscapes where there is a larger number of small, though less isolated, fragments. Testing hypothesis regarding the actual role of fragmentation in promoting or preventing hybridization requires the study of syntopic populations in both fragmented and extensive forest. The *A. caraya* × *A. guariba* hybrid zones portrayed here may provide a unique opportunity within primates, with cases of natural hybridization occurring in both highly fragmented areas of Brazil and the mostly pristine Atlantic Forest of Argentina.

5.6.3 Effect of Hybridization in the Vocal Communication of Hybridizing Species

One characteristic feature of howler monkeys is their conspicuous, loud vocalizations (Whitehead 1995). Although nonhuman primate vocalizations have long been considered genetically determined, some studies have questioned this assumption based on the existent variation among individuals and populations within taxa (Sun et al. 2011). This question can be addressed by analyzing vocalizations from purebred and hybrid individuals with different levels of admixture in the hybrid zone.

During a study on social behavior in one of the Brazilian hybrid zones, Aguiar (2010) detected that loud vocalizations tended to occur more frequently between conspecific males than during heterospecific interactions (including interactions with hybrids). A similar observation has been reported in Argentina (Holzmann et al. 2012) for syntopic *A. caraya* and *A. guariba*. These observations may support the argument of a genetic basis of vocalizations. However, Aguiar (2010) also observed that one hybrid female modified her vocalizations according to the species that she was interacting with. This plasticity could be either ecologically or genetically determined. An ongoing study of vocalizations integrating genetic, behavioral, and morphological data (Kitchen et al. unpubl. data; see also Kitchen et al. 2015) in the Mexican hybrid zone is starting to provide insights into the influence of genetics on the vocalizations of howler monkeys.

5.6.4 Interaction Between Social Dynamics and Hybridization

Hybrid zones have been considered natural laboratories for the study of the characters and processes leading to divergence and speciation (Hewitt 1988), which include behavioral strategies to acquire mates by the two parental populations and their hybrid offspring. However, despite a continuously growing number of studies dedicated to understanding the social and reproductive dynamics in primates, very little work has been focused on reproductive strategies of individuals within primate hybrid zones (e.g., Bergman and Beehner 2004; Bergman et al. 2008). Hybrid zones provide the opportunity to explore reproductive strategies of individuals with very different genetic backgrounds (both pure and admixed) in the same ecological and social context (Bergman et al. 2008). Ongoing studies on the social dynamics in the Mexican hybrid zone (e.g., Ho et al. 2014) will allow us to evaluate the competitive abilities of hybrid versus purebred individuals. In Brazil, Aguiar (2010) conducted a study on social interactions in two mixed groups composed of pure *A. caraya*, *A. guariba* and putative hybrids. Although the two groups were very different in composition, his analyses suggested that heterospecific associations confer some competitive advantages when facing other groups. He also found that affiliative and sexual interactions mostly included putative hybrids and were less frequent between apparently pure heterospecific individuals. Furthermore, he found that one hybrid female had a higher rank in the group than the putatively purebred *A. caraya* female. Although the sample size in his study is very small, these observations suggest the presence of assortative mating and a possible reproductive advantage in hybrids (Aguiar 2010). Behavioral studies comparing social interactions and dynamics of a larger number of groups with different compositions (*A. caraya* and *A. guariba*, as well as mixed and hybrid groups) within the area of contact between these species may allow the understanding of the interaction between hybridization and social dynamics.

Furthermore, the integrated genetic and behavioral study of primate populations in different hybrid zones can provide important information on the genetic composition of reproductively successful individuals and inform the relative effects of genetics and social dynamics on the overall fitness of hybrid versus purebred individuals. The study of social dynamics in the howler monkey hybrid zones would be especially insightful given the relatively good knowledge of different aspects of the social systems of the hybridizing species, due to a large and growing number of basic studies on social and sexual behavior of these taxa (see Van Belle and Bicca-Marques 2015). These studies can serve as a basis to conduct comparative observations between purebred and hybrid individuals in the same ecological and social context. There are important differences in social structure and mating systems between the hybridizing taxa. For example, while *A. pigra* has an average group size of ~6.3 individuals (range 2–16) with an adult sex ratio between 0.7 and 1.3 females per male, *A. palliata* has an average group size of ~15 individuals (range 2–45) with an adult sex ratio between 1.2 and 4.2 females per male (Di Fiore et al. 2010). In both species there is bisexual dispersal, but it is reported that *A. pigra* females commonly stay in natal groups (Van Belle et al. 2011) whereas most *A. palliata* females disperse (Glander 1992). Immigration of *A. pigra* females in well-established groups is rarely observed (Brockett et al. 2000), and females aggressively chase away extra-group females (Brockett et al. 2000; Van Belle et al. 2011). In contrast, *A. palliata* females regularly join established groups, first as low-ranking individuals, and gradually become dominant (Glander 1992). In *A. pigra* alpha or "central" males have almost exclusive access to fertile females, whereas "noncentral" males have few or no mating opportunities (Van Belle et al. 2008), but in *A. palliata* mating opportunities among group males are more evenly distributed (Jones and Cortés-Ortiz 1998; Ellsworth 2000; Milton et al. 2009).

These and other differences in social systems between the two parental species likely affect the genetic structure of individuals within the hybrid zone and will enable evaluations of the success of reproductive strategies of pure versus admixed individuals. These studies would require systematic long-term data collection on behavior, demography, and genetics for a large number of groups with distinct compositions within and outside the hybrid zone, using concordant methodologies. Despite the inherent difficulties of maintaining long-term studies given the costs and demands of field work (Strier 2010), the maintenance of long-term research in primate hybrid zones and the comparative studies across primate hybrid systems is critical to develop a holistic understanding of the evolutionary consequences of hybridization in primates.

5.6.5 *Studies of Hybridization in the Genomic Era*

The advent of the newer technologies to sequence entire genomes opens an exciting possibility in the genetic study of primate hybrid zones. Currently, there is a number of primate genome sequencing projects underway and within the next several years it is likely that genome sequence data will become available for most, if not all, primate genera (Bradley and Lawler 2011). The use of a larger number of genetic markers across the genome that characterize parental taxa will dramatically increase the power and accuracy of detecting admixed individuals. Also, polymorphism of these markers in conjunction with behavioral observations will allow us to establish kin relationships in hybrid populations to evaluate aspects such as individual reproductive success, and the possible effect of kinship in structuring social relationships within a hybrid zone. Furthermore, the understanding of patterns of introgression of different regions of the genome of each of the parental species will potentially enable the identification of genes that contribute to various levels of reproductive isolation, such as those observed in the *A. palliata* × *A. pigra* hybrid system, and the maintenance of species boundaries despite gene flow.

Acknowledgments This work has been supported by NSF grant # BCS-0962807 to LCO. Genetic, behavioral, and morphologic work presented in this study for the Mexican hybrid zone was supported by the University of Michigan (OVPR #U014374), Universidad Veracruzana, PROMEP UVER 98-11-019 and 103.5/03/1154EXB-9, and NSF grants DEB-0640519 and BCS-0962807 to LCO. Research in Argentina was supported by a Conservation Grant of the International Primatological Society to IA and from Idea Wild. Behavioral and morphological studies in Paraná Brazil by LMA were supported by CNPq and Universidade Federal do Paraná (UFPR). Research in the State of Rio Grande do Sul, Brazil, was supported by a CAPES fellowship to FES and by CNPq research grants (#306090/2006-6 and 303154/2009-8) to JCBM.

References

Ackermann RR (2010) Phenotypic traits of primate hybrids: recognizing admixture in the fossil record. Evol Anthropol 19:258–270
Ackermann RR, Bishop JM (2010) Morphological and molecular evidence reveals recent hybridization between gorilla taxa. Evolution 64:271–290
Ackermann RR, Rogers J, Cheverud JM (2006) Identifying the morphological signatures of hybridization in primate and human evolution. J Hum Evol 51:632–645
Agostini I, Holzmann I, Di Bitetti MS (2008) Infant hybrids in a newly formed mixed-species group of howler monkeys (*Alouatta guariba clamitans* and *Alouatta caraya*) in northeastern Argentina. Primates 49:304–307
Agostini I, Holzmann I, Di Bitetti MS (2010) Are howler monkeys ecologically equivalent? Trophic niche overlap in syntopic *Alouatta guariba clamitans* and *Alouatta caraya*. Am J Primatol 72:173–186
Aguiar LM (2010) Sistema social de grupos mistos de espécies de bugios (*Alouatta caraya* e *A. clamitans*) e potenciais híbridos no Alto Rio Paraná, sul do Brasil. Ph.D. thesis. Universidade Federal do Paraná, Curitiba
Aguiar LM, Mellek DM, Abreu KC, Boscarato TG, Bernardi IP, Miranda JMD, Passos FC (2007) Sympatry of *Alouatta caraya* and *Alouatta clamitans* and the rediscovery of free-ranging potential hybrids in Southern Brazil. Primates 48:245–248
Aguiar LM, Pie MR, Passos FC (2008) Wild mixed groups of howler species (*Alouatta caraya* and *Alouatta clamitans*) and new evidence for their hybridization. Primates 49:149–152
Aguiar LM, Tonetto J, Bicca-Marques JC (2014) Novas zonas de contato entre *Alouatta caraya* e *A. guariba clamitans* no sul do Brasil. In: Miranda JMD, Passos FC (eds) A primatologia no Brasil, vol. 13. Sociedade Brasileira de Primatologia, Curitiba
Arnold ML (1997) Natural hybridization and evolution. Oxford University Press, Oxford
Arnold ML (2009) Reticulate evolution and humans origins and ecology. Oxford University Press, Oxford

Arnold M, Meyer A (2006) Natural hybridization in primates: one evolutionary mechanism. Zoology 109:261–276

Barton NH (2001) The role of hybridisation in evolution. Mol Ecol 10:551–568

Barton NH, Hewitt GM (1989) Adaptation, speciation and hybrid zones. Nature 341:497–503

Bergman TJ, Beehner JC (2004) The social system of a hybrid baboon group (*Papio hamadryas anubis* × *P. h. hamadryas*). Int J Primatol 25:1313–1330

Bergman TJ, Phillips-Conroy JE, Jolly CJ (2008) Behavioral variation and reproductive success of male baboons (*Papio anubis* × *Papio hamadryas*) in a hybrid social group. Am J Primatol 70:136–147

Bernstein IS (1966) Naturally occurring primates hybrid. Science 154:1559–1560

Bicca-Marques JC, Calegaro-Marques C (1995) Locomotion of black howlers in a habitat with discontinuous canopy. Folia Primatol 64:55–61

Bicca-Marques JC, Prates HM, Aguiar FRC, Jones CB (2008) Survey of *Alouatta caraya*, the black-and-gold howler monkey, and *Alouatta guariba clamitans*, the brown howler monkey, in a contact zone, State of Rio Grande do Sul, Brazil: evidence for hybridization. Primates 49:246–252

Bradley BJ, Lawler RR (2011) Linking genotypes, phenotypes and fitness in wild primate populations. Evol Anthropol 20:104–119

Brockett RC, Horwich RH, Jones CB (2000) Female dispersal in the Belizean black howling monkey *Alouatta pigra*. Neotrop Primates 8:32–34

Büntge ABS, Pyritz LW (2007) Sympatric occurrence of *Alouatta caraya* and *Alouatta sara* at the Río Yacuma in the Beni Department, Northern Bolivia. Neotrop Primates 14:82–83

Bynum EL (2002) Morphological variation within a macaque hybrid zone. Am J Phys Anthropol 118:45–49

Cheverud J, Jacobs S, Moore A (1993) Genetic differences among subspecies of the saddle-back tamarin (*Saguinus fuscicollis*): evidence from hybrids. Am J Primatol 31:23–39

Chiarelli B (1973) Check-list of catarrhina primate hybrids. J Hum Evol 2:301–305

Coimbra-Filho AF, Silva RR, Pissinatti A (1984) Heterose em fêmea híbrida de *Callithrix* (Callitrichidae—Primates). In: Mello MT (ed) A primatologia no Brasil 1. Sociedade Brasileira de Primatologia, Belo Horizonte

Cortés-Ortiz L (2003) Evolution of howler monkeys, Genus *Alouatta*. Ph.D. dissertation. University of East Anglia, Norwich, Norfolk

Cortés-Ortiz L, Bermingham E, Rico C, Rodríguez-Luna E, Sampaio I, Ruiz-García M (2003) Molecular systematics and biogeography of the Neotropical monkey genus, *Alouatta*. Mol Phylogenet Evol 26:64–81

Cortés-Ortiz L, Duda TJ Jr, Canales-Espinosa D, García-Orduna F, Rodríguez-Luna E, Bermingham E (2007) Hybridization in large-bodied New World primates. Genetics 176:2421–2425

Cortés-Ortiz L, Rylands AB, Mittermeier R (2014) The taxonomy of howler monkeys: integrating old and new knowledge from morphological and genetic studies. In: Kowalewski M, Garber P, Cortés-Ortiz L, Urbani B, Youlatos D (eds) Howler monkeys: adaptive radiation, systematics, and morphology. Springer, New York

Cruz Lima E (1945) Mammals of Amazônia. I. General Introduction and Primates. Contrib. Mus. Paraense Emílio Goeldi Hist. Nat., Ethnogr., Belém do Pará, Rio de Janeiro

Cruzan MB, Arnold ML (1993) Ecological and genetic associations in an Iris hybrid zone. Evolution 47:1432–1445

da Cunha RGT, Oliveira DAG, Holzmann I, Kitchen DM (2015) Production of loud and quiet calls in howler monkeys. In: Kowalewski M, Garber P, Cortés-Ortiz L, Urbani B, Youlatos D (eds) Howler monkeys: adaptive radiation, systematics, and morphology. Springer, New York

da Silva BTF, Sampaio MIC, Schneider H, Schneider MPC, Montoya E, Encarnacion F, Salzano FM (1992) Natural hybridization between *Saimiri* taxa in the Peruvian Amazonia. Primates 33:107–113

Jesus AS, Schunemann HE, Müller J, da Silva MA, Bicca-Marques JC (2010) Hybridization between *Alouatta caraya* and *Alouatta guariba clamitans* in captivity. Primates 51:227–230

de Oliveira EHC, Neusser M, Figueiredo WB, Nagamachi CY, Pieczarka JC, Sbalqueiro IJ, Wienberg J, Müller S (2002) The phylogeny of the howler monkeys, (*Alouatta*, Platyrrhini): reconstruction by multicolor or cross species chromosome painting. Chromosome Res 10:669–683

Defler TR (1994) La conservación de primates en Colombia. Trianea (Act Cien INDERENA) 5:255–287

Detwiler KM (2002) Hybridization between red-tailed monkeys (*Cercopithecus ascanius*) and blue monkeys (*C. mitis*) in East African forests. In: Glenn ME, Cords M (eds) The guenons: diversity and adaptation in African monkeys. Plenum, New York

Detwiler KM, Burrell AS, Jolly CJ (2005) Conservation implications of hybridization in African cercopithecine monkeys. Int J Primatol 26:661–684

Di Bitetti MS (2005) Perspectivas para a conservacao de primatas em Misiones. In: Galindo-Leal C, Camara IG (eds) Mata Atlantica: biodiversidade, ameacas e perspectivas. Fundacao SOS Mata Atlantica, Sao Paulo, Conservacao Internacional, Belo Horizonte

Di Fiore A, Link A, Campbell CJ (2010) The Atelines: behavior and socioecological diversity in a New World monkey radiation. In: Campbell CJ, Fuentes AF, MacKinnon KC, Stumpf R, Bearder S (eds) Primates in perspective, 2nd edn. Oxford University Press, Oxford

Dias PAD, Alvarado-Serrano D, Rangel-Negrín A, Canales-Espinosa D, Cortés-Ortiz L (2013) Landscape attributes affecting the natural hybridization of Mexican howler monkeys. In: Marsh L, Chapman C (eds) Primates in Fragments II. Springer, New York.

Dowling TE, Secor CL (1997) The role of hybridization in the evolutionary diversification of animals. Annu Rev Ecol Syst 28:593–619

Dunbar RIM, Dunbar P (1974) On hybridization between *Theropithecus gelada* and *Papio anubis* in the wild. J Hum Evol 3:187–192

Ellsworth JA (2000) Molecular evolution, social structure and phylogeography of the mantled howler monkey (*Alouatta palliata*). Ph.D. dissertation, University of Nevada. Reno, Nevada

Field DL, Ayre DJ, Whelan RJ, Young AG (2011) The importance of pre-mating barriers and the local demographic conditions for contemporary mating patterns in hybrid zones of *Eucalyptus aggregata* and *Eucalyptus rubida*. Mol Ecol 20:2367–2379

Ford SM (2006) The biogeographic history of Mesoamerican primates. In: Estrada A, Garber PA, Pavelka MSM, Luecke L (eds) New perspectives in the study of Mesoamerican primates: distribution, ecology, behavior and conservation. Developments in primatology: progress and prospects. Tuttle RA (series ed). Kluwer/Springer, New York

Gaubert P, Taylor PJ, Fernandes CA, Bruford MW, Veron G (2005) Patterns of cryptic hybridization revealed using an integrative approach: a case study on genets (Carnivora, Viverridae, *Genetta* spp.) from the southern African subregion. Biol J Linn Soc 86:11–33

Glander KE (1992) Dispersal patterns in Costa Rican mantled howling monkeys. Int J Primatol 13:415–436

Gompert Z, Parchman TL, Buerkle CA (2012) Genomics of isolation in hybrids. Phil Trans R Soc B 367:439–450

Grant PR, Grant BR (2010) Conspecific versus heterospecific gene exchange between populations of Darwin's finches. Philos Trans R Soc B Biol Sci 365:1065–1076

Grant PR, Grant BR, Markert JA, Keller LF, Petren K (2004) Convergent evolution of Darwin's finches caused by introgressive hybridization and selection. Evolution 58:1588–1599

Green RE, Krause J, Briggs AW, Maricic T, Stenzel U, Kircher M, Patterson N, Li H, Zhai W, Fritz MHY, Hansen NF, Durand EY, Malaspinas AS, Jensen JD, Marques-Bonet T, Alkan C, Prufer K, Meyer M, Burbano HA, Good JM, Schultz R, Aximu-Petri A, Butthof A, Hober B, Hoffner B, Siegemund M, Weihmann A, Nusbaum C, Lander ES, Russ C, Novod N, Affourtit J, Egholm M, Verna C, Rudan P, Brajkovic D, Kucan Z, Gusic I, Doronichev VB, Golovanova LV, Lalueza-Fox C, De La Rasilla M, Fortea J, Rosas A, Schmitz RW, Johnson PLF, Eichler EE, Falush D, Birney E, Mullikin JC, Slatkin M, Nielsen R, Kelso J, Lachmann M, Reich D, Pääbo S (2010) A draft sequence of the Neandertal genome. Science 328:710–722

Gregorin R (2006) Taxonomia e variacao geografica das especies do genero *Alouatta* Lacepede (Primates, Atelidae) no Brasil. Rev Bras Zool 23:64–144

Groves CP (2001) The taxonomy of primates. Smithsonian Institution Press, Washington, DC
Haldane JBS (1922) Sex ratio and unisexual sterility in hybrid animals. J Genet 12:101–109
Hernández-Camacho J, Cooper R (1976) The nonhuman primates of Colombia. In: Thorington R Jr, Heltne P (eds) Neotropical primates. National Academy of Sciences, Washington, DC
Hewitt G (1988) Hybrid zones—natural laboratories for evolution studies. Trends Ecol Evol 3:158–166
Hill WCO (1962) Primates comparative anatomy and taxonomy IV—Cebidae, Part B. Edinburgh University Pubs Science & Maths, Edinburgh
Ho L, Cortés-Ortiz L, Dias PA, Canales-Espinosa D, Kitchen DM, Bergman TJ (2014) Effect of ancestry on behavioral variation in two species of howler monkeys (*Alouatta pigra* and *A. palliata*) and their hybrids. Am J Primatol 76:855–867
Holzmann I (2012) Distribución geográfica potencial y comportamiento vocal de dos especies de mono aullador (*Alouatta guariba clamitans* y *Alouatta caraya*). Ph.D. dissertation. Universidad Nacional de La Plata, Argentina
Holzmann I, Agostini I, Di Bitetti M (2012) Roaring behavior of two syntopic howler species (*Alouatta caraya* and *A. guariba clamitans*): evidence supports the mate defense hypothesis. Int J Primatol 33:338–355
Horwich RH, Johnson EW (1986) Geographic distribution of the black howler monkey (*Alouatta pigra*) in Central America. Primates 27:53–62
Isabelle A (1983) Viagem ao Rio Grande do Sul, 1833-1834. Tradução e notas de Dante de Laytano. Martins Livreiro, Porto Alegre
Iwanaga S, Ferrari SF (2002) Geographic distribution of red howlers (*Alouatta seniculus*) in Southwestern Brazilian Amazonia, with notes on *Alouatta caraya*. Int J Primatol 23:1245–1256
Jasinska AK, Wachowiak W, Muchewicz E, Boratyn´ Ska K, Montserrat JM, Boratyn´ Ski A (2010) Cryptic hybrids between *Pinus uncinata* and *P. sylvestris*. Bot J Linn Soc 163:473–485
Jolly CJ (2001) A proper study for mankind: analogies from the Papionin monkeys and their implications for human evolution. Yearb Phys Anthropol 44:177–204
Jolly CJ, Woolley-Barker T, Beyene S, Disotell TR, Phillips-Conroy JE (1997) Intergeneric hybrid baboons. Int J Primatol 18:597–627
Jones CB, Cortés-Ortiz L (1998) Facultative polyandry in the howling monkey (*Alouatta palliata*): Carpenter was correct. Bol Primatol Lat 7:1–7
Kelaita M, Cortés-Ortiz L (2013) Morphology of genetically-confirmed *Alouatta pigra* × *A. palliata* hybrids from a natural hybrid zone in Tabasco, Mexico. Am J Phys Anthropol 150:223–234
Kelaita MA, Dias PAD, Aguilar-Cucurachi MS, Canales-Espinosa D, Cortés-Ortiz L (2011) Impact of intrasexual selection on sexual dimorphism and testes size in the Mexican howler monkeys *Alouatta palliata* and *A. pigra*. Am J Phys Anthropol 146:179–187
Kinzey WG (1982) Distribution of primates and forest refuges. In: Prance GT (ed) Biological diversification in the tropics. Columbia University Press, New York
Kitchen DM, da Cunha RGT, Holzmann I, Oliveira DAG (2015) Function of loud calls in howler monkeys. In: Kowalewski M, Garber P, Cortés-Ortiz L, Urbani B, Youlatos D (eds) Howler monkeys: adaptive radiation, systematics, and morphology. Springer, New York
Kohn L, Langton L, Cheverud J (2001) Subspecific genetic differences in the saddle-back tamarin (*Saguinus fuscicollis*): postcranial skeleton. Am J Primatol 54:41–56
Lorini ML, Persson VG (1990) A contribuicao de Andre Mayer a historia natural no Parana (Brasil) II. Mamíferos do terceiro planalto paranaense. Arq Bras Biol Tecnol 33:117–132
Mallet J (2005) Hybridization as an invasion of the genome. Trends Ecol Evol 20:229–237
Mandujano S, Estrada A (2005) Detección de umbrales de área y distancia de aislamiento para la ocupación de fragmentos de selva por monos aulladores, *Alouatta palliata*, en Los Tuxtlas, Mexico. Universidad y Ciencia Número Especial II 11–21
Merker S, Driller C, Perwitasari-Farajallah D, Pamungkas J, Zischler H (2009) Elucidating geological and biological processes underlying the diversification of Sulawesi tarsiers. Proc Natl Acad Sci U S A 106:8459–8464
Milton K, Lozier JD, Lacey EA (2009) Genetic structure of an isolated population of mantled howler monkeys (*Alouatta palliata*) on Barro Colorado Island, Panama. Conserv Genet 10:347–358

Mudry MD, Nieves M, Steinberg ER (2015) Cytogenetics of howler monkeys. In: Kowalewski M, Garber PA, Cortés-Ortiz L, Urbani B, Youlatos D (eds) Howler monkeys: adaptive radiation, systematics, and morphology. Springer, New York

Nagel U (1973) A comparison of anubis baboons, hamadryas baboons and their hybrids at a species border in Ethiopia. Folia Primatol 19:104–165

Napier PH (1976) Catalogue of primates in the British Museum (Natural History), part 1: families Callitrichidae and Cebidae. British Museum (Natural History), London

Neaves LE, Zenger KR, Cooper DW, Eldridge MDB (2010) Molecular detection of hybridization between sympatric kangaroo species in south-eastern Australia. Heredity 104:502–512

Peres CA, Patton JL, Da Silva MNF (1996) Riverine barriers and gene flow in Amazonian saddle-back tamarin monkeys. Folia Primatol 67:113–124

Pinto LP, Setz EZF (2000) Sympatry and new locality for *Alouatta belzebul discolor* and *Alouatta seniculus* in the Southern Amazon. Neotrop Primates 8:150–151

Pozo-Montuy G, Serio-Silva JC (2007) Movement and resource use by a group of *Alouatta pigra* in a forest fragment in Balancán, México. Primates 48:102–107

Rheindt FE, Edwards SV (2011) Genetic introgression: an integral but neglected component of speciation in birds. Auk 128:620–632

Rohwer S, Bermingham E, Wood C (2001) Plumage and mitochondrial DNA haplotype variation across a moving hybrid zone. Evolution 55:405–422

Samuels A, Altmann J (1986) Immigration of a *Papio anubis* male into a group of cynocephalus baboons and evidence for anubis-cynocephalus hybrid zone in Amboseli, Kenya. Int J Primatol 7:131–138

Schillaci MA, Froehlich JW, Supriatna J, Jones-Engel L (2005) The effects of hybridization on growth allometry and craniofacial form in Sulawesi macaques. J Hum Evol 49:335–369

Silva FE (2010) Extensão da zona de contato e potencial hibridação entre *Alouatta caraya* e *Alouatta guariba clamitans* na região de São Francisco de Assis, RS. M.Sc. dissertation. Porto Alegre, Brazil

Smith JD (1970) The systematic status of the black howler monkey, *Alouatta pigra* Lawrence. J Mammal 51:358–369

Steinberg ER, Cortés-Ortiz L, Nieves M, Bolzán AD, García-Orduña F, Hermida-Lagunes J, Canales-Espinosa D, Mudry MD (2008) The karyotype of *Alouatta pigra* (primates: platyrrhini): mitotic and meiotic analyses. Cytogenet Genome Res 122:103–109

Strier KB (2010) Long-term field studies: positive impacts and unintended consequences. Am J Primatol 72:772–778

Struhsaker TT (1970) Phylogenetic implications of some vocalizations of Cercopithecus monkeys. In: Napier JR, Napier PH (eds) Old World monkeys: evolution, systematics, and behavior. Academic, London

Sun GZ, Huang B, Guan ZH, Geissmann T, Jiang XL (2011) Individuality in male songs of wild black crested gibbons (*Nomascus concolor*). Am J Primatol 73:431–438

Tenaza R (1984) Songs of hybrid gibbons (*Hylobates lar×H. muelleri*). Am J Primatol 8:249–253

Tung J, Charpentier M, Garfield D, Altmann J, Alberts S (2008) Genetic evidence reveals temporal change in hybridization patterns in a wild baboon population. Mol Ecol 17:1998–2011

Van Belle S, Bicca-Marques JC (2015) Insights into reproductive strategies and sexual selection in howler monkeys. In: Kowalewski M, Garber P, Cortés-Ortiz L, Urbani B, Youlatos D (eds) Howler monkeys: behavior, ecology and conservation. Springer, New York

Van Belle S, Estrada A, Strier KB (2008) Social relationships among male *Alouatta pigra*. Int J Primatol 29:1481–1498

Van Belle S, Estrada A, Strier KB (2011) Insights into social relationships among female black howler monkeys *Alouatta pigra* at Palenque National Park, Mexico. Current Zool 57:1–7

Wallace RB, Painter RLE, Rumiz DI, Taber AB (2000) Primate diversity, distribution and relative abundances in the Rios Blanco y Negro Wildlife Reserve, Santa Cruz Department, Bolivia. Neotrop Primates 8:24–28

Whitehead JM (1995) Vox Alouattinae—a preliminary survey of the acoustic characteristics of long distance calls of howling monkeys. Int J Primatol 16:121–144

Wirtz P (1999) Mother species—father species: unidirectional hybridization in animals with female choice. Anim Behav 58:1–12

Wyner YM, Johnson SE, Stumpf RM, DeSalle R (2002) Genetic assessment of a white-collared x red-fronted lemur hybrid zone at Andringitra. Madagascar. Am J Primatol 67:51–66

Zinner D, Groeneveld L, Keller C, Roos C (2009) Mitochondrial phylogeography of baboons (*Papio* spp.): indication for introgressive hybridization? BMC Evol Biol 9:83

Zinner D, Arnold ML, Roos C (2011) The strange blood: natural hybridization in primates. Evol Anthropol 20:96–103

Zunino G, Kowaleski M, Oklander L, Gonzalez V (2007) Habitat fragmentation and population trends of the black and gold howler monkey (*Alouatta caraya*) in a semideciduous forest in northern Argentina. Am J Primatol 69:966–975

Chapter 6
Morphology of Howler Monkeys: A Review and Quantitative Analyses

Dionisios Youlatos, Sébastien Couette, and Lauren B. Halenar

Abstract Recognition of a particularly derived eco-behavioral strategy for the genus *Alouatta* has been crucial for studying and understanding its equally derived cranial and postcranial morphology. The unique architecture of the skull and mandible has very likely evolved in relation to both masticatory correlates associated with an increasingly folivorous diet as well as the use of vocal communication associated with social behavior and an energy-minimizing strategy. Comparisons of cranial morphology using three-dimensional geometric morphometrics have highlighted significant interspecific shape differences. *Alouatta seniculus* is the most divergent in both cranial and hyoid morphology and exhibits the most pronounced levels of sexual dimorphism in those areas. Cranial variability is expressed in facial prognathism and airorhynchy, basicranial flexure, and zygomatic height. Inter- and intraspecific differences based on these axes of variation are very likely linked to interspecific variations in diet, behavior, and life history. This is further evident in the dental anatomy of the genus, indicating adaptations to a shift to a more folivorous diet. In addition, recent studies provide further evidence for significant inter- and intraspecific variations in hyoid size and shape. *Alouatta seniculus* possesses the largest and most inflated hyoid bulla, and the species that occupy distributional extremes (*A. palliata*, *A. caraya*) are differentiated by highly distinct hyoid shapes. These data indicate a complex relationship between morphology and behavior, with possible biogeographic implications. In terms of postcranial morphology,

D. Youlatos (✉)
Department of Zoology, School of Biology, Aristotle University of Thessaloniki, Thessaloniki, Greece
e-mail: dyoul@bio.auth.gr

S. Couette
Ecole Pratique des Hautes Etudes, Laboratoire Paléobiodiversité et Evolution & UMR uB/CNRS 6282 "Biogéosciences", Dijon, France
e-mail: sebastien.couette@u-bourgogne.fr

L.B. Halenar
Department of Biological Sciences, Bronx Community College, The City University of New York, 2155 University Avenue, Bronx, NY 10453, USA

New York Consortium in Evolutionary Primatology (NYCEP), The City University of New York, New York, NY, USA
e-mail: lauren.halenar@gmail.com

the forelimb bears a mixture of features that favor quadrupedalism with restricted abduction and overhead extension, providing stable contact and support along branches. In contrast, the hind limb appears to allow for a wider range of movement in all joints, with an emphasis on thigh extension and abduction, leg rotation, and powerful grasping with a habitually inverted foot. Interspecific variation reveals traits that can be ultimately related to subtle differences in the frequency of use of different positional modes, associated with variable eco-social factors. These results, deriving from different anatomical regions, provide evidence for understanding morphological variation across howler species in terms of morphofunctional adaptations, environmental pressure, and niche partitioning.

Resumen Reconocer la particular estrategia eco-comportamental del género *Alouatta* ha sido crucial para estudiar y entender su morfología craneal y post-craneal. La arquitectura única de su cráneo y mandíbula indican que muy probablemente evolucionaron en relación con su aparato masticatorio particular que se encuentra asociado a un incremento de una dieta folívora, así como con su comunicación vocal asociada a su comportamiento social y de estrategia de minimización energética. Comparaciones de la morfología craneal utilizando geometría morfométrica tridimensional resaltan diferencias significativas inter-específicas en la forma del cráneo. *Alouatta seniculus* es la especie más divergente en cuanto a la morfología del cráneo y del hueso hioides, exhibiendo los niveles más pronunciados de dimorfismo sexual en esas áreas. La variabilidad craneal es expresada en el prognatismo facial y airorrinchia, flexión basocraneal y altura zigomática. Diferencias inter- e intra-específicas basadas en estos ejes de variación están muy posiblemente vinculadas a variaciones interespecíficas en dieta, comportamiento e historias de vida. Por otra parte, evidencia adicional de la anatomía dental de este género indica adaptaciones de cambio a una dieta más folívora. Además, estudios recientes proporcionan evidencia adicional de variación inter- e intraespecífica significativas en la forma y el tamaño del hueso hioides. *Alouatta seniculus* posee la bula más larga e inflada del hueso hioides y las especies que ocupan los extremos de distribución del género (*A. palliata*, *A. caraya*) se distinguen por diferencias acentuadas en la forma del hueso hioides. Estos datos indican una compleja interrelación entre morfología y comportamiento, con posibles implicaciones biogeográficas. En términos de la morfología craneal, los miembros anteriores muestran una combinación de atributos que favorecen el cuadrupedalismo con abducción restringida y extensión amplia que proporciona contacto estable y soporte en las ramas. En contraste, los miembros posteriores parecieran permitir un mayor rango de movimiento de todas las articulaciones, con énfasis en la extensión y abducción del muslo, rotación de la pierna y una mayor capacidad de asir con una inversión común del pie. Variación interespecífica revela rasgos que ultimadamente podrían estar vinculadas con diferencias sutiles en la frecuencia de uso de diferentes posturas, asociadas con factores eco-sociales. Estos resultados, que derivan de diferentes regiones anatómicas, proporcionan evidencia para entender la variación morfológica a través de las especies de monos aulladores en términos de adaptaciones morfo-funcionales, presiones ambientales y partición de nichos.

Keywords *Alouatta* • Cranium • Hyoid • Morphometrics • Postcranium • Teeth

Palabras clave *Alouatta* • Cráneo • Hueso hioides • Morfometría • Postcráneo • Dientes

6.1 Introduction

The monophyly of atelines as a group is unquestioned, but the interrelationships of the genera within the group remain unclear. The phylogenetic signals from molecular and morphological data disagree on the relationships among *Lagothrix*, *Brachyteles*, and *Ateles*, but all phylogenetic hypotheses distinguish *Alouatta* from the rest: howlers apparently diverged first from the group around 15 MA and evolved unique morphobehavioral adaptations related to increased folivory, energy-minimizing ecological strategies, cautious above-branch quadrupedal positional behavior, and enhanced sound production (Rosenberger and Strier 1989; Hartwig 2005 for a review). The distinctiveness of *Alouatta* was pointed out very early by Lacépède (1799) and was later highlighted by several authors (e.g., Lönnberg 1941; Hershkovitz 1949; Hill 1962). In this chapter, we attempt to point out the unique morphology of the genus *Alouatta* focusing on cranial, dental, hyoid, and postcranial traits. For these purposes, two approaches were adopted: review of the literature on cranial, dental, hyoid, and postcranial characters as well as quantitative multivariate analyses of new data on cranial and hyoid morphology. Reviews will help the reader to appreciate the derived morphology of *Alouatta*, while the analyses of original data will further highlight their remarkably different anatomy from that of other atelines and other platyrrhines and will address issues of intra- and interspecific morphological variations within a functional, phylogenetic, and, where available, biogeographic framework.

6.2 Cranium

6.2.1 The Uniqueness of Alouatta Cranial Morphology

Among atelines and other platyrrhines, *Alouatta* can be differentiated by the peculiar form of its skull (Hill 1962): the pyramidal shape of the relatively small brain case, the posteriorly directed nuchal plane, the prognathic face anterior to the braincase, and the extended and unflexed basicranial axis. This morphological organization of the skull has been described as airorhynchy and corresponds to the upward rotation of the functional axis of the splanchnocranium on the neurocranium (see Tattersall 1972; Bruner et al. 2004). The braincase, in comparison to the face, is small (Schultz 1941), a relationship probably related to the early cessation of cranial growth and the obliteration of the cranial sutures (Höfer 1954, 1969). Additionally, the foramen magnum and the occipital condyles face posteriorly, an orientation displaying individual variation, but also undergoing changes during ontogeny (Bolk 1915; Senyurek 1938; Schultz 1955).

The development and form of the face, mandible, and braincase have been interpreted as primitive traits (Anthony et al. 1949). However, a long narrow cranial base with an airorynchous face, a flat, posteriorly facing nuchal plane, and expanded gonial angles of the mandible are most often argued to be associated with opening the subbasal space for an enlarged hyoid (Rusconi 1935; Biegert 1963; Miller and Begun 1998). This functional interpretation is partly supported by the shared airorynchous face and well-developed anatomical structures for sound production in *Alouatta* and *Pongo* (Biegert 1963; Shea 1985; Bruner et al. 2004). Alternatively, this reorganization has been also functionally linked to folivory. For example, in the mandible, the wide bigonial breadth of *Alouatta* appears to both favor folivory and to accommodate the enlarged hyoid bone. The development and posterolateral displacement of the gonial angles is intrinsically related to the enlarged hyoid in males (Watanabe 1982), but their flaring allows the mandible to pass lateral to the enlarged hyoid bone during maximum gape, and their expansion also provides a larger area of attachment for the masseter muscles used during chewing (Herring 1975; Bruner et al. 2004). An airorynchous face and a vertical nuchal plane change the orientation of the head on the neck, which has been suggested to make it easier to crop leaves (Tattersall 1972); this functional link has also been used in interpreting the cranial morphology of the large subfossil lemur *Megaladapis*, other folivorous subfossil lemurs (Godfrey and Jungers 2002), and the folivorous koala *Phascolarctos*.

This reorganization is accompanied by analogous modifications to the arrangement of various cranial muscles. Thus, *Alouatta* is characterized by the pronounced size of the masticatory muscles, which are linked both to the expansion of the hyolaryngeal organs and the mandible. Furthermore, howlers possess a modified insertion of *m. digastricus* on the mandible that does not extend posteriorly, as in other platyrrhines (Leche 1912). The arrangement of facial muscles is comparable to that of the other atelines, but *Alouatta* displays overdeveloped *mm. platysma colli* and *faciei*, which reach the orbitozygomatic plate and are probably linked to the growth of the mandible. The *mm. stylopharyngeus* and the muscles of the pharynx have undergone a large increase in size, especially the pharyngeal middle constrictor and the membranous parts of the pharynx which are implicated in the emptying of the air sacs, as well as the pharyngeal inferior constrictor that is involved in the control of the vocal cords (Schön 1968). *M. pterygoideus medialis* attaches to the enlarged gonial angle and helps elevate and protract the mandible during howling (Schön 1968). Finally, *mm. stylohyoideus, genioglossus, hyoglossus, styloglossus, sternohyoideus*, and *thyrohyoideus*, which operate on the hyoid, are all well developed.

These morphological modifications differentiate *Alouatta* from the other atelines. However, their functional link to folivory and sound production, which both vary across howler species and sexes (Di Fiore and Campbell 2007), would imply some degree of intra- and interspecific differences. With this in mind, quantitative analytical approaches were used in this study to examine potential morphological variations.

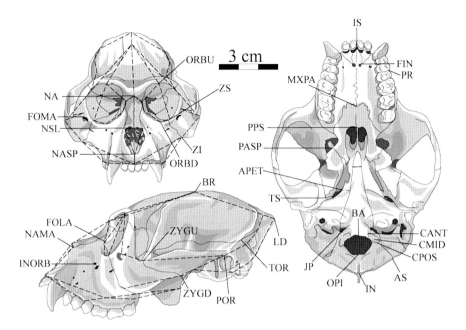

Fig. 6.1 Set of the 55 landmarks defined on the skull. Bilateral landmarks were digitized on the left side only and mirrored (see Table 6.1 for landmark definitions). *Dashed lines* illustrate wireframes used for shape variation visualizations

6.2.2 Cranial Material and Analyses

For the current study, we collected data from the skulls of 14 *Lagothrix* (three species), 48 *Ateles* (seven species), 78 *Alouatta* (seven species) adult males and females, and one male *Brachyteles*. For our analysis at the genus level, we did not consider sex or species. In contrast, we considered both variables in our analysis of *Alouatta* cranial diversity at the species level. The *Alouatta* sample was composed of 15 *A. paliatta* (10M, 5F), 4 *A. pigra* (3M, 1F), 9 *A. guariba* (5M, 4F), 14 *A. belzebul* (7M, 7F), 10 *A. caraya* (6M, 4F), 15 *A. seniculus* (8M, 7F), and 11 *A. nigerrima* (6M, 5F). All specimens were wild-shot individuals housed in the Laboratoire des Mammifères et Oiseaux, Muséum national d'Histoire naturelle (Paris, France), Museu Paraense Emílio Goeldi (Bélem, Pará, Brasil), and the American Museum of Natural History (New York, USA). A set of 55 landmarks were used (Fig. 6.1, Table 6.1), identical to those from previous studies and shown to successfully differentiate small shape variations (Couette 2002, 2007). The landmarks were digitized using a MicroScribe G2X (Immersion Corporation, San José, California). Bilateral landmarks have been digitized on the left side only, and mirroring was performed using the R software (2008). Each specimen was digitized twice in order to estimate measurement error, which varied between 1.9 and 2.3 %.

Table 6.1 List of the 55 landmarks used for the analyses of cranial variations (consult Fig. 6.1 for visualizations)

Landmark	Description	Position(s)
IS	Intradental superior	Midline
NASP	Base of the nasal aperture	Midline
NSL	Nasale	Midline
NA	Nasion	Midline
BR	Bregma	Midline
LD	Lambda	Midline
IN	Inion	Midline
OPI	Opisthion	Midline
BA	Basion	Midline
PPS	Posterior point of the palatine suture	Midline
MXPA	Maxillary/palatine point	Midline
FIN	Incisive foramen	Right, left
PASP	Palatine/ sphenoid suture	Right, left
PR	P1/P2 point	Right, left
ZS	Zygomaxillare superior	Right, left
ORBU	Upper point of the orbit	Right, left
ORBD	Lower point of the orbit	Right, left
INORB	Upper point of the infraorbital foramen	Right, left
NAMA	Nasal/maxillary suture on the nasal aperture	Right, left
FOMA	Upper point of the malar foramen	Right, left
FOLA	Lower point of the lachrymal foramen	Right, left
AS	Asterion	Right, left
TOR	Parietal/occipital suture on the occipital torus	Right, left
POR	Porion	Right, left
ZI	Zygomaxillare inferior	Right, left
ZYGU	Jugal/squamosal suture on the upper ridge of the zygomatic arch	Right, left
ZYGD	Jugal/squamosal suture on the lower ridge of the zygomatic arch	Right, left
TS	Temporosphenoidal junction at the petrous	Right, left
APET	Anterior petrous temporal	Right, left
JP	Jugal process	Right, left
CANT	Anterior point of the occipital condyle	Right, left
CMID	Medial point of the occipital condyle	Right, left
CPOS	Posterior point of the occipital condyle	Right, left

For our analysis, we used landmark-based 3DGM methods, such as generalized Procrustes analysis (GPA; Rohlf and Slice 1990), which can involve a great number of variables for the quantification of variation. A good solution to deal with such numerous variables is the use of multivariate statistics such as principal components analysis (PCA). Common parts of variation described by the variables are summarized and expressed by a new series of orthogonal (independent) axes that constitute new morphological variables. Each PC axis describes only a part of the total varia-

6 Morphology of Howler Monkeys: A Review and Quantitative Analyses

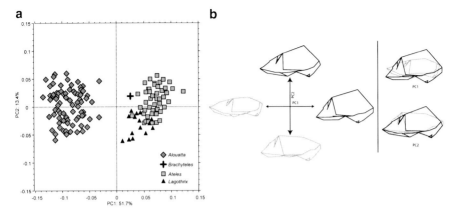

Fig. 6.2 (**a**) PCA results for the analysis of cranial variation at the genus level. *Alouatta* cranial morphology is clearly different from that of the other atelines; (**b**) polarization of the morphologies described by the negative (*gray*) and positive (*black*) values of PC1 and PC2 for the analysis of cranial variation at the genus level (on the *left*). Superimposition of the extreme morphologies of each axis (on the *right*)

tion within the sample, and the contribution of the initial variables on each axis constitutes a key for the understanding of morphological differences between specimens. However, with Procrustes techniques, a great number of variables and the nature of the variables themselves (*X*, *Y*, and *Z* coordinates) usually complicate the polarization of morphological variations described by the PC axes. Therefore, wireframe diagrams and geometric reconstruction of cranial shape are frequently used to visualize PCA results. The GPAs and PCAs were performed using *morphologika²* v. 2.5 (O'Higgins and Jones 2006), which also calculated centroid sizes, PC scores, wireframe diagrams, and the geometric reconstruction of cranial shape. The PAST 1.89 software package (Hammer et al. 2001) was used to perform and output graphical representation of the PCAs. Finally, we used Statistica 7 (Statsoft Inc., USA) for other statistical analyses such as linear regressions to check for allometric effects on the pattern of variation described on each PC axis.

6.2.3 Cranial Variation

In the PCA at the generic level (Fig. 6.2a), the first two PCs accounted for 65.1 % of total variation (51.7 % for PC1 and 13.4 % for PC2). Multivariate regressions (reduced major axis) of the PC scores onto the logarithm of centroid size indicate a significant linear correlation between PC1 scores and size ($a=-0.91$; $b=5.15$; $R^2=0.32$, $p<0.001$). Size has a significant influence on cranial morphology, and larger specimens lie on the negative side of PC1. Cranial distinction is mainly allometric, with *Alouatta* exhibiting the higher scores ($R^2=0.71$; $p<0.001$), compared to *Ateles* ($R^2 = 0.28$, $p = 0.02$). *Alouatta* are clearly separated from the other atelines and occupy the negative side of the morphospace on PC1 but spread along PC2 with a plot centered at the origin of this axis. *Ateles* and *Lagothrix* overlap on the positive side of PC1 but are distinguishable on PC2. The latter are located on the negative side of the axis, while *Ateles* are mainly on the negative side, with significant overlap in the center.

Regarding morphology, negative values on PC1 are associated with a relatively developed face and a high degree of prognathism, while positive values describe rounded skulls with a relatively small face compared to the neurocranium. On PC1, the main morphological variation is a downward and backward rotation and relative shortening of the face. This rotation implies verticalization of the snout and the orbital plane. The relative volume of the neurocranium increases, and the occipital undergoes a downward and forward rotation, a consequence of an increase in the flexion of the basicranium associated with the facial rotation (Fig. 6.2b). On PC2, the negative values describe skulls with a relatively flattened neurocranium, while the positive values are associated with a high braincase. The main morphological variations are linked to the position of the bregma and the relative height of the braincase and the frontal part of the face (Fig. 6.2b).

The main morphological differences between *Alouatta* and the other atelines are in the degree of prognathism, airorhynchy, the oblique position of the orbital plane, the relative reduction of the neurocranium (essentially in height), the posterior orientation of the occipital condyles and the foramen magnum, the unflexed basicranium, and the robustness of the face and the zygomatic arches of howlers. As described above, functionally, these derived features have been associated with both diet and vocalizations. Although *Alouatta* and *Brachyteles* share dental adaptations to their semi-folivorous diet, the latter have skulls that are most similar to those of their ateline relatives, without an airorynchous face, posteriorly directed occipital, or a relatively small brain case (Rosenberger et al. 2011). Lack of nutrients in the mainly folivorous diet in *Alouatta* may account for a small brain, which may in turn be directly linked to an unflexed cranial base, as the "spatial packing" hypothesis suggests (Biegert 1963; Gould 1977; Ross and Ravosa 1993; Lieberman et al. 2000). Additionally, this morphology may be further associated with the necessity for opening of the subbasal space in order to accommodate the morphological modifications related to sound production. Therefore, it seems that the craniomandibular apomorphies of *Alouatta* are the result of complex adaptations to increased and frequent vocalization and also partly to folivory. However, this is variably expressed within the genus, and size seems to play a significant role suggesting intra- (sexual) and interspecific (phylogeographic) variations.

In the analysis at the species level, the first two PCs accounted for 61.1 % of total variation (37.4 % for PC1 and 23.7 % for PC2; Fig. 6.3a). Morphological variations described by PC1 are very similar to the ones described by PC1 at genus level but with lower intensity. Negative values characterize skulls with a very prognathic face, while positive ones characterize skulls with a less well-developed face. This axis describes a reduction in the degree of prognathism accompanied by a verticalization of the nasal and frontal bones as well as the orbital plane. Variation along

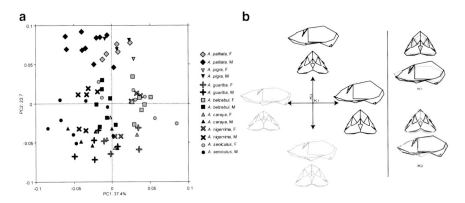

Fig. 6.3 (**a**) PCA results for the analysis of cranial variation at the species level. Morphological variations along PC1 describe differences explained by sexual dimorphism. Differences along PC2 describe geographical differences, distinguishing the Central American (*A. paliatta* and *A. pigra*) from the South American species. (**b**) Polarization of the morphologies described by the negative (*gray*) and positive (*black*) values of PC1 and PC2 for the analysis of cranial variation at the species level (on the *left*). Superimposition of the extreme morphologies of each axis (on the *right*)

PC1 also includes a shortening of the palate, a relative increase of the size of the neurocranium, and a downward rotation of the occipital which leads to increased basicranial flexion (Fig. 6.3b). PC1 also separates males from females. In the morphospace, males occupy the negative side of the axis, while females are clustered in the positive side, with few overlapping. Gender differences and overall variation seem to be more pronounced in *A. seniculus* than for any other species, and are not in accord with body size dimorphism (Ravosa and Ross 1994). These differences in cranial morphological traits do, however, seem to agree with those from previous descriptions of cranial dimorphism [e.g., *A. palliata* (Blanco and Godfrey 2006), *A. arctoidea* (Braza 1980)], although population level differences may be important (Jones et al. 2000). The differences in cranial morphology between the sexes may be related to proximally hypermorphosis during male growth (Ravosa and Ross 1994; Masterson and Hartwig 1998; Jones et al. 2000; Flores and Casinos 2011) but may also be linked evolutionarily to trophic and masticatory functions that are very important for male competition (Flores and Casinos 2011).

Along PC2, the morphological variation represents a difference in the height of the cranial vault associated with a difference in the rotation of the occipital. Compared to the negative side of the axis, positive values are associated with a taller braincase, higher occipital condyles, thicker zygomatic arches, a downward directed muzzle, and a more vertical orientation of the orbital plane. PC2 partly distinguishes between certain species, and the positive side is occupied by the divergent Central American clade (*A. palliata*, *A. pigra*). The rest are rather mixed in the center (*A. belzebul*, *A. seniculus*, *A. nigerrima*), with the southernmost species, *A. caraya* and *A. guariba*, differing slightly. *Alouatta belzebul*, *A. seniculus*, and *A. nigerrima* are distributed widely across the morphospace, displaying higher intraspecific variation (on both PCs). This is noticeable for *A. nigerrima*, with specimens of both sexes positioned on both sides, implying two probable morphological groups for this species. The cranial particularities of the Central American species have been previously reported by Watanabe (1982), and their differentiation within the genus is also supported by molecular data (e.g., Cortés-Ortiz et al. 2003). Watanabe (1982) detected differences between *A. palliata* and *A. seniculus* in basicranium length, height of the nasal bones, bigonial breadth, and the development of the pterygoid wing, which is associated with *m. pterygoideus medialis*, an elevator and a protractor of the mandible (Schön 1968).

6.3 Dental Morphology

In terms of overall proportions, the *Alouatta* dentition can be described as molar centric, where the incisors at the front of the jaw are much smaller than the molars and are often offset from the canines by a diastema (Fig. 6.4). This is the opposite condition seen in many frugivorous primates, including the atelines *Ateles* and *Lagothrix*, which tend to have larger, bladelike incisors for processing the large, potentially tough-skinned fruit into bite-sized pieces (Hylander 1975; Kay and Hylander 1978; Eaglen 1984; Anthony and Kay 1993). In contrast, the narrow width of the incisor row of *Alouatta*, as well as the mediolateral tapering of each incisor itself, reflects the more concentrated force per tooth used to bite and pull leaves (Ungar 1990; Anapol and Lee 1994). The upper incisors are also oriented at an angle due to the howler's airorynchous face and are therefore in an advantageous

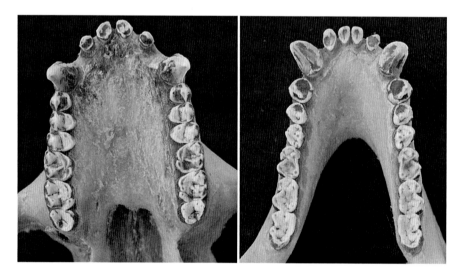

Fig. 6.4 Occlusal views of the upper dentition (*left*) and lower dentition (*right*) of male *Alouatta* (third upper molar length = 5.8 mm; third lower molar length = 7.5 mm)

position for cropping leaves off tree branches (Tattersall 1972). *Alouatta* is highly sexually dimorphic in body size, hyoid size, and canine size so male canines are considerably more prominent than those of females of the same species (Ford 1994).

The molars are large and square with complex surfaces made up of long crests of thin enamel (Fig. 6.4); *Alouatta* has the largest postcanine tooth area of all the ateline primates (Rosenberger and Strier 1989). Their distinct morphology is marked by high crowns, deep basins, deep intercuspal notches on the buccal and lingual sides of the crown, and a long cristid obliquid (Fig. 6.4; Cooke 2011). This is particularly clear when comparing the "shearing quotient" of folivorous howler monkeys to frugivorous spider monkeys (Kay 1978; 1984; Anthony and Kay 1993). The high-crested molars of the howlers are adapted to slicing through the tough structural carbohydrates, like cellulose, that make up their leafy diet, while the low-relief molars of the spider monkeys are better suited to mashing up softer fruit pulp (Kay 1975; Rosenberger and Kinzey 1976; Anthony and Kay 1993; Anapol and Lee 1994). More recent analyses of the overall surface relief and topography of the molars, not just shearing crest length, also illustrate this difference between howler monkeys and their more frugivorous relatives (i.e., Cooke 2011).

The shearing capability of *Alouatta* molars has been shown to be convergent upon the superficially similar surface topography of the molars of another ateline, *Brachyteles* (Rosenberger and Strier 1989; Rosenberger 1992). The folivorous woolly spider monkeys also have relatively small incisors and large, high-relief molars that have a high shearing quotient, but the crests appear on a different aspect of the molar surface. *Alouatta*'s shearing crests are concentrated on the buccal side of the molars which exhibit a large paracone, metacone, and ectoloph in the upper tooth row and an elongate talonid and cristid obliquid in the lower (Fig. 6.4; Rosenberger and Strier 1989). *Brachyteles*, on the other hand, shears on the lingual side of the molars with a large protocone and a tall metaconid, which occludes into the gap between the protocone and hypocone; they are also unique among atelines in possessing a strong lingual notch between the metaconid and entoconid (Rosenberger and Strier 1989). Genetic evidence points to a sister-taxon relationship between *Brachyteles* and *Lagothrix*, while skeletal morphology favors a *Brachyteles-Ateles* pair; no evidence for a close relationship between *Alouatta* and *Brachyteles* exists besides their shared dental morphology, which has convincingly been shown to be a convergent adaptation to their folivorous diet (Anthony and Kay 1993).

Some differences in relative incisor size have been described between various species of *Alouatta* (i.e., Anthony and Kay 1993), which is to be expected based on their differing degrees of folivory (Di Fiore and Campbell 2007), but no systematic study of interspecific dental morphology exists to date [but see the dissertation work of Orlosky (1973)]. At least one species, *Alouatta palliata*, shows a characteristic species-wide wear sequence as their high-relief molars are flattened over time with use, reducing the shearing crest length but preserving the "angularity," or jaggedness, of the surfaces necessary for processing leaves (Dennis et al. 2004). This is in agreement with Rosenberger and Kinzey's (1976) insights on the ontogenetic maintenance of molar shear in folivores such as *Alouatta*, as well as more insectivorous taxa, like *Callithrix*. The thin enamel and large molar surface area assure that the howler monkey teeth will retain their shearing functions while they wear down (see also Kay 1984; Anapol and Lee 1994).

6.4 Hyoid

6.4.1 A Unique Hypertrophied Hyoid

Alouatta is said to make the loudest vocalizations of any terrestrial vertebrate and has the largest, most inflated hyoid bone within the primate order. While the basic anatomy of the hyoid bone, thyroid and epiglottic cartilages, and air sacs of *Alouatta* have been described by many authors, inter- and intraspecific variations are not as well known, and there is debate about how their morphology relates to diet, vocalizations, body size, respiration, and the functional morphology of the skull. Unlike in humans, the hyoid bone of *Alouatta* is always attached to the cranial base through a chain of cartilaginous, or sometimes even ossified, elements, i.e., the more common "integro-cornuate" condition (Howes 1896). It also sits wedged between the gonial angles of the mandible. The hyoid serves as an attachment point for muscles and ligaments that also serve the mandible, tongue, laryngeal cartilages, pharynx, sternum, and cranial base, hence its proposed functions involving several systems including respiration, swallowing, and vocalizations. In most mammals, the hyoid apparatus lies at or above the entrance to the larynx and consists of two basic parts: inferiorly, the unpaired basihyal and paired thyrohyal "horns" which connect the basihyal to the thyroid cartilage and superiorly, the paired suspensory stylohyoid chains which are made up of four cartilaginous or ossified links, the ceratohyal, epihyal, stylohyal, and tympanohyal, that connect the basihyal to the temporal bone (e.g., Howes 1896; Negus 1949).

It is the size and shape of the basihyal portion that vary across primate species. Some Old World monkeys have a caudally expanded basihyal that covers the top of the thyroid cartilage (Negus 1949; Hilloowala 1975; Nishimura 2003), but none are as extremely modified as that of *Alouatta*. Howler monkeys are the only living primates that have evolved a pneumatized hyoid bone with a large, hollow balloon-like basihyal, which is argued to serve, along with their air sacs, as a resonating chamber (Chapman 1929; Kelemen and Sade 1960; Schön 1971; Hewitt et al. 2002). As the larynx of howlers, and the other nonhuman primates, is positioned so high in the throat, the supralaryngeal dimensions of the pharynx are small, limiting the range of sounds that can be made (Laitman and Reidenberg 2009). The expanded hyoid bone and air sacs positioned above the larynx in *Alouatta* increase the volume of this space and, along with stretching out the neck and manipulations of the lips (Schön 1986; Whitehead 1995), shape the sound and volume of their loud long calls.

The laryngeal cartilages are also enlarged in *Alouatta*, especially the thyroid, cuneiforms, and epiglottis, which are also sometimes ossified (Kelemen and Sade 1960; Hill 1962). The basihyal is joined to the thyroid cartilage with a membranous attachment, and the two are freer to move more independently of one another in *Alouatta* than in other primates (Schön-Ybarra 1995). As mentioned above, the stylohyoid chain is of the common mammalian integro-cornuate type, connecting to the cranial base anterior to the stylomastoid foramen (Howes 1896; Hilloowala 1975). Not enough is known about variation in the associated hyoid musculature, hence how the hyoid may move during howling in the different species, as only a few specimens of *A. seniculus* (Hill 1962; Schön 1968, 1971), *A. palliata* (Kelemen and Sade 1960; Hilloowala 1975), and *A. caraya* (Schön 1971) have been dissected. It is worth emphasizing that the size of the hyoid and the anatomical space it requires in the head are not trivial. For example, an average endocranial volume for a mixed sex sample of 16 *A. belzebul* individuals from the American Museum of Natural History was 55 ml, and the average hyoid volume for four males of that species was 57 mL; in two individuals, hyoid volume was approximately 140 % of their endocranial volume.

The impact of this anatomical specialization on the howler skull has yet to be adequately investigated and synthesized, but various unique aspects of cranial and mandibular morphology in *Alouatta* have been suggested to be related to the large size of these vocal tract structures (e.g., Biegert 1963; Watanabe 1982). While the unique morphology of the bone was mentioned by many early authors (e.g., Grew 1681), variation in hyoid size and shape among the species of *Alouatta* was first systematically described by Hershkovitz in 1949. The morphology of this region has been used as a diagnostic character for species recognition since then by Hill (1962) and many others. This section will quantitatively explore hyoid size and shape within *Alouatta* in order to better understand interspecific differences that in the past have been used as a basis for taxonomic, phylogenetic, and functional interpretations about the genus.

6.4.2 Hyoid Material and Analyses

Over 100 undeformed *Alouatta* hyoids from collections at the American Museum of Natural History (New York, USA), the United States National Museum (Washington, DC, USA), and the Field Museum of Natural History (Chicago, IL, USA) were laser scanned using either a portable Minolta Vivid 910 laser surface scanner or a NextEngine desktop 3D scanner (Table 6.2). As the bony walls of many of the smaller hyoids were too thin to be "seen" properly by the lasers, the hyoid bullae were stuffed with paper towels to make the bone more opaque. The scans were edited using *Geomagic Studio* and ScanStudio HD v. 1.3.0, and a .ply file was created for each specimen (Harcourt-Smith et al. 2008). The *Landmark Editor* program (Wiley 2006) was then used to place homologous landmark points across the superior aspect of the tentorium as well as at the four corners of the posterior opening of the basihyal. These landmarks were used as anchors for drawing automated curves that generate semi-landmarks which quantify the bulbous aspect of the anterior portion of the bone (Bookstein 1997; Gunz et al. 2005; Fig. 6.5). This procedure

Table 6.2 Extant sample for hyoid measurements

Genus	Species	Subspecies[a]	Country	Hyoid M	F
Alouatta	*belzebul*	*belzebul*	Brazil	6	7
			N	6	7
Alouatta	*caraya*		Brazil	7	6
			Bolivia	5	6
			N	12	12
Alouatta	*guariba*	*guariba*	Brazil (NE)	3	2
Alouatta	*guariba*	*clamitans*	Brazil (SE)	4	4
			N	7	6
Alouatta	*palliata*	*palliata*	El Salvador	2	1
Alouatta	*palliata*	*aequatorialis*	Panama	1	–
			N	3	1
Alouatta	*pigra*		Guatemala	2	1
			N	2	1
Alouatta	*seniculus*		Bolivia	2	4
			Brazil	1	–
Alouatta	*seniculus*	*seniculus*	Colombia	26	22
			Peru	3	2
Alouatta	*seniculus*	*straminea*[b]	Guyana	2	2
			N	34	30
Lagothrix	*lagotricha*	*lagotricha*	Colombia	1	–
Lagothrix	*lagotricha*	*lugens*	Colombia	2	–
			N	3	0

[a]Taxonomic allocation and provenance taken from specimen boxes
[b]Currently recognized as *A. macconnelli*

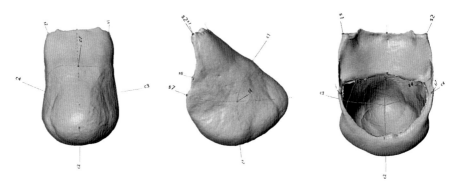

Fig. 6.5 Landmarks recorded on the hyoid. Images are from *Landmark Editor*, showing points and curves on a typical male *A. seniculus* hyoid in (from *left* to *right*) anterior, lateral, and posterior views

captured not only the size of the hyoid, which has been approximated in previous studies by measuring its length and/or width (see, for example, Watanabe 1982; Halpern 1987), but also the bone's contours, which contribute to the unique shape of the hyoid in each species. Several *Lagothrix* hyoids, which are much smaller and less concave than those of *Alouatta*, were also scanned and landmarked, functioning as an "outgroup."

The data were subjected to a GPA (Rohlf and Slice 1990) to translate, rotate, and scale all specimens to unit centroid size using the *morphologika*^{2} v2.5 software program (O'Higgins and Jones 2006). The curves of semi-landmarks on the hyoids were also subjected to a "sliding" protocol using a program implemented in MATLAB in order to minimize Procrustes distance between the specimens and the calculated reference specimen (Bookstein 1997; Gunz et al. 2005). Sliding the semi-landmarks along the outline curves is an extension of the standard GPA superimposition procedure and is necessary to properly account for type III semi-landmarks having fewer degrees of freedom than those of type I or type II (Adams et al. 2004). PCA was then used to visualize the morphological variation in the sample. The PAST 1.89 software package (Hammer et al. 2001) was used to perform and output graphical representation of the PCAs.

6.4.3 Hyoid Variation

The development of the tentorium superiorly and the size and shape of the posterior opening of the basihyal are the most important variables for distinguishing the species from one another (Hershkovitz 1949; Hill 1962; Fig. 6.6a, b). Male *A. seniculus* have the largest, most inflated hyoid bullae with relatively small posterior openings and wide convex tentoria. *A. guariba* has a larger posterior opening with a less inflated tentorium and no cornua. The hyoid of *A. belzebul* is also in a larger size class with the largest posterior opening and a slightly concave tentorium. Of the larger species, *A. caraya* has the smallest tentorium but a very large posterior opening and a uniquely shortened anteroposterior dimension that gives the bone a "snub-nosed" appearance. The hyoid bullae of *A. palliata* are the smallest, with a flared opening, extremely reduced tentorium, and broad attachment sites for the thyrohyal proximally and the stylohyoid ligament distally. While *A. pigra* has been considered a subspecies of *A. palliata* and shares with it a Central American distribution, the female hyoid looks more similar to that of the South American species, while the bone in the male has a very unique, almost two-dimensional appearance.

These qualitative descriptions led Hershkovitz (1949) to divide the genus into three groups: the *seniculus* group which includes *A. seniculus*, *A. belzebul*, and *A. guariba*; the *palliata* group which includes *A. palliata* and *A. pigra*; and the *caraya* group for the snub-nosed *A. caraya*. For Hershkovitz, these groups have evolutionary significance, with *A. palliata* retaining the ancestral condition and *A. seniculus*, especially the males, exhibiting the most derived hyoid shape. Interestingly, while *Alouatta* males do have larger, more inflated hyoid bullae than females of the same species, body size differences between the species do not seem to correspond to differences in hyoid size or the length of the supralaryngeal vocal tract (Hilloowala 1975; Schön-Ybarra 1995). In other words, the species with the largest average body size does not have the absolutely largest hyoid volume

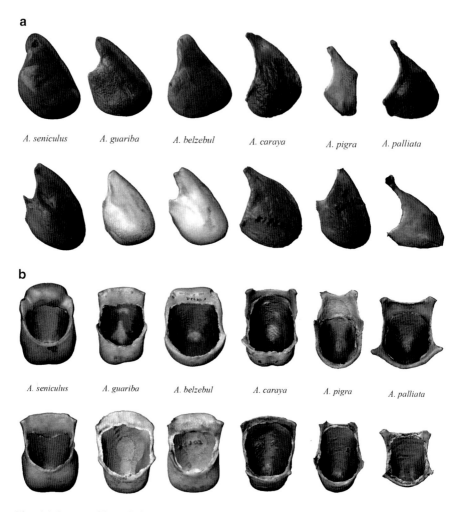

Fig. 6.6 Interspecific variation in hyoid shape across *Alouatta*, lateral (**a**, *top*) and posterior (**b**, *bottom*) views. Images are .ply files created from laser surface scans of individual specimens (see Materials and Methods—ragged edges around the posterior opening of the hyoid are an artifact of the scan editing process) chosen from the hyoid sample as typical representatives of each species and scaled to similar size to emphasize shape differences. *Top row* in each view are males, *bottom row* are females (*A. pigra* female is mislabeled as a male)

(Table 6.3). These differences in howler monkey hyoid size and shape do, however, seem to correlate with various acoustic features of their calls (Table 6.3): *A. palliata* has been shown to vocalize for shorter periods of time over a narrower bandwidth and higher frequency (Sekulic and Chivers 1986; Whitehead 1995), they have the smallest hyoids, and they only have one type of air sac (Schön-Ybarra 1995; Hewitt et al. 2002).

Table 6.3 Relationship between body size, hyoid size, and long call acoustics in *Alouatta*

Taxon	Body weight (kg)[a] Males	Body weight (kg)[a] Females	Dimorphism index[b]	Hyoid size[c] Males	Hyoid size[c] Females	Complete rules	Emphasized frequency (Hz)[d]	Call duration (s)[e]
A. belzebul	7.3	5.5	1.327	212	130	1.631	480.6	4.9
A. caraya	6.8	4.6	1.478	164	116	1.414	439.7	–
A. guariba	6.2	4.5	1.378	197	122	1.615	616.6	–
A. palliata	7.1	5.3	1.339	117	74	1.581	413.3	3.5
A. pigra	11.3	6.4	1.766	124	120	1.033	598.9	5–6
A. seniculus	7.2	5.6	1.286	220	131	1.679	475.7	19, 8, 120

[a]Averages taken from Ford and Davis (1992)
[b]Dimorphism index = average male centroid size/average female centroid size (Ford 1994)
[c]Male and female average centroid size calculated in this study
[d]From Whitehead (1995)
[e]From Sekulic and Chivers (1986) and Whitehead (1995)

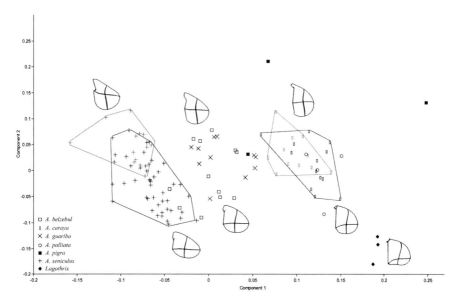

Fig. 6.7 PCA results for the hyoid. Sample includes males and females of all species of *Alouatta* and three *Lagothrix* individuals. PC1 (37 % total variance) separates *A. seniculus* from the other species as the most bulbous with the smallest posterior opening. PC2 (18 %) separates males, individuals towards the positive end of the axis, from females, individuals towards the negative end (also see Fig. 6.8). PC2 is also somewhat related to geographic differences within the *A. seniculus* and *A. caraya* samples. The gray polygon bounds the subset of the *A. seniculus* sample not from Colombia. The *A. caraya* individuals bounded by the *gray* polygon are from Bolivia, while those bounded in *black* are from Brazil. Wireframes show the curves on the basihyal in lateral view with the anterior portion of the bone facing to the right to more easily visualize the shape change along PC1 and the difference between males and females of each species (see Fig. 6.5)

In the PCA results, the *A. palliata* and non-*palliata* groups are not as distinct as one might expect based on the descriptions given above or quantitative studies of interspecific variation of the skull, but this could be due to the much smaller sample size of *A. palliata* compared to the other species (Figs. 6.7 and 6.8). These groupings do, however, generally agree with those described in previous studies of interspecific hyoid variation (i.e., Hershkovitz 1949; Halpern 1987; Gregorin 2006). While having a slightly concave basihyal, the three *Lagothrix* hyoids are clearly separated from the *Alouatta* sample; *Ateles* hyoids look similar to those of *Lagothrix*, but unfortunately none could be laser scanned for this analysis. The anterior-posterior dimensions of the basihyal, along with the position of the four corners surrounding the posterior opening, are the main variables driving separation of the species clusters seen across PC1. The phenetic similarities in hyoid shape suggested by the overlap of the *A. guariba* and *A. belzebul* samples agree with their linkage in molecular phylogenetic studies of the genus (e.g., Cortés-Ortiz et al. 2003). The molecular evidence, however, also suggests a sister-taxon relationship between

6 Morphology of Howler Monkeys: A Review and Quantitative Analyses

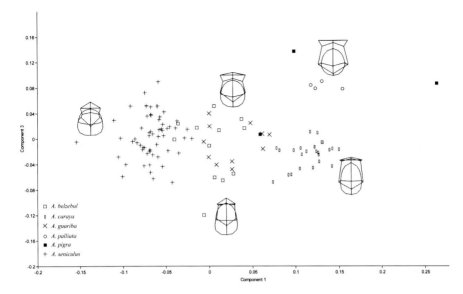

Fig. 6.8 PCA results for the hyoid using the *Alouatta* sample, PC1 vs. PC3. When the *Lagothrix* individuals are excluded from the sample, PC3 (9 % total variance) separates *A. palliata* from *A. caraya*. PC1 (39 %) is still driven by the contours of the inferior portion of the basihyal as well as the dimensions of the posterior opening and tentorium. Wireframes show the hyoid in posterior view, again to visualize the shape differences between the species

A. seniculus and *A. caraya*, two species that are shown here to be quite distinct in their hyoid morphology. Interestingly, there is some suggestion of geographic separation within the *A. caraya* and *A. seniculus* samples; Brazilian and Bolivian *A. caraya* form relatively distinct clusters as do Colombian and non-Colombian *A. seniculus* (Fig. 6.7). Perhaps hyoid size and shape are more influenced by socioecological factors related to long call behavior, like group size and habitat type (e.g., Sekulic and Chivers 1986; Chiarello 1995), rather than phylogenetic relatedness.

PC2 separates *Alouatta* male hyoid shape from female hyoid shape within each species cluster reasonably well, and the dimorphism index calculated by dividing male hyoid centroid size by female hyoid centroid size confirms a relatively high degree of sexual dimorphism in *Alouatta* hyoid size (Fig. 6.9; Table 6.3). The species with the largest hyoids as measured by centroid size, *A. belzebul*, *A. seniculus*, and *A. guariba*, are also the most dimorphic. But as noted above, these are not the species with the largest body sizes nor the highest body size dimorphism indices (Table 6.3). Male hyoid centroid sizes also vary more widely from species to species than those of females. An explanation for this, as well as why there are some males on the "female side" of the clusters in Fig. 6.7, could be that non-dominant but fully adult males are vocalizing less than the dominant males and therefore have smaller musculature moving their smaller, more female-like hyoids.

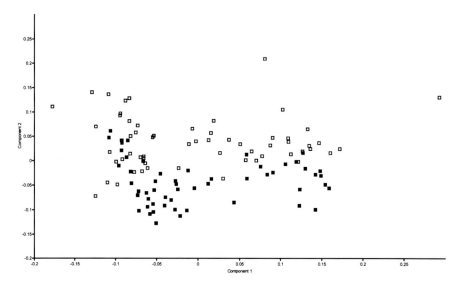

Fig. 6.9 Sexual dimorphism in the *Alouatta* hyoid. This shows the same distribution of specimens as in Fig. 6.6, but with male individuals symbolized by *open squares* and females by *closed squares*. Graphically, the separation between males and females for most species looks distinct, but regressing in centroid size on PC1 scores gives R^2 values for PC1 and PC2 that are both relatively low at 0.4. This would indicate that hyoid shape differs more than hyoid size between the sexes in *Alouatta*

6.5 A Derived Postcranium

This section reviews the current evidence on postcranial apomorphies and provides a functional interpretation of the *Alouatta* skeleton.

6.5.1 *Forelimb*

In *Alouatta*, the forelimb is relatively short compared to other atelines, representing only 91 % of the length of the vertebral column, whereas it is much longer in *Lagothrix* and *Ateles* (109 % and 150 % respectively, Erikson 1963). However, it is almost equal in length compared to the hind limb, as indicated by the intermembral index (IMI), which ranges between 97 and 99 (Erikson 1963; Youlatos 1994; Jones 2004). A similar condition is encountered in *Lagothrix* (IMI=97.6), whereas *Ateles* is very different (IMI=105; Erikson 1963).

The scapula is positioned relatively cranially compared to most other quadrupeds (Youlatos 1994). It is craniocaudally wide and reminiscent of climbing and suspensory mammals, compared to the more dorsoventrally elongated scapular shape of

6   Morphology of Howler Monkeys: A Review and Quantitative Analyses   153

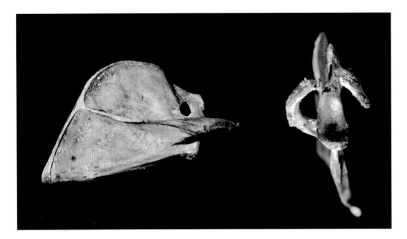

Fig. 6.10 Lateral view (*left*) and proximal view (*right*) of the scapula of *A. seniculus* (glenoid-midaxillary border length is 71.4 mm) showing the enlargement of the suprascapular fossa, the development of the acromion, the distal extension of the coracoid process joining the cranial end of the suprascapular fossa via a large ligament (*left*), and the oblong shape of the glenoid fossa (*right*)

both arboreal and terrestrial quadrupeds (Fig. 6.10; Roberts 1974). The scapular spine is only slightly oblique, compared to the more oblique position in *Lagothrix* and very oblique in *Ateles*, facilitating the rotatory action of the powerful *m. trapezius* during quadrupedal activities (Erikson 1963; Anapol and Fleagle 1988).

The suprascapular fossa is particularly enlarged (Fig. 6.10), a condition also encountered in many climbing and suspensory mammals and associated with the substantial arm raising and abducting action of *m. supraspinatus* (Oxnard 1963; Schön 1968; Larson and Stern 1989; Youlatos 2000). However, the greater relative surface area in *Alouatta* (33–42 % of total scapular surface area) compared to climbing primates (24–33 %, Youlatos 1994) may be related to a more cranial origin and the overdevelopment of *m. atlantoscapularis posterior* providing additional support for the neck (Schön 1968; Youlatos 1994).

The acromion is large and protuberant facing moderately dorsad (Fig. 6.10), whereas it is slightly less prominent and dorsally oriented in *Ateles* (Erikson 1963). This provides a better leverage for *m. deltoideus* during powerful arm protraction and abduction. Medially, the coracoid process is moderately developed, protruding distally over the glenoid fossa in a way reminiscent of *Lagothrix* and arboreal quadrupeds. In contrast, *Ateles* possess an enlarged coracoid, oriented cranio-distally, contributing to a better leverage for *mm. coracobrachialis* and *biceps brachii caput breve* (Schön 1968; Anapol and Fleagle 1988).

The glenoid fossa faces cranially, in contrast to the more lateral and completely lateral direction in *Lagothrix* and *Ateles*, respectively. This is related to the position of the scapula on the dorsoventrally elongated rib cage of *Alouatta*, compared to the more mediolaterally wide ribcage of *Ateles* (Erikson 1963; Rosenberger and Strier

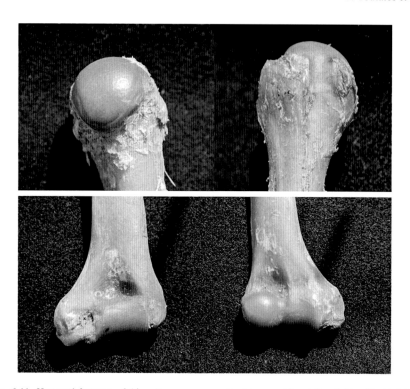

Fig. 6.11 Humeral features of *Alouatta*: narrow proximal humeral head (*top left*), wide bicipital groove and reduced humeral tuberosities (*top right*), protruding medial condyle and shallow olecranon fossa (*bottom left*), wedged trochlea, and ovoid capitulum (*bottom right*) (humeral head width = 16.4 mm; distal epiphysis width = 28.5 mm)

1989; Youlatos 1994). Its shape is oblong in *Alouatta* (Fig. 6.10) with a relatively prominent cranial buttress with dense trabecular bone distribution, unlike the more spherical facet in *Lagothrix* and *Ateles* with a reduced buttress (Schön-Ybarra and Schön 1987; Anapol and Fleagle 1988; Fleagle and Meldrum 1988; Jones 2004; Kagaya 2007).

The humeral head is ovoid in shape, relatively mediolaterally narrow and faces posteriorly (Fig. 6.11; Erikson 1963; Schön-Ybarra and Schön 1987; Kagaya 2007). Its mediolateral convexity is shorter and more pronounced than the anteroposterior one (Schön-Ybarra and Schön 1987; Jones 2004). The head is more circular and larger in *Ateles*, and its surface also expands and faces medially (Erikson 1963; Larson 1988; Jones 2004). This has been metrically expressed by the intertuberosity angle, which is quite reduced in *Alouatta* (63.5°), compared to the intermediate position in *Lagothrix* (76.7°) and the very wide one in *Ateles* (111.2°; Jones 2004). The head projects proximad to the tuberosities (Fig. 6.11). The latter are rather well developed, compared to their reduced development in *Lagothrix* and *Ateles* (Fleagle and Simons 1982; Jones 2004). The lesser tuberosity, the insertion point of *m. subscapularis*, is also well marked, whereas it is quite reduced in *Ateles* and *Brachyteles* and may be related to ample arm rotations (Fleagle and Simons 1982; Jones 2004). This combination of humeral head features allows a greater degree of arm protraction and retraction, as well as increased stability during arm abduction necessary for arm raising movements during climbing and suspensory locomotion, as well as in foraging activities (Fleagle and Simons 1982; Harrison 1989).

The bicipital groove is wide and shallow in *Alouatta* (Fig. 6.11), somewhat more defined in *Lagothrix*, and narrow and deep in *Ateles*; this keeps the tendon of the biceps in place during arm movements (Fleagle and Simons 1982; Schön-Ybarra and Schön 1987; Harrison 1989; Jones 2004). The deltopectoral crest on the humeral shaft is moderately developed and similar to climbing and suspensory forms, unlike the prominent and distally situated crest in more quadrupedal primates (Fleagle and Simons 1982; Schön-Ybarra and Schön 1987; Ford 1988).

Alouatta, and to a lesser extent *Lagothrix*, possesses straight and quite robust humeri (robusticity index = 6.9) similar to most other quadrupedal platyrrhines able to withstand the frequent action of compressive forces (Schön-Ybarra and Schön 1987; Jones 2004). In contrast, *Ateles* and *Brachyteles* are characterized by equally straight but more slender shafts (5.2 and 5.0, respectively; Jones 2004).

The distolateral crest, where *m. brachialis* originates, is noticeable and seems to extend rather proximally on the lateral side of the humeral shaft as in other primates which habitually flex their elbow (Conroy 1976; Fleagle and Simons 1982). The crest is less evident and more distally located in *Ateles* (Jones 2004). At its distal end, the lateral epicondyle is unreduced in *Alouatta*, but small in *Ateles* (Erikson 1963; Rosenberger and Strier 1989).

The humeral trochlea is shallow and spool-like, with a well-developed medial trochlear lip, and is relatively mediolaterally extended (Fig. 6.11), a morphology that facilitates arboreal quadrupedal movements by allowing a certain degree of mediolateral translation (Jenkins 1973; Feldesman 1982; Fleagle and Simons 1982; Schön-Ybarra and Schön 1987; Ford 1988; Rose 1988; Harrison 1989; Jones 2004). In contrast, *Lagothrix* and *Ateles* possess more cylindrical trochlea with relatively reduced medial lips, morphology associated with well-guided elbow flexion and extension during suspensory activities (Jones 2004).

Alouatta possesses a transversely ovoid and proximodistally high capitulum (Fig. 6.11), which seems to be functionally associated with habitual elbow flexion and forearm pronation during arboreal quadrupedal movements (Feldesman 1982; Schön-Ybarra and Schön 1987; Rose 1988; Harrison 1989; Schön-Ybarra 1998). On the other hand, *Lagothrix* and *Ateles* possess a low, well-developed, and more spherical capitulum that provides ample pronosupinatory rotations during elbow excursions (Rose 1988; Jones 2004).

The medial epicondyle of *Alouatta* is well developed and faces slightly posteromedially but is prominent and medially oriented in *Ateles* (Fig. 6.11; Erikson 1963; Feldesman 1982; Schön-Ybarra and Schön 1987; Harrison 1989). The overall projection of the condyle increases the leverage of the forearm flexors and pronators that contribute to arboreal quadrupedal locomotion, and a more posterior direction further enhances the moment of the acting forces in quadrupedalism (Jenkins 1973; Fleagle and Simons 1982; Schön-Ybarra 1998).

Fig. 6.12 Radial view of the ulna of *Alouatta* (ulna length = 176.3 mm) showing the gentle anterior curvature of the shaft, the robust straight olecranon proximally, and the prominent styloid distally (*left*); the distally inclined shallow sigmoid and the gently concave radial notch in closer views (*right*)

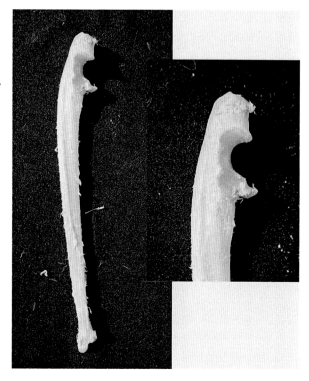

In the ulna, the olecranon is large and faces anteriorly (Fig. 6.12; Youlatos 1994; Jones 2004) providing enhanced mechanical advantage to the action of *mm. triceps* during elbow extension and supporting the weight during habitual elbow flexion (Schön-Ybarra and Conroy 1978; Rodman 1979; Harrison 1989; Schön-Ybarra 1998; Rein et al. 2011; Drapeau 2004). In contrast, it is moderately developed in *Lagothrix* and relatively reduced in *Ateles*, as in most forelimb suspensory mammals (Drapeau 2004).

The sigmoid notch is relatively wide, proximodistally concave and its axis angles medially and distally, with both proximal borders flaring proximally (Schön-Ybarra and Schön 1987; Youlatos 1994). These characters indicate the weight-bearing role of the ulna and elbow excursions that are mediolaterally stable with controlled deviation of the forearm for its placement on arboreal substrates (Conroy 1976; Schön-Ybarra and Conroy 1978; Schön-Ybarra and Schön 1987; Fleagle and Meldrum 1988; Harrison 1989; Schön-Ybarra 1998). Additionally, over the coronoid, the surface is flat and slopes distally; this is related to the habitually flexed postures of the elbows (Fig. 6.12; Schön-Ybarra and Schön 1987; Anapol and Fleagle 1988). The sigmoid of *Ateles* is slightly different, characterized by a weaker angle and being wide, deep, and semicircular with a reduced coronoid process facing anteriorly.

The radial notch is reduced, gently concave, and facing anterolaterally, a condition encountered in most arboreal quadrupedal primates (Fig. 6.12; Youlatos 1994;

Halenar 2011). This morphology allows a certain degree of forearm rotation and provides increased stability during pronation (Rose 1988). In contrast, *Ateles* possess a large concave notch facing anterolaterally, as in other suspensory primates.

In *Alouatta*, and to a lesser extent *Lagothrix*, the ulnar shaft is straight, very robust, and gently concave anteriorly (Fig. 6.12). It is ovoid in section and similar to arboreal quadrupedal platyrrhines (Schön-Ybarra and Schön 1987; Youlatos 1994; Jones 2004). This morphology withstands the action of high compressive forces, resists the action of the elbow flexors, and provides space for radial rotations (Conroy 1976; Fleagle and Meldrum 1988). In contrast, *Ateles* and *Brachyteles* are characterized by quite slender shafts that are T-shaped in section (Jones 2004). Distally, the *Alouatta* ulna is characterized by a robust styloid process, which protrudes distally within the wrist joint. This morphology establishes an extended articulation between the ulna and the lateral part of the wrist, assuring wrist stability and resistance to compressive forces at ulnar deviation during quadrupedal activities (Lewis 1989).

Alouatta, and to a lesser extent *Lagothrix*, possess very robust radii similar to other quadrupedal platyrrhines, implying the action of frequent compressive forces (Schön-Ybarra and Schön 1987; Fleagle and Meldrum 1988; Youlatos 1994; Jones 2004). In contrast, *Ateles* and *Brachyteles* are characterized by quite slender shafts (Jones 2004).

The head of the radius is slightly elliptical with a shallow articular facet that is laterally inclined as in most arboreal quadrupedal primates (Fig. 6.13; Conroy 1976; Rose 1988; Youlatos 1994; Jones 2004). A similar morphology indicates limits to the range of forearm rotation and extended radioulnar and radiohumeral contact during forearm pronation, establishing a locking mechanism for elbow stability (Jenkins 1973; Conroy 1976; Rose 1988; Harrison 1989; Schön-Ybarra 1998). *Lagothrix* and *Ateles* possess horizontal, more cylindrical radial heads with deep facets that permit articular contact throughout a wide range of forearm rotations (Jenkins 1973; Conroy 1976; Rose 1988).

The bicipital tuberosity, that hosts the insertion of *m. biceps brachii*, is distally located as in most quadrupedal platyrrhines (Jones 2004). In contrast, it is rather proximally located in *Lagothrix* and even more so in *Ateles*, indicating a short lever arm, which favors rapid and wider forearm movements (Harrison 1989; Jones 2004).

The distal radioulnar joint is a distally restricted and anteroposteriorly wide facet hosting a syndesmosis that allows a certain range of forearm rotations (Ford 1988; Lewis 1989; Youlatos 1994). In *Ateles*, the facet is practically absent, reflecting the wide range of pronosupination that occurs in the forearm of this suspensory platyrrhine. Distally, *Alouatta* and *Lagothrix* have relatively wide radiocarpal articular surfaces (Fig. 6.13) that provide wide contact for the carpus to withstand compressive forces. Nevertheless, this facet is still narrower than that of other platyrrhines. In contrast, *Ateles* possess very narrow distal articular surface, very likely associated with the reduced presence of compressive forces during their habitual suspensory locomotion (Jones 2004).

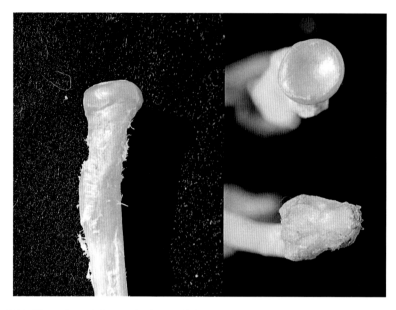

Fig. 6.13 Ulnar view of the proximal part of the radius (radial length = 160.9 mm), showing the inclination of the head and the extended articular facet for the radial notch (*left*), proximal view of the ovoid and shallow radial head (radial head length = 12.8 mm) (*right top*), and distal view of the broad radiocarpal facet (*bottom right*)

Alouatta's wrist is rather short and its relative proportions do not depict any functional specializations, other than generalized quadrupedal activities (Godinot 1992). The proximal carpal row, composed of the scaphoid, lunate, triquetrum, and pisiform, is mediolaterally extended and gently curved (Fig. 6.14). In *Alouatta* and *Lagothrix*, the proximal facets of the scaphoid and lunate are similar to most quadrupedal primates and assure an extended contact with the wide distal radial facet allowing flexion and extension movements with limited abduction and adduction (Jenkins and Fleagle 1975; Youlatos 1994, 1996; Kivell and Begun 2009).

The proximal facets of the triquetrum and pisiform are concave and articulate with the corresponding well-developed distal facets of the styloid process of the ulna (Fig. 6.14; Youlatos 1994). This morphology, also encountered in *Lagothrix* and other quadrupedal primates, assures a stable joint during quadrupedal stances when the hand is ulnarly deviated (O'Connor 1975; Beard and Godinot 1988; Lewis 1989). This stability is further enhanced by the long and robust pisiform which is supported by two ligaments that connect with the palmar tubercle of the hamate and the styloid process of metacarpal V (Grand 1968a; Ziemer 1978; Youlatos 1994).

The distal carpal row is composed of the trapezium, trapezoid, capitate, and hamate, which articulate proximally with the distal facets of the proximal carpal row and distally with the metacarpals (Fig. 6.14). At the level of the midcarpal joint, the proximal facets of the capitate and hamate of *Alouatta* and *Lagothrix* are smoothly curved allowing a certain degree of midcarpal rotation (Jenkins and Fleagle 1975;

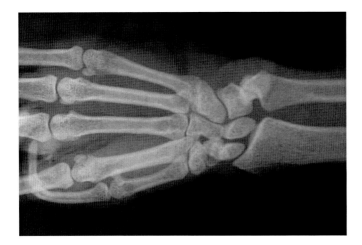

Fig. 6.14 Radiograph of the hand of *Alouatta*, showing the arrangement of the proximal and distal carpal rows, with the well-developed heads of the capitate and hamate, the well-developed pisiform ulno-proximally, and the extended contact between the distal capitate, and proximal ends of metacarpal II and metacarpal III that accommodate the zygodactylous grasp (metacarpal III length = 33.7 mm)

Kivell and Begun 2009). Additionally, the capitate bears a dorsal expansion of the proximal facet for the os centrale, similar to that found in *Lagothrix*, *Ateles*, and some other suspensory primates, which seem to enhance midcarpal pronosupination (Ziemer 1978). Furthermore, the dorsal facet of the hamate bears a facet for the triquetrum, similar to that of other atelines. This facet allows for the radial rotation of the os centrale over the hamate and increases midcarpal supination (Jenkins 1981; Sarmiento 1988; Kivell and Begun 2009). However, a palmar process on the head of the capitate, non-existent in *Lagothrix* and *Ateles*, would eventually restrain the extended rotations permitted by the above morphology (Youlatos 1994).

The ulnoradial direction and proximodistal elongation of the ulnar facet of the hamate, for the triquetrum, is similar to *Lagothrix* and other quadrupedal primates and is indicative of restricted mobility and enhanced stability for weight bearing (Ziemer 1978; Beard and Godinot 1988; Youlatos 1994). The corresponding facet on the distal surface of the triquetrum is ulnoradially elongated establishing an interlocking mechanism with the hamate in the ulnar side of the wrist (Youlatos 1994).

Intrinsic hand proportions do not show any functional specializations and fall within the ranges of most arboreal quadrupedal primates (Fig. 6.14; Jouffroy et al. 1991; Godinot 1992). *Alouatta*'s metacarpals and phalanges are comparably short and very gently curved, morphology associated with pronograde quadrupedal habits and contrasted with the particularly long and especially curved metacarpals of *Ateles* and other suspensory primates (Erikson 1963; Jouffroy et al. 1991; Hamrick et al. 1995; Stern et al. 1995; Halenar 2011; Rein 2011). *Alouatta* also possess a paraxonal hand with digits III and IV having similar lengths, while the functional axis of the hand usually passes through digit III (Fig. 6.14; Grand 1968a; Youlatos 1999).

The metacarpal I and trapezium facets are rather flat, indicating limited excursions of the pollex (Napier 1961; Rose 1992). The pollex is capable of being positioned against the lateral digits mainly because of the arrangement of digits II–V, but advanced rotatory movements, similar to the ones observed in catarrhines are practically impossible (Rose 1992). However, strong manual prehension upon substrates is achieved by frequent use of a zygodactylous grasp (Fig. 6.14; Grand 1968a). Morphologically, this is assured by the convex joint between the capitate and metacarpal II and between the proximal ends of the latter and metacarpal III, which provide enhanced mobility in both the dorsopalmar and radioulnar planes (Youlatos 1999). Furthermore, the functional separation of the tendons of *m. flexor digitorum profundus and m. extensor digitorum profundus* towards the two pincer arms and the well-developed adductor of digit II probably assure a strong grasp (Youlatos 1999).

6.5.2 Hind Limb

In the pelvis, the iliosacral joint is located cranially in *Alouatta* and the other atelines (Youlatos 1994). The particularly cranial location of the joint increases the distance from the hip joint, providing better leverage for the transmission of the hind limb forces during quadrupedal progression (Rodman 1979). More cranially, the surface of the insertion of *mm. erector spinae* is significantly reduced in *Alouatta*, as in other atelines (Youlatos 1994). A reduced insertion area is functionally associated with the lesser development of these muscles in *Alouatta* and other climbing primates, as these muscles have been shown instead to contribute to pelvic movements during running (Waterman 1929; Stern 1971; Grand 1968b, 1977).

The gluteal surface of the iliac blade is long, wide, concave, and oriented dorsolaterally, presenting an intermediate morphology between the quadrupedal platyrrhines and the more suspensory atelines. The wider and more concave blades in the latter very likely host a well-developed *m. gluteus medius*, which also contributes to hind limb adduction during quadrupedal climbing and frequent suspensory postures (Stern 1971; Zuckerman et al. 1973).

The sciatic area, between the acetabulum and the iliosacral joint, is comparably elongated in *Alouatta*, as in other atelines, and the distribution of trabecular bone is not very dense, as in most quadrupedal primates (Zuckerman et al. 1973; Leutenegger 1974). This condition may be related to the action of tensile forces applied during tail and hind limb hanging postures (Grand 1968b; Zuckerman et al. 1973).

Alouatta possess a large and shallow acetabulum, similar to that found in other atelines and suspensory/climbing mammals, which allows for a wide range of femoral movements (Schultz 1969; Jenkins and Camazine 1977). In addition, it is located rather dorsally and caudally and has equally developed dorsal and ventral buttresses (Youlatos 1994). This morphology very likely reflects resistance to forces applied equally during pronograde quadrupedal actions and hind limb suspension (Fleagle and Simons 1979).

6 Morphology of Howler Monkeys: A Review and Quantitative Analyses

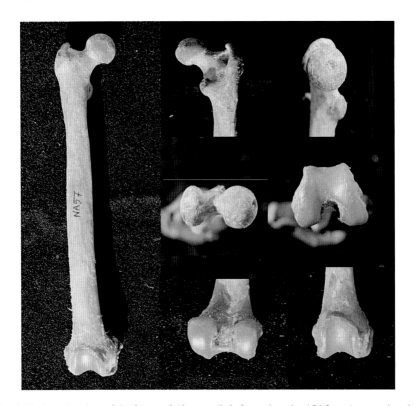

Fig. 6.15 Anterior view of the femur of *Alouatta* (*left*; femur length = 154.2 mm); posterior view of proximal femur showing the head and the greater and lesser trochanters (*top middle*); medial view of the femoral head showing the position of the fovea capitis (*top right*); proximal view of the head and trochanter showing their arrangement (*center middle*); distal view of the knee joint, showing its shape, the depth of the trochlea, and the relative extent and width of the condyles (*center right*); posterior view of the distal femur showing the condyles (*bottom middle*); anterior view of the trochlea showing its shape and its borders (*bottom right*)

The ischium of *Alouatta* is intermediate in length, between the short ischia of the other atelines and the longer ones of quadrupedal platyrrhines. As relatively long ischia provide a better leverage for rapid retraction of the thighs during quadrupedal activities (Smith and Savage 1956), the intermediate condition of *Alouatta* probably reflects slower and more controlled retraction associated with a quadrupedal and climbing repertoire.

The femoral head is more proximally located than the greater trochanter (Fig. 6.15). This condition is encountered in most arboreal mammals and allows a greater range of femoral movements at the hip joint (Dagosto 1983). However, this position is also related to the femoral neck angle which is a variable character among atelines (Ford 1988). In all cases, scansorial and climbing mammals tend to possess higher angles, thus placing the head over the trochanter than more quadrupedal forms (Halaczek 1972; Fleagle 1977; Ford 1988).

The femoral head is semispherical and extends over the neck of the femur (Fig. 6.15). This posterior extension is reduced in *Ateles* and almost lacking in *Lagothrix*, and its expansion in *Alouatta* may be related to its climbing habits as it provides contact between the acetabulum and the head during increased abduction (Jenkins and Camazine 1977; Dagosto 1983). The fovea capitis is large and deep and located either on or near the intersection of the meridian and equator (Fig. 6.15; Schön-Ybarra and Schön 1987; Youlatos 1994). Its morphology and position favor abducted and laterally rotated postures of the hind limb used during climbing and hind limb suspension (Jenkins and Camazine 1977; Rose and Walker 1985).

Alouatta is characterized by a large, deep, and medially oriented trochanteric fossa and a wide intertrochanteric line (Fig. 6.15). This arrangement deviates the tendons of *mm. obturatores* and *gemelli* to an acute angle that favors wide femoral excursions with rotations and abduction (Grand 1968b; Jenkins and Camazine 1977; Schön-Ybarra and Schön 1987; Bacon 1992). Additionally, the increased depth of the fossa enhances the power of the tendons by elongating their lever arm providing controlled and powerful hip rotations (Bacon 1992).

The lesser trochanter is large as in other atelines and is medio-posteriorly directed, as in *Lagothrix* (Fig. 6.15). In contrast, *Ateles* possess a more medially oriented trochanter. This morphology provides a better leverage for the flexion and lateral rotation of the femur by the action of *m. iliopsoas* during climbing activities (Ford 1988, 1990).

The femoral shaft is relatively robust, gently convex anteriorly, and anteroposteriorly compressed (Fig. 6.15; Schön-Ybarra and Schön 1987; Youlatos 1994). This morphology, also shared by the other atelines, is functionally related to the action of forces in various planes, as during climbing and suspensory activities (Halaczek 1972; Godfrey 1988; Ruff 1988; Ruff et al. 1989).

The distal femur of *Alouatta* and the other atelines is mediolaterally wide and oriented slightly laterally (Fig. 6.15; Youlatos 1994). This morphology and deviation is indicative of ample movements in different planes and a more abducted position of the knee joint and is also shared with hominoids and lorisoids (Halaczek 1972; Ciochon and Corruccini 1975; Dagosto 1983; Tardieu 1983). The patellar groove of *Alouatta*, *Ateles*, and *Lagothrix* is relatively deep and wide, similar to that of climbing and suspensory mammals (Fig. 6.15; Schön-Ybarra and Schön 1987; Ford 1988; Youlatos 1994). This morphology allows mediolateral excursions at the knee joint, permitting analogous rotations of the tibia to place the hindfoot on the branches (Rose 1983; Ford 1988; Bacon 1992; Madar et al. 2002).

Both femoral condyles are rather circular in profile implying an extensive range of flexion and extension in contact with the tibial plateau (Rose 1983; Tardieu 1983). However, the medial condyle is slightly longer anteroposteriorly, narrower, and angled medially and posteriorly (Fig. 6.15). This condition reflects the action of stresses through the medial side of the joint and is indicative of mediolateral rotations and a more abducted knee posture (Tardieu 1983; Ford 1988; Bacon 1992; Rafferty and Ruff 1994). Finally, *Alouatta* possesses a very wide and deep intercondylar fossa, a character shared with other climbing primates and mammals. This morphology, in association with the very low intercondylar tibial tubercles, enables

6 Morphology of Howler Monkeys: A Review and Quantitative Analyses 163

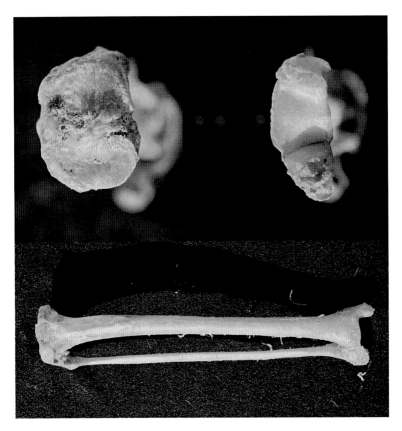

Fig. 6.16 Anterior view of the tibia and fibula of *Alouatta* (left; tibia length = 139.3 mm), proximal view of the rectangular tibial plateau (*top right*), and distal view of the wide talocrural joint and the direction of the medial malleolus (*bottom right*)

mediolateral rotations at the knee joint while keeping the knee flexed (Tardieu 1983; Schön-Ybarra and Schön 1987; Youlatos 1994).

In the tibia, the shape of the plateau is rectangular, similar to arboreal quadrupedal primates and not mediolaterally elongated as in most climbing mammals (Fig. 6.16). The plateau strongly retroflexes and overhangs the shaft posteriorly, a condition indicating habitual flexion of the knee joint (Grand 1968b; Schön-Ybarra and Schön 1987; Youlatos 1994). The tibial facets for the femoral condyles are gently concave with the lateral being longer and narrower than the medial (Fig. 6.16). These features reflect the action of stresses through the medial side of the joint and would favor flexion and medial rotations of the knee (Grand 1968b; Tardieu 1983; Schön-Ybarra and Schön 1987).

Additionally, the anterior tibial tuberosity, insertion of the distal tendon of *mm. vasti* and *rectus femoris*, is relatively reduced and characterized by the presence of dissipated trabecular fibers (Youlatos 1994). This morphology is functionally associated with a controlled role of knee extensors with a reduced lever arm for the tendon's insertion, as well as the presence of multiplanar forces during flexion and extension of the knee joint.

The tibial shaft is straight, relatively robust, and moderately mediolaterally compressed, as in the other atelines (Fig. 6.16). This morphology indicates that major compressive forces act through the medial side of the hind limb and that their action is multidirectional, as encountered during climbing activities (Schön-Ybarra and Schön 1987; Youlatos 1994). The fibula is slender and situated laterally and away from the tibial shaft. This condition is reminiscent of other climbing and hind limb suspensory mammals, providing a wide interosseous membrane for the origin of *mm. flexor digitorum tibialis* and *flexor digitorum fibularis* that flex the pedal digits for powerful hindfoot grasping (Grand 1968b; Youlatos 1994).

The distal tibiofibular facet is low and anteroposteriorly wide as in most platyrrhines (Ford 1988). This condition hosts a synovial articulation that allows a degree of tibiofibular movement that assures a relatively flexible talocrural joint (Ford 1988). The medial malleolus is slightly reduced and fairly medially oriented, similar to the other atelines and most suspensory primates and other mammals (Fig. 6.16; Youlatos 1994; Desilva et al. 2010). Additionally, the anterior surface of the tibiotalar facet is wide and lacks a pronounced distal projection in the middle, as in other atelines (Fig. 6.16). The combination of these features contributes to ample talocrural movements along multiple planes, assuring wide contact with the talus during flexion/extension and inversion/eversion (Ford 1988; Gebo 1989; Harrison 1989; Bacon 1992; DeSilva et al. 2010).

Alouatta has a moderately high talus, higher than *Ateles*, indicating comparably less mobility associated with foot inversion and suspensory habits (Gebo 1986, 1989; Meldrum 1990; Jones 2004). The talar trochlea is large, shallow, and proximodistally wedged as in other atelines. These features provide wide contact of the talus with the tibia throughout flexion, extension, inversion, and eversion of the foot during climbing activities (Dagosto 1983; Langdon 1986; Gebo 1989; Lewis 1989).

The talar neck and head deviate slightly medially, to enable a powerful foot-hold during climbing or pedal suspension against the action of medially directed forces during foot supination (Langdon 1986; Gebo 1989). Furthermore, the talar head exhibits a moderate lateral inclination associated with ample midtarsal movements positioning the navicular and entocuneiform, and thus the hallux, more vertically and enhancing habitual foot supination for powerful foot grasping (Conroy 1976). Powerful grasping is achieved by the leverage of the well-developed *m. flexor digitorum fibularis* (Youlatos 1994), which is evident through the deep groove at the posterior side of the talus and at the plantar surface of the sustentaculum tali of the calcaneus. Additionally, midtarsal supination is further facilitated by the confluent and convex distal talocalcaneal facets on the plantar surface of the head which articulate with the corresponding concave and extended common calcaneal facet on the dorsal surface of the sustentaculum (Dagosto 1983; Ford 1988; Gebo 1989). The proximal talocalcaneal facet is long, wide, and gently curved and associates with a long and equally curved proximal calcaneal facet. This morphology, also shared by the other atelines, facilitates extensive talocalcaneal movements and enhances ankle

Fig. 6.17 Dorsal view of the foot of *A. seniculus* showing the divergent hallux (*bottom*; metatarsal III length = 35.4 mm) and the metatarsal I medial cuneiform joint (*top*)

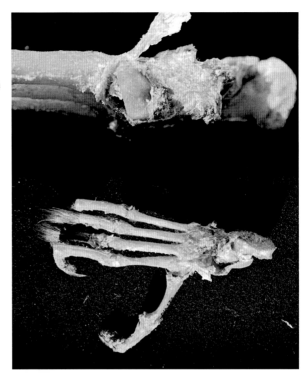

flexibility during pedal grasping in foot hanging postures (Ford 1988; Gebo 1989; Lewis 1989; Meldrum 1990).

The proximal part of the calcaneus is relatively short, as in most medium-sized arboreal primates, and reflects the leverage for the action of the *mm. triceps surae* (Fig. 6.17; Strasser 1988). Their action is further improved by the elongated tuber calcanei, which further increases their insertion angle for increased power and controlled plantar flexion. Furthermore, the well-developed calcaneal tubercle on the plantar side (Fig. 6.17), where *m. flexor digitorum brevis* originates, reflects its powerful action assisting in digital flexion during foot hanging activities (Sarmiento 1983).

On the medial surface of the calcaneal body, the sustentaculum tali is protruding and medially inclined. These features contribute to frequent subtalar inversion and midtarsal supination related to habitual climbing activities (Gebo 1989). On the lateral surface of the calcaneal body, the peroneal tubercle of *Alouatta* is large, compared to that of other atelines, reflecting the important action of *mm. peroneus longus* and *peroneus brevis*, responsible for foot eversion during quadrupedal activities (Langdon 1986).

Distally, the calcaneocuboid facet is semicircular in shape and relatively shallow, with a well-developed lateral surface. It articulates with a mediolaterally wide calcaneocuboid facet with an equally developed lateral surface on the proximal surface of the cuboid. This morphology enables a certain degree of midtarsal excursions that facilitate mediolateral rotations contributing to foot inversion (Langdon 1986; Ford 1988; Strasser 1988; Gebo 1989).

Alouatta and *Ateles* possess a proximodistally short cuboid, which suggests frequent hindfoot reversal associated with hind limb hanging activities (Gebo 1989; Meldrum 1990; Jones 2004). The proximodistally wide and robust navicular reflects the action of compressive forces on the medial side of the foot during climbing and quadrupedal activities (Langdon 1986). In *Alouatta*, and the other atelines, these two bones articulate via a proximodistally extended facet which enables a wide range of movements between them and contributes to overall midtarsal mobility of the foot (Langdon 1986). On the distal surface of the navicular, the facets reserved for the cuneiforms are transversely arranged, in a manner similar to most quadrupedal primates.

The facet between the medial cuneiform and metatarsal I is similar to most platyrrhines. The dorsomedial part is ovoid in shape reflecting an alignment of the hallux in relation to the lateral digits. Additionally, the plantar part is reduced but retains a saddle-shaped facet that stabilizes the articulation during hallucal prehension (Fig. 6.17; Szalay and Dagosto 1988). This grasping performance is further assured by the relatively elongated and robust metatarsal I, which is achieved by well-developed hallucal musculature, of which the adductors and flexors represent almost 70 % (Youlatos 1994).

6.5.3 Vertebral Column

The vertebral column of *Alouatta* is longer than that of the other atelines in both number of vertebrae in the thoracic and lumbar region, as well as in relative length, and seems similar in these metrics to quadrupedal *Cebus* (Fig. 6.18; Erikson 1963; Johnson and Shapiro 1998). This morphology likely provides more flexibility and lengthens the stride during quadrupedal walking but may also contribute to bridging and clambering maneuvers commonly used by both taxa (Slijper 1946).

Alouatta possesses relatively long lumbar vertebral bodies for their size, contrasting with the rather craniocaudally short vertebrae of *Ateles* (Johnson and Shapiro 1998). Furthermore, the transverse processes of these vertebrae are oriented approximately perpendicular to the sagittal plane, a character shared with other atelines (Ankel 1972; Johnson and Shapiro 1998). These features contribute to a relative firmness and resistance to flexing moments and assure a lumbar region that helps support a hanging animal by the tail and hind limbs (Johnson and Shapiro 1998).

The sacral vertebrae of *Alouatta* are large with wide neural arches, whose proximal and distal openings are more or less the same height, in contrast to most nonprehensile-tailed primates. However, in *Ateles* and *Lagothrix*, the distal opening is even longer than the proximal one, a condition metrically expressed by the sacral index and scoring as high as 112–120 in the latter forms, compared to the 94.4 in

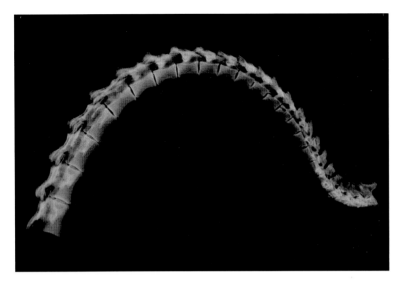

Fig. 6.18 Radiograph of the vertebral column of *Alouatta* showing the different vertebral regions (length of last sacral vertebra = 21.6 mm)

Alouatta (Ankel 1972). This morphology may be associated with the bulkier nerves that provide a fine motor control in the Atelini, compared to that in *Alouatta*, which uses its tail comparatively less (Ankel 1972).

More distally, at the caudal region, *Alouatta* and the other atelines possess several apomorphies that are functionally related to tail prehensility and its extensive use in hanging postures and locomotion. *Alouatta* possesses a very heavy tail, representing ~6 % of total body weight, where caudal vertebrae represent the major component (Grand 1977). Additionally, while atelines have a relatively long tail for their body size, it is least elongate in *Alouatta* (Rosenberger 1983). In effect, howlers have tails that are the same length as their bodies (tail to head body ratio = 0.99 (Youlatos 1994; see also Braza 1980), while there is a gradient to comparably longer tails from *Lagothrix* (1.25) towards *Ateles* (1.38; see also Rosenberger 1983). Interestingly, the relative extension of the naked tactile area shows some variation between atelines. *Alouatta* and *Brachyteles* possess the greatest relative naked areas (0.41 of total tail length) compared to the shorter areas of *Lagothrix* and *Ateles* (0.35 and 0.34, respectively).

In terms of caudal vertebral morphology, *Alouatta* possesses the least number of caudal vertebrae (~25–26) and total caudal vertebral length, compared to *Ateles* (~29–30 (Ankel 1972; German 1982; Youlatos 1994; Organ 2010)). The caudal region is divided into proximal and distal regions, delineated by the transitional vertebra (Ankel 1972). Despite the variability in caudal vertebrae number, all atelines possess a relatively long proximal region characterized by short and high vertebrae articulated by lumbar-like joints that facilitate intrinsic mobility and favor elevation and dorsal flexion of the tail (Ankel 1972). Compared to nonprehensile-tailed primates, the transitional vertebra is located more or less in the same place along tail length in all atelines, indicating a constant relative position (German 1982; Organ 2010). Thus, the relatively long proximal caudal region would favor increased mobility and is associated with its frequent engagement in tail-assisted behaviors (Ankel 1972).

Alouatta, like all atelines and *Cebus*, possess robust caudal vertebrae (German 1982; Lemelin 1995; Organ 2010), a condition also shared by other prehensile-tailed mammals (Youlatos 2003). However, within atelines, *Ateles* has even shorter, more robust vertebrae; this difference is particularly evident in the distal region and more particularly towards the tip of the tail (Ankel 1972; German 1982; Lemelin 1995; Organ 2010). These adaptations reflect the action of frequent and high reaction stresses involving the substrate during extended tail contact and the forces exerted by the powerful *m. flexor caudae longus*.

In effect, *mm. flexor* and *extensor caudae longii* are powerful in prehensile-tailed atelines and span fewer caudal vertebrae (4–7 and 7–9, respectively) compared to nonprehensile-tailed primates. This condition assures greater precision in flexion and extension and more controlled actions (Lemelin 1995). Additionally, the *mm. intertransversarii caudae*, which are the primary lateral tail flexors and rotators, are bulkier in mass and show higher physical cross section area (PCSA) in prehensile-tailed primates, which reflects greater force while maximizing tail contact with the substrate. In *Alouatta*, however, this muscle is not as developed as in other atelines (Lemelin 1995; Organ et al. 2009).

This difference within atelines is skeletally evident in the transverse processes of the caudal vertebrae, whereupon insert *mm. flexor caudae longus* and *intertransversarii caudae*. They are large in atelines compared to those of other platyrrhines, but within the group, *Alouatta* possess the narrowest processes (Ankel 1972; German 1982; Lemelin 1995; Jones 2004). These muscles flex the tail and provide for an enhanced grasp between the distal part of the tail and the substrate while resisting the torsion exerted during tail-assisted locomotion (Lemelin 1995). Their morphology in *Alouatta* may reflect their comparably reduced role in such tail-assisted suspensory activities.

6.6 *Alouatta* Anatomy

The above review of the cranial, dental, hyoid, and postcranial anatomy of *Alouatta* highlights the distinctive morphology of the genus, compared to other atelines and the other platyrrhines as well. The skull of *Alouatta* is quite unique among platyrrhines and reveals strong adaptations to a combination of increased folivory and enhanced sound production. The former is strongly indicated by the arrangement and morphology of the teeth, while the latter is even more prominently evident in the highly modified hyoid bone. Finally, the postcranial skeleton reveals a body plan adapted to above-branch quadrupedal movements with relative arm protraction, abducted forelimbs, and relative stability at the forearm and wrist joints to withstand compressive forces from the substrate. On the other hand, the hind limb reflects adaptations to increased flexion, extension, and abduction, with the capacity for knee rotation and strong flexion and powerful foot grasping. These movements accommodate propulsion as well as a frequent use of suspensory postural behavior, aided by the modifications of the lumbar and caudal parts of the axial skeleton.

This suite of morphological modifications was further substantiated by quantitative analyses of the skull and hyoid bone, which represent the most noticeably derived areas of the skeleton. *Alouatta* is the only extant representative of a group of alouattines that has changed significantly since their divergence from the basal ateline stock at least 15.5 MA, as indicated by molecular and paleontological data (Rosenberger and Strier 1989; Cortés-Ortiz et al. 2003; Hartwig 2005; Jones 2008). This group followed a very distinct evolutionary path from the ancestral stock and the rest of the atelines, which resulted in a very different activity strategy and remarkable morphobehavioral adaptations that currently characterize the genus as a whole. These adaptations are functionally linked to four major behavioral axes: variable but generally increased rates of folivory, enhanced sound production, decreased daily ranges, and emphasis on an above-branch quadrupedal positional repertoire (Rosenberger and Strier 1989; Rosenberger et al. 2011, 2014). These axes are strongly interrelated and very likely evolved in parallel. The simultaneous shift of *Alouatta*'s diet to one focused on leaves, a nutrient-poor food source that imposes important developmental and ontogenetic constraints and also benefits from an energy-minimizing strategy with extended periods of resting for digestion and cautious locomotion, as well as avoidance of costly intergroup violence by the development of an impressive sound production mechanism, probably accounts for the particularly derived skeletal morphology of *Alouatta*. Given the more flexible and ateline-like anatomy and behavior of the last common ancestor of *Alouatta* and the rest of the atelines (Jones 2008; Halenar 2012; Rosenberger et al. 2014), it appears that *Alouatta* evolved a mosaic of cranial and postcranial features that promoted this novel morphobehavioral strategy. Adaptations to these unusual patterns very likely evolved prior to the wide geographic dispersal of howlers in Central and South America and probably contributed to the successful invasion of divergent forested habitats from the northern Mesoamerican wet and dry forests to the seasonal forests of the northern parts of the southern cone (Rosenberger et al. 2009).

Acknowledgments All authors wish to thank the editors for inviting us to participate in this exciting volume, thus giving us the opportunity to contribute our knowledge on specific issues of the morphology of howler monkeys. This work would not have been possible without the funding provided by the Muséum national d'Histoire naturelle and the Centre National de la Recherche Scientifique, France; the Aristotle University of Thessaloniki and the State Scholarship Foundation, Greece; and the National Science Foundation. Advice provided by Jeffrey Laitman, Alfred Rosenberger, and Eloïse Zoukouba is much appreciated by the authors.

References

Adams D, Rohlf F, Slice D (2004) Geometric morphometrics: ten years of progress following the "revolution". Ital J Zool 71:5–16

Anapol F, Fleagle JG (1988) Fossil platyrrhine forelimb bones from the early Miocene of Argentina. Am J Phys Anthropol 76:417–426

Anapol F, Lee L (1994) Morphological adaptation to diet in platyrrhine primates. Am J Phys Anthropol 94:239–261

Ankel F (1972) Vertebral morphology of fossil and extant primates. In: Tuttle RH (ed) The functional and evolutionary biology of primates. Aldine, Chicago

Anthony MRL, Kay RF (1993) Tooth form and diet in ateline and alouattine primates: reflections on the comparative method. Am J Sci 293-A:356–382

Anthony J, Serra O, Serra RG (1949) La surface de la voûte palatine rapportée à la capacité crânienne chez les Singes Platyrhiniens. Bull Mém Soc Anthrop Paris 10:129–145

Bacon AM (1992) Les os longs du membre pelvien chez les primates Miocènes et plio-pleistocènes: morphologie fonctionnelle, taxinomie, phylogénie. PhD Dissertation, Université Paris 7, Paris

Beard KC, Godinot M (1988) Carpal anatomy of *Smilodectes gracilis* (Adapiformes, Primates) and its significance for lemuriform phylogeny. J Hum Evol 17:71–92

Biegert J (1963) The evaluation of characteristics of the skull, hands, and feet for primate taxonomy. In: Washburn S (ed) Classification and human evolution. Aldine, Chicago

Blanco MB, Godfrey LR (2006) Craniofacial sexual dimorphism in *Alouatta palliata*, the mantled howling monkey. J Zool 270:268–276

Bolk L (1915) Ueber lagerung, versehiebung und neigung des foramen magnum am schädel der Primaten. Z Morph Anthropol 17:611–692

Bookstein F (1997) Landmark methods for forms without landmarks: localizing group differences in outline shape. Med Image Anal 1:225–243

Braza F (1980) El araguato rojo (*Alouatta seniculus*). Doñana Acta Vertebr 7:1–175

Bruner E, Mantini S, Manzi G (2004) A geometric morphometric approach to airorhynchy and functional cranial morphology in *Alouatta* (Atelidae, Primates). J Anthropol Sci 82:47–66

Chapman F (1929) My tropical air castle. D. Appleton, New York

Chiarello A (1995) Role of loud calls in brown howlers, *Alouatta fusca*. Am J Primatol 36:213–222

Ciochon RL, Corruccini RS (1975) Morphometric analysis of platyrrhine femora with taxonomic implications and notes on two fossil forms. J Hum Evol 4:193–217

Conroy GC (1976) Primate postcranial remains from the Oligocene of Egypt. Contrib Primatol 8:1–134

Cooke SB (2011) Paleodiet of extinct platyrrhines with emphasis on the Caribbean forms: three-dimensional geometric morphometrics of mandibular second molars. Anat Rec 294: 2073–2091

Cortés-Ortiz L, Bermingham E, Rico C, Rodriguez-Luna E, Sampaio L, Ruiz-Garcia M (2003) Molecular systematics and biogeography of the Neotropical monkey genus, *Alouatta*. Mol Phylogenet Evol 26:64–81

Couette S (2002) Quantitative characterisation of the cranial development of *Cebus apella* and *Alouatta seniculus* (Primates, Platyrrhines). Interest of geometrical morphometric methods. Folia Primatol 73:153

Couette S (2007) Différenciation morphologique et génétique des populations de douroucoulis (*Aotus infulatus*, Primates, Platyrrhiniens, Cebidae) provenant des rives droite et gauche du rio Tocantins (Brésil). C R Biol 330:148–158

Dagosto M (1983) Postcranium of *Adapis parisiensis* and *Leptadapis magnus* (Adapiformes, Primates). Folia Primatol 41:49–101

Dennis JC, Ungar PS, Teaford MF, Glander KE (2004) Dental topography and molar wear in *Alouatta palliata* from Costa Rica. Am J Phys Anthropol 125:152–161

DeSilva JM, Morgan ME, Barry JC, Pilbeam D (2010) A hominoid distal tibia from the Miocene of Pakistan. J Hum Evol 58:147–154

R Development Core Team (2008) R: a language and environment for statistical computing. R Foundation for Statistical Computing, Vienna, Austria. ISBN 3-900051-07-0. http://www.R-project.org

Di Fiore A, Campbell CJ (2007) The atelines: Variation in ecology, behavior, and social organization. In: Campbell CJ, Fuentes A, MacKinnon KC, Panger M, Beader SK (eds) Primates in perspective. Oxford University Press, New York

Drapeau SMS (2004) Functional anatomy of the olecranon process in hominoids and Plio-Pleistocene hominins. Am J Phys Anthropol 124:297–314

Eaglen RH (1984) Incisor size and diet revisited: the view from a platyrrhine perspective. Am J Phys Anthropol 64:263–275

Erikson GE (1963) Brachiation in New World monkeys and in anthropoid apes. Symp Zool Soc Lond 10:135–164

Feldesman MR (1982) Morphometric analysis of the distal humerus of some Cenozoic catarrhines: the late divergence hypothesis revisited. Am J Phys Anthropol 58:73–95

Fleagle JG (1977) Locomotor behavior and skeletal anatomy of sympatric Malaysian leaf monkeys (*Presbytis obscura* and *Presbytis melalophos*). Yearb Phys Anthropol 20:440–453

Fleagle JG, Meldrum DJ (1988) Locomotor behavior and skeletal morphology of two sympatric pithecine monkeys, *Pithecia pithecia* and *Chiropotes satanas*. Am J Primatol 16:227–249

Fleagle JG, Simons EL (1979) Anatomy of the bony pelvis in parapithecid primates. Folia Primatol 31:176–186

Fleagle JG, Simons EL (1982) The humerus of *Aegyptopithecus zeuxis*: a primitive anthropoid. Am J Phys Anthropol 59:175–193

Flores D, Casinos A (2011) Cranial ontogeny and sexual dimorphism in two New World monkeys: *Alouatta caraya* (Atelidae) and *Cebus apella* (Cebidae). J Morphol 272:744–757

Ford SM (1988) Postcranial adaptations of the earliest platyrrhine. J Hum Evol 17:155–192

Ford SM (1990) Locomotor adaptations of fossil platyrrhines. J Hum Evol 19:141–173

Ford SM (1994) Evolution of sexual dimorphism in body weight in platyrrhines. Am J Primatol 34:221–244

Ford SM, Davis LC (1992) Systematics and body size: implications for feeding adaptations in New World monkeys. Am J Phys Anthropol 88:415–468

Gebo DL (1986) Anthropoid origins—the foot evidence. J Hum Evol 15:421–430

Gebo DL (1989) Locomotor and phylogenetic considerations in anthropoid evolution. J Hum Evol 18:201–233

German RZ (1982) The functional morphology of caudal vertebrae in New World monkeys. Am J Phys Anthropol 58:453–459

Godfrey LR (1988) Adaptive diversification of Malagasy strepsirrhines. J Hum Evol 17:93–134

Godfrey LR, Jungers WL (2002) Quaternary fossil lemurs. In: Hartwig W (ed) The primate fossil record. Cambridge University Press, New York

Godinot M (1992) Early euprimate hands in evolutionary perspective. J Hum Evol 22:267–283

Gould SJ (1977) Ontogeny and phylogeny. Belknap, Cambridge

Grand TI (1968a) Functional anatomy of the upper limb. In: Malinow MR (ed) Biology of the howler monkey (*Alouatta caraya*). Karger, Basel

Grand TI (1968b) The functional anatomy of the lower limb of the howler monkey (*Alouatta caraya*). Am J Phys Anthropol 28:163–182

Grand TI (1977) Body weight: its relation to tissue composition, segment distribution, and motor function. I. Interspecific comparisons. Am J Phys Anthropol 47:211–240

Gregorin R (2006) Taxonomy and geographic variation of species of the genus *Alouatta* Lacépède (Primates, Atelidae) in Brazil. Rev Bras Zool 23:64–144

Grew N (1681) Museum regalis societatis or a catalogue and description of the natural and artificial rarities belonging to the Royal Society and preserved at Gresham College. Rawlins W, London

Gunz P, Mitteroecker P, Bookstein F (2005) Semilandmarks in three dimensions. In: Slice D (ed) Modern morphometrics in physical anthropology. Kluwer, Dordrecht

Halaczek B (1972) Die langknochen der hinterextremität bei simichen primate. Juris Druck und Verlag, Zurich

Halenar LB (2011) Reconstructing the locomotor repertoire of *Protopithecus brasiliensis*. II. Forelimb morphology. Anat Rec 294:2048–2063

Halenar LB (2012) Paleobiology of *Protopithecus brasiliensis*, a plus-size Pleistocene platyrrhine from Brazil. PhD dissertation, City University of New York, New York

Halpern I (1987) The relationship between the hyoid and cranial and mandibular morphology in *Alouatta*: a factor analytic approach. PhD dissertation, New York University, New York

Hammer Ø, Harper DAT, Ryan PD (2001) PAST: Paleontological Statistics software package for education and data analysis. Palaeontol Electron 4(1):9

Hamrick MW, Meldrum DJ, Simons EL (1995) Anthropoid phalanges from the Oligocene of Egypt. J Hum Evol 28:121–145

Harcourt-Smith W, Tallman M, Frost SR, Wiley D, Rohlf F, Delson E (2008) Analysis of selected hominoid joint surfaces using laser scanning and geometric morphometrics: a preliminary report. In: Sargis E, Dagosto M (eds) Mammalian evolutionary morphology: a tribute to Frederick S. Szalay. Springer, New York

Harrison T (1989) New postcranial remains of *Victoriapithecus* from the Middle Miocene of Kenya. J Hum Evol 18:3–54

Hartwig W (2005) Implications of molecular and morphological data for understanding ateline phylogeny. Int J Primatol 26:999–1015

Herring SW (1975) Adaptations for gape in the hippopotamus and its relatives. Forma Functio 8:85–100

Hershkovitz P (1949) Mammals of Northern Colombia. Preliminary report n° 4: monkeys (Primates), with taxonomic revisions of some forms. Proc U S Natl Mus 98:323–427

Hewitt G, MacLarnon A, Jones K (2002) The functions of laryngeal air sacs in primates: a new hypothesis. Folia Primatol 73:70–94

Hill W (1962) Primates: comparative anatomy and taxonomy V Cebidae Part B. Edinburgh University Press, Edinburgh

Hilloowala R (1975) Comparative anatomical study of the hyoid apparatus in selected primates. Am J Anat 142:367–384

Höfer H (1954) Die craniocerebrale topographie bei affen und ihre bedeutung fur die menschliche schadelform. Homo 5:52–72

Höfer H (1969) On the evolution of the craniocerebral topography in primates. Ann NY Acad Sci 162:341–356

Howes G (1896) On the mammalian hyoid, with especial reference to that of *Lepus*, *Hyrax*, and *Choloepus*. J Anat Physiol 30:513–526

Hylander WL (1975) Incisor size and diet in anthropoids with special reference to the Cercopithecidae. Science 189:1095–1098

Jenkins FA Jr (1973) The functional anatomy and evolution of the mammalian humeroulnar articulation. Am J Anat 137:281–298

Jenkins FA Jr (1981) Wrist rotation in primates: a critical adaptation for brachiators. Symp Zool Soc Lond 48:429–451

Jenkins FA Jr, Camazine SW (1977) Hip structure and locomotion in ambulatory and cursorial carnivores. J Zool Lond 181:351–370

Jenkins FA Jr, Fleagle JG (1975) Knuckle-walking and the functional anatomy of the wrists in living apes. In: Tuttle RH (ed) Primate Functional Morphology and Evolution. Mouton; The Hague

Johnson SE, Shapiro LJ (1998) Positional behavior and vertebral morphology in atelines and cebines. Am J Phys Anthropol 105:333–354

Jones AL (2004) The evolution of brachiation in atelines: a phylogenetic comparative study. PhD dissertation, University of California, Davis

Jones AL (2008) The evolution of brachiation in ateline primates, ancestral character states and history. Am J Phys Anthropol 137:123–144

Jones AL, Degusta D, Turner S, Campbell CJ, Milton K (2000) Craniometric variation in a population of mantled howler monkeys (*Alouatta palliata*): evidence of size selection in females and growth in dentally mature males. Am J Phys Anthropol 113:411–434

Jouffroy FK, Godinot M, Nakano Y (1991) Biometrical characteristics of primate hands. Hum Evol 6:269–306

Kagaya M (2007) Glenohumeral joint surface characters and its relation to forelimb suspensory behavior in three ateline primates, *Ateles*, *Lagothrix*, and *Alouatta*. Anthropol Sci 115:17–23

Kay RF (1975) The functional adaptation of primate molar teeth. Am J Phys Anthropol 42:195–215

Kay RF (1978) Molar structure and diet in extant Cercopithecidae. In: Butler PM, Joysey KA (eds) Development, function, and evolution of teeth. Academic, London

Kay RF (1984) On the use of anatomical features to infer foraging behavior in extinct primates. In: Rodman PS, Cant JGH (eds) Adaptations for foraging in nonhuman primates: contributions to an organismal biology of prosimians, monkeys and apes. Columbia University Press, New York

Kay RF, Hylander WL (1978) The dental structure of mammalian folivores with special reference to primates and phalangeroids (Marsupialia). In: Montgomery GG (ed) The ecology of arboreal folivores. Smithsonian Institution Press, Washington, DC

Kelemen G, Sade J (1960) The vocal organ of the howling monkey (*Alouatta palliata*). J Morphol 107:123–140

Kivell TL, Begun DR (2009) New primate carpal bones from Rudabanya (late Miocene, Hungary): taxonomic and functional implication. J Hum Evol 57:697–709

Lacépède BGE de (1799) Mémoire sur une nouvelle table méthodique des animaux à mamelles. Mém Inst Nat Sci Art Sci Math Phys Paris 3:469–502

Laitman J, Reidenberg J (2009) Evolution of the human larynx: nature's great experiment. In: Fried M, Ferlito A (eds) The larynx. Plural, San Diego

Langdon JH (1986) Functional morphology of the Miocene hominoid foot. Contrib Primatol 22:1–226

Larson SG (1988) Subscapularis function in gibbons and chimpanzees: implications for interpretation of humeral head torsion in hominoids. Am J Phys Anthropol 76:449–462

Larson SG, Stern JT (1989) The use of supraspinatus in the quadrupedal locomotion of vervets (*Cercopithecus aethiops*): implications for interpretation of humeral morphology. Am J Phys Anthropol 79:369–377

Leche W (1912) Über Beziehungen zwischen Gehirn und Schadel bei den Affen. Zool Jahrb Suppl 15:1–106

Lemelin P (1995) Comparative and functional myology of the prehensile tail in New World monkeys. J Morphol 224:351–368

Leutenegger W (1974) Functional aspects of pelvic morphology in simian primates. J Hum Evol 3:207–222

Lewis OJ (1989) Functional morphology of the evolving hand and foot. Clarendon, Oxford

Lieberman DE, Pearson OM, Mowbray KM (2000) Basicranial influence on overall cranial shape. J Hum Evol 38:291–315

Lönnberg E (1941) Notes on members of the genera *Alouatta* and *Aotus*. Arkiv Zool 33A(10):1–44

Madar SI, Rose MD, Kelley J, MacLatchy L, Pilbeam D (2002) New *Sivapithecus* postcranial remains from the Siwaliks of Pakistan. J Hum Evol 42:705–752

Masterson TJ, Hartwig WC (1998) Degrees of sexual dimorphism in *Cebus* and other New World monkeys. Am J Phys Anthropol 107:243–256

Meldrum DJ (1990) New fossil platyrrhine tali from the early Miocene of Argentina. Am J Phys Anthropol 83:403–418

Miller JR, Begun DR (1998) Facial flexion and craniometric variation in the genus *Alouatta*. Am J Phys Anthropol 105(S26):163

Napier JR (1961) Prehensility and opposability in the hands of primates. Symp Zool Soc Lond 5:115–132

Negus V (1949) The comparative anatomy and physiology of the larynx. Grune and Stratton, New York

Nishimura T (2003) Comparative morphology of the hyo-laryngeal complex in anthropoids: two steps in the evolution of the descent of the larynx. Primates 44:41–49

O'Connor BL (1975) The functional morphology of the cercopithecoid wrist and inferior radioulnar joints and their bearing on some problems in the evolution of Hominoidea. Am J Phys Anthropol 43:113–121

O'Higgins P, Jones N (2006) *morphologika*². Functional Morphology and Evolution Research Group, The Hull York Medical School, Heslington

Organ JM (2010) Structure and function of platyrrhine caudal vertebrae. Anat Rec 293:730–745

Organ JM, Teaford MK, Taylor AB (2009) Functional correlates of fiber architecture of the lateral caudal musculature in prehensile and nonprehensile tails of the platyrrhini (Primates) and Procyonidae (Carnivora). Anat Rec 292:827–841

Orlosky F J (1973) Comparative dental morphology of extant and extinct Cebidae. Ph.D. dissertation, University Microfilms, Ann Arbor

Oxnard CE (1963) Locomotor adaptations in the primate forelimb. Symp Zool Soc Lond 10:165–182

Rafferty J, Ruff C (1994) Articular structure and function in *Hylobates*, *Colobus*, and *Papio*. Am J Phys Anthropol 94:395–408

Ravosa MJ, Ross CF (1994) Craniodental allometry and heterochrony in two howler monkeys: *Alouatta seniculus* and *A. palliata*. Am J Primatol 33:277–299

Rein TR (2011) The correspondence between proximal phalanx morphology and locomotion: implications for inferring the locomotor behavior of fossil catarrhines. Am J Phys Anthropol 146:435–445

Rein TR, Harrison T, Zollikofer CPE (2011) Skeletal correlates of quadrupedalism and climbing in the anthropoid forelimb: implications for inferring locomotion in Miocene catarrhines. J Hum Evol 61:564–574

Roberts D (1974) Structure and function of the primate shoulder. In: Jenkins FA, Jr (ed) Primate Locomotion. Academic Press; New York

Rodman PS (1979) Skeletal differentiation of *Macaca fascicularis* and *Macaca nemestrina*, in relation to arboreal and terrestrial quadrupedalism. Am J Phys Anthropol 51:51–62

Rohlf F, Slice D (1990) Extensions of the Procrustes method for the optimal superimposition of landmarks. Syst Zool 39:40–59

Rose MD (1983) Miocene Hominoid postcranial morphology: monkey-like, ape-like, neither or both? In: Ciochon RL, Corruccini RS (eds) New Interpretations of Ape and Human Ancestry. Plenum Press, New York

Rose KD, Walker A (1985) The skeleton of early Eocene *Cantius*, oldest lemuriform primate. Am J Phys Anthropol 66:73–89

Rose MD (1988) Another look at the anthropoid elbow. J Hum Evol 17:193–224

Rose MD (1992) Kinematics of the trapezium-1st metacarpal joint in extant anthropoids and Miocene hominoids. J Hum Evol 22:255–266

Rosenberger AL (1983) Tale of tails: parallelism and prehensility. Am J Phys Anthropol 60:103–107

Rosenberger AL (1992) Evolution of feeding niches in New World monkeys. Am J Phys Anthropol 88:525–562

Rosenberger AL, Kinzey WG (1976) Functional patterns of molar occlusion in platyrrhine primates. Am J Phys Anthropol 45:281–298

Rosenberger AL, Strier KB (1989) Adaptive radiation of the ateline primates. J Hum Evol 18:717–750

Rosenberger AL, Tejedor MF, Cooke SB, Pekar S (2009) Platyrrhine ecophylogenetics in space and time. In: Garber P (ed) South American primates. Comparative perspectives in the study of behavior, ecology, and conservation. Springer, New York

Rosenberger AL, Halenar L, Cooke SB (2011) The making of platyrrhine semifolivores: models for the evolution of folivory in primates. Anat Rec 294:2112–2130

Rosenberger AL, Cooke SB, Halenar L, Tejedor MF, Hartwig WC, Novo NM, Munoz-Saba Y (2014) Fossil Alouattines and the origin of Alouatta: craniodental diversity and interrelationships. In: Kowalewski M, Garber PA, Cortés-Ortiz L, Urbani B, Youlatos D (eds) Howler monkeys: adaptive radiation, systematics and morphology. Springer, New York

Ross CF, Ravosa MJ (1993) Basicranial flexion, relative brain size, and facial kyphosis in nonhuman primates. Am J Phys Anthropol 91:305–324

Ruff CB (1988) Hindlimb articular surface allometry in Hominoidea and *Macaca*, with comparisons to diaphyseal scaling. J Hum Evol 17:687–714

Ruff CB, Walker AC, Teaford MF (1989) Body mass, sexual dimorphism and femoral proportions of *Proconsul* from Rusinga and Mfangano Islands, Kenya. J Hum Evol 18:515–536

Rusconi C (1935) Sobre morfogenesis basicraneana de algunos primates actuales y fosiles. Rev Argent Paleont Antropol Ameghina 1:3–23

Sarmiento EE (1983) The significance of the heel process in anthropoids. Int J Primatol 4:127–152

Sarmiento EE (1988) Anatomy of the hominoid wrist joint: Its evolutionary and functional implications. Int J Primatol 9:281–345

Schön MA (1968) The muscular system of the red howling monkey. Smithsonian Institution Press, Washington, DC

Schön MA (1971) The anatomy of the resonating mechanism in howling monkeys. Folia Primatol 15:117–132

Schön MA (1986) Loud calls of adult male red howling monkeys (*Alouatta seniculus*). Folia Primatol 47:204–216

Schön-Ybarra MA (1995) A comparative approach to the nonhuman primate vocal tract: implications for sound production. In: Zimmermann E, Newman J, Jurgens U (eds) Current topics in primate vocal communication. Plenum, New York

Schön-Ybarra MA (1998) Arboreal quadrupedism and forelimb articular anatomy of red howling monkeys. Int J Primatol 19:599–613

Schön Ybarra MA, Conroy GC (1978) Non-metric features in the ulna of *Aegyptopithecus, Alouatta, Ateles*, and *Lagothrix*. Folia Primatol 29:178–195

Schön-Ybarra MA, Schön MA III (1987) Positional behavior and limb bone adaptations in red howling monkeys (*Alouatta seniculus*). Folia Primatol 49:70–89

Schultz AH (1941) The relative size of the cranial capacity in primates. Am J Phys Anthropol 28:273–287

Schultz AH (1955) The position of the occipital condyles and of the face relative to the skull base in primates. Am J Phys Anthropol 13:97–120

Schultz AH (1969) Observation on the acetabulum of primates. Folia Primatol 11:181–199

Sekulic R, Chivers D (1986) The significance of call duration in howler monkeys. Int J Primatol 7:183–190

Senyurek MS (1938) Cranial equilibrium index. Am J Phys Anthropol 24:23–41

Shea BT (1985) On aspects of skull form in African apes and orangutans, with implications for hominoid evolution. Am J Phys Anthropol 68:329–342

Slijper E (1946) Comparative biologic-anatomical investigations on the vertebral column and spinal musculature of mammals. Verh K Ned Akad Wet 42:1–128

Smith JM, Savage SJG (1956) Some locomotor adaptations in mammals. J Linn Soc Zool 42:603–622

Stern JT (1971) Functional myology of the hip and thigh of Cebid monkeys and its implications for the evolution of erect posture. Bibl Primatol 14:1–318

Stern JT, Jungers WL, Susman RL (1995) Quantifying phalangeal curvature: an empirical comparison of alternative methods. Am J Phys Anthropol 97:1–10

Strasser E (1988) Pedal evidence for the origin and diversification of cercopithecid clades. J Hum Evol 17:225–245

Szalay FS, Dagosto M (1988) Evolution of hallucial grasping in the primates. J Hum Evol 17:1–33

Tardieu C (1983) L'articulation du genou. CNRS, Paris

Tattersall I (1972) The functional significance of airorhynchy in *Megaladapis*. Folia Primatol 18:20–26

Ungar PS (1990) Incisor microwear and feeding behavior in *Alouatta seniculus* and *Cebus olivaceus*. Am J Primatol 20:43–50

Watanabe T (1982) Mandible/basihyal relationships in red howler monkeys (*Alouatta seniculus*): a craniometrical approach. Primates 23:105–129

Waterman HC (1929) Studies on the evolution of the pelvis of man and other primates. Bull Am Mus Nat Hist 58:585–642

Whitehead J (1995) Vox Alouattinae: a preliminary survey of the acoustic characteristics of long-distance calls of howling monkeys. Int J Primatol 16:121–143

Wiley D (2006) *Landmark editor* 3.0. University of California, Institute for Data Analysis and Visualization, Davis

Youlatos D (1994) Maîtrise de l'espace et accès aux ressources chez le singe hurleur roux (*Alouatta seniculus*) de la Guyane française: étude morpho-fonctionnelle. PhD Dissertation, Muséum national d'Histoire naturelle, Paris

Youlatos D (1996) Atelines, apes and wrist joints. Folia Primatol 67:193–198

Youlatos D (1999) The schizodactylous grasp of the howling monkey. Z Morph Anthropol 82:187–198

Youlatos D (2000) Functional anatomy of forelimb muscles in Guianan atelines (Platyrrhini: Primates). Ann Sci Nat 21:137–151

Youlatos D (2003) Osteological correlates of tail prehensility in carnivorans. J Zool Lond 259:423–430

Ziemer LK (1978) Functional morphology of the forelimb joints in the woolly monkey *Lagothrix lagothricha*. Contrib Primatol 14:1–130

Zuckerman S, Ashton EH, Flinn RM, Oxnard CE, Spence TF (1973) Some locomotor features of the pelvic girdle in primates. Symp Zool Soc Lond 33:71–165

Part III
Physiology

Chapter 7
Hematology and Serum Biochemistry in Wild Howler Monkeys

Domingo Canales-Espinosa, María de Jesús Rovirosa-Hernández, Benoit de Thoisy, Mario Caba, and Francisco García-Orduña

Abstract Hematological and blood biochemistry parameters are valuable tools for determining the health of free-ranging primate populations. However, baseline data on these parameters are needed to discriminate between healthy and unhealthy individuals. This type of information is currently limited for wild primate populations and especially for those that cannot be easily kept in captivity. This is particularly true for howler monkeys. The aim of this chapter is twofold. First, we review the hematological and serum biochemistry values of free-ranging individuals of three howler monkey species, *Alouatta pigra* and *A. palliata* from Mexico and *A. macconnelli* from French Guiana, in order to establish reference values for these species. We also obtain published data for two populations of black and gold howler monkeys (*A. caraya*). Second, we infer the health status of each population highlighting the benefits of blood screening as a tool to evaluate the responses of howler monkeys to the disturbance of their habitats. We found the following patterns: (a) females have higher concentration of white blood cell (WBC) count than males with the exception of *A. caraya*, (b) *A. palliata* and *A. caraya* have higher concentration of WBC count with respect to the other *Alouatta* species, (c) Mexican howler monkeys (*A. palliata* and *A. pigra*) have low total protein concentration with respect to other *Alouatta* species, and d) creatinine concentration is higher in males possibly due to their higher body mass. Overall, the present study will help to monitor blood parameters in threatened wild howler monkey populations as well as in captive individuals.

D. Canales-Espinosa (✉) • M. de Jesús Rovirosa-Hernández • F. García-Orduña
Instituto de Neuroetología, Universidad Veracruzana, Xalapa, Veracruz, Mexico
e-mail: dcanales@uv.mx; jrovirosa@uv.mx; fragarcia@uv.mx

B. de Thoisy
Laboratoire des Interactions Virus Hôtes, Institut Pasteur de la Guyane,
Cayenne, French Guiana
e-mail: bdethoisy@pasteur-cayenne.fr

M. Caba
Centro de Investigaciones Biomédicas, Universidad Veracruzana, Xalapa, Veracruz, Mexico
e-mail: mcaba@uv.mx

Resumen Los parámetros bioquímicos y hematológicos son una herramienta valiosa para determinar el estado de salud de poblaciones de primates en vida libre. Sin embargo, se requiere de datos base que sirvan de referencia para discriminar la calidad de la salud de los distintos individuos en una población. Este tipo de información no está disponible para la mayoría de las especies en vida libre, especialmente para especies que son difíciles de mantener en cautiverio, como es el caso de los monos aulladores. En este capítulo nos propusimos dos objetivos. Primero revisamos los valores hemáticos y bioquímicos de tres poblaciones de monos aulladores en vida libre: *Alouatta pigra* y *A. palliata* de Mexico y *A. macconnelli* de la Guiana Francesa, con la finalidad de establecer valores de referencia para estas especies. También los comparamos con valores publicados para *A. caraya*. Segundo, inferimos el estado de salud de estas poblaciones con el fin de resaltar las ventajas de los análisis hemáticos como una herramienta para evaluar la respuesta de los monos aulladores a la perturbación de su hábitat. Como resultado encontramos los siguientes patrones: (a) las hembras tienen mayor concentración en el conteo de leucocitos que los machos excepto en *A. caraya*, (b) *A. palliata* y *A. caraya* tiene mayor concentración en el conteo de leucocitos con respecto a otras especies de *Alouatta*, (c) los monos aulladores Mexicanos (*A. palliata* y *A. pigra*) tienen baja concentración de proteínas totales con respecto a otras especies de *Alouatta*, y d) los machos tienen una mayor concentración de creatinina lo cual probablemente está relacionado a su mayor masa corporal. En general, el presente estudio ayudará a monitorear patrones hematológicos en especies de monos aulladores amenazadas en su medio ambiente natural así como en cautiverio.

Keywords *Alouatta* • Biochemical parameters • Blood parameters • Health disease • Hematology • Howler biochemistry • Blood parameters

7.1 Introduction

Physiological, nutritional, and pathological conditions of wild individuals are usually evaluated using clinical procedures such as hematological and biochemical analyses (Jain 1986; Bush 1991). This type of information is useful to characterize changes in the health of individuals in response to their natural habitat (de Thoisy and Richard-Hansen 1997; de Thoisy et al. 2001). Information on baseline values for all the variables analyzed using this approach is necessary to serve as reference on these evaluations. These parameters may vary from individual to individual based on age, sex (Martin et al. 1973; Davy et al. 1984; Larsson et al. 1999; Riviello and Wirz 2001; Rovirosa-Hernández et al. 2012), and reproductive status (Rosenblum and Coulston 1981), as well as the levels of hydration (Gotoh et al. 2001), stress (Gotoh et al. 1987; Dinh 2002), nutrition (Pronkuk et al. 1991; Divillard et al. 1992; Klosla and Hayes 1992), and temperature at the moment of sampling (Ogunrinade et al. 1981; Bush 1991; Ogunsanmi et al. 1994; Večerek et al. 2002; Blahová et al. 2007). Hematological and biochemical values in the blood can also be affected by the animal restrain methods (Hassimoto et al. 2004; Lambeth et al.

2006), type of anesthetic used (Lugo-Roman et al. 2009; Rovirosa-Hernández et al. 2011), time of day when sampled, anticoagulant used to preserve the samples, and sample processing and storage (Thrall et al. 2006). Therefore, it is important to obtain reference blood biochemistry in wild primate populations.

For hematological studies, the complete blood count (CBC) is used as a screening procedure to give an overview of the health of the individuals. There are three elements that are regularly analyzed to determine quantitative variations of peripheral blood: erythrocyte or red blood cell (RBC), leukocyte or white blood cells (WBC), and platelet series. RBC includes the parameters hemoglobin (Hb), hematocrit (Hct), mean corpuscular volume (MCV), mean corpuscular hemoglobin (MCH), and mean corpuscular hemoglobin concentration (MCHC). These parameters can reveal disorders such as anemia, hemorrhagic syndromes, and decreased vascular fluid due to dehydration, hemoconcentration, and hypovolemia (Jain 1993; Brockus 2011). WBC includes monocytes, lymphocytes, eosinophil, basophil, segmented neutrophil, and band neutrophil, and these parameters indicate infectious, inflammatory, toxic, and stress conditions (Jain 1993; Webb and Latimer 2011). The platelet series are essential to the coagulation process and for clot retraction (Morrison 1995) and therefore are good indicators of the cellular composition of the blood for health assessments. On the other hand, the serum chemistry profile indicates the function of the organs and systems that are involved in the metabolism (food transformed into energy) and help to determine the amount of electrolytes in the serum (Sirois and Hendrix 2007; Evans 2011).

Despite the importance of hematological parameters to evaluate the health of individuals, the physiological data available for wild primates is limited, particularly for howler monkey species. This is probably due to the difficulties in capturing wild individuals (Vié et al. 1998a). However, a number of conservation strategies throughout the distribution range of howler monkeys have included the rescue and translocation of individuals (Porter 1971; Vié et al. 1998a; de Thoisy et al. 2001; Richard-Hansen et al. 2000; Crissey et al. 2003; Schmidt et al. 2007; Flaiban et al. 2008; Rovirosa-Hernández et al. 2011, 2012), providing researchers with an excellent opportunity to acquire blood samples and analyze hematological parameters of individuals in different contexts.

Howler monkeys, genus *Alouatta*, are the most widely distributed Neotropical primates. Twelve species have been recognized (see Cortés-Ortiz et al. 2014) and most of them are included in some of the categories of the IUCN Red List of threatened (IUCN 2013), mainly due to the loss and alteration of their habitats. In order to understand how these and other anthropogenic activities affect the health of natural populations of howler monkeys, it is important to establish chemical and hematological profiles, to understand the natural variation of those values in wild populations. This would allow the development of population monitoring as part of the conservation strategies aimed at maintaining healthy populations in the wild. In this chapter, we present biochemical and hematological values for four wild populations of howler monkeys, two of *A. pigra*, one of *A. palliata* (subspecies *mexicana*), and one of *A. macconnelli*. We also revise the information reported for *A. caraya* and compared these values across all the species.

The two Mexican species of howler monkeys are distributed in the southeast part of the country. *Alouatta pigra* is distributed in the state of Tabasco, Campeche, Chiapas, Yucatán, and Quintana Roo (Horwich and Johnson 1986; Estrada et al. 2002; Barrueta-Rath et al. 2003; García-Orduña 2005), whereas *A. palliata* inhabits the states of Veracruz, Tabasco, and Chiapas (Horwich and Johnson 1986). In Tabasco both *Alouatta* species are sympatric. In South America, the Guiana red howler monkey (*A. macconnelli*) occurs throughout a large part of the Guiana Shield, north of the Amazonas River, east of the Negro River, and east and south of the Orinoco River (Gregorin 2006). Howler monkeys are considered folivorous-frugivorous or frugivorous-folivorous, depending on the habitat and the season. However, they are extremely selective feeders (Milton 1987), preferring to eat young rather than mature leaves, but also consuming fruits, flowers, petioles, buds, seeds, stems, and twigs (Milton 1980; Crockett and Eisenberg 1987 Among the group of plants that *A. pigra* eat are Leguminosae, Moraceae, and Sapotaceae families (Coyohua 2008), while *A. palliata* eat Moraceae, Lauraceae, and Fabaceae-Mimosoideae families (Estrada and Coates-Estrada 1984; Cristobal-Azkarate and Arroyo-Rodríguez 2007; Dias and Rangel-Negrin 2014). *Alouatta macconnelli* shows a marked preference for fruits of the Sapotaceae and Moraceae, leaves of Fabaceae-Mimosoideae and Caesalpiniaceae families, and flowers of Caesalpiniaceae, Fabaceae, Sapotaceae, and Apocynaceae families (Julliot and Sabatier 1993).

In Mexico, the two species of howler monkeys can be found in highly fragmented habitats, living next to (and within) agricultural plantations (e.g., cacao, Muñoz et al. 2006), and within small human settlements (Estrada et al. 2006; García-Orduña, pers. obs.), as well as in highly disturbed areas with very small fragments (Cristobal-Azkarate et al. 2005; Rivera and Calmé 2006). In those environments, they may even travel on the ground and drink water from rivers and ponds (Bravo and Sallenave 2003; Pozo-Montuy and Serio-Silva 2007). The steady increase in human activities and settlements in natural areas has resulted in the reduction of many populations of howler monkeys. As a consequence, the IUCN has listed *A. pigra* as "endangered" (Marsh et al. 2008) and *A. palliata* (subspecies *mexicana*) as "critically endangered" (Cuarón et al. 2008).

The Guianan red howler monkey is found most often in tall rain forests but also occurs in many other forest types, like riverbank and ridge forests, marshes and swamps, and savanna forests (Muckenhirn et al. 1975; Mittermeier 1977; Norconk et al. 1996). This species is also found in liana forest and mangroves (B. de Thoisy pers. obs.). The species is considered of "least concern" by the IUCN (Boubli et al. 2008), because there are no major range-wide threats. However, Guianan howler monkeys are affected by local hunting (de Thoisy et al. 2005, 2009) and, in some parts of its range, by habitat loss.

One strategy to protect and potentially preserve primate populations threatened in their natural habitat is through translocations (de Vries 1991; Ostro et al. 1999, 2000, 2001). During the last decades, translocation of primate groups and populations have been used for howler monkeys in Mexico (e.g., Aguilar-Cucurachi et al. 2010; Shedden-González and Rodríguez-Luna 2010) and in French Guiana (Vié 1999). During those translocation events and through population monitoring, we collected blood samples of *A. pigra*, *A. palliata*, and *A. macconnelli* in order to determine their biochemical and hematological reference values and to better understand the success of post-release stages (Richard-Hansen et al. 2000).

In this chapter, we present hematologic and biochemical data for these three howler monkey species. The hematological and biochemical analyses of individuals of these species are relevant: (1) to gather baseline data for each species and determine whether there are interspecific differences; (2) to compare the values with other populations of the same species, both free-ranging and in captivity, in order to gain some insights regarding the factors that could affect these parameters; (3) to determine the health status of individuals and populations in order to develop better conservation programs for these primates; and (4) to compile hematological and biochemical information of these three species and compare it with published values of *A. caraya*, another species of the genus.

7.2 Methods

7.2.1 Sites and Data Collection

Individuals of *A. pigra* and *A. palliata* were captured in evergreen forest fragments from Campeche and Tabasco states in Mexico (Fig. 7.1). Sixteen *A. pigra* groups were sampled in the municipalities of Ciudad del Carmen (18°37′16″N, 90°41′11″W)

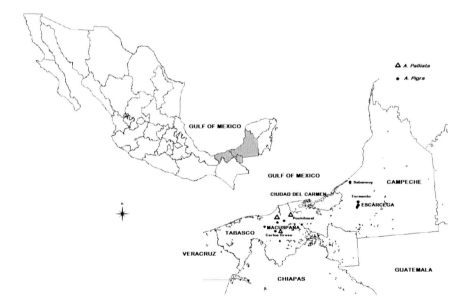

Fig. 7.1 Location of study area and sampled sites for Mexican howler monkeys, at Tabasco and Campeche states (Municipalities of the study in *greay with black dots* indicating sampling sites for *A. pigra* and *A. palliata*)

and Escarcega (18°51′00″ N, 90°43′55 W) and 13 groups in the municipality of Macuspana, Tabasco (17°38′ 14.2″ N, 92°29′ 35″ W). *Alouatta palliata* groups were only sampled in Tabasco, when they are sympatric with populations of *A. pigra* (Smith 1970).

For the present study, we considered only adult subjects, 26 males and 30 females of *A. pigra* from Campeche, 12 males and 14 females of *A. pigra* from Tabasco, and 12 males and 16 females of *A. palliata* from Tabasco. All individuals were immobilized using ketamine hydrochloride (10–15 mg/kg—Inoketam® 1000 Virbac, S.A. Lab. Guadalajara, Jal., Mexico) as anesthetic, following procedures described by Rodríguez-Luna and Cortés-Ortíz (1994). Each individual was weighed in a hanging scale (LightLine® Spring Scales, Forestry Suppliers, Inc. Jackson, MS, USA); its size was registered from the top of the bregma to the base of the tail in a dorsal posture with a flexometer (Truper® EN 8 México, DF). The samples were taken within the first 30 min of immobilization.

Individuals of *A. macconnelli* were captured during the filling of the Petit-Saut hydroelectric dam, along the Sinnamary River, French Guiana (4°45′–05°04′ N, 52°55′–53°15′ W, Fig. 7.2.). Complete groups were captured and transported by boat (trip duration <2 h) to the veterinary facility. After arrival, monkeys were located in a quiet place overnight, with water and fruits provided in the cages. For clinical examination and sampling, animals were anesthetized with an intramuscular combination of medetomidine (Domitor©, Pfizer, France) and ketamine (Ketamine UVA©, UVA, France), with 0.15 and 4 mg/kg, respectively (Vié et al. 1998b). All animals were apparently healthy, although body condition was variable. Sixty percent were in good condition, and 40 % were thin. Seventy-four individuals (26 males and 48 females) were adults and 29 (20 males and 9 females) were juveniles (weights <6 kg for males, <4 kg for females). Nine females were pregnant, and 17 were lactating. During clinical examinations, hemoparasite infection was also carefully investigated on thin blood smear stained with a 4 % May-Grünwald-Giemsa solution (de Thoisy et al. 2000).

We provide a synopsis on blood parameters found in gold and black howler monkeys (*A. caraya*) in two sites: a fragmented forest of Alto Rio Paraná, southern Brazil (Flaiban et al. 2008), and a flooded forest of northern Argentina (Schmidt et al. 2007).

7.2.2 Blood Sampling

Blood samples of *A. pigra* and *A. palliata* were collected by ventromedial venipuncture of the tail (2.5 mL) with a syringe (BD-Vacutainer® 21Gx 32 mm, Mexico City, Mexico). For hematological analyses 0.5 mL of blood was quickly transferred into Microtainer® EDTA tubes (Brand Becton Dickinson™, Franklin Lakes, New Jersey, USA), kept on ice, and analyzed within 8 h of sampling. These samples were quantified with a Coulter Beckman equipment ACT-5-DIF. For serum biochemistry, the remaining 2 mL of blood was deposited in tubes with serum separator

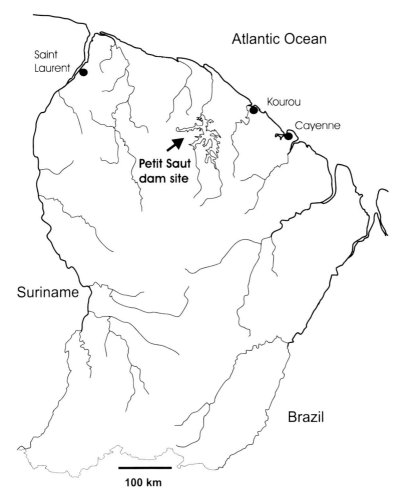

Fig. 7.2 Location of study area for *A. macconnelli* at the Petit-Saut hydroelectric dam, Sinnamary River, French Guiana

(BD-Vacutainer® SST™) and centrifuged at 3,400 rpm for 10 min within 1 h of collection (Cole Parmer®). The serum was collected and placed in clean plastic vials that were transported on ice for less than 8 h and stored at −20 °C for a maximum of 1 month before biochemical analyses were conducted. The serum was quantified with a Johnson & Johnson™ Vitros 250 autoanalyzer. Blood samples from 110 animals were evaluated for 14 hematology and 18 serum biochemistry parameters.

Blood samples of *A. macconnelli* were collected from the femoral vein in both dry and EDTA-filled tubes. The latter were kept at 4 °C for a maximum of 15 h before hematological analyses were conducted. Serum was separated from coagulated blood by centrifugation (10 min; 1,500 rpm; 4 °C) and then preserved in a freezer at −80 °C for a maximum of 6 months before biochemical analyses were conducted. On the basis of the of the quantity of blood available, hematological and biochemical profiles were established for two sample sets, one based on 42 males and 54 females and another based on 40 males and 48 females. An electronic cell counter (STKS Coutronics, France) was used for hematological analyses, although differential WBC counts were made visually on thin blood smears. Chemistry parameters were determined with a Synchron™ CX5 analyzer (Beckman, USA). Blood samples from 103 animals were evaluated for 13 hematology and 22 serum biochemistry parameters.

This study complies with the institutional animal care guidelines of the Mexican and French governments, which follow the principles for the Ethical Treatment of Non-human Primates of the American Society of Primatologists.

7.2.3 Statistical Analyses

Descriptive statistics were applied to obtain the average weight and size of the specimens. The Kolmogorov-Smirnov test was performed to examine the normal distribution of data in each of the parameters of blood count and chemistry by sex, and Mann–Whitney test was used to determine differences of each parameter. The GB-STAT statistical package V 6.0 (Dynamic Microsystems, Inc., Silver Spring, MD, USA) was used for nonparametric datasets. In *A. macconnelli*, data were also analyzed according to the reproductive status ("not in reproduction," "pregnant," or "lactating") of individuals using the Kruskal-Wallis test, and differences related to body condition (good body condition, animals well-muscled, with good hair coats, vs. poor body condition, animals thin and angular, with protruding bones, taut and unpadded skin, and signs of dehydration) were tested with the Mann–Whitney test. In both studies probability levels of $P<0.05$ were considered significant. Results are expressed as means (mean±SD), with the maximum and minimum values for each parameter.

7.3 Results

7.3.1 Physical Parameters

Males of *A. pigra* from Campeche had an average weight of 6.99±0.88 kg and size of 49.76±2.04 cm, while the females had an average weight of 5.05±0.62 kg and size of 44.88±3.04 cm. *A. pigra* males from Tabasco weighed 7.09±2.5 kg and has an average size of 44.14±10.34 cm, while females weighed 5.88±1.5 kg and measured 39.82±5.4 cm. Males of *A. palliata* from Tabasco had a weight average of 5.18±0.874 kg and a size of 41.97±12.08 cm, while females had a weight of 3.94±0.99 kg and a size of 38.39±11.7 cm.

For *A. macconnelli*, weights ranged from 0.80 to 10.06 kg (mean± SD=5.65±2.15 kg). Based on our own body dimension data and dental eruptions, we defined adults as body weights for males and females that range above 6 kg and 4 kg, respectively. Weight and size of males was significantly higher than the weight of females (weight=8.27 kg vs. 5.67 kg; head and body length=63.9 cm vs. 54.8 cm, respectively; both differences were significant, $p<0.0001$, one-tailed t test) (Richard-Hansen et al. 1999). Nine females were found to be pregnant by rectal palpation, and 18 were lactating.

7.3.2 Hematology

Table 7.1 shows the hematological parameters for males and females of both Mexican species. For *A. pigra* individuals from Campeche, counts of hemoglobin ($P=0.006$), hematocrit ($P=0.001$), and RBC ($P\leq0.001$) were higher in males than in females. However, in *A. pigra* individuals from Tabasco, WBC count was higher in females ($P\leq0.034$) than in males, and all other parameters were not significantly different between sexes ($P>0.05$). For *A. palliata*, MCV ($P=0.019$) and MCHC ($P=0.023$) were significantly higher in males than in females. There were no significant differences between sexes in this species for the rest of the parameters.

Table 7.2 shows the hematological parameters of the howler monkeys from French Guiana. Adult females of *A. macconnelli* had lower RBC counts ($P<0.05$), MCV ($P<0.01$), and hemoglobin ($P<0.05$) level than males. MCV was higher in lactating females ($P<0.05$). Adults had lower WBC ($P<0.05$), lymphocyte ($P<0.01$), and platelet counts ($P<0.001$) than juveniles.

7.3.3 Serum Biochemistry

Table 7.3 shows blood chemistry values for *A. pigra* and *A. palliata*. Cholesterol ($P<0.02$) and triglycerides ($P<0.006$) were higher in males than in females in *A. pigra* individuals from Campeche, whereas *A. pigra* individuals from Tabasco showed significantly higher levels of creatinine ($P<0.001$) in males than in females, but females showed higher concentrations of potassium ($P<0.025$) and phosphorous ($P<0.009$) than males. There were no significant differences between sexes in all other parameters. For *A. palliata*, creatinine ($P<0.008$) was also significantly higher in males than in females; globulin ($P<0.003$) was significantly higher in females than in males. There were no significant differences between sexes in all other parameters.

Table 7.4 shows the blood chemistry parameters of *A. macconnelli*. Adult males had higher creatinine ($P<0.05$), cholesterol ($P<0.01$), and calcium ($P<0.05$) levels than females. Juveniles had higher alkaline phosphatase ($P<0.001$) than adults. Cholesterol was lower in adult females than in juveniles ($P<0.05$), and cholesterol was lower in pregnant females ($P<0.005$). In males, creatinine and calcium were lower ($P<0.01$ and $P<0.05$, respectively) in juveniles than in adults.

Table 7.1 Hematological values (mean ± SD) of free-ranging black howler monkeys (*Alouatta pigra*) and mantled howler monkey (*A. palliata*) adults from Campeche and Tabasco states, respectively

Parameters	*Alouatta pigra* Campeche free-ranging adult male (n = 26) Mean ± SE	Max–Min	*Alouatta pigra* Campeche free-ranging adult female (n = 30) Mean ± SE	Max–Min	P	*Alouatta pigra* Tabasco free-ranging adult male (n = 12) Mean ± SE	Max–Min	*Alouatta pigra* Tabasco free-ranging adult female (n = 14) Mean ± SE	Max–Min	P	*Alouatta palliata* Tabasco free-ranging adult male (n = 12) Mean ± SE	Max–Min	*Alouatta palliata* Tabasco free-ranging adult female (n = 16) Mean ± SE	Max–Min	P
Hemoglobin (g/dL)	10.91 ± 1.21	13.4–8.1	10.21 ± 1.78	17.8–7.5	**0.006**[a]	10.89 ± 1.62	13.3–8.3	10.31 ± 0.94	13.3–7.9	0.318	10.24 ± 0.90	13.3–7.9	9.91 ± 0.91	12.3–7.9	0.344
Hematocrit (%)	34.48 ± 3.93	40.5–25.5	30.32 ± 5.75	39.1–7.1	**0.001**[a]	35.18 ± 4.93	41.2–26.5	33.67 ± 3.35	41.2–23.3	0.405	32.50 ± 2.73	41.2–23.3	31.65 ± 2.78	42.2–23.3	0.419
Red blood cell count (× 10⁶/μL)	3.97 ± 0.40	4.8–3.08	3.49 ± 0.64	4.45–0.9	**0.001**[a]	3.91 ± 0.64	4.9–2.81	3.67 ± 0.36	5.17–2.54	0.292	3.61 ± 0.38	5.17–2.54	3.73 ± 0.31	5.17–2.54	0.378
Mean corpuscular volume (fL)	84.43 ± 12.90	97.5–27.5	86.39 ± 4.30	97.5–77.8	0.837	90.02 ± 5.57	100–67.7	93.17 ± 7.67	100–69.5	0.270	89.69 ± 4.10	100–67.7	84.99 ± 5.53	100–65.5	0.019[a]
Mean corpuscular hemoglobin (pg)	29.7 ± 11.26	84.5–23.4	28.11 ± 2.11	36.4–25.7	0.973	27.88 ± 1.33	34.3–21.3	29.20 ± 2.42	37.3–23.3	0.118	28.24 ± 1.68	34.3–21.3	26.60 ± 1.78	34.0–20.3	0.023[a]
Mean corpuscular hemoglobin concentration (g/dL)	31.69 ± 1.33	34.4–30	31.92 ± 1.17	33.9–30	0.426	30.96 ± 1.38	33.0–24.5	30.45 ± 1.14	33.7–24.4	0.347	31.52 ± 0.78	33.7–24.5	31.2 ± 0.88	31.7–24.0	0.315
Platelet (× 10³/μL)	150.88 ± 67.21	319–43	132.32 ± 84.12	310–20	0.538	236.41 ± 105.31	399.0–159.9	237.09 ± 57.90	399–158	0.985	219.66 ± 41.48	399–158.9	193.58 ± 69.46	369–150	0.257
White blood cell (WBC) count (× 10³/μL)	6.29 ± 1.68	10.6–3.1	7.95 ± 2.45	15.9–3.3	0.203	6.04 ± 1.32	24.1–5.4	7.57 ± 1.88	24.9–3.2	**0.034**[b]	12.49 ± 3.04	24.1–3.2	12.49 ± 4.08	24.1–4.2	0.834
Monocyte % of WBC	9.6 ± 12.20	21–3	12.67 ± 11.60	26.4–4.1	0.201	5.75 ± 4.04	10–3	5.63 ± 3.50	10–1	0.975	5.66 ± 3.42	10–3	4.88 ± 3.42	10–1	0.549
Lymphocyte % of WBC	38.63 ± 19.65	75–13	34.96 ± 21.82	1.43–12	0.440	30.83 ± 9.0	75–15	38.18 ± 13.0	55.0–12	0.128	28.83 ± 13.59	55–15	30.81 ± 10.23	55.0–12	0.663
Eosinophil % of WBC	0.39 ± 0.09	0.50–0.03	0.32 ± 0.06	0.49–0.03	0.440	0.66 ± 0.37	1.6–0.01	1.63 ± 0.33	2.5–0.01	0.183	1.41 ± 0.50	1.82–0.01	1.35 ± 0.83	2.5–0.03	0.726
Basophil % of WBC	0.62 ± 1.07	1.76–0.4	0.49 ± 0.09	1.79–0.3	0.838	0.5 ± 0.11	0.70–0.01	0.57 ± 0.09	0.71–0.01	0.432	0.05 ± 0.09	0.70–0.01	0.0 ± 0.0	0	0.268
Segmented neutrophil % of WBC	20.57 ± 18.76	52–5	18.97 ± 18.38	49.0–5.2	0.691	61.91 ± 10.72	82–23	52.72 ± 14.87	72–18	0.102	58.16 ± 22.11	72–20	62.94 ± 9.80	72–18	0.440
Band neutrophil % of WBC	0.51 ± 1.37	1.3–0.56	0.60 ± 1.3	1.19–0.52	0.313	0.83 ± 1.4	1.4–0.50	0.27 ± 0.9	0.89–0.5	0.201	0.55 ± 1.66	1.3–0.50	0.66 ± 1.28	1.89–0.5	0.463

[a]Significant difference between adult males and adult females
[b]Significant difference between adult females and adult males

Table 7.2 Hematological values (mean±SD) of free-ranging Guiana red howler monkeys (*Alouatta macconnelli*) from French Guiana

Parameters	*Alouatta macconnelli* French Guiana free-ranging male adult (n=23) Mean±SD	Max–Min	*Alouatta macconnelli* French Guiana free-ranging female adult (n=46) Mean±SD	Max–Min	P	*Alouatta macconnelli* French Guiana free-ranging male juvenile (n=19) Mean±SD	Max–Min	*Alouatta macconnelli* French Guiana free-ranging female juvenile (n=8) Mean±SD	Max–Min	P
Hemoglobin (g/dL)	10.90±0.40	13.1–8.9	10.11±0.21	13.6–7.4	**0.020**[a]	10.30±0.42	11.9–83	10.00±0.48	11.4–8.4	0.213
Hematocrit (%)	34.80±0.95	41.6–27.2	31.70±1.05	45.5–20.1	0.100	33.00±1.12	38.5–27.5	31.60±2.15	36.0–26.6	0.600
Red blood cell count (×10⁶/μL)	4.40±0.15	5.6–3.3	4.00±0.10	6.6–2.8	**0.049**[a]	4.30±0.12	5.5–3.4	4.10±0.10	4.6–3.5	0.251
Mean corpuscular volume (fL)	79.10±2.10	94.7–69.1	80.30±1.10	93.6–70.8	0.007[b]	77.60±1.45	83.4–66.4	77.30±3.50	85.6–69.3	0.256
Mean corpuscular hemoglobin (pg)	24.9±0.60	29.5–22.0	25.40±0.30	31.9–21.0	0.945	24.10±0.35	26.0–23.0	24.60±0.75	26.1–21.9	0.114
Mean corpuscular hemoglobin concentration (g/dL)	31.50±0.25	33.0–29.5	31.60±0.15	34.3–29.1	0.342	31.00±0.32	32.9–27.8	31.80±0.60	34.1–30.4	0.412
Platelet (×10³/μL)	146.00±25.00	273–22	137.00±12.00	290–21	0.231	221.00±29.00	385–22	195.00±12.00	222.0–162.0	**0.001**[c]
White blood cell count (WBC) (×10³/μL)	8.60±1.80	22.6–3.0	8.9±0.45	14.9–2.9	0.871	11.50±2.00	21.3–40	12.30±2.8	18.9–4.8	**0.03**[c]
Monocyte % of WBC	5.4±2.0	20–0	4.30±1.11	19–0.0	0.255	4.20±1.15	11.0–0.0	5.30±1.60	10.0–1.0	**0.543**
Lymphocyte % of WBC	35.90±7.00	82–2	38.10±5.12	77–10	0.821	42.90±7.10	78.0–14.0	46.30±9.30	72.0–18.0	**0.005**[c]
Eosinophil % of WBC	0.70±0.36	5.0–0.0	1.20±0.95	25–0	0.321	1.90±1.10	13.0–0.0	0.60±0.25	2.0–0.0	0.562
Basophil % of WBC	0	0	0.30±0.08	3.0–0.0	0.346	0.10±0.08	1.0–0.0	0	0	0.231
Neutrophil % of WBC	57.90±7.2	92–12	55.20±5.4	84–8	0.212	50.90±8.10	85.0–10.0	47.60±9.50	76.0–26.0	0.126

[a]Significant difference between adult males and adult females
[b]Significant difference between adult females and adult males
[c]Significant difference between juveniles and adults

Table 7.3 Blood biochemical values (mean ± SD) of free-ranging males and females black howler monkeys (*A. pigra*) and mantled howler monkey (*A. palliata*) from Campeche and Tabasco states

Parameters	*Alouatta pigra* Campeche free-ranging adult male (n=26) Mean±SE	Max–Min	*Alouatta pigra* Campeche free-ranging adult female (n=30) Mean±SE	Max–Min	P	*Alouatta pigra* Tabasco free-ranging adult male (n=12) Mean±SE	Max–Min	*Alouatta pigra* Tabasco free-ranging adult female (n=14) Mean±SE	Max–Min	P	*Alouatta palliata* Tabasco free-ranging adult male (n=12) Mean±SE	Max–Min	*Alouatta palliata* Tabasco free-ranging adult female (n=16) Mean±SE	Max–Min	P
Glucose (mmol/L)	5.43±0.38	8.65–3.33	4.73±0.20	6.27–3.21	0.090	4.29±1.0	8.76–3.10	4.39±1.46	8.76–3.10	0.863	5.19±1.55	6.43–2.16	5.75±1.46	6.76–2.10	0.346
Blood urea nitrogen (mmol/L)	5.76±0.67	9.63–1.42	4.92±0.46	8.92–0.71	0.301	5.3±1.42	12.5–3.92	5.6±1.98	8.56–3.92	0.698	4.64±2.72	8.55–3.9	4.17±2.73	8.56–3.92	0.736
Urea (mg/dL)	34.42±5.73	57.0–8.6	29.47±2.77	53.3–0.72	0.307	31.80±8.60	74.9–23.5	33.63±11.95	51.4–32.1	0.701	27.69±16.23	51.4–23.5	24.98±16.30	51.4–32.1	0.721
Total protein (g/L)	65.42±3.42	78–40	67.37±1.85	84–74	0.589	74.5±7.0	82–52	75.9±4.8	84–53	0.605	74.4±7.5	80–56	78.8±5.0	84–53	0.066
Albumin (g/dL)	4.05±.27	5.4–2.0	4.06±0.17	5.4–2.3	0.986	4.41±0.36	6.9–2.9	4.55±0.32	5.7–2.8	0.366	4.55±0.62	5.6–3.7	4.60±0.43	5.6–3.5	0.599
Globulin (g/dL)	2.48±0.13	3.5–1.7	2.6±0.11	4.2–1.8	0.292	3.04±0.50	3.6–1.7	3.04±0.35	3.49–1.16	0.988	2.89±0.32	3.2–1.7	3.28±0.31	3.94–1.7	0.003[a]
Albumin/globulin relation (mg/dL)	1.59±0.12	2.25–0.86	1.56±0.10	5.3–2.3	0.870	1.47±0.24	2.41–0.94	1.50±0.20	2.41–0.94	0.746	1.52±0.29	2.41–1.32	1.40±0.18	1.96–1.32	0.155
Creatinine μmol/L	80.44±5.30	114.9–44.2	68.06±3.53	114.9–35.36	0.061	107.84±11.49	114.9–35.36	74.25±11.49	114.9–44.2	0.001[a]	88.4±15.2	168.0–84.0	73.37±10.60	106.1–44.2	0.008[a]
Cholesterol (mmol/L)	3.03±0.20	4.94–1.91	2.31±0.19	4.48–1.16	0.021[a]	2.54±0.61	3.49–1.16	2.34±0.70	3.48–1.16	0.490	2.69±0.72	4.35–2.17	2.74±1.12	4.35–2.20	0.753
Triglyceride (mmol/L)	1.08±0.09	1.76–0.57	0.77±0.06	1.43–0.40	0.006[a]	1.42±0.63	1.79–0.70	1.34±0.51	1.79–0.70	0.754	1.7±0.61	1.96–0.99	1.78±0.70	2.04–0.80	0.952
Alkaline phosphatase (U/L)	399.85±66.04	904.0–113.0	424.66±46.22	805.0–72.0	0.754	387.66±400.09	1,700.0–120.0	371.2±226.	858.0–198.0	0.348	274.46±181.34	700.0–125.0	256.87±90.64	575.0–125.0	0.539
Calcium (mmol/L)	2.39±0.07	2.76–1.62	2.42±0.06	2.79–1.79	0.773	2.46±0.14	2.57–1.87	2.5±.06	2.97–1.8	0.318	2.46±0.15	2.81–2.24	2.5±.12	2.97–1.8	0.274
Phosphorous (mmol/L)	0.98±0.09	1.6–0.45	1.09±0.24	1.9–0.58	0.361	1.02±0.58	2.51–0.71	2.7±4.02	4.1–1.5	0.009[a]	1.06±.63	2.0–0.64	0.82±0.66	2.0–0.64	0.252
Magnesium (μmol/L)	0.84±0.04	1.06–0.53	0.90±0.03	1.19–0.65	0.313	0.87±0.47	1.08–0.53	0.97±0.14	1.09–0.55	0.285	0.73±.98	1.06–0.52	0.49±0.57	1.03–0.43	0.585
Chloride (mmol/L)	92.92±3.42	108–66.0	99.54±1.96	114.0–79.0	0.079	97.77±5.23	112.0–94.0	96.41±3.47	100.0–93.0	0.482	101.23±5.37	105.0–92.0	102.05±4.57	105.9–92.0	0.652
Potassium (mmol/L)	5.42±0.28	7.8–3.8	5.45±0.14	7.0–4.4	0.94	4.78±0.48	6.4–4.1	5.89±1.01	6.0–4.1	0.025[b]	5.37±1.05	7.7–3.9	5.45±0.91	7.8–3.9	0.867
Sodium (mmol/L)	134.35±5.60	158.0–92.0	141.29±2.93	163.0–104.0	0.234	140.66±8.20	148.0–134.0	142.91±3.47	160.9–134.0	0.476	144.76±6.69	155.0–130.0	140.17±6.40	153.0–130.0	0.067
Iron (μmol/L)	15.10±1.00	19.69–9.48	17.55±1.04	27.57–8.41	0.126	–	–	–	–	–	–	–	–	–	–

[a]Significant difference between adult males and adult females
[b]Significant difference between adult females and adult males

Table 7.4 Blood biochemistry values (mean ± SD) of free-ranging Guiana red howler monkeys (*A. maccornelli*) from French Guiana

Parameters	*Alouatta maccornelli* French Guiana, free-ranging male adult (n = 23) Mean ± SD	Max–Min	*Alouatta maccornelli* French Guiana, free-ranging female adult (n = 44) Mean ± SD	Max–Min	P	*Alouatta maccornelli* French Guiana, free-ranging juvenile male (n = 17) Mean ± SD	Max–Min	*Alouatta maccornelli* French Guiana, free-ranging juvenile female (n = 4) Mean ± SD	Max–Min	P
Glucose (mmol/L)	5.90 ± 1.60	24.3–0.6	5.60 ± 0.70	19.0–0.3	0.091	6.40 ± 1.80	19.1–1.1	7.50 ± 2.25	10.5–2.3	0.663
Blood urea nitrogen (mmol/L)	9.40 ± 1.80	26.3–1.4	9.00 ± 0.85	19.3–1.9	0.310	7.00 ± 1.85	12.7–2.6	6.00 ± 0.60	7.9–4.5	0.650
Total protein (g/L)	96.50 ± 5.45	126–75	86.90 ± 3.00	123–62	0.085	85.20 ± 3.00	97–75	90.20 ± 8.90	103–71	0.605
Creatinine µmol/L	140.00 ± 8.00	202–97	119.00 ± 4.50	181–79	0.041[a]	93.00 ± 11.00	154–53	94.00 ± 18.00	140–69	**0.015**[c]
Cholesterol (mmol/L)	3.10 ± 0.15	5.7–1.6	2.40 ± 0.08	4.3–1.1	0.021[a]	2.90 ± 0.42	5.2–0.9	4.30 ± 0.45	5.3–3.7	0.060
Triglyceride (mmol/L)	0.70 ± 0.06	1.8–0.2	0.80 ± 0.07	1.3–0.2	0.065	0.80 ± 0.09	1.6–0.2	1.40 ± 0.75	3.0–0.7	0.095
Alkaline phosphatase (U/L)	111.00 ± 38.00	502–58	86.00 ± 22.00	639–15	0.068	231.00 ± 63.00	617–80	279.00 ± 82.00	437–156	**0.001**[b]
Calcium (mmol/L)	2.70 ± 0.06	3.5–2.3	2.50 ± 0.75	3.2–1.9	0.021[a]	2.50 ± 0.06	3.1–1.5	2.70 ± 0.05	3.0–2.5	**0.004**[c]
Phosphorous (mmol/L)	2.60 ± 0.40	6.0–1.0	2.50 ± 0.22	5.7–0.4	0.321	2.60 ± 0.15	4.5–1.7	2.40 ± 0.53	3.4–1.3	0.093
Magnesium (µmol/L)	1.50 ± 0.12	1.9–1.3	1.40 ± 0.12	2.1–1.0	0.113	1.50 ± 0.20	1.3–1.4	–	–	0.285
Chloride (mmol/L)	110.00 ± 3.00	129–90	105.00 ± 4.15	134–11	0.079	106.00 ± 3.52	125–93	110.00 ± 7.50	128–93	0.482
Potassium (mmol/L)	55.00 ± 0.40	8.2–3.8	5.10 ± 0.21	7.9–3.8	0.281	5.10 ± 0.45	7.4–3.8	5.30 ± 0.69	71–4.5	0.025
Sodium (mmol/L)	156.00 ± 5.00	187–130	149.00 ± 3.20	174–132	0.076	149.00 ± 0.40	175–130	155.00 ± 11.50	182–135	0.476
Iron (µmol/L)	13.40 ± 2.10	25.8–3.7	15.30 ± 2.15	41.1–4.0	0.056	11.80 ± 2.20	25.0–3.7	16.60 ± 3.55	21.0–9.1	0.065

[a]Significant difference between adult males and adult females
[b]Significant difference between adults and juveniles
[c]Significant difference between adult males and juvenile males

7.3.4 Relations Between Health Status, Infections, and Hematological Parameters in A. macconnelli

In *A. macconnelli*, due to the flooding and subsequent habitat degradation (de Thoisy and Richard-Hansen 1997) a few months before the collection of samples, some animals were in poor health condition. Thin animals had lower basophil ($P<0.00005$) and monocyte counts ($P<0.00001$) than animals in good condition. Animals with poor body condition also had lower total protein ($P=0.02$), sodium ($P<0.005$), chloride ($P=0.02$), and lactate dehydrogenase ($P<0.05$) but higher levels of creatinine ($P<0.05$) than animals in good condition. We found high prevalence of *Plasmodium brasilianum* (18.4 %), *Trypanosoma cruzi* like (16 %), and microfilaria (*Mansonella, Dipetalonema*, global prevalence = 21 %); however, this infection had no significant effect on the hematological profiles (all paired *t*-test comparisons, for each parameter, between infected and noninfected animals, were not significant).

7.3.5 Comparing Blood Parameters: The Case of Alouatta caraya

Table 7.5 shows the hematological parameters for adult males and females, as well as adults and juveniles of Brazilian *A. caraya* (Flaiban et al. 2008). The count of hematocrit ($P \leq 0.05$) was higher in adult females than in males. However, adult males had higher count of MCV than females ($P \leq 0.05$). Adults showed higher values in MCV with respect to juveniles ($P \leq 0.05$). Eosinophil count was higher in adult males than in females ($P \leq 0.05$). On the other hand, Table 7.6 shows the blood chemistry parameters for Argentinean *A. caraya* (Schmidt et al. 2007). The adults showed higher values in total protein ($P \leq 0.05$) with respect to juvenile. The HDL cholesterol and copper concentrations were significantly different between adult males and females ($P \leq 0.05$).

7.4 Discussion

There is limited information documenting hematological and chemical values to allow a clinical diagnosis of the health status in populations of wild Neotropical primates. The current study is therefore important to assess those parameters in wild populations of the endangered *A. pigra* and *A. palliata*, inhabiting forest fragments in southern Mexico and *A. macconnelli* from French Guiana.

In terms of length and weight, as expected based on previous studies (e.g., Schmidt et al. 2007; Rovirosa-Hernández et al. 2012), we found that males are larger and heavier than females in the three studied species. In relation to hematological parameters between sexes, we found significantly lower values in hemoglobin, hematocrit, and

Table 7.5 Hamatological values (mean ± SD) of free-ranging adult females, adult males, and juvenile black and gold howler monkeys (*A. caraya*) from Alto Rio Paraná, south of Brazil (Flaiban et al. 2008)

Parameters	*Alouatta caraya* Alto Rio Paraná free-ranging male adult (n = 20) Mean ± SD	*Alouatta caraya* Alto Rio Paraná free-ranging female adult (n = 16) Mean ± SD	P	*Alouatta caraya* Alto Rio Paraná Free-ranging adults (n = 23) Mean ± SD	*Alouatta caraya* Alto Rio Paraná free-ranging juvenile (n = 13) Mean ± SD	P
Hemoglobin (g/dL)	9.96 ± 2.77	11.18 ± 3.82	NS	10.66 ± 3.29	10.22 ± 3.40	NS
Hematocrit (%)	39.00 ± 3.49	37.78 ± 4.03	NS	39.46 ± 3.53	36.69 ± 3.54	**<0.05**[c]
Red blood cell count (×10⁶/μL)	3.58 ± 0.91	4.48 ± 1.36	**<0.05**[a]	4.15 ± 1.36	3.67 ± 0.73	NS
Mean corpuscular volume (fL)	118.36 ± 44.63	90.99 ± 27.65	**<0.05**[b]	107.21 ± 46.07	104.40 ± 27.71	NS
Mean corpuscular hemoglobin (pg)	30.06 ± 12.66	25.58 ± 7.65	NS	27.44 ± 10.08	29.18 ± 8.62	NS
Mean corpuscular hemoglobin concentration (g/dL)	25.42 ± 6.34	29.40 ± 9.03	NS	26.91 ± 7.48	27.69 ± 8.62	NS
Platelet (×10³/μL)	ND	ND	NS	ND	ND	NS
White blood cell (WBC) count (×10³/μL)	12.17 ± 7.70	10.18 ± 6.47	NS	11.33 ± 7.47	11.21 ± 6.84	NS
Monocyte % of WBC	ND	ND	NS	ND	ND	NS
Lymphocyte % of WBC	31.38 ± 2.55	27.79 ± 2.21	NS	28.42 ± 2.50	32.64 ± 2.48	NS
Eosinophil % of WBC	4.74 ± 1.10	2.94 ± 0.37	**<0.05**[b]	5.29 ± 1.04	4.28 ± 0.65	NS
Basophil % of WBC	ND	ND	NS	ND	ND	NS
Neutrophil % of WBC	61.05 ± 7.58	68.56 ± 9.14	NS	65.04 ± 7.10	62.26 ± 2.48	NS

[a]Significant difference between adult females and adult males
[b]Significant difference between adult males and adult females
[c]Significant difference between adults and juvenile

Table 7.6 Biochemical blood values (mean ± SD) of free-ranging adult females, adult males, and juvenile black and gold howler monkeys (*A. caraya*) from San Cayetano, Province of Corrientes, in northern Argentina (from Schmidt et al. 2007)

Parameters	*Alouatta caraya* (Flaiban et al. 2008) Alto Rio Paraná free-ranging male adult (n=20) Mean±SD	*Alouatta caraya* (Flaiban et al. 2008) Alto Rio Paraná free-ranging female adult (n=16) Mean±SD	P	*Alouatta caraya* (Flaiban et al. 2008) Alto Rio Paraná free-ranging adults (n=23) Mean±SD	*Alouatta caraya* (Flaiban et al. 2008) Alto Rio Paraná free-ranging juvenile (n=13) Mean±SD	P	*Alouatta caraya* (Schmidt et al. 2007) Male (n=13) free-ranging Mean±SEM	*Alouatta caraya* (Schmidt et al. 2007) Female (n=13) free-ranging Mean±SEM	P
Total protein (mg/L)	75.70±6.4	79.10±5.4	NS	79.10±5.3	74.00±6.3	<0.05[a]	ND	ND	
Cholesterol (mmol/L)							4.9±0.4	4.0±0.3	
Triglyceride (mmol/L)							1.1±0.1	1.1±0.1	
HDL-cholesterol (mmol/L)							0.03±.002	0.02±.002	**<0.05**[b,c]
LDL-cholesterol (mmol/L)							0.08±.005	0.06±.007	
Calcium (mmol/L)							2.3±.05	2.3±.05	
Phosphorous (mmol/L)							3.8±.2	3.7±0.2	
Magnesium (μmol/L)							0.8±0.02	0.8±0.04	
Potassium (mmol/L)							4.2±.2	4.5±0.2	
Sodium (mmol/L)							134.5±.6	136.8±2.1	
Iron (μmol/L)							50.1±7.2	50.1±5.4	
Copper (μmol/L)							14.2±1.6	22.0±3.1	**<0.05**[b,c]
Zinc (mmol/L)							16.8±1.5	16.8±1.5	

[a]Significant difference between adults and juvenile
[b,c]Means in the same column within the current study with different superscripts are significantly different

RBC count in females of the three species as compared to males. These same differences have been reported for other Neotropical primates such as *Sapajus apella* (Larsson et al. 1999; Riviello and Wirz 2001; Núñez et al. 2008; Wirz et al. 2008). Although not significant, a similar trend was also observed in *Cebus capucinus* and *A. palliata* in captivity (Porter 1971). It has been suggested that these hematological differences by sex might be in accordance with sexual dimorphism as related to the larger muscular mass of adult males (Larsson et al. 1999). However, contrary to this trend, *A. caraya* females in Brazil showed higher RBC count than males (Flaiban et al. 2008). Flaiban et al. (2008) argue that these differences could be due to habitat degradation.

In *A. palliata* from Tabasco, we found significantly higher values of MCV and MCH in males than in females. In *A. caraya*, MCV is also significantly higher in males than in females (Flaiban et al. 2008). We did not find a similar tendency in *A. pigra* or *A. macconnelli*; so it is likely that those differences are not related to the sexual dimorphism of howlers but rather to the age of the individuals. These parameters vary between juveniles, adults, and older adults (Morrison 1995) and likely among the sampled individuals with older adults. This age-related difference can be observed when we compared adults and juveniles in *A. macconnelli* (present study) and is also confirmed in *A. caraya* (Flaiban et al. 2008).

The WBC counts in *A. pigra* from Tabasco were significantly higher in females than in males. It is possible that this difference may indicate an infection, as the percentages of lymphocytes and eosinophils were also higher (even though they did not reach statistical significance), suggesting a possible helminthic infection (Webb and Latimer 2011). This could also occur in males and females of *A. caraya*, which are reported to have a higher leukocyte count (not significant) and a significant increase in eosinophil count (Flaiban et al. 2008). However, these authors suggest that this may be due to the condition of the environment and the capturing process in their study. Other factors that can also affect the WBC counts are the social (Alexander 1974; Freeland 1976; Moller et al. 1993; Dobson and Meagher 1996) and ecological contexts (Hausfater and Meade 1982), as well as sexual contacts (Cates and Meheus 1990). Nevertheless a similar tendency in WBC counts was reported previously in *A. pigra* in a different location (Rovirosa-Hernández et al. 2012), and a similar trend is observed in *A. macconnelli* (present study). On this basis, we suggest that the increase in WBC counts in females is probably a characteristic of the genus *Alouatta*.

Alouatta palliata from Tabasco did not show any difference between females and males in WBC count. Moreover, mean values of WBC counts are considerably higher in this species with respect to *A. pigra* and those reported in other *Alouatta* species (Porter 1971; Vié et al. 1998b; de Thoisy et al. 2001; Flaiban et al. 2008), as well as other species of New World monkeys like *Sapajus apella* (Riviello and Wirz 2001; Núñez et al. 2008; Wirz et al. 2008), *Pithecia pithecia*, and *Saguinus midas* (de Thoisy et al. 2001) and *Saguinus leucopus* (Fox et al. 2008). Thus, we propose that high WBC counts in males and females of *A. palliata* in Tabasco could be characteristic for this species or that *A. palliata* is more susceptible than other New World monkeys species to the stress of capture and handling (see: Webb and Latimer 2011; Morrison 1995; Flaiban et al. 2008). Further studies are required to explore the effects of these factors on the levels of WBC in Neotropical primates.

Platelet counts in *A. pigra* and *A. palliata* from Tabasco were higher compared with *A. pigra* from Campeche and *A. macconnelli* from French Guiana. Probably this difference could be due to the time lapse between ketamine administration and blood sampling. More specifically the time between capture and sampling was longer in *A. pigra* from Campeche in comparison to *A. pigra* and *A. palliata* from Tabasco. Previously, we reported that the time elapsed between capture and sampling affected significantly platelet counts (Rovirosa-Hernández et al. 2011). For humans it has been reported that ketamine inhibits platelet aggregation possibly by suppression of cytosolic free calcium concentration, likely via receptor-coupled mechanisms, including G protein (Nakagawa et al. 2002). Thus, we suggest that the differences in platelets counts between populations might be due to handling procedures.

Regarding blood chemistry, total protein levels did not differ between sexes in *A. pigra* and *A. palliata*; both species showed similar values to those reported previously for *A. palliata mexicana* (Crissey et al. 2003). In general, these values are lower in comparison with other wild New World monkeys such as *Callithrix jacchus* (McNees et al. 1984), *Sapajus apella* (Riviello and Wirz 2001), and *Lagothrix lagotricha* (Heugten et al. 2008) in captivity. This suggests that the levels of this parameter are the baseline for free-ranging howler monkeys in Mexico, contrasting with much higher values recorded in *A. macconnelli*. This serves to highlight the importance of having particular profiles for each species and cautions against making generalizations about these types of parameters, even in closely related species. Other causes of these low total protein values in Mexican howlers could be the poor habitat quality where these species forage (Cristobal-Azkarate et al. 2005; Haugaasen and Peres 2005). Similar values were reported in adults of both sexes and juveniles of *A. caraya*; however, Flaiban et al. (2008) found only statistical differences in the total protein of adults with respect to the juvenile, suggesting that this difference is due to nutritional and energy requirements of juveniles. More studies are needed to test this hypothesis.

In general, values of cholesterol and triglycerides in both species from Tabasco were similar to those previously reported for *A. palliata mexicana* in Veracruz, Mexico (Crissey et al. 2003), and *A. caraya* in Argentina (Schmidt et al. 2007). In contrast, the cholesterol values were higher in *A. pigra* males from Campeche in comparison to *A. pigra* and *A. palliata* from Tabasco. This suggests that these *A. pigra* groups have access to different food sources, or perhaps they have disorders of cholesterol metabolism (Evans 2011). Interestingly, *A. macconnelli* have cholesterol values close to the later *A. pigra* groups. Thus, the chemistry profiles do not show only specific signatures but also regional signatures likely in response to ecological conditions. Moreover Schmidt et al. (2007) reported that *A. caraya* males showed significant concentration of HDL cholesterol regarding females and argue that these differences could be due to small sample size.

Males of *A. pigra* and *A. palliata* from Tabasco showed higher creatinine concentration than females. A similar tendency, although not significant, was found in *A. pigra* males from Campeche and was also reported for *A. macconnelli* (Vié et al. 1998b; de Thoisy et al. 2001; Crissey et al. 2003; Schmidt et al. 2007). Also, creatinine values were significantly higher in adult *A. macconnelli* males in comparison to adult females. Similar results are reported for other *Alouatta* species (Crissey et al. 2003; Schmidt et al. 2007); higher creatinine values in males are likely related to their larger body mass (Vié et al. 1998b).

In regard to ions and minerals, the values are similar in both Mexican species and also to those reported for other species of *Alouatta* (Vié et al. 1998b; de Thoisy et al. 2001; Crissey et al. 2003; Schmidt et al. 2007). Only phosphorous and potassium values for *A. pigra* females from Tabasco were significantly higher as compared to the males. But also these values were higher than those of animals from both Campeche and Tabasco. It is known that a delay in the separation of serum from the blood cells could result in an increase in these parameters and also of creatinine (Vié et al. 1998b). In this regard, it is interesting that creatinine was also significantly higher in the *A. pigra* males from Tabasco in comparison to the females and also higher than in remaining subjects from both Campeche and Tabasco. As already mentioned the time between capture and sampling could have eventually affected the mentioned parameters. Also, increase of creatinine can indicate muscle damage (Kock et al. 1990), which could be the result of male aggressive behaviors during capture. In *A. caraya* females, a significant concentration of copper with respect to the males was found without apparent reason (Schmidt et al. 2007).

Hematological data highlighted slight anemia in most of Guianan howlers. Five of the animals in good condition had a low protein level (60–80 g/L vs. 90.8 g/L for the whole sample), as well as modified sodium, chloride, and lactate dehydrogenase. During the flooding of the dam reservoir, with the progressive disappearance of their foliage resources, the red howlers showed an ability to adapt both their social behavior and their diet (de Thoisy and Richard-Hansen 1997). But over a long period, this substitute diet would not provide them with their high energy needs (Milton 1996), as shown by protein levels recorded on those animals after capture (de Thoisy et al. 2001). Thus, some hematological parameters can be considered relevant descriptors of short-term health status, once other spurious effects such as stress and anesthesia are considered. During the translocation process of *A. macconnelli*, four animals died soon after the release. Interestingly, malnutrition and dehydration had not been detected at clinical examination, but those animals showed lower glucose and iron levels and higher chloride and sodium. This reinforces the idea of monitoring hematology and blood biochemistry, as some parameters can be considered markers of poor physical condition. In contrast, it has to be noted that charges of hemoparasites, if any, detected from blood smears (*Trypanosoma*, *Filaria*, *Plasmodium*) did not influence hematological and biochemical profiles (de Thoisy et al. 2001). This may reflect the ability of howlers to control endemic infection pressures by microorganisms. The results of hematology and blood biochemistry in *A. pigra*, *A. palliata*, and *A. macconnelli* are similar to those reported for MCV and different for RBC parameters in *A. caraya* reported by Flaiban et al. (2008).

In summary, the hematological and serum biochemical studies are useful for understanding the adaptation of different howler species to their habitats. In this particular case, the hematological and serum biochemical data of *A. pigra* and *A. palliata* are valuable not only as a reference, but will also be useful to understand the physiological adaptation of these species to their changing environment. In the case of *A. macconnelli*, although the species exhibit an ability to support habitat disturbance by modifying diet and social structure (de Thoisy and Richard-Hansen 1997), those data clearly show the physiological consequences of lower habitat quality. In the case of hematological values of *A. caraya*, they seem to be similar to those reported as physiological adaptation to changing habitat.

In conclusion, we found the following patterns on hematology and blood biochemistry parameters for the four species of *Alouatta* here explored: (a) females have higher concentrations in WBC count than males, except in *A. caraya*, (b) *A. palliata* and *A. caraya* have higher concentrations in WBC count with respect to other *Alouatta* species, (c) Mexican howler monkeys (*A. palliata* and *A. pigra*) have low total protein quantification with respect to other *Alouatta* species, and (d) creatinine concentration is higher in males probably due to their higher body mass. The information presented in this chapter contributes significantly to establish reference values of hematology and blood biochemistry parameters for the genus *Alouatta* in the wild. This information will also be useful as a baseline for diagnosis and treatment of howler monkey individuals maintained in ex situ facilities.

Acknowledgments The authors would like to thank all the individuals who helped to collect samples for this project; a contribution that was essential to its success. Thanks to the reviewers and editors for the constructive comments. This research was supported by a grant of the Federal Electricity Commission of Mexico (CFE RGCPTT-UV-001/04). We thank Dr. Carlos Rafael Blazquez-Domínguez, Director of the Hospital, School of Gynecology, Universidad Veracruzana, for his valuable help with blood analyses.

References

Aguilar-Cucurachi S, Duarte-Dias PA, Rangel-Negrín A, Chavira R, Boek L, Canales-Espinosa D (2010) Preliminary evidence of accumulation of stress during translocation in mantled howlers. Am J Primatol 72:805–810

Alexander RD (1974) The evolution of social behavior. Annu Rev Ecol Syst 5:325–383

Barrueta-Rath T, Estrada A, Pozo C, Calmé S (2003) Reconocimiento demográfico de *Alouatta pigra* y *Ateles geoffroyi* en la reserva El Tormento, Campeche, México. Neotrop Primates 11:165–169

Blahová J, Dobšíková R, Straková E, Suchý P (2007) Effect of low environmental temperature on performance and blood system in broiler chickens (*Gallus domesticus*). Acta Vet Brno 76:S17–S23

Boubli JP, Di Fiore A, Mittermeier RA (2008) *Alouatta macconnelli*. In: IUCN 2012. IUCN red list of threatened species. Version 2013.2. <www.iucnredlist.org>. Downloaded on 7 February 2013

Bravo SP, Sallenave A (2003) Foraging behavior and activity patterns of *Alouatta caraya* in the northeastern Argentinean flooded forest. Int J Primatol 24:825–846

Brockus CW (2011) Erythrocytes. In: Latimer KS (ed) Duncan & Prasse's veterinary laboratory medicine: clinical pathology, 5th edn. Wiley-Blackwell, New York

Bush BM (1991) Interpretation of laboratory results for small animal clinicians. Blackwell Scientific Publication, London

Cates W Jr, Meheus S (1990) Strategies for development of sexually transmitted diseases control program. In: Holmes KK, Mardh P-A, Sparling PF, Wiesner PJ (eds) Sexually transmitted diseases, 2nd edn. McGraw-Hill, New York

Cortés-Ortiz L, Rylands AB, Mittermeier RA (2014) Howler monkeys: adaptive radiation, systematics, and morphology. In: Kowalewski M, Garber P, Cortés-Ortiz L, Urbani B, Youlatos D (eds) Howler monkeys: examining the adaptive radiation, systematics and morphology of the most widely distributed genus of Neotropical primate. Springer, New York

Coyohua FA (2008) La dieta de los monos aulladores negros (*Alouatta pigra*) en el Estado de Campeche: Descripción general y variaciones en función del grado de conservación del hábitat BSc. Biology, Universidad Veracruzana, México

Crissey SD, Serio-Silva JC, Meehan T, Slifka KA, Bowen PE, Stacewicz-Sapuntzakis M, Holick MF, Chen TC, Mathieu J, Meerdlink G (2003) Nutritional status of free-ranging (*Alouatta palliata mexicana*) in Veracruz, Mexico: serum chemistry; lipoprotein profile; vitamins D, A, and E; carotenoids; and minerals. Zoo Biol 22:239–251

Cristobal-Azkarate J, Arroyo-Rodríguez V (2007) Diet and activity pattern of howler monkeys (*Alouatta palliata*) in Los Tuxtlas and implications for conservation. Am J Primatol 69:1013–1029

Cristobal-Azkarate J, Veà JJ, Asensio N, Rodríguez-Luna E (2005) Biogeographical and floristic predictors of the presence and abundance of mantled howlers (*Alouatta palliata mexicana*) in rainforest fragments at Los Tuxtlas, Mexico. Am J Primatol 67:209–222

Crockett CM, Eisenberg JF (1987) Howlers: variation in group size and demography. In: Smuts BB, Cheney DL, Seyfarth RM, Wrangham RW, Struhsaker TT (eds) Primates societies. The University of Chicago Press, Chicago

Cuarón AD, Shedden A, Rodríguez-Luna E, de Grammont PC, Link A (2008) Alouatta palliata ssp. mexicana. In: IUCN 2011. IUCN red list of threatened species. Version 2013.2. <www.iucnredlist.org>. Downloaded on 7 February 2013

Davy MW, Jackson MR, Walter J (1984) Reference intervals for some clinical chemical parameters in the marmoset (*Callithrix jacchus*): effect of age and sex. Lab Anim 18:135–142

de Thoisy B, Richard-Hansen C (1997) Diet and social behavior changes in a red howler monkey (*Alouatta seniculus*) troop in a highly degraded rainforest. Folia Primatol 68:357–361

de Thoisy B, Michel JC, Vogel I, Vié JC (2000) A survey of hemoparasite infections in free-ranging mammals and reptiles in French Guiana. J Parasitol 86:1035–1040

de Thoisy B, Vogel I, Reynes JM, Pouliquen JF, Carme B, Kazanji M (2001) Health evaluation of translocated free-ranging primates in French Guiana. Am J Primatol 54:1–16

de Thoisy B, Renoux F, Julliot C (2005) Hunting in northern French Guiana and its impacts on primates communities. Oryx 39:149–157

de Thoisy B, Richard-Hansen C, Peres CA (2009) Impacts of subsistence game hunting on Amazonian primates. In: Garber PA, Estrada A, Bicca-Marques JC, Heymann EW, Strier KB (eds) South American primates: comparative perspectives in the study of behavior, ecology, and conservation. Book series developments in primatology: progress and prospects. Springer, New York

de Vries A (1991) Translocation of mantled howler monkeys (*Alouatta palliata*) in Guanacaste, Costa Rica, Thesis M.A. University of Calgary, Alberta, Canada

Dias P, Rangel-Negrin A (2014) Diets of howler monkeys. In: Kowalewski M, Garber P, Cortés-Ortiz L, Urbani B, Youlatos D (eds) Howler monkeys: behavior, ecology and conservation. Springer, New York

Dinh VO (2002) Haematological values of macaques (*Macaca fascicularis*) in a mangrove forest Vietnam. J Zoo 10:15

Divillard SP, Stucchi AF, Terstra AHM, Nicolosi RJ, Von Duvillard SP (1992) The effects of dietary casein and soybean protein on plasma lipid levels in *Cebus* monkeys fed cholesterol-free or cholesterol-enriched semi purified diets. J Nutr Biochem 3:71–74

Dobson A, Meagher M (1996) The population dynamics of brucellosis in the Yellowstone National Park. Ecology 77:1026–1036

Estrada A, Coates-Estrada R (1984) Fruit eating and seed dispersal by howler monkey (*Alouatta palliata*) in the tropical rain forest of Los Tuxtlas Mexico. Am J Primatol 6:77–91

Estrada A, Castellanos L, Mendoza A, Pacheco R (2002) Población, ecología y comportamiento de monos aulladores (*Alouatta pigra*) en Palenque, Chiapas, México. Lakamha Boletín Informativo del Museo y Zona Arqueológica de Palenque (CONACULTA-INAH) 3:9–15

Estrada A, Saenz J, Harvey C, Naranjo E, Muñoz D, Rosales-Meda M (2006) Primates in agroecosystems: conservation value of some agricultural practices in Mesoamerican landscapes. In: Estrada A, Garber PA, Pavelka MSM, Luecke L (eds) New perspectives in the study of Mesoamerican primates: distribution, ecology, behavior and conservation. Springer, New York

Evans EW (2011) Proteins, lipids and carbohydrates. In: Latimer KS (ed) Duncan & Prasse's veterinary laboratory medicine: clinical pathology, 5th edn. Wiley-Blackwell, Ames

Flaiban KKMC, Spohr KAH, Malanski LS, Svoboda WK, Shiozawa MM, Hilst CLS, Aguilar LM, Ludwig G, Passos FC, Navarro IT, Lisbôa JAN, Balarin MRS (2008) Valores hematológicos de burgio pretos (*Alouatta caraya*) de vida livre da regiao do Alto Río Paraná, sul do Brasil. Arq Bras Med Vet Zootec 61:628–634

Fox M, Brieva C, Moreno C, Mac Williams P, Thomas C (2008) Hematologic and serum biochemistry reference values in wild-caught white-footed tamarins (*Saguinus leucopus*) housed in captivity. J Zoo Wildl Med 39:548–557

Freeland WJ (1976) Pathogens and the evolution of primate sociality. Biotropica 8:12–24

García-Orduña F (2005) Los primates del sureste de México y la fragmentación de su habitat. II Mexican Congress of Primatology Xalapa Veracruz-México

Gotoh S, Matsubayashi K, Nozawak R (1987) Reports on crab eating monkey in Anguar 11. Result of clinical examination, Kyoto University. Overseas research reports of studies 6:91

Gotoh S, Takennako O, Watanabe K, Kawamoto R, Watanabe T, Surgobroto B, Sajuthi O (2001) Haematological values and parasitic fauna in free ranging *Macaca hecki* and the *Macaca tonkeana*/*Macaca hecki* hybrid group of Sulawesi Island, Indonesia. Primates 42:27–34

Gregorin R (2006) Taxonomia e variação geográfica das espécies do gênero *Alouatta* Lacépède (Primates, Atelidae) no Brasil. Rev Bras Zool 23:64–144

Hassimoto M, Harada T, Harada T (2004) Changes in hematology, biochemical values, and restraint ECG of rhesus monkeys (*Macaca mulatta*) following 6-month laboratory acclimation. J Med Primatol 33:175–186

Haugaasen T, Peres CA (2005) Primate assemblage structure in Amazonian flooded and unflooded forests. Am J Primatol 67:243–258

Hausfater G, Meade BJ (1982) Alteration of sleeping groves by yellow baboons (*Papio cynocephalus*) as a strategy for parasite avoidance. Primates 23:287–297

Heugten KA, Verstegen M, Ferket PR, Stoskopf M, van Heugten E (2008) Serum chemistry concentration of captive wooly monkeys (*Lagothrix lagotricha*). Zoo Biol 27:188–199

Horwich RH, Johnson EW (1986) Geographic distribution of the black howler monkey (*Alouatta pigra*) in Central America. Primates 27:53–62

Jain NC (1986) Schalm veterinary haematology, 4th edn. Lea & Febiger, Philadelphia

Jain NC (1993) Essentials of veterinary hematology. Lea & Febiger, Philadelphia

Julliot C, Sabatier D (1993) Diet of the red howler monkey (*Alouatta seniculus*) in French Guiana. Int J Primatol 14:527–550

Klosla P, Hayes KC (1992) Comparison between the effects of dietary saturated (16:0), monosaturated (18:1) and polyunsaturated fatty acids on plasma lipoprotein metabolism. Am J Clin Nutr 55:51–62

Kock MD, Du Toit R, Kock N, Morton D, Foggin C, Paul B (1990) Effects of capture and translocation on biological parameters in free-ranging black rhinoceroses (*Diceros bicornis*) in Zimbabwe. J Zoo Wildl Med 21:414–424

Lambeth SP, Hau J, Perlman JE, Martino M, Schapiro SJ (2006) Positive reinforcement training affects hematologic and serum chemistry values in captive chimpanzees (*Pan troglodytes*). Am J Primatol 68:245–256

Larsson MHMA, Birgel EH, Benesi FJ, Birgel EH, Lazaretti P, Fedullo JDL, Larsson CE, Molina SR, Guerra PPCA, Prada CS (1999) Hematological values of *Cebus apella* anesthetized with ketamine. Braz J Vet Res Anim Sci 36:3–5

Lugo-Roman LA, Rico PJ, Sturdivant R, Burks R, Settle TL (2009) Effects of serial anesthesia using ketamine or medetomidine on hematology and serum biochemical values in rhesus macaques (*Macaca mulatta*). J Med Primatol 39:41–49

Marsh LK, Cuarón AD, Cortés-Ortiz L, Shedden A, Rodríguez-Luna E, de Grammont PC (2008) *Alouatta pigra*. In: IUCN 2011. IUCN red list of threatened species. Version 2013.2. <www.iucnredlist.org>. Downloaded on 7 February 2013.

Martin DP, McGowan MJ, Loeb WF (1973) Age related changes of hematologic values in infant *Macaca mulatta*. Lab Anim Sci 23(2):194–200

McNees DW, Lewis RW, Ponzio BJ, Sis RF, Stein FJ (1984) Blood chemistry of the common marmoset (*Callithrix jacchus*) maintained in an indoor-outdoor environment: primate comparisons. Primates 25:103–109

Milton K (1980) The foraging strategy of howler monkeys. A study in primate economics. Columbia University Press, New York

Milton K (1987) Physiological characteristics of the genus *Alouatta*. Int J Primatol 8:428

Milton K (1996) Dietary quality and demographic regulation in a howler monkey population. In: Ceigh EG, Rands AS, Windsor DM (eds) The ecology of a tropical rainforest: seasonal rhythms and long-term changes, 2nd edn. Smithsonian Institution Press, New York

Mittermeier RA (1977) Distribution, synecology, and conservation of Surinam monkeys. Ph.D. dissertation, Harvard University, Boston

Moller AP, Dufva R, Allander K (1993) Parasites and the evolution of host social behavior. Adv Study Behav 22:65–101

Morrison TK (1995) Haematological screening. In: Lemus GA (ed) Clinical laboratory and diagnostic test. El Manual Moderno, Mexico

Muckenhirn NA, Mortensen S, Vessey S, Fraser CEO, Singh B (1975) Report on a primate survey in Guyana. Pan American Health Organization, Washington, DC

Muñoz D, Estrada A, Naranjo E, Ochoa S (2006) Foraging ecology of howler monkeys in a cacao (*Theobroma cacao*) plantation in Comalcalco, Mexico. Am J Primatol 68:127–142

Nakagawa T, Hirakata H, Sato M, Nakamura K, Hatano Y, Nakamura T, Kazuhiko F (2002) Ketamine suppresses platelet aggregation possibly by suppressed inositol triphosphate formation and subsequent suppression of cytosolic calcium increase. Anesthesiology 96:1147–1152

Norconk MA, Sussman RW, Phillips-Conroy J (1996) Primates of Guayana shield forests: Venezuela and the Guianas. In: Norconk MA, Rosenberger AL, Garber PA (eds) Adaptive radiations of neotropical primates. Plenum Press, New York

Núñez H, Araya M, Cisternas F, Arredondo M, Méndez M, Pizarro F, Ortíz A, Ortíz R, Olivares M (2008) Blood biochemical indicators in young and adult *Cebus apella* of both sexes. J Med Primatol 37:12–17

Ogunrinade A, Fajimi J, Adenike A (1981) Biochemical indices in the White Fulani (Zebu) cattle in Nigeria. Rev Elev Med Vet Pays Trop 34:41

Ogunsanmi AO, Akpavie SO, Anosa VO (1994) Serum biochemical changes in West African Dwarf sheep experimentally infected with *Trypanosoma brucei*. Rev Elev Med Vet Pays Trop 47:195

Ostro LET, Silver SC, Koontz FW, Young TP, Horwich RH (1999) Ranging behavior of translocated and established groups of black howler monkeys *Alouatta pigra* in Belize, Central America. Biol Conserv 87:181–190

Ostro LET, Silver SC, Koontz FW, Young TP (2000) Habitat selection by translocated black howler monkeys in Belize. Anim Conserv 3:175–181

Ostro LET, Silver SC, Koontz FW, Horwich RH, Brockett R (2001) Shifts in social structure of black howler (*Alouatta pigra*) groups associated with natural and experimental variation in population density. Int J Primatol 22:733–748

Porter JA (1971) Hematologic values of the black spider monkey (*Ateles fusciceps*), red spider monkey (*Ateles geoffroyi*), white face monkey (*Cebus capucinus*), and black howler monkey (*Alouatta villosa*). Lab Anim Sci 21:426–433

Pozo-Montuy G, Serio-Silva JC (2007) Movement and resource use by a group of *Alouatta pigra* in a forest fragment in Balancán, México. Primates 48:102–107

Pronkuk A, Patton GM, Stephan ZF, Hayes KC (1991) Species variation in the atherogenic profile of monkeys: relationship between dietary fats lipoprotein and platelets aggregation. Lipids 26:213–222

Richard-Hansen C, Vié JC, Vidal N, Kéravec J (1999) Biometrical data on 40 species of mammals from French Guiana. J Zool (Lond) 247:419–428

Richard-Hansen C, Vié JC, de Thoisy B (2000) Translocation of red howler monkeys *Alouatta seniculus* in French Guiana. Biol Conserv 93:247–253

Rivera A, Calmé S (2006) Forest fragmentation and its effects on the feeding ecology of black howlers (*Alouatta pigra*) from the Calakmul area in Mexico. In: Estrada A, Garber PA, Pavelka MSM, Luecke L (eds) New perspectives in the study of Mesoamerican primates: distribution, ecology, behavior and conservation. Springer, New York

Riviello MC, Wirz A (2001) Haematology and blood chemistry of *Cebus paella* in relation to sex and age. J Med Primatol 30:308–312

Rodríguez-Luna E, Cortés-Ortíz L (1994) Translocacion y seguimiento de un grupo de monos *Alouatta palliata* liberado en una isla (1988-1994). Neotrop Primates 2:1–4

Rosenblum IY, Coulston F (1981) Normal range and longitudinal bloom chemistry and hematology values in juvenile and adult rhesus monkeys (*Macaca mulatta*). Ecotoxicol Environ Saf 5:401–411

Rovirosa-Hernández MJ, García-Orduña F, Caba M, Canales-Espinosa D, Hermida-Lagunes J, Torres-Pelayo VR (2011) Blood parameters are little affected by time of sampling after the application of ketamine in black howler monkey (*Alouatta pigra*). J Med Primatol 40:294–299

Rovirosa-Hernández MJ, Caba M, García-Orduña F, López-Muñoz JJD, Canales-Espinosa D, Hermida-Lagunes J (2012) Hematological and biochemical blood values in wild populations of black howler monkey (*Alouatta pigra*) of Campeche, México. J Med Primatol 4:309–316

Schmidt DA, Kowalewski MM, Ellersieck MR, Zunino GE, Stacewicz-Sapuntzakis M, Chen TC, Holick MF (2007) Serum nutritional profiles of free-ranging *Alouatta caraya* in northern Argentina: lipoproteins; amino acid; vitamins A, D, and E; carotenoids; and minerals. Int J Primatol 28:1093–1107

Shedden-González A, Rodríguez-Luna E (2010) Responses of a translocated howler monkey *Alouatta palliata* group to new environmental conditions. Endang Species Res 12:25–30

Sirois M, Hendrix CHM (2007) Clinical chemistry. In: Hendrix CHM, Sirois M (eds) Laboratory procedures for veterinary technicians, 5th edn. Mosby Elsevier, St. Louis

Smith JD (1970) The systematic status of the black howler monkey, *Alouatta pigra* Lawrence. J Mammal 51:358–369

Thrall MA, Baker DC, Campbell TW, De Nicola D, Fettman MJ, Duane Lassen E, Rebar A, Weiser G (2006) Clinical chemistry of mammals: laboratory animals and miscellaneous species. In: Thrall MA, Baker DC, Campbell TW, De Nicola D, Fettman MJ, Duane Lassen E, Rebar A, Weiser G (eds) Veterinary hematology and clinical chemistry. Blackwell-Wiley, Ames

Večerek V, Straková E, Suchý P, Voslářová E (2002) Influence of high environmental temperature on production and haematological and biochemical index in broiler chickens. Czech J Anim Sci 47:176–182

Vié JC (1999) Wildlife rescues: the case of the Petit Saut hydroelectric dam in French Guiana. Oryx 33:115–126

Vié JC, de Thoisy B, Fournier P, Fournier-Chambrillon C, Genty C, Keravec J (1998a) Anesthesia of wild red howler monkeys (*Alouatta seniculus*) with medetomidine-ketamine and reversal by atipamezole. Am J Primatol 45:399–410

Vié JC, Moreau B, de Thoisy B, Fournier P, Genty C (1998b) Hematology and serum biochemistry values of free-ranging red howler monkeys (*Alouatta seniculus*) from French Guiana. J Zoo Wildl Med 29:142–149

Webb JL, Latimer KS (2011) Leukocytes. In: Latimer KS (ed) Duncan & Prasse's veterinary laboratory medicine: clinical pathology, 5th edn. Wiley-Blackwell, Ames

Wirz A, Truppa V, Riviello MC (2008) Hematological and plasma biochemical values for captive tufted capuchin monkeys (*Cebus apella*). Am J Primatol 70:463–472

Chapter 8
Endocrinology of Howler Monkeys: Review and Directions for Future Research

Sarie Van Belle

Abstract Endocrine studies that investigate the interplay between hormones, behavior, and social and ecological environment are critical for our understanding of proximate, physiological mechanisms underlying the biology and sociality of a species. Nonetheless, only recently have endocrine studies been incorporated into research on howler monkeys (*Alouatta* spp.), and only few aspects of endocrinology in 6 out of 12 species have been addressed. These include androgen and estrogen profiles of juvenile *A. palliata*, and progestin and estrogen profiles of the ovarian cycle of *A. arctoidea*, *A. caraya*, and *A. pigra*. In addition, socioendocrine studies in *A. pigra* and *A. palliata* have investigated how male androgen levels and male and female glucocorticoid levels are influenced by intra- and extragroup male–male competition, whereas ecologically oriented endocrine studies have revealed that in *A. pigra*, *A. palliata*, *A. belzebul*, and *A. seniculus* male and female glucocorticoid levels are influenced by a scarcity of high-quality food resources, habitat fragmentation, human disturbance, and translocation. Endocrine studies have shed light on howler monkey biology and sociality that were not anticipated based on behavioral data alone. This includes a nonaggressive form of intragroup male–male competition over access to females, a more prominent reliance on high-quality food resources such as fruits for these primarily folivorous primates, and an apparent higher sensitivity to social and ecological stress in females than in males. Additional endocrine studies across howler monkey species are needed to further elucidate relationships among diet, mating competition, and social interactions.

Resumen Estudios endocrinológicos que investigan la interacción entre las hormonas, el comportamiento, y el contexto social y ecológico son fundamentales para nuestra comprensión de los mecanismos fisiológicos subyacentes a la biología y al sistema social de una especie. Sin embargo, sólo recientemente estudios

S. Van Belle (✉)
Instituto de Biología, Universidad Nacional Autónoma de México,
Circuito Exterior s/n, Ciudad Universitaria, Copilco, Coyoacan,
Distrito Federal 04510, Mexico
e-mail: sarievanbelle@primatesmx.com

endocrinológicos han sido incorporado en la investigación de los monos aulladores (*Alouatta*), y sólo unos pocos aspectos de la endocrinología en seis de las 12 especies han sido abordado. Estos incluyen investigar los perfiles de andrógenos y estrógenos de juveniles de *A. palliata*, y los perfiles de progesterona y estrógeno de los ciclos ováricos en *A. arctoidea, A. caraya,* y *A. pigra*. Además, los estudios socio-endocrinológicos de *A. pigra* y *A. palliata* han investigado como los niveles de andrógenos en machos y los niveles de glucocorticoides en machos y hembras se ven influidos por la competencia entre machos adentro y entre grupos, mientras que los estudios endocrinológicos en *A. pigra, A. palliata, A. belzebul,* y *A. seniculus* con un enfoque en factores ecológicos han revelado que los niveles de glucocorticoides de machos y hembras son afectados por la disponibilidad de alimentos de alta calidad, la fragmentación del hábitat, perturbación humana y el estrés acumulado durante eventos de translocación. Estudios endocrinológicos en el mono aullador han revelado aspectos de su biología y su sistema social que no fueron anticipados en base a sólo datos de comportamiento. Esto incluye una forma de competición no agresivo entre machos por el acceso a hembras, una dependencia más prominente en los recursos alimenticios de alta calidad como las frutas para estos primates principalmente folívoros, y una sensibilidad superior al estrés social y ecológico en hembras en comparación con machos. Otros estudios endocrinológicos entre más especies de monos aulladores serán necesarios para elucidar aún más el vínculo entre la dieta, la competición reproductiva, y las interacciones sociales.

Keywords Progesterone • Estrogen • Testosterone • Cortisol • Ovarian cycle • Social system

8.1 Introduction

Endocrine studies that investigate the interplay between hormones, behavior, and social and ecological environment are critical for our understanding of proximate, physiological mechanisms underlying individual, age, and sex-based variation in behavior. Studies of primate socioendocrinology provide a framework for identifying factors that regulate differential reproductive success among individuals (Bercovitch and Ziegler 2002). Recent advances in noninvasive techniques for measuring hormonal profiles in feces and urine have provided primatologists with the opportunity to investigate a wide range of endocrine systems and their influence on behavior in both captive and wild primates (Strier and Ziegler 2005). For example, endocrine studies are central in our understanding of reproductive maturation (i.e., puberty), which is driven by an increase in gonadal activity (Plant and Witchel 2006; Saltzman et al. 2011). In addition, reproductive endocrinology is essential for understanding the basic biology of reproduction, because "the occurrence and timing of largely concealed reproductive events such as ovulation, conception, pregnancy, and natural abortions can only be accurately and reliably detected through hormone analyses" (Lasley and Savage 2007:357). Similarly, socioendocrinology examines how the social environment influences the feedback loop between hormones and behavior, and has the power to enhance our understanding of male and female reproductive strategies (Bercovitch and Ziegler 2002; Anestis 2010), whereas ecological endocrinology investigates how the physical environment modulates behavioral and hormonal responses to fluctuating ecological factors such as seasonality in food availability, photoperiod, or weather conditions (Bradshaw 2007). Furthermore, endocrine studies may contribute importantly to the conservation of endangered primate species by using hormone studies on wild populations living in undisturbed habitats as baseline information to evaluate how the interplay between behavior and hormones is affected under conditions of demographic, ecological, and behavioral disruption resulting from habitat alteration due to human activity (Cockrem 2005).

Despite the important contributions that endocrine studies have made to our understanding of primate behavior and ecology, such studies only recently have been incorporated into research on howler monkeys (*Alouatta* spp.). Hormonal studies have been conducted on six howler monkey species: the ursine howler monkey (*A. arctoidea*), the red-handed howler monkey (*A. belzebul*), the black-and-gold howler monkey (*A. caraya*), the mantled howler monkey (*A. palliata*), the black howler monkey (*A. pigra*), and the red howler monkey (*A. seniculus*), and these have focused principally on questions of pubertal development, female reproduction, male competition, the influence of ecological stressors, and conservation management (Table 8.1).

Table 8.1 Review of endocrine studies in howler monkeys

Species	Hormones	Context	Male/female	Captive/wild	Reference
A. arctoidea	P	Ovarian cycles	F	W	Herrick et al. (2000)
A. belzebul	A, E, GC, P	Ecological stress	M, F	W	Monteiro et al. (2013)
A. caraya	A	Male reproduction	M	C	Moreland et al. (2001)
A. caraya	E, P	Ovarian cycles	F	C	Kugelmeier et al. (2011)
A. palliata	A, E	Puberty	M, F	W	Clarke et al. (2007)
A. palliata	GC	Male reproductive strategies	M, F	W	Cristóbal-Azkarate et al. (2007)
A. palliata	A	Male reproductive strategies	M	W	Cristóbal-Azkarate et al. (2006)
A. palliata	GC	Translocation	M, F	C, W	Aguilar-Cucurachi et al. (2010)
A. palliata	I	Method validation	M, F	C	Wasser et al. (2010)
A. palliata	GC	Ecological stress	M, F	Semi-wild	Aguilar-Melo et al. (2013)
A. palliata	GC	Habitat fragmentation	M, F	W	Dunn et al. (2013)
A. palliata	GC	Social and ecological stress	M, F	W	Gómez-Espinosa et al. (2013)
A. pigra	GC	Habitat fragmentation	M, F	W	Martínez-Mota et al. (2007)

(continued)

Table 8.1 (continued)

Species	Hormones	Context	Male/female	Captive/wild	Reference
A. pigra	GC	Method validation	M, F	C	Martínez-Mota et al. (2008)
A. pigra	E, P	Ovarian cycles	F	W	Van Belle et al. (2009a)
A. pigra	A, GC	Male reproductive strategies	M	W	Van Belle et al. (2009b)
A. pigra	GC	Ecological stress	M, F	W	Behie et al. (2010)
A. pigra	A	Male reproductive strategies	M	W	Rangel-Negrín et al. (2011)
A. pigra	E, P	Method validation	F	W	Torres-Pelayo et al. (2011)
A. pigra	GC	Ecological stress	M, F	W	Behie and Pavelka (2013)
A. seniculus	GC	Method validation	M, F	W	Rimbach et al. (2013a)
A. seniculus	GC	Habitat fragmentation	M, F	W	Rimbach et al. (2013b)

P progestins, *E* estrogens, *GC* glucocorticoids, *A* androgens, *I* thyroid hormones

8.2 Method Validation

Of the 22 endocrine studies on howler monkeys (Table 8.1), 21 studies used noninvasively collected fecal samples, and 1 study used noninvasively collected urine samples (Herrick et al. 2000). Both fecal and urine samples contain very low or no concentrations of active hormones but are characterized by high concentrations of hormone metabolites (Bahr et al. 2000; Heistermann et al. 2006). This presents an opportunity for biased results and therefore samples need to be validated to ensure that the substance measured from urine or feces is a metabolite derived from the hormone in question, and not a substance that cross-reacts with the antibody of the respective immunoassay. This would render the hormone profile meaningless (Whitten et al. 1998; Ziegler and Wittwer 2005). The procedures required for validation include providing a stimulus to an individual that will cause an increase in circulatory levels of the hormone in question, such as injecting adrenocorticotropic hormone (ACTH) or gonadotropin-releasing hormone (GnRH) when interested in glucocorticoids or gonadal steroids, respectively, or injecting radioactive-labeled forms of the particular hormone, and subsequently assess whether concentrations of the metabolites in feces and urine vary in parallel with changes in blood hormone concentrations (Palme 2005).

When such physiological validations are impossible to perform because of a lack of captive individuals or ethical restraints, biological validations are suitable alternatives. For example, patterns in production of gonadal steroids can be biologically validated by comparing steroid concentrations from samples collected during different reproductive stages (e.g., juveniles versus adults). Similarly, Martínez-Mota et al. (2008) validated the detection and measurement of glucocorticoids in fecal samples using enzyme immunoassays (EIA) in black howler monkeys before and after the application of a stressor, i.e., anesthesia. Levels of fecal glucocorticoid metabolites measured in four adults, residing at the Chapultepec Zoo in Mexico City, reached peak concentrations 24–96 h after anesthesia, parallel to an increase in circulating glucocorticoid concentrations, suggesting that the assay correctly measured glucocorticoid metabolites. Similarly, fecal glucocorticoid metabolites increased 21 and 24 h after the anesthesia of wild mantled howler monkeys (one male and two females, Gómez-Espinosa et al. 2013) and wild red howler monkeys (one male, Rimbach et al. 2013a), respectively. Wasser et al. (2010) validated the measurement of thyroid hormones in fecal samples using radioimmunoassay methods in two adult male and three adult female mantled howler monkeys that were temporarily kept in captivity during a translocation project. Thyroid concentrations decreased significantly post-capture in females but not in males. The authors argued that endocrine differences documented in this study corresponded to the limited food intake exhibited by females compared to males, suggesting that the measured thyroid levels accurately represented to metabolic state of these individuals.

Besides biological validations, the efficiency by which hormonal metabolites are recovered from samples using different extraction methods may need to be assessed to optimize endocrine studies, as has been done for estrogen and progesterone in black howler monkeys (Torres-Pelayo et al. 2011). These authors assessed whether the type of substrate (moist versus lyophilized feces), organic solvent (80 or 100 % ethanol versus methanol), and extraction method (agitation versus ebullition) affected the extraction efficiency of estrogens and progestins in fecal samples of black howler monkeys. Their findings suggested considerable variation in percentages of recovered steroids according to the substrate, solvent, and extraction method, with an ebullition extraction method as the most efficient (Torres-Pelayo et al. 2011).

8.3 Endocrinology of Puberty

Puberty in primates is characterized by behavioral, morphological, neurological, and endocrine changes, along with the development of distinctive secondary sex characteristics (reviewed in Plant and Witchel 2006; Saltzman et al. 2011). The onset of puberty in both males and females is marked by the increase in the pulsative release of GnRH from the hypothalamus, due to the diminishing inhibitory effect of the neurotransmitter GABA on the GnRH neurons. This causes a dramatic elevation in circulatory levels of luteinizing hormone (LH), and to a lesser extent follicle-stimulating hormone (FSH), released from the pituitary gland. The surge in LH stimulates testicular growth, the production of gonadal androgens, and the initiation of spermatogenesis in males, and the initiation of cyclic ovarian activity and the production of ovarian steroids in females (Plant and Witchel 2006). The onset of ovarian cycles is typically associated with a period of "adolescent infertility" characterized by anovulatory or irregular cycles prior to the onset of fertile ovarian cycles. Limited information is available on the onset of sexual maturation in howler monkeys. Based on behavioral observations, without the underlying endocrine confirmation, male howler monkeys are believed to be sexually mature between 2 and 5.5 years of age, while females are believed to be sexually mature between 2.9 and 4.5 years of age and experience their first parturition at 3.4–5.0 years of age (Table 8.2).

In mantled howler monkeys that usually live in groups with a large number of resident adult males, male puberty is also characterized by the descent of the testes from the inguinal canal into the scrotum around 3 years of age (Glander 1980), while in other howler monkey species, testes descend during infancy (Crockett and Eisenberg 1987). Prior to the descent of the testes, the non-pendulous scrotum of juvenile male mantled howlers cannot be visually differentiated from female juvenile genitalia, and juveniles can only be reliably sexed when they are newborn or by external palpation of captured individuals (Glander 1980; Clarke et al. 2007). In order to determine whether developmental differences in hormone concentrations can be used to differentiate between juvenile males and females, Clarke et al. (2007) collected fecal samples from 31 mantled howler juveniles (0.8–3.5 years old) of known sex (19 males and 12 females; juveniles were captured or positively sexed as infants) from five social groups at Hacienda La Pacifica, Costa Rica, during three 1-month field studies across 3 years. Mean fecal androgen and estrogen concentrations did not differ significantly between 1-year-old, 2-year-old, and 3-year-old males and females. Male fecal androgen levels increased during puberty with 3-year-old males ($N=6$) having significantly higher fecal androgen levels than younger males ($N=13$). However, three 3-year-old males who were observed to be actively evicted from their natal group by other group members had lower fecal androgen levels than 3-year-old males who remained in their natal group ($N=3$), and these hormonal differences could have resulted from increased aggression and harassment received and a decrease in competitive ability prior to eviction (Bernstein et al. 1979).

Clarke et al. (2007) argued that juvenile monomorphic genitalia in mantled howler monkeys might serve to prolong the period that juveniles remain in their natal group because aggressive eviction from the natal group frequently occurred immediately after the appearance of adult genitalia. Because other howler monkeys do not exhibit juvenile genitalia monomorphy, comparative data on hormonal profiles underlying juvenile development, puberty, and eviction from their natal group across other howler monkey species are needed to better understand whether hormonal profiles observed in mantled howler juveniles are unique or represent a common pattern in *Alouatta*. Furthermore, given that dichromatism has evolved in four

Table 8.2 Parameters of sexual maturation in howler monkeys

Species	Male age (months) of sexual maturity	Female age (months) of sexual maturity	Female age (months) of first parturition	Reference
A. arctoidea	58–66	43–54	48–60	Crockett and Eisenberg (1987)
A. caraya	24–37	35–42	41–48	Shoemaker (1982)
A. palliata	42	36	43	Glander (1980)

howler taxa (*A. caraya*, *A. guariba clamitans*, *A. seniculus puruensis*, and *A. ululata*) and males change to their adult coat color starting at 6 months of age (Bicca-Marques and Calegaro-Marques 1998), it would be important to study the interplay between hormonal profiles and adult–juvenile social interactions during changes from natal to adult coat coloration in these sexually dichromatic howler species, to assess whether natal pelage patterns serve as sexual mimicry during puberty. This is especially true when considering that a delay in the onset of coat color change observed in two black-and-gold howler juvenile males might have been linked to the increased probability of being evicted from their natal group due to the reduced size (0.3 ha) of the forest fragment they resided in compared to other groups with lower ecological and demographic pressures (Bicca-Marques and Calegaro-Marques 1998). The authors suggested that the delay in color change might be a mechanism allowing a juvenile male to remain in their natal group for a longer period attaining larger body sizes before being forcefully evicted, similar to mantled howler juvenile genitalia mimicry.

8.4 Reproductive Endocrinology

Information on estrogen and progestin profiles of the howler ovarian cycle is limited to three species (*A. arctoidea*, *A. caraya*, and *A. pigra*, Table 8.1). Steroids were measured from urinary (*A. arctoidea*) or fecal samples (*A. caraya* and *A. pigra*), and the steroid profiles of the ovarian cycle followed the pattern unique to platyrrhines. In all platyrrhines examined to date, the urinary and fecal estrogen and progestin profiles do not match the profiles present in circulating steroid hormones in that the urinary and fecal estrogen profiles do not exhibit a follicular surge prior to ovulation (reviewed in Ziegler et al. 2009a). Instead, urinary and fecal estrogen levels show a sustained increase during the luteal phase of the cycle, similar to urinary and fecal progestin profiles. Several studies have indicated that the onset of the sustained increase in urinary and fecal progestin concentrations occurs shortly after the serum luteinizing hormone (LH) peak, which triggers ovulation, while the delay in the excretion of estrogens is more variable (Fig. 8.1, Ziegler et al. 2009a). Therefore, in New World monkeys urinary and fecal progestin profiles are more reliable to estimate days of ovulation than estrogen profiles, and days of ovulation are estimated as the sample preceding the first sustained rise in urinary and fecal progestin levels. The length of the ovarian cycle is calculated as the time interval between two consecutive estimated days of ovulation (Van Belle et al. 2009a). A representative hormonal profile of a black howler female is shown in Fig. 8.2.

Following these methods, the mean ovarian cycle length for three adult black-and-gold howler females living in captivity was $19.1 \pm SD$ 2.1 days (range = 9–47 days, $N = 18$; Kugelmeier et al. 2011). For four wild multiparous black howler females, mean ovarian cycle length was $18.4 \pm SE$ 1.4 days (range = 13–25 days, $N = 12$; Van Belle et al. 2009a), and for two wild adult ursine howler females, mean ovarian cycle length was $29.5 \pm SE$ 1.5 days (range = 28–31 days, $N = 2$; Herrick et al. 2000). The average ovarian cycle length of 18–19 days observed in black

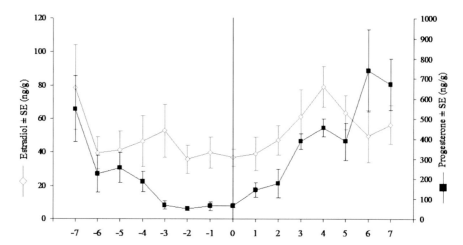

Fig. 8.1 Mean ± SE of fecal estrogen and progesterone levels across the ovarian cycle, aligned relative to estimated day of ovulation (=day 0, $n=18$) in four wild female black howler monkeys (*A. pigra*). The average profile shows that fecal progesterone concentrations decrease to baseline levels during the follicular phase when the ovum is growing. The rise in fecal progesterone levels indicates that ovulation has occurred. Fecal estrogen levels remain relatively constant during ovulation but show an increase about 2–4 days later

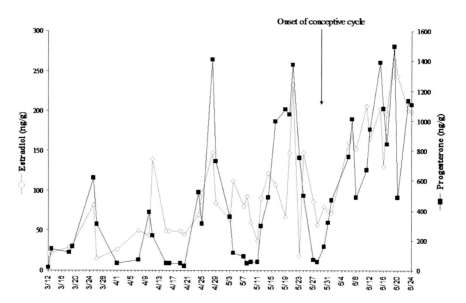

Fig. 8.2 The fecal hormonal profile for a female black howler monkey (*A. pigra*) at Palenque National Park, Mexico Van Belle et al. 2009a

howler and black-and-gold howler monkeys is similar to the average ovarian cycle lengths estimated from patterns of observed copulations in ursine howler monkeys (mean = 17 days, range = 11–27 days, Crockett and Sekulic 1982) and mantled howler monkeys (mean = 15.8 days, range = 10–24 days, Glander 1980; Jones 1985). The longer estimates of ovarian cycles from urinary progestin profiles in ursine howler monkeys (Herrick et al. 2000) could reflect an artifact of limited sample size.

The fecal steroid profiles of three black howler females who had lost their young infants showed a resumption in cyclicity 1–3 weeks later, indicating an almost immediate physiological switch from lactational amenorrhea to ovarian cyclicity (Van Belle et al. 2009a). Nonetheless, the first cycle these females experienced after losing their infant may have been anovulatory because they were noticeably shorter (9–11 days) and had lower fecal estrogen and progesterone levels (Van Belle et al. 2009a). This delay in returning to reproductive condition might be linked to the replenishment of maternal energetic reserves as has been noted for several other primates (reviewed in Ziegler et al. 2009a).

Mean ovarian cycle length has also been estimated from cytological profiles of vaginal swabs in captive black-and-gold howler females. Based on the recurrence of squamous epithelial cells, Colillas and Coppo (1978) estimated a mean cycle length of $19.7 \pm SD$ 1.0 days (range = 17–24 days), with ovulation estimated to occur immediately after the surge of squamous cells. Kugelmeier et al. (2011) estimated a mean ovarian cycle length of $19.8 \pm SD$ 0.9 days (range = 18–22 days, $N=4$) based on profiles of erythrocytes in vaginal swabs with a mean bleeding period of $4.1 \pm SD$ 1.0 days (range = 1–7 days), coinciding with basal levels of both fecal progesterone and estrogen.

A reliable estimate of the day of ovulation and peak fertility based on hormonal profiles allows for mapping copulations and male and female sexual solicitations onto profiles of the ovarian cycle to provide critical insights into male and female reproductive strategies and reproductive success. Examining the occurrence of these behaviors during the periovulatory period, when conception is most likely to occur, compared to outside the periovulatory periods, allows one to assess the degree to which dominant or central howler males successfully mate-guard fertile females and the extent to which cycling females might undermine a male's ability to monopolize reproductive opportunities through female mate choice. This type of study has only been conducted in black howler monkeys, and revealed that female sexual solicitations, male mate-guarding efforts, male's monitoring of a female's reproductive state by sniffing her genitalia, and copulations were largely confined to the periovulatory period (Van Belle et al. 2009a, reviewed in Van Belle and Bicca-Marques 2014).

8.5 Socioendocrinology

Socioendocrine studies in howler monkeys have investigated how male androgen levels and male and female glucocorticoid levels are influenced by intra- and extragroup male–male competition. The challenge hypothesis posits that an increase in adult male androgen levels, above baseline levels required for spermatogenesis and full display of sexual behavior, are closely associated with levels of male intrasexual competition for access to fertile females (Wingfield et al. 1990). Evidence of the challenge hypothesis in primates has been found in species in which males aggressively compete for social rank resulting in dominant males having higher androgen levels than subordinate males throughout the year (e.g., chimpanzees, *Pan troglodytes*, Muehlenbein et al. 2004; Muller and Wrangham 2004a; mandrills, *Mandrillus sphinx*, Setchell et al. 2008). Because of the high costs and potentially detrimental effects (e.g., suppression of immune system, suppression of parental care, reduced survival) associated with chronically elevated androgen levels (Braude et al. 1999; Muehlenbein and Bribiescas 2005; Hau 2007), rank-related differences in androgen levels may be expected to only coincide with periods of heightened male–male aggression during social instability due to rank reversal, immigration events, or reproductive competition (e.g., Verreaux's sifakas, *Propithecus verreauxi*, Brockman et al. 2001; chacma baboons, *Papio ursinus*, Bergman et al. 2005). Consistent with the challenge hypothesis, androgen levels were found not to differ among males in species with limited aggressive competition over access to fertile females (e.g., Northern muriquis, *Brachyteles hypoxanthus*, Strier et al. 1999; tufted capuchin monkeys, *Sapajus nigritus*, Lynch et al. 2002; moustached tamarins, *Saguinus mystax*, Huck et al. 2005).

The stress response is characterized by a marked increase in circulating glucocorticoids released from the adrenal gland that accelerates carbohydrate metabolism and leads to an increased availability of glucose in the bloodstream, which in turn enables individuals to deal with short- and long-term threats (Sapolsky 2002). Short-term stress responses are generally thought to be beneficial in that they mobilize energy reserves. In contrast, chronic stress can result in the suppression of reproduction, the immune system, growth, and muscle wasting (reviewed in Sapolsky 2005). Potential threats or stressors include predation, nutritional deficiencies, and conspecific agonism. The frequency and degree of aggressive interactions among conspecifics, as well as the degree of social support available to individuals, are expected to contribute to variation in glucocorticoid concentrations within and among individuals (Abbott et al. 2003; Goymann and Wingfield 2004). As such, dominant individuals are expected to have higher glucocorticoid levels than subordinates when they are frequently challenged by others and need to regularly assert their high social status (e.g., female ring-tailed lemurs, *Lemur catta*, Cavigelli et al. 2003; male chimpanzees, Muller and Wrangham 2004b), or subordinates might have higher glucocorticoid levels than dominant individuals when subordinates endure frequent attacks by higher ranking individuals and have limited social support such as alliances and grooming partners to cope with these social stressors (e.g., chacma baboons, Sapolsky 1993). In other species, dominant and subordinate individuals might not differ in their glucocorticoid levels because the stressors faced by dominants are similar to those faced by subordinates (e.g., long-tailed macaques, *Macaca fascicularis*, van Schaik 1991; mountain gorillas, *Gorilla beringei*, Robbins and Czekala 1997) or social status may not be directly related to frequent intragroup aggression (e.g., northern muriquis, Strier et al. 1999; tufted capuchin monkeys, Lynch et al. 2002; moustached tamarins, Huck et al. 2005).

The social system of howler monkeys is characterized by very low levels of intragroup male–male competition with low frequencies of resident male–male aggression (*A. pigra*: mean rate=0.007 interactions/dyad/h, Van Belle et al. 2008; mean rate=0.04

interactions/ind/h, Rangel-Negrín et al. 2011; *A. palliata*: mean rate = 0.018 interactions/ind/h, Wang and Milton 2003; reviewed in Kowalewski and Garber 2014). Overt male–male aggression is principally confined to immigration events and intergroup encounters, which may result in injuries or death (*A. arctoidea*: Sekulic 1983; Crockett and Pope 1988; *A. palliata*: Glander 1992; Cristóbal-Azkarate et al. 2004; Dias et al. 2010; *A. pigra*: Horwich et al. 2000; Van Belle et al. 2008; *A. seniculus*: Izawa and Lozano 1991; Izawa 1997). Both single males or pairs of males have been observed to successfully immigrate into established groups during which none, one, several, or all of the resident males might be evicted (*A. arctoidea*: Pope 1990; Crockett and Pope 1993; Agoramoorthy and Rudran 1995; Crockett and Sekulic 1984; *A. palliata*: Clarke 1983; Glander 1992; Dias et al. 2010; *A. pigra*: Horwich et al. 2000; Van Belle et al. 2008; *A. seniculus*: Izawa and Lozano 1991; Kimura 1992; Izawa 1997). Take-over attempts may be accompanied by infanticidal attacks (reviewed in Crockett 2003) and both take-over attempts and intergroup encounters may involve extragroup copulations (reviewed in Van Belle and Bicca-Marques 2014), indicating that encounters with extragroup males may pose substantial threats to the reproductive success of resident males and females.

Cristóbal-Azkarate et al. (2006, 2007) investigated how male fecal androgen and male and female fecal glucocorticoid levels were influenced by the threat of solitary males in ten mantled howler monkey groups living in six forest fragments at Los Tuxtlas Biosphere Reserve, Mexico. Groups had, on average, 3.4 adult males (range = 1–6) and 5.4 adult females (range = 2–9). A total of 31 samples were collected from 17 males, with a mean of 3.1 (range = 1–7) male fecal samples per group, while a total of 35 samples were collected from 18 different adult females and 12 unidentified adult females, with a mean of 3.5 (range = 1–10) female fecal samples per group. Hormonal levels were averaged across all sampled males and all sampled females per group. No encounters between social groups and solitary males were observed during the study. Nonetheless, mean male fecal androgen levels per group were positively correlated with the number of extragroup males living in the same forest fragment, suggesting that mantled howler males on average exhibited a hormonal response proportional to the potential threat posed by solitary males (Cristóbal-Azkarate et al. 2006). Fecal glucocorticoid levels averaged across all sampled males per group were not significantly correlated, while those averaged across all sampled females per group were positively correlated with the number of solitary males living in the same forest fragment (Cristóbal-Azkarate et al. 2007). The authors suggested that differences in the ways in which males and females cope with stressful, unpredictable situations posed by the presence of extragroup males in their forest fragment may account for different mean fecal glucocorticoid levels among resident males and females. The passive response displayed by resident mantled howler females towards solitary males might be associated with an activation of the hypothalamic–pituitary–adrenal (HPA) axis resulting in chronically elevated fecal glucocorticoid levels, whereas the more active and aggressive response displayed by resident males seemed not to be associated with the activation of the HPA axis resulting in lower fecal glucocorticoid levels (Cristóbal-Azkarate et al. 2007), similar to differential glucocorticoid levels associated with passive versus active coping styles in captive rodents (Koolhaas et al. 1999; Ebner et al. 2005).

Similarly, a study of five black howler monkey groups living in five different sites in the state of Campeche, Mexico that were each observed for two sampling periods of 4 weeks separated by 3–8 months provided evidence that extragroup male–male competition also may be reflected in male fecal androgen levels in this species (Rangel-Negrín et al. 2011). In this study, mean fecal androgen levels averaged across samples per male were significantly higher for males living in one-male groups ($N=2$ adult males) compared to males living in two-male or three-male groups ($N=7$ adult males). Furthermore, two of the three multimale groups changed from multimale to unimale groups between the first and the second observation period, and mean fecal androgen levels of both males who remained in their respective groups increased significantly when being the sole male in a unimale group compared to when part of a multimale group. Because the probability of resident males being evicted by extragroup males during take-over attempts might be higher for unimale compared to multimale groups (*A. arctoidea*, Agoramoorthy and Rudran 1995; Pope 2000; *A. pigra*, Horwich et al. 2000; Van Belle et al. 2008), Rangel-Negrín et al. (2011) argued that unimale groups might be more attractive targets to dispersing males, and resident males of unimale groups might manifest elevated androgen levels in response to possible confrontations with extragroup males and increased risk of being ousted from their groups.

In contrast, a 14-month study of two multimale–multifemale black howler groups in Palenque National Park, Mexico, suggested that intragroup male–male competition, but not extragroup male–male competition, modulated male fecal androgen and glucocorticoid concentrations (Van Belle et al. 2009b). During this study, both focal groups underwent several changes in male group membership. These included (1) three take-over events during which three coalitions of two extragroup males each took over one of the study groups and evicted all adult resident males ($N=2$ males per event) and (2) three male immigration events during which a single extragroup male joined either of the two study groups that had one, two, or three resident males at the time of immigration. This did not result in the eviction of resident males, except for one case in which one of the three adult resident males was evicted (for more details see Van Belle et al. 2008). Immigrant males ($N=9$) had no significant differences in their fecal glucocorticoid and androgen levels during week 1 and week 2 following the take-over/immigration compared to week 3 and week 4. Similarly, resident males ($N=5$) did not differ in their fecal glucocorticoid and androgen levels 2 weeks before and 2 weeks after a male immigration event (Van Belle et al. 2009b). Also, the five resident females in the two study groups did not have consistently higher fecal glucocorticoid levels 2 weeks after compared to 2 weeks before changes in male membership in their social group (Van Belle, unpubl. data; see Appendix 1). Furthermore, biweekly rates of encounters with either adjacent social groups or extragroup males were not correlated with changes in male hormonal levels (Van Belle et al. 2009b). These findings suggest that actual events of male–male competition over group membership, as opposed to potential threat, are not readily reflected in the fecal hormonal levels of black howler males and females in Palenque National Park. It is possible, however, that the unusually rapid turnover of male group membership in one of the study groups (7 changes in 6 months) might have affected the researchers' ability to evaluate the influence of male immigration on male and female hormonal levels. Additional studies are needed to elucidate and compare changes in hormonal profiles during extragroup male–male competition across howler monkey species and groups living in both unimale and multimale social groups to better understand the social and demographic factors mediating hormonal responses to extragroup male–male competition.

Instead of extragroup male–male competition, Van Belle et al. (2009b) found that intragroup male–male competition is reflected in male fecal steroid levels in black howler monkeys. Although, resident black howler males seldom engaged in intrasexual agonistic (mean rate = 0.007 interactions/h/dyad) or affiliative interactions (mean rate = 0.009 interactions/h/dyad) and no agonistic dominance hierarchy could be discerned (Van Belle et al. 2008), one resident adult male per group, referred to as the "central" male, was found to monopolize almost all mating opportunities, spent significantly more time in close proximity to females, and engaged in affiliation at significantly higher rates with cycling females than did "noncentral" males (Van Belle et al. 2009b). Noncentral males had very limited mating opportunities and accounted for only 4 % of copulations (Van Belle et al. 2008). In addition, central males had significantly higher fecal androgen and glucocorticoid levels compared to noncentral males, suggesting that their efforts of fostering social relationships with females might represent a nonaggressive form of male–male competition over sexual access to females. This nonaggressive form of male–male competition might be socially challenging to central males as indicated by their higher fecal glucocorticoid levels (Van Belle et al. 2009b).

During this study, central males had significantly higher hormonal levels than noncentral males during resident female periovulatory and nonperiovulatory periods of ovarian cycles, as well as when none of the resident females were cycling (Van Belle et al. 2009b). This suggests that central males had elevated steroid levels throughout the year, even during periods when resident females were not sexually active. Furthermore, central male hormonal levels did not increase during the times when at least one resident female was cycling or during periovulatory periods of cycling females, despite higher copulation rates and heightened efforts by central males to spend time close to and groom cycling females compared to noncycling females (Van Belle et al. 2009a, b). In contrast, noncentral males had significantly lower mean fecal androgen levels during resident female periovulatory periods, which might be indicative of some suppression of testicular endocrine function at times when resident females are most likely to conceive (Van Belle et al. 2009b). Yet, it is unlikely that these lower androgen levels fully suppressed the sexual function of noncentral males because at least one noncentral male was observed copulating during the periovulatory period of a cycling female (Van Belle et al. 2009a). Furthermore, a study that investigated male reproductive physiology and sperm production in three adult male and three juvenile male black-and-gold howler monkeys in captivity revealed that even low levels of fecal androgens were sufficient for normal sperm count, quality, and motility (Moreland et al. 2001). Additionally, small testes size relative to body size characteristic for black howler monkeys suggests an overall weak level of sperm competition in this howler monkey species (Kelaita et al. 2011).

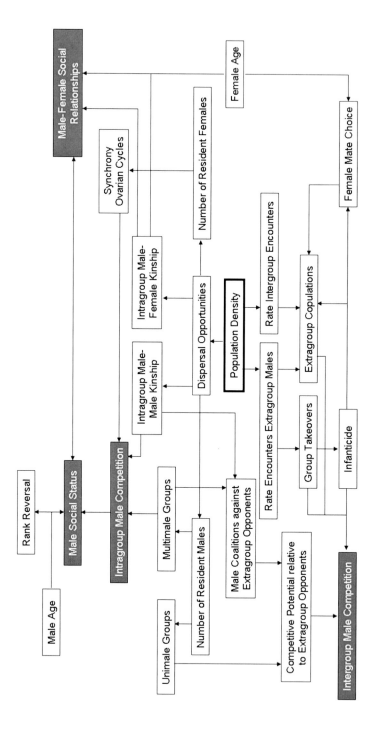

Fig. 8.3 Hypothetical, social, and demographic context of male hormonal profiles underlying male and female reproductive strategies in howler monkeys. Population density is considered central as it mediates both group composition and encounter rates with opponents. The *gray boxes* indicate social factors that have been observed to influence male hormonal profiles in howler monkeys

In Rangel-Negrín et al.'s (2011) study of *A. pigra*, mean fecal androgen levels in males of two focal groups for which copulations were observed were not significantly higher in weeks when copulations were observed compared to weeks without observed copulations. One of these two study groups contained only one adult male, while the other study group had two adult males. In the latter group, both males were observed to copulate with the resident females. In contrast to the findings of Van Belle et al. (2009b), Rangel-Negrín et al. (2011) did not observe significant differences in mean fecal androgen levels between central and noncentral males in three multimale–multifemale groups. These different results could reflect variable hormonal responses underlying individual male reproductive strategies associated with group size, group composition, number of reproductively active females, female mate choice, or demographic factors such as the number of solitary males in the local area.

Further studies are needed to elucidate hormonal profiles underlying alternative male and female reproductive strategies in this and other howler monkey species. This is especially true when considering that the genus *Alouatta* is characterized by a highly flexible social system including great variability in group size and composition with a mixture of unimale and multimale bisexual groups in most populations regardless of the species (Di Fiore et al. 2011). As such, the reproductive strategies and the underlying endocrine correlates of individual males and females may be influenced by the number of resident adult males and females, population density, the number of extragroup males, the intensity of intra- and intergroup male–male competition over access to the fertile females, individual social status, age, kinship patterns, the degree to which females exert mate choice, and dispersal opportunities (Fig. 8.3). It remains unclear exactly how each of these social and demographic factors differentially affect male and female reproductive strategies across howler monkeys species. What is clear is that endocrine studies will help elucidate subtle distinctions not revealed by behavioral observations alone.

8.6 Ecological Endocrinology

The physiological stress response of increased circulating glucocorticoids and subsequent elevated glucose levels is also affected by a wide range of environmental factors. For example, elevated glucocorticoid levels is characteristic of individuals facing ecological challenges such as food scarcity (e.g., ring-tailed lemurs, Pride 2005), cold temperatures (e.g., chacma baboons, Weingrill et al. 2004), prolonged daylight duration (e.g., gray mouse lemurs, *Microcebus murinus*, Génin and Perret 2000), high gastrointestinal parasite load (red colobus monkeys, *Piliocolobus tephrosceles*, Chapman et al. 2006), increased risk of predation (e.g., gray-cheeked mangabey, *Lophocebus albigena*, Arlet and Isbell 2009), and habitat fragmentation and degradation (e.g., red colobus monkeys, Chapman et al. 2006; spider monkeys, *Ateles geoffroyi yucatanensis*, Rangel-Negrín et al. 2009). In addition, elevated glucocorticoid levels can be experienced by individuals residing in atypically large or small social groups (e.g., ring-tailed lemurs, Pride 2005) or by females during the energetically demanding period of lactation (e.g., chacma baboons, Weingrill et al. 2004).

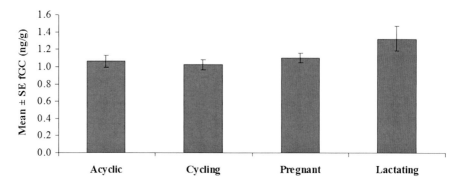

Fig. 8.4 Mean fecal glucocorticoid levels (fGC) of female black howler monkeys (*A. pigra*, n = 5) during different reproductive states during a 14-month study at Palenque National Park, Mexico

Female primates face different energetic demands across different reproductive stages, with lactation exerting the highest energetic demands (Dufour and Sauther 2002). To offset the energetic demands of lactation, black howler females are reported to increase caloric intake, rest less, and range further distances during the first 10 months of lactation compared to adult non-lactating females (Dias et al. 2011, see also Serio-Silva et al. 1999). In parallel with these behavioral changes, black howler females also may increase their metabolic rate by experiencing significantly higher fecal glucocorticoid levels, and therefore purportedly higher level of circulating glucose, when lactating compared to less energetic demanding reproductive stages (Fig. 8.4, Van Belle, unpubl. data; see Appendix 1). Elevated fecal glucocorticoid metabolites in pregnant and lactating females compared with nonreproducing females were also reported for mantled howler monkeys (Dunn et al. 2013). These elevated glucocorticoid levels in pregnant and lactating females might be partially a result of an increased stress response towards psychosocial stressors such as agonistic interactions during periods when energetic demands are high, as was observed for eight mantled howler females from two social groups (Gómez-Espinosa et al. 2013).

Using a large set of fecal samples (n = 350) collected from ten social groups during a 21-month study spread across 4 years, Behie et al. (2010) investigated variation in fecal glucocorticoid levels in the black howler population at Monkey River, Belize in relation to group size, presence of tourists, and maximum monthly temperature. Only lactating females were considered in this study to control for the effect that different reproductive stages have on glucocorticoid concentrations. These authors found that fecal glucocorticoid levels of both males and females were not influenced by group size and maximum monthly temperature, but were positively influenced by the presence of tourists. The two study groups that were frequently visited by tourists had significantly higher fecal glucocorticoid levels than other groups that experienced limited human contact. Behie et al. (2010) suggested that either the unpredictability of tourist visits or the intrusive and noisy behavior by tourists might account for the elevated fecal glucocorticoid levels in these groups. However, a group of red-handed howler monkeys exposed to loud noises from nearby mining practices did not have significantly higher fecal glucocorticoid levels compared with a nearby group living away from these mining practices (Monteiro et al. 2013). Similarly, in two mantled howler groups residing in a recreational forest reserve of which a section is a zoological park open to tourism and that had been exposed to humans throughout their lives, daily number of people visiting the park did not influence fecal glucocorticoid levels (Aguilar-Melo et al. 2013). In addition, fecal glucocorticoid levels in 31 red howler monkey groups residing in 10 different forest fragments also were not influenced by the level of human disturbance (i.e., minimal, logging, hunting, and logging and hunting), suggesting that howler monkeys might have a lower sensitivity to anthropogenic disturbance compared to other species such as spider monkeys (Rimbach et al. 2013b).

During their 21-month study, Behie and colleagues also measured fruit production of 200 trees belonging to the top 12 food species for the black howler monkeys at Monkey River and assessed parasite load in fecal samples. Infection by multiple parasite species, but not parasite prevalence, negatively influenced fecal glucocorticoid levels, suggesting greater physiological stress on the body when it is parasitized by more than one species that compete with each other for the host's nutrients and energy (Behie and Pavelka 2013). Fruit availability also influenced male and female fecal glucocorticoid levels with elevated levels when fruits were scarce. These data suggest that black howler monkeys may be more dependent on fruit than previously believed, resulting in a subtle form of feeding competition in this primarily folivorous primate species (Behie et al. 2010; Behie and Pavelka 2013).

Similarly based on data collected during an 8-month study, limited fruit availability in particular, or low food quality in general, may have partially accounted for the higher mean fecal glucocorticoid levels found in two black howler groups inhabiting small forest fragments (<2 ha, Ejido Leona Vicario, Balancán, Mexico) compared to two black howler groups living in a more continuous forest (>1,400 ha, Calakmul Biosphere Reserve, Mexico; Martínez-Mota et al. 2007). Groups inhabiting the forest fragments rested, fed, and socially interacted with group members at similar rates as those in the continuous forest, but traveled more frequently. Members of one of the howler groups living in forest fragments spent 11 % of their traveling time walking on the ground, which may have increased their exposure to terrestrial predators including dogs or coyotes, or their contact with fecal material and increased their susceptibility to infection from anthropogenic diseases such as gastrointestinal parasites (Martínez-Mota et al. 2007). However, the precise factors that resulted in elevated fecal glucocorticoid levels in these howler groups remain unclear.

Fecal glucocorticoid levels in two groups of mantled howler monkeys residing in forest fragments of 244 ha and 7 ha, respectively, at Los Tuxtlas Biosphere Reserve were significantly and positively correlated to the percentage of time traveling, providing critical insight into the proximate mechanisms mediating elevated glucocorticoid levels in primates in fragments (Dunn et al. 2013). Although fecal glucocorticoid levels were not significantly influenced by fruit consumption per se, increased travel time was associated with reduced fruit consumption, decreased time feeding in large trees, and decreased time feeding on primary food resources resulting in increased feeding efforts in the study group living in the small and more disturbed forest fragment than the study group living in the more conserved fragment

(Dunn et al. 2009, 2010). As such, in 31 groups of red howler monkeys living in 10 forest fragments ranging in size from 4 to 500 ha in Colombia, fragment size was not a significant predictor of fecal glucocorticoid levels (Rimbach et al. 2013b). The authors suggested that the apparent absence of physiological stress in the small forest fragments may be due to the fact that the study region has been undergoing fragmentation recently (<10 years), and hence that drastic changes in food availability might not have occurred yet.

A subsequent 4-month study spread evenly over the dry and wet season observing two additional mantled howler groups living in forest fragments of 230 ha and 15 ha, respectively, at the Los Tuxtlas Biosphere Reserve demonstrated that the group in the small fragment had higher fecal glucocorticoid levels during the dry season, but not the wet season, than the study group in the larger forest fragment (Gómez-Espinosa et al. 2013). However, this was not due to decreased food availability in the small forest fragment, nor were time spent traveling and ranging distance significant predictors of weekly glucocorticoid levels in these study groups, contrasting with the results by Dunn et al. (2013). Instead, Gómez-Espinosa et al. (2013) hypothesized that anthropogenic disturbance concentrated during the dry season when humans visited the river running through the small forest fragment more regularly might better explain the seasonal pattern of glucocorticoid levels in the group members residing in the small forest fragment. These inconsistent patterns of glucocorticoid profiles in howler monkeys inhabiting forest fragments reveal that multiple factors, including fragment size and history, food availability and associated feeding efforts, and the level of human disturbance, mediate the stress response in howler monkeys in fragments.

Endocrine studies also contribute directly to conservation programs, especially when researchers actively monitor glucocorticoid levels during translocation projects. One Mexican mantled howler group, composed of two adult males and two adult females, was translocated from a 4.9 ha forest fragment that was scheduled to be converted to agricultural land. The group was moved to an 80 ha protected forest 50 km away from their pre-translocation area (Aguilar-Cucurachi et al. 2010). The translocation followed a soft-release protocol in which the four adults were first held in captivity close to the post-translocation site and were provided with food for 1 month. They were then released into a 0.18 ha outdoor enclosure where the monkeys could forage on natural vegetation for 1 month, and finally released into the wild at the post-translocation site and monitored for an additional month. Each stage of the process from their pre-translocation site, to captivity, to semi-captivity, to their post-translocation site involved recapturing the four individuals, and fecal samples were collected during all four stages. Male fecal glucocorticoid levels remained stable throughout the translocation process, whereas female fecal glucocorticoid levels progressively increased from pre-translocation, to captivity, to semi-captivity, after which their glucocorticoid levels dropped below those recorded at the pre-translocation site once released at the post-translocation site and human handling was minimal. Females had significantly higher fecal glucocorticoid levels than males during all translocation stages, except during the pre-translocation stage, suggesting that female mantled howler monkeys were more sensitive to stressors. In addition, the data suggest that the increase in female howler stress response during translocation to a new habitat were lower than their response to being captured, handled, confined, and in close proximity to humans (Aguilar-Cucurachi et al. 2010). Possible ways to reduce stress during translocation include minimizing human handling and the period of confinement by curtailing the period in captivity in favor of semi-captivity.

8.7 Conclusion

Howler monkey studies that included the examination of hormonal profiles underlying behavioral strategies have revealed several important insights into their biology, social interactions, and mating strategies that were not anticipated based on behavioral data alone. For example, endocrine studies have suggested a nonaggressive form of intragroup male–male competition over access to fertile females in black howler monkeys based on elevated androgen and glucocorticoid levels in central compared to noncentral males (Van Belle et al. 2009b), despite otherwise neutral male–male social relationships based on tolerance and avoidance (Van Belle et al. 2008). A direct relationship between male steroid levels and male–female social and sexual interactions also has been observed in tufted capuchin monkeys (Lynch et al. 2002), Japanese macaques (*Macaca fuscata*, Barrett et al. 2002), and bonobos (*Pan paniscus*, Surbeck et al. 2012), suggesting that social relationships of males with cycling females may be a more pervasive driver of male endocrine function than generally thought. In this regard, howler monkeys are not unique but fit a broader pattern and may serve as an important instructive model for examining questions regarding male endocrinology underlying male–female relationships. In addition, the two endocrine studies investigating the influence of inter- and intragroup male–male competition on male androgen profiles in black howler monkeys (Van Belle et al. 2009b; Rangel-Negrín et al. 2011) revealed considerable differences in hormonal profiles, most likely reflecting distinct male reproductive strategies in different social and demographic settings. This suggests that howler endocrine function is highly adaptable and responds to subtle proximate changes in the social environment.

Endocrine studies also revealed that glucocorticoid levels were higher during periods of fruit scarcity compared to periods of fruit abundance in a black howler population recovering from a population collapse and habitat destruction caused by a hurricane in Belize. This suggests more prominent reliance on fruits as a staple food resource than was previously thought for howler monkeys (Behie et al. 2010; Behie and Pavelka 2013; also see Behie and Pavelka 2014 and Garber et al. 2014 on the importance of fruit in the diet of several howler species). At a proximate level, increased feeding effort by spending more time traveling between feeding sites may result in elevated glucocorticoid levels during periods of fruit scarcity or in degraded forest fragments (Dunn et al. 2013). Such findings can be incorporated into conservation programs for howler monkeys by examining fruit species diversity and seasonal availability in howler monkey habitats that have been degraded and fragmented because of human activity. For example, effective conservation policies could include the planting of particular fruiting species in fragmented habitats occupied by howler monkeys.

Finally, comparison of glucocorticoid levels between males and females has revealed that females of mantled howler monkeys appear to be more sensitive than males to either social (e.g., threat of extragroup males in habitat; Cristóbal-Azkarate et al. 2007) or ecological stressors (e.g., translocation and human handling; Aguilar-Cucurachi et al. 2010), information that could only have been identified by combining behavioral observations with endocrine studies.

I advocate for additional endocrine–behavioral–ecological studies on other howler monkey species and populations encompassing a wider range of environmental conditions, such as environments exhibiting prominent seasonality in food resources, rainfall, or temperatures, and a wider range of social/demographic conditions, including groups containing one male versus multiple males, groups with adults coresiding with kin versus nonkin, or well-established groups versus newly formed groups. Such studies are needed to assess how social, demographic, and ecological fluctuations affect male and female hormonal profiles and reproductive strategies. Although a limited number of studies of howler monkeys have focused on ovarian cycles, the effects of male–male competition, fruit availability, fragmentation, and translocation, information is extremely limited on hormonal profiles during infant development, the onset of sexual maturation, pregnancy, and patterns of hormonal fluctuation in response to seasonal changes in food availability across disturbed and undisturbed habitats. In addition, endocrine studies exploring the nature and formation of social bonding among male–male coalitions, mother–offspring bonds, or the role of paternal care during infant social development focusing on peptide hormones like oxytocin, prolactin, and vasopressin (e.g., Schradin and Anzenberger 2004; Seltzer and Ziegler 2007; Ziegler et al. 2009b; Anestis 2010; Moscovice and Ziegler 2012) are needed, especially in order to compare differences between howler groups or species in which collective action, male–male tolerance, and female–female tolerance, and female mate choice have been reported (Pope 1990, 2000; Van Belle et al. 2008, 2009a, b, 2011; Kowalewski and Garber 2010; Garber and Kowalewski 2011).

Appendix: The Effect of Male Migration and Reproductive Status on Black Howler Female Fecal Glucocorticoid Levels

Methods

Two multimale–multifemale black howler (*A. pigra*) groups were studied in Palenque National Park, Mexico from June 2006 through July 2007. The Balam group had three adult females at the onset of the study, but one female (MI) disappeared on October 18, 2006. The Motiepa group had two adult females throughout the study period. Both groups underwent several changes in adult male group membership (Balam, $n=7$; Motiepa, $n=2$; for detailed description, see Van Belle et al. 2008). Fresh fecal samples were collected from each adult female, on average, every 4.1 ± 1.4 days, resulting in a total of 246 samples. Methods used for sample storage, hormone extraction, assay validation, and glucocorticoid EIA are described in detail in Van Belle et al. (2009b). Values of hormone concentrations were $\log_{10}$ transformed to normalize the distribution and equalize the variance (Kolmogorov–Smirnov tests and Levene's tests, $P > 0.05$), allowing the use of parametric tests.

General linear mixed models (GLMM) were used to analyze whether glucocorticoid levels of females differed between the 2 weeks prior to and the 2 weeks after male migration events in their groups. Random factors in the GLMM included female identity to account for the repeated sampling of the same individual, which was nested within groups to account for the possibility that coresiding females had correlated hormone levels and migration events nested within groups to account for the possibility that coresiding females changed hormonal levels similarly to the same migration events. To examine whether female glucocorticoid levels changed according to their reproductive status, a GLMM was used with female identity nested within groups as random factors and reproductive status as predictor variable. Female reproductive status was classified as acyclic, cycling, pregnant, or lactating based on their fecal estrogen and progestin profiles and presence of young offspring (Van Belle et al. 2009a).

Results

Adult females did not differ in their fecal glucocorticoid levels between the 2 weeks before and after male migrations in their groups (Mean$_{Before}$ = 0.99 ± SE 0.09, Mean$_{After}$ = 1.05 ± 0.07, $F_{2,39.6} = 0.650$, $P = 0.527$). However, female fecal glucocorticoid levels across different reproductive states were significantly different with females having higher fecal glucocorticoid levels when lactating compared to other reproductive states (Fig. 8.3, $F_{3,240.7} = 10.48$, $P < 0.001$)

Acknowledgments I thank Martín M. Kowalewski, Paul A. Garber, Liliana Cortés-Ortiz, Bernardo Urbani, and Dionisios Youlatos for the invitation to contribute to this volume and to Martín M. Kowalewski, Paul A. Garber, and an anonymous reviewer for constructive suggestions that improved the quality of the chapter. I was supported by a postdoctoral fellowship from Universidad Nacional Autónoma de México (UNAM) during the writing of this manuscript.

References

Abbott DH, Keverne EB, Bercovitch FB, Shively CA, Mendoza SP, Saltzman W, Snowdon CT, Ziegler TE, Banjevic M, Garland TJ, Sapolsky RM (2003) Are subordinates always stressed? A comparative analysis of rank differences in cortisol levels among primates. Horm Behav 43:67–82

Agoramoorthy G, Rudran R (1995) Infanticide by adult and subadult males in free-ranging red howler monkeys, *Alouatta seniculus*, in Venezuela. Ethology 99:75–88

Aguilar-Cucurachi MAS, Dias PAD, Rangel-Negrín A, Chavira R, Boeck L, Canales-Espinosa C (2010) Preliminary evidence of accumulation of stress during translocation in mantled howlers. Am J Primatol 72:805–810

Aguilar-Melo AR, Andresen E, Cristóbal-Azkarate J, Arroyo-Rodríguez V, Chavira R, Schondube J, Serio-Silva JC, Cuarón AD (2013) Behavioral and physiological responses to subgroup size and number of people in howler monkeys inhabiting a forest fragment used for nature-based tourism. Am J Primatol 75:1108–1116

Anestis SF (2010) Hormones and social behavior in primates. Evol Anthropol 19:66–78

Arlet ME, Isbell LA (2009) Variation in behavioral and hormonal responses of adult male gray-cheeked mangabeys (*Lophocebus albigena*) to crowned eagles (*Stephanoaetus coronatus*) in Kibale National Park, Uganda. Behav Ecol Sociobiol 63:491–499

Bahr NI, Palme R, Möhle U, Hodges JK, Heistermann M (2000) Comparative aspects of the metabolism and excretion of cortisol in three individual non-human primates. Gen Comp Endocrinol 117:427–438

Barrett GM, Shimizu K, Bardi M, Asaba S, Mori A (2002) Endocrine correlates of rank, reproduction, and female-directed aggression in male Japanese macaques (*Macaca fuscata*). Horm Behav 42:85–96

Behie AM, Pavelka MSM (2013) Interacting roles of diet, cortisol levels, and parasites in determining population density of Belizean howler monkeys in a hurricane damaged forest fragment. In: March LK, Chapman CA (eds) Primates in fragments: complexity and resilience. Springer Press, New York

Behie AM, Pavelka SM (2014) Fruit as a key factor in howler monkey population density: conservation implications. In: Kowalewski M, Garber P, Cortés-Ortiz L, Urbani B, Youlatos D (eds) Howler monkeys: behavior, ecology and conservation. Springer, New York

Behie AM, Pavelka MSM, Chapman CA (2010) Sources of variation in fecal cortisol levels in howler monkeys in Belize. Am J Primatol 71:1–7

Bercovitch FB, Ziegler TE (2002) Current topics in primate socioendocrinology. Annu Rev Anthropol 31:45–67

Bergman TJ, Beehner JC, Cheney DL, Seyfarth RM, Whitten PL (2005) Correlates of stress in free-ranging male chacma baboons, *Papio hamadryas ursinus*. Anim Behav 70:703–713

Bernstein IS, Rose RM, Gordon TP, Grady CL (1979) Agonistic rank, aggression, social context, and testosterone in male pigtail monkeys. Aggress Behav 5:329–339

Bicca-Marques JC, Calegaro-Marques C (1998) Behavioral thermoregulation in a sexually and developmentally dichromatic Neotropical primate, the black-and-gold howling monkey (*Alouatta caraya*). Am J Phys Anthropol 106:533–646

Bradshaw D (2007) Environmental endocrinology. Gen Comp Endocrinol 152:125–141

Braude S, Tang-Martinez Z, Taylor G (1999) Stress, testosterone, and the immunoredistribution hypothesis. Behav Ecol 10:345–350

Brockman DK, Whitten PL, Richard AF, Benander B (2001) Birth season testosterone levels in male Verreaux's sifaka, *Propithecus verreauxi*: insight into socio-demographic factors mediating seasonal testicular function. Behav Ecol Sociobiol 49:117–127

Cavigelli SA, Dubovick T, Levash W, Jolly A, Pitts A (2003) Female dominance status and fecal corticoids in a cooperative breeder with low reproductive skew: ring-tailed lemurs (*Lemur catta*). Horm Behav 43:166–179

Chapman CA, Wasserman MD, Gillespie TR, Speirs ML, Lawes MJ, Saj TJ, Ziegler TE (2006) Do food availability, parasitism, and stress have synergistic effects on red colobus populations living in forest fragments? Am J Phys Anthropol 131:525–534

Clarke MR (1983) Infant-killing and infant disappearance following male takeovers in a group of free-ranging howling monkeys (*Alouatta palliata*) in Costa Rica. Am J Primatol 5:241–247

Clarke MR, Zucker EL, Ford RT, Harrison RM (2007) Behavior and endocrine concentrations do no distinguish sex in monomorphic juvenile howlers (*Alouatta palliata*). Am J Primatol 69:477–484

Cockrem JF (2005) Conservation and behavioral neuroendocrinology. Horm Behav 48:492–501

Colillas O, Coppo J (1978) Breeding *Alouatta caraya* in Centro Argentino de Primates. In: Chivers DJ, Lane-Petter W (eds) Recent advances in primatology, 2. Conservation. Academic, London

Cristóbal-Azkarate J, Dias PAD, Veà JJ (2004) Causes of intraspecific aggression in *Alouatta palliata mexicana*: evidence from injuries, demography, and habitat. Int J Primatol 25:939–953

Cristóbal-Azkarate J, Chavira R, Boeck L, Rodríguez-Luna E, Veà JJ (2006) Testosterone levels of free-ranging resident mantled howler monkey males in relation to the number and density of solitary males: a test of the challenge hypothesis. Horm Behav 49:261–267

Cristóbal-Azkarate J, Chavira R, Boeck L, Rodríguez-Luna E, Veà JJ (2007) Glucocorticoid levels in free ranging resident mantled howlers: a study of coping strategies. Am J Primatol 69:866–876

Crockett CM (2003) Re-evaluating the sexual selection hypothesis for infanticide by Alouatta males. In: Jones CB (ed) Sexual selection and reproductive competition in primates: new perspectives and directions, vol 3, American Society of Primatology: Special Topics in Primatology. The American Society of Primatologists, Norman, Oklahoma

Crockett CM, Eisenberg JF (1987) Howlers: variations in group size and demography. In: Smuts BB, Cheney DL, Seyfarth RM, Wrangham RW, Struhsaker TT (eds) Primate societies. The University of Chicago Press, Chicago

Crockett CM, Pope TR (1988) Inferring patterns of aggression from red howler monkey injuries. Am J Primatol 15:289–308

Crockett CM, Pope TR (1993) Consequences of sex differences in dispersal for juvenile red howler monkeys. In: Pereira MA, Fairbanks LA (eds) Juvenile primates: life history, development and behavior. Oxford University Press, New York

Crockett CM, Sekulic R (1982) Gestation length in red howler monkeys. Am J Primatol 3:291–294

Crockett CM, Sekulic R (1984) Infanticide in red howler monkeys (*Alouatta seniculus*). In: Hausfater G, Hrdy SB (eds) Infanticide: comparative and evolutionary perspectives. Aldine, New York

Di Fiore A, Link A, Campbell CJ (2011) The atelines: behavioral and socioecological diversity in a New World radiation. In: Campbell CJ, Fuentes A, MacKinnon KC, Panger M, Bearder SK (eds) Primates in perspective, 2nd edn. Oxford University Press, Oxford

Dias PAD, Rangel-Negrín A, Veà JJ, Canales-Espinosa D (2010) Coalitions and male–male behavior in *Alouatta palliata*. Primates 51:91–94

Dias PAD, Rangel-Negrín A, Canales-Espinosa D (2011) Effects of lactation in the time-budgets and foraging patterns of female black howlers (*Alouatta pigra*). Am J Phys Anthropol 145:137–146

Dufour DL, Sauther ML (2002) Comparative and evolutionary dimensions of human pregnancy and lactation. Am J Hum Biol 14:585–602

Dunn JC, Cristóbal-Azkarate J, Veà J (2009) Differences in diet and activity pattern between two groups of *Alouatta palliata* associated with the availability of big trees and fruit of top food taxa. Am J Primatol 71:654–662

Dunn JC, Cristóbal-Azkarate J, Veà J (2010) Seasonal variations in the diet and feeding effort of two groups of howler monkeys in different sized forest fragments. Int J Primatol 31: 887–903

Dunn JC, Cristóbal–Azkarate J, Schulte-Herbrüggen B, Chavira R, Veà JJ (2013) Travel time predicts fecal glucocorticoid levels in free-ranging howlers (*Alouatta palliata*). Int J Primatol 34:246–259

Ebner K, Wotjak CT, Landgraf R, Engelmann M (2005) Neuroendocrine and behavioral response to social confrontation: residents versus intruders, active versus passive coping styles. Horm Behav 47:14–21

Garber PA, Kowalewski MM (2011) Collective action and male affiliation in howler monkeys (*Alouatta caraya*). In: Sussman RW, Cloninger CR (eds) Origins of altruism and cooperation. Springer Publishers, New York

Garber P, Righini N, Kowalewski M (2014) Evidence of alternative dietary syndromes and nutritional goals in the genus *Alouatta*. In: Kowalewski M, Garber P, Cortés-Ortiz L, Urbani B, Youlatos D (eds) Howler monkeys: behavior, ecology and conservation. Springer, New York

Génin F, Perret M (2000) Photoperiod-induced changes in energy balance in gray mouse lemurs. Physiol Behav 71:315–321

Glander KE (1980) Reproduction and population growth in free-ranging mantled howling monkeys. Am J Phys Anthropol 53:25–36

Glander KE (1992) Dispersal patterns in Costa Rican mantled howling monkeys. Int J Primatol 13:415–436

Gómez-Espinosa E, Rangel-Negrín A, Chavira R, Canales-Espinosa D, Dias PAD (2013) The effect of energetic and psychosocial stressors in glucocorticoids in mantled howler monkeys (*Alouatta palliata*). Am J Primatol 76(4):362–373

Goymann W, Wingfield JC (2004) Allostatic load, social status and stress hormones: the costs of social status matter. Anim Behav 67:591–602

Hau M (2007) Regulation of male traits by testosterone: implications for the evolution of vertebrate life-histories. Bioessays 29:133–144

Heistermann M, Palme R, Ganswindt A (2006) Comparison of different enzyme immunoassays for assessment of adrenocortical activity in primates based on fecal analysis. Am J Primatol 68:257–273

Herrick JR, Agoramoorthy G, Rudran R, Harder JD (2000) Urinary progesterone in free-ranging red howler monkeys (*Alouatta seniculus*): preliminary observations of the estrous cycle and gestation. Am J Primatol 51:257–263

Horwich RH, Brockett RC, Jones CB (2000) Alternative male reproductive behaviors in the Belizean black howler monkey (*Alouatta pigra*). Neotrop Primates 8:95–98

Huck M, Löttker P, Heymann EW, Heistermann M (2005) Characterization and social correlates of fecal testosterone and cortisol excretion in wild male *Saguinus mystax*. Int J Primatol 26:159–179

Izawa K (1997) Social changes within a group of red howler monkeys, VI. Field Stud Fauna and Flora, La Macarena, Colombia 11:19–34

Izawa K, Lozano HM (1991) Social changes within a group of red howler monkeys (*Alouatta seniculus*), III. Field Stud New World Monkeys, La Macarena, Colombia 5:1–16

Jones CB (1985) Reproductive patterns in mantled howler monkeys: estrus, mate choice and copulation. Primates 26:130–142

Kelaita M, Dias PAD, Aguilar-Cucurachi MS, Canales-Espinosa D, Cortés-Ortiz L (2011) Impact of intrasexual selection on sexual dimorphism and testes size in the Mexican howler monkeys *Alouatta palliata* and *A. pigra*. Am J Phys Anthropol 146:179–187

Kimura K (1992) Demographic approach to the social group of wild red howler monkeys (*Alouatta seniculus*). Field Stud New World Monkeys, La Macarena, Colombia 7:29–34

Koolhaas JM, Korte SM, De Boer SF, Van Der Vegt BJ, Van Reenen CG, Hopster H, De Jong IC, Ruis MA, Blokhuis HJ (1999) Coping styles in animals: current status in behavior and stress physiology. Neurosci Biobehav Rev 23:925–935

Kowalewski MM, Garber PA (2010) Mating promiscuity and reproductive tactics in female black and gold howler monkeys (*Alouatta caraya*) inhabiting an island on the Parana River, Argentina. Am J Primatol 72:734–748

Kowalewski M, Garber P (2014) Solving the collective action problem during intergroup encounters: the case of black and gold howler monkeys. In: Kowalewski M, Garber P, Cortés-Ortiz L, Urbani B, Youlatos D (eds) Howler monkeys: behavior, ecology and conservation. Springer, New York

Kugelmeier T, del Rio Do Valle R, de Barros Vaz Guimarães MA, Carneiro Muniz JAP, Barros Monteiro FO, Alvarenga de Oliveira C (2011) Tracking the ovarian cycle in black-and-gold howlers (*Alouatta caraya*) by measuring fecal steroids and observing vaginal bleeding. Int J Primatol 32:605–615

Lasley BL, Savage A (2007) Advances in the understanding of primate reproductive endocrinology. In: Campbell CJ, Fuentes A, MacKinnon KC, Panger M, Bearder SK (eds) Primates in Perspective. Oxford University Press, New York

Lynch JW, Ziegler TE, Strier KB (2002) Individual and seasonal variation in fecal testosterone and cortisol levels of wild male tufted capuchin monkeys, *Cebus apella nigritus*. Horm Behav 41:275–287

Martínez-Mota R, Valdespino C, Sánchez-Ramos MA, Serio-Silva JC (2007) Effects of forest fragmentation on the physiological stress responses of black howler monkeys. Anim Conserv 10:374–379

Martínez-Mota R, Valdespino C, Rivera Rebolledo AR, Palme R (2008) Determination of fecal glucocorticoid metabolites to evaluate stress response in *Alouatta pigra*. Int J Primatol 29:1365–1373

Monteiro FOB, Kugelmeier T, Valle RR, Lima ABF, Silva FE, Martins SS, Pereira LG, Dinucci KL, Viau P (2013) Evaluation of the fecal steroid concentrations in *Alouatta belzebul* (Primates, Atelidae) in the national forest of Tapirape-Aquiri in Pará, Brazil. J Med Primatol 42:325–332

Moreland RB, Richardson ME, Lamberski N, Long JA (2001) Characterizing the reproductive physiology of the male Southern black howler monkey, *Alouatta caraya*. J Androl 22:395–403

Moscovice LR, Ziegler TE (2012) Peripheral oxytocin in female baboons relates to estrous state and maintenance of sexual consortships. Horm Behav 62(5):592–597

Muehlenbein MP, Bribiescas RG (2005) Testosterone-mediated immune functions and male life histories. Am J Hum Biol 17:527–558

Muehlenbein MP, Watts DP, Whitten PL (2004) Dominance rank and fecal testosterone levels in adult male chimpanzees (*Pan troglodytes schweinfurthii*) at Ngogo, Kibale National Park, Uganda. Am J Primatol 64:71–82

Muller MN, Wrangham RW (2004a) Dominance, aggression and testosterone in wild chimpanzees: a test of the "challenge hypothesis". Anim Behav 67:113–123

Muller MN, Wrangham RW (2004b) Cortisol, aggression, and stress in wild chimpanzees (*Pan troglodytes schweinfurthii*). Behav Ecol Sociobiol 55:332–340

Palme R (2005) Measuring fecal steroids: guidelines for practical application. Ann NY Acad Sci 1046:75–80

Plant TM, Witchel SF (2006) Puberty in nonhuman primates and humans. In: Neill JD (ed) Knobil and Neill's physiology of reproduction, vol 2, 3rd edn. Elsevier, St. Louis, MO

Pope TR (1990) The reproductive consequences of male cooperation in the red howler monkey: paternity exclusion in multi-male and single-male troops using genetic markers. Behav Ecol Sociobiol 27:439–446

Pope TR (2000) The evolution of male philopatry in Neotropical monkeys. In: Kappeler PM (ed) Primate males . Cambridge University Press, Cambridge, pp 219–235

Pride ER (2005) Foraging success, agonism, and predator alarms: behavioral predictors of cortisol in *Lemur catta*. Int J Primatol 26:295–319

Rangel-Negrín A, Alfaro JL, Valdez RA, Romano MC, Serio-Silva JC (2009) Stress in Yucatan spider monkeys: effects of environmental conditions on fecal cortisol levels in wild and captive populations. Anim Conserv 12:496–502

Rangel-Negrín A, Dias PAD, Chavira R, Canales-Espinosa D (2011) Social modulation of testosterone levels in male black howlers (*Alouatta pigra*). Horm Behav 59:159–166

Rimbach R, Heymann EW, Link A, Heistermann M (2013a) Validation of an enzyme immunoassay for assessing adrenocortical activity and evaluation of factors that affect levels of fecal glucocorticoid metabolites in two New World primates. Gen Comp Endocrinol 191:13–23

Rimbach R, Link A, Heistermann M, Gomez-Posada C, Galvis N, Heymann EW (2013b) Effects of logging, hunting, forest fragment size on physiological stress levels of two sympatric ateline primates in Colombia. Conserv Physiol 1. doi:10.1093/conphys/cot031

Robbins MM, Czekala NM (1997) A preliminary investigation of urinary testosterone and cortisol levels in wild male mountain gorillas. Am J Primatol 43:51–64

Saltzman W, Tardif SD, Rutherford JN (2011) Hormones and reproductive cycles in primates. In: Norris DO, Lopez KH (eds) Hormones and reproduction of vertebrates, vol 5, Mammals. Academic, San Diego

Sapolsky RM (1993) The physiology of dominance in stable versus unstable social hierarchies. In: Mason WA, Mendoza SP (eds) Primate social conflict. State University of New York Press, Albany

Sapolsky RM (2002) Endocrinology of stress response. In: Becker JB, Breedlove SM, Crews D, McCarthy M (eds) Behavioral endocrinology, 2nd edn. MIT Press, Cambridge

Sapolsky RM (2005) The influence of social hierarchy on primate health. Science 308:648–652

Schradin C, Anzenberger G (2004) Development of prolactin levels in marmoset males: from adult son to first-time father. Horm Behav 46:670–677

Sekulic R (1983) Male relationships and infant deaths in red howler monkeys (*Alouatta seniculus*). Behaviour 81:38–54

Seltzer LJ, Ziegler TE (2007) Non-invasive measurement of small peptides in the common marmoset (*Callithrix jacchus*): a radiolabeled clearance study and endogenous excretion under varying social conditions. Horm Behav 51:436–442

Serio-Silva JC, Hernández-Salazar LT, Rico-Gray V (1999) Nutritional composition of the diet of *Alouatta palliata mexicana* females in different reproductive states. Zoo Biol 18:507–513

Setchell JM, Smith T, Wickings EJ, Knapp LA (2008) Social correlates of testosterone and ornamentation in male mandrills. Horm Behav 54:365–372

Shoemaker AH (1982) Fecundity in the captive howler monkey, *Alouatta caraya*. Zoo Biol 1:149–156

Strier KB, Ziegler TE (2005) Advances in field-based studies of primate behavioral endocrinology. Am J Primatol 67:1–4

Strier KB, Ziegler TE, Wittwer DJ (1999) Seasonal and social correlates of fecal testosterone and cortisol levels in wild male muriquis (*Brachyteles arachnoides*). Horm Behav 35:125–134

Surbeck M, Deschner T, Weltring A, Hohmann G (2012) Social correlates of variation in urinary cortisol in wild male bonobos (Pan paniscus). Horm Behav 62:27–35

Torres-Pelayo VR, Rovirosa-Hernández MJ, García-Orduña F, Chavira-Ramírez RD, Boeck L, Canales-Espinosa D, Rodríguez-Landa JF (2011) Variation in the extraction efficiency of estradiol and progesterone in moist and lyophilized feces of the black howler monkeys (*Alouatta pigra*): alternative methods. Front Physiol 2:97. doi:10.3389/fphys.2011.00097

Van Belle S, Bicca-Marques JC (2014) Insights into reproductive strategies and sexual selection in howler monkeys. In: Kowalewski MM, Garber PA, Cortés-Ortiz L, Urbani B, Youlatos D (eds) Howler monkeys: behavior, ecology and conservation. Springer, New York

Van Belle S, Estrada A, Strier KB (2008) Social relationships among male *Alouatta pigra*. Int J Primatol 29:1481–1498

Van Belle S, Estrada A, Ziegler TE, Strier KB (2009a) Sexual behavior across ovarian cycles in wild black howler monkeys (*Alouatta pigra*): male mate guarding and female mate choice. Am J Primatol 71:153–164

Van Belle S, Estrada A, Ziegler TE, Strier KB (2009b) Social and hormonal mechanisms underlying male reproductive strategies in black howler monkeys (*Alouatta pigra*). Horm Behav 56:355–363

Van Belle S, Estrada A, Strier KB (2011) Insights into social relationships among female black howler monkeys *Alouatta pigra* at Palenque National Park, Mexico. Curr Zool 57:1–7

van Schaik CP (1991) A pilot study of the social correlates of levels of urinary cortisol, prolactin, and testosterone in wild long-tailed macaques (*Macaca fascicularis*). Primates 32:345–356

Wang E, Milton K (2003) Intragroup social relationships of male *Alouatta palliata* on Barro Colorado Island, Republic of Panama. Int J Primatol 24:1227–1243

Wasser SK, Cristóbal-Azkarate J, Booth RK, Hayward L, Hunt K, Ayres K, Vynne C, Gobush K, Canales-Espinosa D, Rodríguez-Luna E (2010) Non-invasive measurement of thyroid hormone in feces of a diverse array of avian and mammalian species. Gen Comp Endocrinol 168:1–7

Weingrill T, Gray DA, Barrett L, Henzi SP (2004) Fecal cortisol levels in free-ranging female chacma baboons: relationship to dominance, reproductive state and environmental factors. Horm Behav 45:259–269

Whitten PL, Brockman DK, Stavisky RC (1998) Recent advances in noninvasive techniques to monitor hormone-behavior interactions. Am J Phys Anthropol 41:1–23

Wingfield JC, Hegner RE, Dufty AMJ, Ball GF (1990) The "challenge hypothesis": theoretical implications for patterns of testosterone secretion, mating systems, and breeding strategies. Am Nat 136:829–846

Ziegler TE, Wittwer DJ (2005) Fecal steroid research in the field and laboratory: improved methods for storage, transport, processing, and analysis. Am J Primatol 67:159–174

Ziegler TE, Strier KB, Van Belle S (2009a) The reproductive ecology of South American primates: ecological adaptations in ovulation and conception. In: Garber PA, Estrada A, Bicca-Marques JC, Heymann EW, Strier KB (eds) South American primates: comparative perspectives in the study pg behavior, ecology, and conservation. Springer, New York

Ziegler TE, Prudom SL, Zahed SR, Parlow AF, Wegner F (2009b) Prolactin's mediative role in male parenting in parentally experienced marmosets (*Callithrix jacchus*). Horm Behav 56:436–443

Chapter 9
The Howler Monkey as a Model for Exploring Host-Gut Microbiota Interactions in Primates

Katherine R. Amato and Nicoletta Righini

Abstract The mammalian gut microbiota is essential to many aspects of host physiology, including nutrition, metabolic activity, and immune homeostasis. Despite the existence of numerous studies of the impact of the gut microbiota on human health and disease, much work remains to be done to improve our understanding of the host-microbe relationship in nonhuman primates. Howler monkeys (*Alouatta* spp.) are highly dependent on the gut microbiota for the breakdown of plant structural carbohydrates, and in this chapter we use new data describing the gut microbiome of captive and wild black howler monkeys (*A. pigra*) to develop and test two models of host-microbe interactions and bioenergetics. Improving our understanding of how spatial and temporal fluctuations in diet affect the nonhuman primate gut microbiota, and how this in turn influences host nutrition and physiology, has important implications for the study of the role that the gut microbiota plays in primate ecology, health, and conservation.

Resumen El papel de la microbiota intestinal es fundamental para muchos aspectos de la fisiología de los mamíferos, incluyendo la nutrición, la actividad metabólica y la homeostasis del sistema inmune. A pesar de la existencia de muchos estudios acerca de la microbiota intestinal humana debido a sus implicaciones para la salud, aún queda mucho por hacer para poder entender la relación huésped-microorganismos en primates no humanos. Los monos aulladores (*Alouatta* spp.) dependen de manera importante de los microbios intestinales para la digestión de los carbohidratos estructurales de las plantas. En este capítulo utilizamos nuevos datos sobre la composición de la microbiota de monos aulladores negros cautivos y silvestres (*A. pigra*) para desarrollar y poner a prueba dos modelos sobre las interacciones

---

K.R. Amato (✉)
Program in Ecology, Evolution, and Conservation Biology, University of Illinois at Urbana-Champaign, Champaign, IL, USA

Department of Anthropology, University of Colorado Boulder, Boulder, CO, USA
e-mail: katherine.amato@colorado.edu

N. Righini
Department of Anthropology, University of Illinois at Urbana-Champaign, Champaign, IL, USA

Instituto de Ecología, A.C., Xalapa, Veracruz, Mexico

huésped-microbios desde un punto de vista ecológico y bioenergético. El análisis del efecto de las fluctuaciones espaciales y temporales de la dieta sobre la microbiota intestinal de los primates, y de cómo esto a su vez se refleja en la nutrición y fisiología del huésped, tiene implicaciones importantes para entender el papel de la microbiota en la ecología, salud y conservación de los primates.

Keywords Gut microbiome • Health • Nutrition • Growth • Reproduction

9.1 Introduction

Mutualistic microbial communities composed of bacteria, ciliate and flagellate protozoa, archaea, anaerobic fungi, and bacteriophages (Mackie 2002) are an essential part of the mammalian gut and play an important role in host physiology by influencing nutrition, metabolic activity, and immune homeostasis (Dethlefsen et al. 2007; Sekirov et al. 2010; Flint et al. 2011). These communities are dominated by bacteria, particularly in the colon, which in the case of humans, is estimated to contain more than 70 % of all of the microbes present in the body, with 10^{11}–10^{12} bacteria per gram of content (Sekirov et al. 2010). These bacteria contribute to host health by regulating xenobiotic metabolism (Bjorkholm et al. 2009), producing vitamins (Hill 1997), excluding pathogenic microbes, attenuating inflammation (Kelly et al. 2003), and affecting immune system development through the formation and modification of the intestinal epithelia and gut-associated lymphoid tissue (GALT) (Bauer et al. 2006; Neish 2009; Hooper et al. 2012). They also are thought to play a role in modulating brain development and function and affect behavior by altering gene expression and neuronal circuits involved in motor control and anxiety (Forsythe et al. 2010; Foster and McVey Neufeld 2013).

While each of these functions is important to the host, the role of the gut bacterial community, or the gut microbiota, in host energy and nutrient acquisition is the most well studied. Because all vertebrates lack the enzyme cellulase, which is required to break down cellulose, their ability to digest fiber is dependent on enzymes either present in their food or produced by intestinal microbes (Stevens and Hume 1995; Barboza et al. 2009). As a result, foods containing high proportions of plant cell wall material and resistant starches can only be digested if hosts maintain rich microbial communities. These microbial communities convert indigestible compounds such as cellulose into short-chain fatty acids (SCFA) such as acetate, butyrate, and propionate, which can be absorbed directly by the host and used as an energy source or stored as glucose in the liver (Mackie 2002). Short-chain fatty acids produced by the gut microbiota can supply hosts with up to 70 % of their daily energy needs (Flint and Bayer 2008) and have been reported to reduce the pH of the intestinal lumen to facilitate nutrient absorption and to prevent the accumulation of potentially toxic metabolic by-products (Neish 2009; Sekirov et al. 2010).

Like all mammals, primates rely on their gut microbiota to process low-quality resources such as woody plants, mature leaves, fungi, and plant exudates that are difficult to digest and require greater handling and processing due to mechanical defenses, limited nutrients, and high concentrations of indigestible material or toxins (Lambert 2011). As a result, in addition to adaptations in dental morphology, specialized features of either the foregut or the hindgut have evolved independently in species of prosimians (indriids, *Lepilemur*), New World monkeys (*Alouatta*, *Callithrix*), Old World monkeys (colobines), and apes (gorillas) that regularly exploit low-quality resources. Specifically, fermentative processes are either pregastric (occurring before hydrolytic/enzymatic digestion), as seen in colobine monkeys, or post-gastric/cecocolic (occurring after hydrolytic/enzymatic digestion), as seen in some prosimians, New World monkeys, cercopithecines, apes, and humans (Chivers and Hladik 1980; Chivers and Langer 1994; Lambert 1998).

Howler monkeys (*Alouatta* spp.) are known for their ability to consume low-quality diets consisting of mostly leaves during some periods of the year (more than 80 % of feeding time in a given month) (Pavelka and Knopff 2004) and are post-gastric, or hindgut, fermenters (Milton 1980; Edwards and Ullrey 1999). They do not possess a particularly specialized gut morphology compared to foregut fermenters such as colobines (Kay and Davies 1994; Edwards and Ullrey 1999), but gut measurements for *A. palliata* (Chivers and Hladik 1980) reveal larger-than-expected cecum and colon volumes given their body mass (positive residuals from the least squares regression of cecum and colon volumes on body weight). Moreover, howlers are characterized by relatively long food transit times compared to other atelines (20.4 h for *A. palliata*) (Milton 1984). A large gut volume is usually associated with a greater production of microbial SCFA, and SCFA absorption and assimilation depend principally on the surface area available and on the length of time food is retained in fermenting chambers (Brourton and Perrin 1991; Kay and Davies 1994). Indeed, howlers are estimated to gain as much as 31 % of required daily energy from SCFA produced by the gut microbiota (Milton and McBee 1983).

Although it is widely accepted that the gut microbiota plays a critical role in howler nutrition, very little is understood about the dynamics of the howler-microbe relationship. For example, as energy minimizers, howler monkeys are able to persist in a wide range of habitats, including highly fragmented or anthropogenically impacted areas (Strier 1992; Phillips and Abercrombie 2003; Bicca-Marques 2003; Behie and Pavelka 2005; Zunino et al. 2007; Pozo-Montuy et al. 2011; Bonilla-Sanchez et al. 2012), and can endure marked seasonal changes in availability of food items such as mature fruit by exploiting hard-to-digest foods such as mature leaves and unripe fruits (Arroyo-Rodriguez et al. 2008), instead of dramatically increasing day range or time spent traveling [as observed in other atelines (Di Fiore et al. 2011)]. However, the role of the gut microbiota in allowing howlers to extract sufficient energy and nutrients from a wide range of resources across seasons and habitats is not well studied, and many questions remain to be answered. Does the composition of the howler gut microbiota shift in response to changes in diet? Do these shifts allow howlers to obtain the energy and nutrients they need under conditions of marked fluctuations in food availability? Do changes in the gut microbiota affect other aspects of howler health and behavior? Similarly, the influence of the gut microbiota on howler life history via nutrition has not been explored. Because howler monkeys have an earlier age at first reproduction (42 vs. 84 months for *Ateles*), shorter gestation length (186 vs. 225 days for *Ateles* and *Lagothrix*), and shorter interbirth intervals (19.9 vs. 34.7 months for *Ateles*) than other atelines (Fedigan and Rose 1995), their daily nutritional demands for growth and reproduction are expected to be greater. Can differences in the composition of juvenile and female gut microbiota help compensate for some of these demands? If so, what triggers the gut microbiota to change? Are these changes important regardless of season and diet?

In this chapter, we begin by reviewing the factors that influence mammalian gastrointestinal microbial community structure and function and the impacts of the gut microbiota on host nutrition, physiology, and health. Using this information we develop two models—a general model of host-microbiota interactions and a revised bioenergetics model that includes gut microbiota effects—and use data from black howler monkeys (*A. pigra*) to test the predictions of these models. Finally, we discuss patterns that correspond to our models within and among other primate species and detail important avenues for future research that integrate gut microbiome analyses with ecological, nutritional, and physiological data to describe interactions between diet, behavior, nutrition, and health in wild primate populations.

9.2 The Mammalian Gut Microbiome

In recent years, the study of microbial communities has benefited from molecular approaches that use the extraction and amplification of microbial DNA to identify patterns in community composition across samples. Given these techniques, it is now possible to overcome the limitations associated with bacterial culturing methods, such as the small number of samples that can be processed at a time and the bias against strict anaerobes, many of which play an important role in the gut microbial community (Sekirov et al. 2010). Instead, analyses such as termination restriction fragment length polymorphism (T-RFLP) (Osborn et al. 2000), denaturing gel gradient electrophoresis (DGGE) (Fischer and Lerman 1979), automated ribosomal intergenic spacer analysis (ARISA) (Fisher and Triplett 1999), and high-throughput pyrosequencing (Ronaghi et al. 1998) allow researchers to describe in detail the taxonomic composition, function, and diversity of the fecal (i.e., mainly colonic) microbiota in a variety of animal species (see Sekirov et al. 2010 for a comparison among these techniques and their respective benefits and limitations).

9.2.1 Evolution of the Mammalian Gut Microbiota

Due to their involvement in host nutrient metabolism, gut microbes are thought to have played a primary role in host evolution by facilitating the adoption of a particular diet and providing specific metabolic pathways for the digestion of that diet

(Neish 2009; Yildirim et al. 2010). However, it is also possible that gut microbial communities co-diversified and coevolved with their hosts, leading to specializations and increased dependence between the host and its microbial colonists (Kau et al. 2011; Yeoman et al. 2011). Recent studies analyzing bacterial 16S ribosomal RNA gene sequences from 60 mammalian species (Ley et al. 2008a, b) indicate that gut bacterial diversity may be affected by host phylogeny since the fecal microbial communities of conspecific hosts are more similar to each other than to the communities of more distantly related hosts. Additionally, data from this study demonstrate an effect of diet on gut bacterial diversity. Herbivorous mammals exhibit a higher diversity of microbial phyla than omnivores, which in turn exhibit a higher diversity than carnivores. Separating these genetic and environmental influences from each other is crucial not only for understanding the role the gut microbiota has played in the evolution of mammalian dietary diversification but also for determining the impact of the microbiota on host diet and nutrition at different time scales (e.g., days, weeks, years).

9.2.2 Factors Affecting Gut Microbiota Composition

In all mammals the fetal gut is generally sterile, and microbial colonization occurs during and after birth via horizontal transfer of microbes from the surrounding environment (Mackie et al. 1999; Donnet-Hughes et al. 2010). In humans, the establishment of the gut microbial community takes approximately 1 year (Mackie et al. 1999). Initially, the gut microbiota exhibits a relatively simple structure (mainly composed of lactic acid bacteria (LAB), enterobacteria, and streptococci) and varies greatly among individuals and across time according to diet (Mackie et al. 1999; Sekirov et al. 2010; Spor et al. 2011). However, as weeks pass, gut microbiota composition stabilizes and begins to include higher numbers of obligate anaerobes (Mackie et al. 1999). Studies have shown that during this process there is a strong maternal influence on the structure of the gut microbial community. For example, among individuals fed the same diet, mouse gut microbiota composition is more similar between mother and weaning offspring than between unrelated individuals, even when the unrelated individuals share the same genotype for obesity traits while the mother and offspring do not (Ley et al. 2005). Increasing evidence suggests that this maternal influence is a result of microbial transfer via colostrum and breast milk. Enteric bacterial translocation and colonization of the mammary tissue have been documented in pregnant and lactating mice, and an analysis of human milk confirmed the presence of autochthonous ileal and colonic microbes (Donnet-Hughes et al. 2010). Therefore, nursing is crucial to gut microbial community development.

Although human microbial community composition stabilizes after about a year, it remains highly dynamic throughout an individual's lifetime. Rapid responses by the microbiota to changes in the selective pressures in the gut result in intra- and interindividual variation according to factors such as host diet, age, nutrition, health status, and genetics (Spor et al. 2011). Of these, diet has been shown to play a particularly strong role in determining gut microbial community composition. For example, within the span of a day, mice that were switched from a low-fat diet rich in plant polysaccharides to a high-fat, high-sugar diet experienced a dramatic increase in the abundance of several classes of bacteria belonging to the Firmicutes phylum (Turnbaugh et al. 2009). Additionally, studies in humans show that there exist at least two stable, broad "enterotypes" which are linked to long-term dietary habits: the *Bacteroides* enterotype, associated with animal protein and saturated fat intake, and the *Prevotella* enterotype, associated with plant-based nutrition (Wu et al. 2011). However, shifting from a primarily plant-based diet to a primarily animal-based diet also affects the gut microbial community by increasing the abundance of bile-tolerant bacteria and reducing the abundance of bacteria from the Firmicutes phylum over a span of one week (David et al. 2014). Together these patterns suggest that host diet exerts strong selective pressure on the mammalian gut microbiota on time scales from hours to years.

In addition to environmental influences such as diet, host genetics appear to impact gut microbiota composition (Benson et al. 2010). This is especially evident when focusing on specific groups of microorganisms within the gut community. For example, variations in fecal abundance of LAB, a group of gram-positive microbes belonging to the phylum Firmicutes, are reported to be associated with particular mouse genetic lines, regardless of the maternal microbiota the mice are exposed to (Buhnik-Rosenblau et al. 2011). Additionally, mice that are genetically predisposed to obesity possess a higher proportion of Firmicutes and a lower proportion of Bacteroidetes compared to mice with a "lean" genotype (Ley et al. 2005; Turnbaugh et al. 2006). Obesity is thought to be a result of the ability of Firmicutes to harvest energy with higher efficiency from a given diet, thereby providing the host with surplus energy (Turnbaugh et al. 2006). As a result, hosts that are genetically predisposed to higher Firmicutes abundances are more likely to become obese. Of course, environmental effects can interact with these genetic effects. In humans, changes in diet leading to weight loss result in decreased proportions of Firmicutes (Ley et al. 2006). Similarly, Zoetendal et al. (2001) argue that genetics strongly influence gut microbiota composition since monozygotic twins living separately show more microbiome similarity than domestic partners, and profiles of domestic partners do not differ in similarity from those of unrelated individuals. However, many environmental effects (e.g., maternal effect, diet, lifestyle, illness) were not controlled for in twin pairs. Therefore, while host genotype appears to have some effect on mammalian gut microbiota composition, in many cases nongenetic effects are equally, if not more, important.

Aside from maternal influences and diet, other environmental factors can interact with the gut microbiota via shifts in host physiology. For example, a study of captive rhesus macaques (*Macaca mulatta*) indicated that physical and psychological stress alter gut microbial community composition (Bailey and Coe 1999). Six- to nine-month-old infants separated from their mothers showed stress-indicative behaviors (e.g., distress calls), increases in plasma cortisol, and a significant reduction in fecal lactobacilli starting the third day after separation (Bailey and Coe 1999). Similarly, rats and chicks exposed to stress from heat and crowding possess distinct gut microbiota compared to individuals not exposed to these stressors (Suzuki et al. 1983) and mouse models of depression also exhibit changes in the gut microbiota (Park et al. 2013). However, the relationship between host stress and gut microbiota composition is not unidirectional. Studies of rodents and humans provide evidence that gut microbiota composition can influence host stress responses. Based on measures of plasma adrenocorticotropic hormone (ACTH) and corticosterone responses, germ-free mice are more susceptible to stress when physically restrained than specific pathogen-free mice (Sudo 2006). As a result, it appears that the gut microbiota has a role in the development of hypothalamic-pituitary-adrenal (HPA) stress responsiveness. Similarly, the administration of certain bacteria strains (such as *Lactobacillus* and *Bifidobacterium*) has been shown to have beneficial effects on stress, anxiety, and depression in rats and humans, as indicated by reductions in anxiety-like behaviors and cortisol levels (Messaoudi et al. 2010). Thus, there may exist a positive feedback loop relating stress to depauperate gut microbiota.

Interactions between host physiology and gut microbiota composition also appear to occur via immune system function. The adaptive immune system constantly monitors the gut microbiota and stimulates the secretion of local strain-specific immunoglobulin A (IgA) across mucous membranes (Macpherson et al. 2008; Neish 2009). Secretion of IgA influences gut microbiota composition and protects beneficial microbiota from host immune attacks since IgA is used by the mammalian humoral immune system (i.e., mediated by antibodies produced by B cells) to recognize cells and tag only pathogenic invaders for destruction. However, as with the stress-microbiota relationship, this relationship is not unidirectional. The gut microbiota appears to play an active role in host immune function. IgA-secreting cells are significantly reduced (1–2 orders of magnitude lower) in germ-free animals and absent in neonates suggesting that intestinal IgA levels are regulated by the presence of gut microflora (Benveniste et al. 1971a, b; Macpherson et al. 2008). Furthermore, the gut microbiota is thought to contribute to the development of the host intestinal mucosal and systemic (i.e., peripheral) immune systems (Neish 2009; Forsythe et al. 2010; Sekirov et al. 2010; Hooper et al. 2012). For example, germ-free mice lack immune activity, and only colonization of their guts with specifically selected bacteria provokes the complete restoration of immune activity (Talham et al. 1999). Similarly, in humans, the Hygiene Hypothesis suggests that reduced exposure to microorganisms suppresses the normal development of the immune system, resulting in the increased rates of allergies or immune/inflammatory conditions associated with sanitation, antibiotic use, and other "Western" habitats (Strachan 1989; Sekirov et al. 2010).

9.3 General Model of Host-Microbiota Interactions

The mammalian studies described above indicate a wide variety of interactions between mammals and their gut microbiota. In general, host physiology and diet impart strong selective pressures on the gut microbiota. Therefore as host

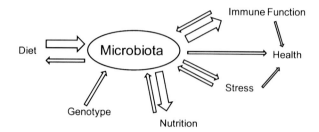

Fig. 9.1 General model of host-gut microbiota interactions. Size of arrows indicates relative size of effect

physiology or diet changes, selective pressures also change, influencing the competitive abilities of microbial taxa and inciting shifts in gut microbial community composition. In turn, these shifts can affect host nutrition and health. Based on these dynamics, we have developed a general model to predict host-gut microbiota interactions in mammals (Fig. 9.1). Because diet appears to play a stronger role in determining microbiota composition than host genotype or physiology (Friswell et al. 2010), in this model, host diet is presented as the main influence on gut microbiota composition. Likewise, because host nutrition and immune development and function depend so heavily on the gut microbiota, these factors in our model are the most strongly affected by changes in gut microbiota composition.

Using this framework as a guide, a series of predictions regarding host-gut microbiota interactions within mammalian species can be made. First, because diet varies spatially across habitats and temporally according to seasonal food availability (Ostfeld and Keesing 2000; Meserve et al. 2003), gut microbiota composition should vary among individuals of a species occupying distinct habitats or within individuals across seasons. These differences should be associated with differences in overall diet diversity or composition. For example, bacteria such as *Clostridium* and *Ruminococcus* have high cellulolytic capability and can outcompete other microbes in the presence of cellulose (Cavedon et al. 1990; Ohara et al. 2000; Louis et al. 2007). Therefore, we would expect individuals consuming a leaf-heavy diet to have higher abundances of these two genera compared to individuals consuming a fruit-heavy or lower-fiber plant-based diet. However, as diet shifts and leaf eating decreases, *Clostridium* and *Ruminococcus* may no longer be able to outcompete other microbes and survive in the gut, and microbiota composition should change as other microbes increase in abundance or invade the gut community. Similarly, a diverse host diet delivering a large array of nutrients and different types of carbohydrate substrates to the gut provides a variety of feeding niches to support microbial taxa or functional groups (Louis et al. 2007). Therefore, for an herbivore/frugivore, we would expect that the more plant species an individual is able to utilize in a particular habitat, the richer and more diverse its gut microbiota. In other cases, a specific food item, plant species, or set of plant species may determine the amount of one or two key macro- or micronutrients in the gut and strongly influence the composition of the gut microbiota.

Once changes in gut microbiota composition occur, our model predicts that host nutrition should be affected. Because previous studies suggest that certain gut microbial communities are specialized for the digestion of certain host diets (Ley et al. 2008a; Wu et al. 2011), food digestion by the microbiota adapted to that diet should be highly efficient. Moreover, adaptation of the gut microbiota to changes in host diet can occur within days (Turnbaugh et al. 2009), meaning that increased digestive efficiency can be achieved rapidly and should aid hosts in acquiring sufficient energy and nutrients to meet metabolic demands despite variability in diet across habitats and seasons. If this is the case, we would expect host energy balances, body masses, and/or body conditions not to vary dramatically spatially or temporally, while the products of microbial fermentation such as SCFA should.

Finally, our model predicts that shifts in gut microbiota composition should have strong effects on host immune function. The literature suggests that depletion of the gut microbiota leads to decreased immune function, specifically decreased IgA secretion (Benveniste et al. 1971a, b; Moreau et al. 1978). Therefore, regardless of their nutritional status, we would expect individuals with low gut microbiota diversity and/or richness to exhibit low IgA levels compared to the rest of the population. Additionally, if depleting the microbiota reduces overall immune function, we would also expect individuals with reduced gut microbiota diversity and/or richness to contract more illnesses and to serve as hosts to more parasites. This has critical implications for primate conservation and survivorship. Similarly, reducing gut microbiota diversity is believed to facilitate colonization of the gut by gastrointestinal pathogens and parasites by reducing the number of feeding niches occupied by mutualistic microbes, increasing the number of niches available for colonization, and/or eliminating those bacterial taxa that actively exclude pathogens (Fons et al. 2000; Servin 2004; Costello et al. 2012). As a result, we would expect individuals with reduced gut microbial diversity to have higher occurrences and abundances of gastrointestinal pathogens and parasites. Although it is not directly linked to immune function, we would also expect gut microbial diversity to interact with host glucocorticoid levels as suggested in the literature (Bailey and Coe 1999; Sudo 2006). Whether microbiota composition influences glucocorticoid levels or vice versa is difficult to distinguish. However, individuals with reduced microbial diversity should exhibit higher glucocorticoid levels.

To explore the validity of our model, we used behavioral and gut microbiome data from an 8-week study of wild, black howler monkeys (*A. pigra*) in southeastern Mexico (Amato et al. 2013) to test the hypotheses that (1) differences in diet across habitats result in differences in gut microbiome composition and (2) differences in gut microbiome composition affect host health. Specifically, we expected that howlers consuming diets with a relatively higher proportion of leaves during the study period would exhibit relatively higher abundances of cellulose-degrading bacteria such as *Ruminococcus* compared to other howlers. We also expected howlers consuming a relatively more diverse diet in terms of plant species to exhibit a relatively more diverse microbiome. Finally, for those howlers with the lowest microbial diversity, we expected to find higher abundances of pathogenic bacteria.

Fecal samples analyzed in this study were collected from five groups of black howler monkeys occupying four habitats—a continuous evergreen rainforest; an evergreen rainforest fragment; a continuous, semi-deciduous forest; and a rehabilita-

Table 9.1 Percent of total feeding time spent consuming plant parts by howlers in the continuous evergreen rainforest and semi-deciduous forest (adapted from Amato et al. 2013)

Food type	Feeding time (%) Continuous evergreen	Continuous semi-deciduous
Mature fruit	53.03	30.68
Immature fruit		0.79
Mature leaf	21.25	11.70
Young leaf	13.22	56.83
Stem	9.17	
Other	3.34	

Howlers in the evergreen rainforest fragment consumed less mature fruits than those in the continuous forest, and the captive howlers were fed a mixture of mature fruits, cereal, and monkey chow

Table 9.2 Percent of total feeding time spent consuming plant species by howlers in the continuous rainforest and semi-deciduous forest (adapted from Amato et al. 2013)

Plant species	Feeding time (%) Continuous evergreen	Continuous semi-deciduous
Acacia usumacintensis (Fabaceae)		35.32
Alseis yucatanensis (Rubiaceae)		1.19
Brosimum alicastrum (Moraceae)		35.32
Bursera simaruba (Burseraceae)		9.92
Cecropia peltata (Urticaceae)	2.47	
Ficus americana (Moraceae)	23.99	
Ficus aurea (Moraceae)	11.37	
Ficus sp. (Moraceae)	8.34	
Ficus yoponensis (Moraceae)	14.61	
Lonchocarpus castilloi (Fabaceae)		2.28
Manilkara zapota (Sapotaceae)		1.19
Metopium brownei (Anacardiaceae)		26.58
Monstera sp. (Araceae)		0.26
Poulsenia armata (Moraceae)	15.58	
Schizolobium parahyba (Fabaceae)	1.82	
Simarouba glauca (Simaroubaceae)		1.78
Vitex gaumeri (Verbenaceae)		7.27
Unknown sp. 1		1.24
Unknown sp. 2	0.13	
Unknown sp. 3	1.04	
Unknown sp. 4	0.17	
Unknown sp. 5	3.38	
Unknown sp. 6 (Fabaceae)	3.6	
Unknown sp. 7 (Araceae)	2.70	
Unknown sp. 8	4.94	
Unknown sp. 9	3.04	
Unknown sp. 10		0.59
Vines	3.24	2.17

Species consumed by the howlers in the evergreen rainforest fragment were a subset of those consumed by the howlers in the continuous rainforest and also included several distinct species. Captive howlers were fed melon, mango, papaya, and banana in addition to cereal and monkey chow

tion center—in southeastern Mexico. Quantitative diet data collected using focal sampling were available for two of these habitats, and qualitative diet data were available for the other two (Tables 9.1 and 9.2). Bacterial community fingerprinting (ARISA) was used to detect broad patterns in overall microbiome composition while pyrosequencing provided information regarding which bacterial taxa were driving the patterns. Functional genes associated with the production of VFA's and other microbial fermentation products were measured using quantitative real-time PCR.

Analyses revealed that howler gut microbiome richness, diversity, and composition differed by habitat. Captive howlers ($N=8$) exhibited the lowest microbial richness and diversity (Chao1 = 9,821, Shannon = 6.82), and howlers in the continuous rainforest ($N=14$) exhibited the highest microbial richness and diversity (Chao1 = 1,549, Shannon = 4.78). Because the captive howlers came from distinct geographic regions of Mexico and were not genetically related, these patterns must be an effect of their captive environment. Indeed, gut microbiome variation was strongly correlated with howler diet both in terms of plant parts (Spearman's $\rho=0.54$, $p<0.001$; Table 9.1) and plant species (Spearman's $\rho=0.34$, $p<0.005$; Table 9.2), which differed according to habitat. Howlers consuming a more diverse diet also exhibited more diverse gut microbiome.

Diet composition also influenced gut microbiome composition. Cellulolytic *Ruminococcus* increased with the proportion of fiber-rich, mature leaves in the howler diet (Table 9.3). *Prevotella*, which degrades the monosaccharide xylose (Yildirim et al. 2010), was found in higher abundances in the captive howlers (Table 9.3). Simple sugars like xylose are typical of fruits, and since the captive howlers consumed a fruit-heavy diet, it is likely that these sugars favored the presence of *Prevotella*. Similarly, *Lachnospira pectinoschiza*, a Clostridia that utilizes pectin, was found in captive howlers in higher abundances (Nakamura et al. 2011). Pectin is a complex polysaccharide contained in many fruits such as apple and guava. *Lactobacilli* are benefited by dietary calcium (Bovee-Oudenhoven et al. 1999), and the captive howlers with high calcium content in their manufactured diet (Mazuri Leaf-Eater Primate Chow: 1.12 % Ca vs. 0.40 % and 0.30 % for young leaves and fruits, respectively (Righini, 2014) had the highest levels of *Lactobacilli* (Table 9.3). *Ficus* is also known to have high calcium content compared to other fruit species (O'Brien et al. 1998), and *Ficus* trees were present in every wild habitat except the semi-deciduous forest. Likewise, all howlers outside of the semi-deciduous forest possessed *Lactobacilli*.

Measures of howler health showed less clear patterns. In general, there was a low occurrence of pathogenic bacteria in all of our sampled howlers, and the presence of a pathogen was not related to gut microbiome diversity. However, the eight individuals (all captives and one continuous forest howler) with the lowest microbiome richness and diversity died within 6 months following the sampling period. Although

Table 9.3 Percent of total bacterial sequences sampled belonging to bacterial genera for each howler group sampled

Bacterial genus	Cont. evergreen	Cont. semi-deciduous	Frag. evergreen	Captive
Ruminococcus	0.42	0.19	0.70	0.10
Prevotella	0.0060	0.0070	0.0030	12.98
Lactobacillus	0.0100	0.0040	0.029	0.28

this relationship cannot be assumed to be causative, it suggests a potential connection between the gut microbiome and howler health.

Shifts in metabolic functional genes across habitats reinforce this connection. The butyryl-CoA:acetate CoA-transferase gene involved in the microbial production of health-promoting butyrate was more prevalent in howlers in the continuous evergreen rainforest than in howlers at other sites. Similarly, the number of acetyl-CoA synthase genes used for microbial production of acetate, another important VFA, was significantly higher in the continuous evergreen rainforest than in other habitats. Increased hydrogen production is associated with increased microbial fermentation, and (Ni-Fe)-hydrogenase ((NF)*hyd*) genes for hydrogen production were most abundant in the continuous evergreen rainforest. Finally, hydrogen sulfide is a toxic gas produced by the consumption of hydrogen by sulfate-reducing bacteria that affects smooth muscle and has been linked to colonic disease (Medani et al. 2011; Carbonero et al. 2012). The dissimilatory (bi)sulfite reductase gene associated with hydrogen sulfide production was most abundant in the evergreen fragment howlers and the captive howlers.

The results from this initial study confirm that diet diversity and composition play an important role in determining howler gut microbiota composition and indirectly support the hypothesis that reductions in gut microbiota diversity negatively affect howler health. However, measurements of IgA levels, parasite abundance, or glucocorticoid levels would more accurately pinpoint the effect of microbiota depletion on howler health. Studies of wild, nonhuman primates have used fecal measurements of IgA to estimate immune function (Lantz et al. 2011), and this technique could be easily integrated into future microbiome studies. Similarly, measuring gastrointestinal parasite abundance in primates relies on fecal sample collection and could be easily incorporated into future protocols. Studies of several howler species, including black howlers in Palenque, have reported higher gastrointestinal parasite diversity and abundance in primates inhabiting degraded areas compared to those in relatively undisturbed habitats (Eckert et al. 2006; Stoner and Gonzalez Di Pierro 2006; Trejo-Macias et al. 2007; Vitazkova and Wade 2007), suggesting that there may be a connection between reduced gut microbial diversity and parasite abundance in these habitats. However, analyses of parasite diversity and abundance must be carried out simultaneously with analyses of gut microbiota composition to truly test this relationship. Similarly, Martinez-Mota et al. (2007) report that black howler monkeys living in small (<2 ha), highly disturbed forest fragments in Mexico have higher fecal glucocorticoid levels than monkeys inhabiting less-disturbed forest. Again, this relationship provides indirect support for the interaction of the gut microbiota and host stress responses, but fecal samples must be analyzed for glucocorticoids and gut microbiota composition concurrently to confirm the pattern.

Although the results from this study demonstrate that howler microbiota composition differs with diet across habitats, it is unclear whether these differences are associated with the expected shifts in microbial activity and digestive efficiency that would allow howlers to meet nutritional demands in all habitats. Measurements of gut microbiota activity as well as host nutritional status are necessary to clarify this relationship. Fecal volatile fatty acid (VFA, a subset of SCFA) and ammonia content provide an estimate of carbohydrate and protein metabolism by the microbiota (Erwin et al. 1961; Chaney and Marbach 1962). Generally, the more VFA and ammonia detected in fecal samples, the more produced by the gut microbiota, and the more available to the forager. Similarly, the nutritional status of individuals from a variety of primates species has been described using C-peptide analyses from urine samples (Sherry and Ellison 2007; Deschner et al. 2008; Thompson and Knott 2008; Thompson et al. 2008; Harris et al. 2009; Girard-Buttoz et al. 2011). Urinary excretion levels of C-peptide are positively correlated with insulin production in humans (Kruszynskia et al. 1987), and in nonhuman primates, high C-peptide levels are correlated with increased body mass and high food availability, among other factors (Sherry and Ellison 2007; Deschner et al. 2008; Thompson and Knott 2008; Girard-Buttoz et al. 2011). Therefore, if changes in the gut microbiome result in increased microbial activity and provide sufficient energy and nutrients to the host, we would expect spatial and temporal changes in gut microbiota composition to be associated with strong variations in fecal VFA and ammonia content and only weak variations in C-peptide.

9.4 Integrating the Gut Microbiota into Mammalian Bioenergetics Models

In addition to knowing how the mammalian gut microbiota changes with habitat and diet, information regarding how the microbiota differs among individuals and within individuals over time is crucial to understanding the relationship between the gut microbiota and host nutrition, health, and ecology. Foragers face challenges in obtaining sufficient energy and nutrients as food availability varies across habitats in response to disturbance and fragmentation and within habitats across seasons. However, these challenges are compounded for individuals as their energy and nutrient requirements change due to processes like growth and reproduction. In primates, pregnancy and lactation are estimated to increase female daily energy requirements by 20–30 % and 37–39 %, respectively (Aiello and Wells 2002), and lactation is estimated to increase protein requirements by more than a third (Oftedal et al. 1991). Similarly, growth in weaned juveniles can require 50 % more energy and 100 % more protein than basal requirements (Altmann and Alberts 1987; Altmann and Samuels 1992). According to mammalian bioenergetics models, as these nutritional demands increase, individuals must (1) increase energy and nutrient intake, (2) decrease metabolic consumption of energy and nutrients, and/or (3) increase energy and nutrient assimilation efficiency to compensate (McNab 2002; Peles and Barrett 2008). Although a large number of mammalian studies have investigated changes in diet and activity in response to growth and reproduction (e.g., Mellado et al. 2005; Chilvers and Wilkinson 2009; Larimer et al. 2011), few explore differences in assimilation or digestive efficiency (Hammond and Kristan 2000; Jaroszewska and Wilczynska 2006). However, changing digestive efficiency in response to growth and reproduction is likely to be an important mechanism for meeting increased nutritional demands, especially in cases where diet and/or activity are constrained. Although physiological changes that increase intestinal volume, surface area, or permeability can improve primate digestive efficiency, changes in gut microbiota composition, which lead to changes in SCFA production, may represent a faster, less energetically expensive, and more labile mechanism because they do not require host growth or physiological changes. Therefore, we propose a revised bioenergetics model, which incorporates shifts in gut microbiota composition and function.

As in traditional bioenergetics models, this model predicts that as nutritional needs change due to processes such as reproduction and growth, individuals must to consume more energy and nutrients, become less active, or increase gut volume, surface area, and permeability to compensate (Fig. 9.2). However, it also predicts that individuals should exhibit shifts in gut microbiota composition that result in the production of more energy and nutrients. Laboratory studies of mice and humans have demonstrated that the ratio of Bacteroidetes to Firmicutes bacteria influences host digestive efficiency since Firmicutes produce energy more efficiently (Ley et al. 2005; Armougom et al. 2009). Additionally, in humans, increased *Bacteroides* numbers in pregnant women have been associated with increased HDL cholesterol and folic acid, increased *Bifidobacterium* with increased folic acid, and increased *Enterobacteriaceae* and *E. coli* with increased ferritin and reduced transferrin (Santacruz et al. 2010). Therefore, we predict that juvenile and reproductively active female primates should alter the proportion of functional groups of microbes such as Firmicutes or *Bifidobacterium* in the gut to increase digestive efficiency and nutrient production. Although we do not expect changes in the microbiota to replace changes in diet, activity, and/or gut morphology as mechanisms for meeting increased nutritional demands, we do expect them to be most pronounced and most critical when these other mechanisms are constrained.

To test some of the predictions of our revised bioenergetics model, we again use data from wild, black howler monkeys in Mexico. Specifically, we investigate the relationship between diet, activity budget, and gut microbial composition and activity

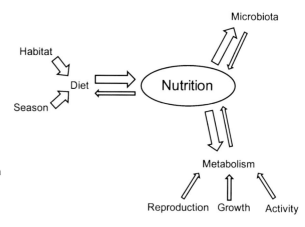

Fig. 9.2 Revised mammalian bioenergetics model. Size of arrows indicates relative size of effect

across howler age and sex classes during an 8-week period. As energy minimizers, howlers exhibit clear behavioral constraints with respect to activity levels. Additionally, howler studies indicate few changes in activity patterns and/or diet for juveniles or reproductively active females (pregnant or lactating) that would suggest compensation for increased nutritional demands (Nagy and Milton 1979; Schoeninger et al. 1997; Serio-Silva et al. 1999; Raguet-Schofield 2009; Dias et al. 2011). Therefore, it is likely that juvenile and adult female howlers rely to some extent on changes in digestive efficiency to meet nutritional demands and provide an excellent system for testing our model.

To compare behavioral and physiological mechanisms for meeting nutritional demands, K. Amato collected data describing diet, activity budget, and gut microbiome composition and activity from black howlers from different age, sex, and reproductive classes (pregnant, lactating) in Palenque National Park, Chiapas, Mexico. Approximately 159 h of behavioral data were collected during May–July 2009 from two neighboring groups of howlers: the Motiepa group ($N=8$ individuals) and the Balam group ($N=6$ individuals). Twenty-minute focal samples with activity recorded instantaneously every 2 min were used to describe feeding (active consumption of food resources), foraging (movement within a feeding tree), resting (periods of inactivity), traveling (movement between tree crowns), and social behavior (aggression, howling, play, sexual activity, etc.) between 6 am and 4 pm each day. During a feeding bout, the food type (young leaves, mature fruit, flowers, etc.) and plant species were recorded. Average daylight hours during the study period were used to calculate the average amount of time spent daily by the howlers in each activity based on the percent of time spent in each activity during focal observations. For feeding data, average ingestion rates collected across seasons (Amato 2013) were used to estimate the number of food items consumed per minute by each individual for each food type and plant species when possible. Average wet masses of food items were used to estimate the average daily amount of grams of food ingested by each individual for each food type and plant species (Amato 2013), and the average kcal and grams of protein ingested by each individual was calculated using general estimates for Neotropical food types (Norconk et al. 2009). Feeding data were standardized by metabolic body weight for each age/sex class before analysis (Kleiber 1975; Kelaita et al. 2011).

To determine whether gut microbial community composition and activity differed across age and sex classes, fecal samples were collected from each individual weekly over the course of 8 weeks (114 samples total, ≈8 samples per individual). Each fecal sample was preserved for the measurement of ammonia concentration and VFA content, as well as for microbial community fingerprinting (automated ribosomal intergenic spacer analysis, ARISA) and sequencing (Chaney and Marbach 1962; Erwin et al. 1961; Mackie et al. 1978; Ronaghi et al. 1998; Fisher and Triplett 1999; Yannarell and Triplett 2005). Fecal ammonia concentration estimates microbial protein metabolism, while VFA content can be used to estimate microbial carbohydrate fermentation. Because these values vary according to body size, all data were standardized by body weight for each age/sex class before analysis (Kelaita et al. 2011).

For both focal data and microbial data, dissimilarity between samples was visualized using nonmetric multidimensional scaling (NMDS) on PRIMER 6 for Windows v 6.1.10 (PRIMER-E, Plymouth, UK). NMDS plots for focal data were created using Euclidean distances, while those for microbial data were created using Bray-Curtis similarity indices. Overall activity budget, diet, and gut microbiome composition were tested for significant differences using analysis of similarity (ANOSIM), and nonparametric similarity percentage analysis (SIMPER) was used to determine which variables accounted most for observed differences in activity budget and diet (Clark and Gorley, 2006). Permutational (nonparametric) multivariate analysis of variance (PerMANOVA) was also used to detect significant patterns in microbial community composition across samples as well as to describe the amount of variation in microbial community composition explained by howler group, age, sex, and reproductive status as well as by sampling week (R software). Differences in kcal and grams of protein ingested, as well as differences in fecal ammonia concentration and VFA content, were tested for significance using ANOVA (R software) with Bonferroni corrections applied to p-values.

The behavioral data revealed few differences in activity budget among individuals. Male and female activity budgets did not differ (ANOSIM $R=0.038$, $p=0.35$), and the activity budgets of reproductively active females (pregnant or lactating) were the same as those of all other adults (ANOSIM $R=-0.068$, $p=0.58$). Activity budgets for juveniles were significantly different from adult activity budgets (ANOSIM $R=0.791$, $p=0.001$). Nonparametric SIMPER analyses revealed that 67.2 % of the variation between juvenile and adult activity budgets is a result of juveniles resting less than adults, while 25.1 % is a result of juveniles spending more time in social behavior (e.g., play) than adults.

Few differences in diet existed among age and sex classes. Overall, there were no differences in the diets of males, females, or juveniles when analyzed by food type or by plant species (ANOSIM $R=-0.069$, $p=0.72$; ANOSIM $R=-0.02$, $p=0.45$). Reproductively active females also showed no overall diet differences when compared to other adults (ANOSIM $R=0.198$, $p=0.16$, ANOSIM $R=0.296$, $p=0.083$). There were no age or sex differences in the number of kcal ingested per day ($F_{2,11}=1.32$, $p=0.3057$). Adult females tended to consume more grams of protein per day than juveniles and adult males ($F_{2,11}=3.65$, $p=0.060$), but this trend was not significant. Reproductively active females showed the same trend as nonreproductively active females.

Microbial analyses of fecal samples identified Firmicutes (68.4 %), Bacteroidetes (13.3 %), and Proteobacteria (0.92 %) in all individuals. Among Firmicutes, Clostridia were the most abundant (64.5 % of the entire microbiota). Microbial community fingerprinting revealed that gut microbiome composition clustered by individual over time, suggesting stability in the microbiota over the 2-month sampling period (ANOSIM $R=0.384$; $p=0.001$). Variation in gut microbiome composition within individuals from week to week was also detected, but there were no significant trends in gut microbiome composition across the study period ($F_{1,72}=3.01$, $p=0.42$). Similarly, while females and juveniles exhibited similar temporal shifts in gut microbiome composition that differed from males, these differences were small (Fig. 9.3).

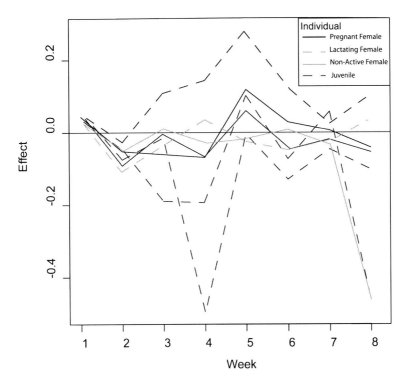

Fig. 9.3 Principal response curves depict weekly variation in gut microbiota composition in adult females and juveniles when compared to adult males (baseline)

PerMANOVA revealed that howler group, age/sex, and individual identity as well as sampling week accounted for approximately 41.2 % of the variation in gut microbiome composition. Of this, approximately 16.0 % of the variation was explained by individual howler identity ($F_{10, 105}=2.09$, $p=0.001$), 4.3 % by howler group ($F_{1, 105}=5.62$, $p=0.001$) and age/sex class ($F_{2, 105}=2.79$, $p=0.001$), and 2.1 % by sampling week ($F_{1, 105}=2.68$, $p=0.001$). When female gestation was incorporated into the age/sex class data, it explained more variation than age/sex class alone ($r^2=0.062$, $F_{3, 105}=2.68$, $p=0.001$). Similarly, when female gut microbiome patterns were depicted across the study period using NMDS, composition varied with reproductive status (ANOSIM $R=0.473$, $p=0.001$; Fig. 9.4). Average gut microbiome composition across the study period was distinct for reproductively active females and other individuals (ANOSIM $R=0.262$, $p=0.023$; Fig. 9.5). However, Firmicutes:Bacteroidetes proportions did not differ across age, sex, or reproductive classes. There were also no strong patterns in *Bifidobacterium*, *Bacteroides*, or *Enterobacteriaceae* (although the lactating female had higher *Enterobacteriaceae* abundances than any other individual sampled). *Oxalobacter*, a bacterium which increases calcium availability (Stuart et al. 2004; Nakata and McConn 2007), was present in both pregnant females and detected in only one other individual in lower abundance (0.007 % vs. 0.033 % of total sequences).

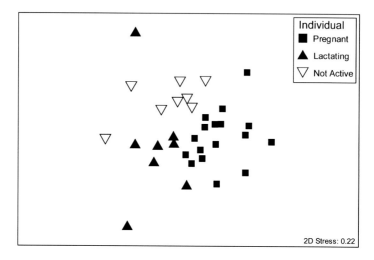

Fig. 9.4 NMDS demonstrates clustering of female gut microbiota composition by reproductive status. Symbols represent all samples collected for each female by week. Pregnant females ($N = 2$) exhibit different gut microbiota composition than the lactating female or the nonreproductively active female

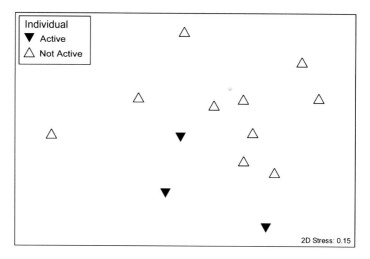

Fig. 9.5 NMDS demonstrates clustering of gut microbiota composition by individual reproductive status. Each symbol represents an average gut microbiota composition for each individual during the study period. Adult males, females, and juveniles that are not reproductively active exhibit distinct microbiota composition from reproductively active females

Gut microbiota activity assays revealed few differences. Fecal ammonia concentration did not vary significantly by age or sex ($F_{2,10} = 2.04$, $p = 0.18$) or by reproductive status ($F_{1,11} = 0.016$, $p = 0.90$). The proportions of fecal volatile fatty acids were similar for all individuals (average molar ratio of acetate to propionate to butyrate = 84.8:8.4:4.4) and generally matched those reported for *A. palliata* (Milton et al. 1980). However, adult females and juveniles had significantly higher total fecal VFA content than adult males (adult female 1,218.96 ± 126.67 µg/g fecal material/kg body weight, adult male 907.70 ± 26.08 mM/kg body weight, juvenile 1,427.34 ± 368.51 µg/g fecal material/kg body weight; $F_{2,7} = 7.64$, $p = 0.017$). Pregnant females also had somewhat lower total fecal VFA content than nonpregnant females (pregnant 1,120.89 ± 97.13 µg/g fecal material/kg body weight, nonpregnant 1,317.03 ± 15.10 µg/g fecal material/kg body weight; $F_{3,6} = 6.02$, $p = 0.031$).

Together these data indicate that the gut microbiota may play an important role in satisfying nutritional demands in juvenile and reproductively active female black howlers. Howlers did not show strong differences in activity patterns or diet according to age, sex, or reproductive status that would imply they are reducing activity or increasing energy and nutrient intake to compensate for differences in nutritional demands. However, variation in gut microbiome composition was in part explained by howler age, sex, and reproductive status, and differences in gut microbial activity were detected across age and sex classes. Low variation in total microbiome composition associated with howler age, sex, and reproductive status is likely a consequence of shifts in only a subset of microbial taxa or functional groups and the somewhat limited resolution of community fingerprinting. However, as indicated by the sequencing results, some microbial taxa may be important for juvenile and reproductively active female nutrition. More research describing microbial function is necessary to understand these patterns as are studies with larger sample sizes and longevity.

Additionally, microbial activity data suggest that juvenile and female howler monkeys are processing the same diet differently (producing different amounts of VFA) even if shifts in microbiota composition are limited. Increases in fecal VFA content may indicate higher energy production by females and juveniles. However, variation in energy absorption among individuals is unknown. As a result, lower VFA levels for pregnant females may not indicate that they produce less energy but rather that they absorb more of it than other individuals. Additional measures of digestive efficiency are necessary to separate these processes. Nevertheless, the production of distinct amounts of VFA by individuals of different age, sex, and reproductive classes is likely to have important consequences for host digestive efficiency and ultimately nutrition.

Although this study provides preliminary evidence validating our revised bioenergetics model, additional data are necessary to confirm the roles of diet, activity, gut morphology, and the gut microbiota in allowing howlers to meet nutritional demands. Furthermore, because behavior and diet may change seasonally depending on climate and/or food availability (Overdorff et al. 1997; Altmann 2009; Grueter et al. 2009; Marshall et al. 2009), cross-seasonal data must be collected to examine how the relationships and importance of these mechanisms vary. Future studies should also integrate C-peptide measurements from juveniles and reproductively active females [or other estimates of host nutritional status since pregnancy is known to interfere with insulin production (Havel 1998)] to test whether changes in diet, activity, and/or digestive efficiency actually allow hosts to meet nutritional demands and maintain body condition.

9.5 Application of Microbiota-Centered Models to Other Primate Taxa

While howler monkeys are an ideal system for exploring host-gut microbe interactions, the models developed in this chapter should extend to other primate taxa as well and allow us to predict both intraspecific and interspecific relationships among gut microbiota composition and host diet, health, and nutrition. Although studies of rodent and human gut microbiomes are still the most numerous due to the implications of gut microbes in human health and disease, some initial investigations of nonhuman primate gut microbiota composition and function exist. These investigations have generally focused on catarrhines (Frey et al. 2006; Fujita and Kageyama 2007; Uenishi et al. 2007; Kisidayova et al. 2009; Nakamura et al. 2009; Szekely et al. 2010; Degnan et al. 2012; Moeller et al. 2012; McCord et al. 2013; Moeller et al. 2013), although other primate taxa, such as the pygmy loris (*Nycticebus pygmaeus*) (Bo et al. 2010; Xu et al. 2013) and the black and mantled howler monkeys (*A. pigra, A. palliata*) (Nakamura et al. 2011; Clayton et al. 2012; Amato et al. 2013) also have been sampled.

In most of these studies, the main goal is to provide data regarding the composition of the gut microbiota. As a result, we know that the three dominant bacterial phyla in human and nonhuman primate gut microbial communities are Firmicutes, Proteobacteria, and Bacteroidetes (Frey et al. 2006; Yildirim et al. 2010). However, because these studies present little to no information regarding host diet, health, age, sex, or reproductive status, patterns and functions of intraspecific gut microbiota variation are difficult to discern. For example, a study of 23 chimpanzees revealed differences in gut microbiome composition between wild and captive individuals (as indicated by different TGGE band profiles), with wild chimpanzees' feces containing more bacteria such as *Clostridium, Ruminococcus*, and *Eubacterium* (Firmicutes), which are known for their sugar-fermenting and cellulolytic activity (Uenishi et al. 2007). Similarly, studies of African apes suggest an influence of host geography on gut microbial community composition that results in convergence of the gut microbial community for individuals of the same host species inhabiting the same area as well as for individuals of distinct host species (Degnan et al. 2012; Moeller et al. 2013). Although these results indicate that diet may play a role in determining ape gut microbiota composition, no data describing differences in microbial richness and diversity or host diet are provided.

Literature-based investigations of the gut microbial influence on individual bioenergetics within primate species are equally limited. Only one study of 12 wild chimpanzees (*P. troglodytes schweinfurthii*), belonging to the same social group and including parent-offspring pairs, has to some extent provided detailed data at the individual level (Szekely et al. 2010). However, this study reported that while the most common bacterial phyla (Firmicutes and Bacteroidetes) were shared by all individuals, indices of microbial community similarity were only high among a few samples. Therefore, while the authors suggest that kinship might play a role in determining microbial community composition, the identification of patterns across age and sex classes in response to growth and reproduction is not possible.

Although the overall scarcity of data limits our ability to more formally test the predictions of our models *within* a given nonhuman primate species, there exists a growing data set which can be used to compare *across* species. For example, a recent survey compared fecal samples from 23 free-ranging great apes (*Pan troglodytes, P. paniscus, Gorilla gorilla,* and *G. beringei*) and two humans (Ochman et al. 2010). Since a phylogeny based on the microbiome composition in these samples matched the great ape species phylogeny (mtDNA) more closely than a phylogeny based on the chloroplast sequence diversity (an indicator of diet) in these samples, the authors argue that host phylogeny is the most likely determinant of primate gut microbiota composition. However, branch length in the mtDNA phylogeny only explained 25 % of the variance in the gut microbiome tree, leaving a large percentage of the variation unexplained. Additionally, because these plant DNA sequences cannot distinguish between leaves and fruits of the same species, and because diet data obtained from plant DNA in fecal material have been shown to differ somewhat from observational dietary data (Bradley et al. 2007), chloroplast diversity may not completely describe host diet. In fact, when the comparison was expanded to include populations of the same species of apes in different habitats, the data suggested that host geography affects the magnitude of differences in the gut microbiota across host species (Moeller et al. 2013). Further evidence is necessary to better understand the influence of diet on interspecific gut microbiota composition and revise our model.

In fact, other studies provide evidence in support of this aspect of our model. Although a molecular analyses of nine fecal samples collected from sympatric wild, foregut-fermenting *Colobus guereza* and *Piliocolobus tephrosceles* and hindgut-fermenting *Cercopithecus ascanius* also revealed an influence of host phylogeny on gut microbiota composition (Yildirim et al. 2010), some effects of diet were clear. In particular, bacterial diversity and community composition analyses showed that the more-folivorous red colobus monkeys were characterized by the highest bacterial richness and highest diversity compared with the other two species. Similarly, a study by Lambert and Fellner (2012) reported significantly higher fecal acetate concentration in *Colobus guereza* (61 mol%) than in *Cercopithecus neglectus* (47 mol%). Although these data do not provide information on gut microbiota composition, they illustrate differential microbial activity possibly associated with dietary and digestive strategies.

Additionally, based on our model, we would predict that primate species specializing on a high-fiber diet such as gorillas or leafy diets such as colobines and indriids would harbor higher abundances of cellulolytic and proteolytic bacteria such as Clostridia and Eubacteria. Studies describing the gut microbiota composition of a variety of nonhuman primates verify these predictions. The fecal bacterial analysis of a wild male gorilla (*G. beringei*) (Frey et al. 2006) indicated a high abundance of Clostridia (51.5 %), as well as the presence *Ruminococcus flavefaciens* (a cellulolytic bacterium) and *Eubacterium oxidoreducens*, which decarboxylates gallate, a phenolic compound found in plant flavonoids, tannins, and lignin. These bacteria could confer an advantage to gorillas when consuming pith containing high cellulose concentrations (17.5–19.8 % of dry matter intake) (Rothman et al. 2007) and condensed tannins (Rothman et al. 2006). Likewise, the pygmy loris, an insectivorous primate which also includes fruits, gums, and small mammals in its diet, possesses more gut bacteria with proteolytic activity such as *Bacteroides* (Bacteroidetes) and carbohydrate-degrading Proteobacteria (34.5 %) than other primate species (0.6–2.2 % reported in the study of the two colobus species and the guenon) (Bo et al. 2010; Yildirim et al. 2010). Gut microbiota analyses also uncovered several species of the genus *Pseudomonas* (Proteobacteria) in the pygmy loris, some of which are known to degrade organic solvents such as toluene (Marques and Ramos 1993) and might play a role in the digestion of toxic insects and other plant material. This capacity of the pygmy loris gut microbiota for processing plant toxins was recently confirmed using metagenomic analyses (Xu et al. 2013).

While results from current primate gut microbiome studies provide some general support for the importance of diet and genotype on gut microbiota composition on an interspecific level, more data from both within and across primate species are necessary to fully test the validity of these relationships in our model. Likewise, intra- and interspecific studies investigating the effects of microbiota composition shifts on host health and nutrition are critical for understanding the role that the gut microbiota plays in host ecology and evolution. Much work remains to be done to improve our understanding of the host-microbe relationship in primates, but we hope that the models presented in this chapter will provide a guide for future project designs.

9.6 Conclusions

The studies reviewed in this chapter provide important baseline data with regard to the interactions between the mammalian gut microbiota and host diet, nutrition, and health. However, many questions remain unanswered regarding the role of the gut microbiota in wild host populations, especially in the case of primates. In even the most comprehensive primate microbiome studies, data reflecting the dynamics of the gut microbiota across time and space are largely absent (but see Amato 2013; Amato et al. 2013; McCord et al. 2013), and studies that integrate behavioral and ecological data with microbiome analyses of primates in their natural habitats simply do not exist. Data obtained from laboratory studies in a variety of taxa have allowed us to develop models to predict patterns in host-gut microbiota interactions in natural habitats, which can then be tested in field settings. Howler monkeys are an excellent system for testing these models due to their presence in a wide variety of habitats with diverse types of plant resources and their nutritional reliance on the gut microbiota. Additionally, improving the understanding of how howlers adjust to changing nutritional demands when diet is constrained has important implications for the study and conservation of these primates. It has recently been suggested that conservation biology could benefit greatly from the research on the human microbiome, by applying its methods and frameworks to improve conservation goals such as reintroductions, captive breeding, and dealing with invasions of nonnative species (Redford et al. 2012). If gut microbiota diversity, composition, and turnover are related to dietary flexibility, conservation of endangered howler species in the face of habitat fragmentation may depend, in part, on a better understanding of host-microbe relationships since fragmentation alters the amounts and types of resources to which the howlers have access (Cristobal-Azkarate and Arroyo-Rodriguez 2007; Dunn et al. 2009; Arroyo-Rodriguez and Dias 2010).

Acknowledgments We would like to thank A. Estrada and Universidad Nacional Autónoma de México for logistic support. Thanks are also due to R. Mackie for use of lab supplies and space at the University of Illinois. Research was carried out under permits from the Mexican environmental agencies, the Secretaría de Medio Ambiente y Recursos Naturales (SEMARNAT), the Comisión Nacional de Áreas Naturales Protegidas (CONANP), and the Instituto Nacional de Investigaciones Forestales, Agrícolas y Pecuarias (INIFAP). The Secretaría de Agricultura, Ganadería, Desarrollo Rural, Pesca y Alimentación (SAGARPA) in Mexico and the Center for Disease Control in the U.S. provided permits for sample transport. We acknowledge the helpful comments of P. Garber and one anonymous reviewer on this manuscript. This project was funded by the NSF grant #0935347 (HOMINID). Fieldwork and preliminary lab work were funded by grants from the University of Illinois (Beckman, Tinker, and the Program in Ecology, Evolution and Conservation Biology) to KRA. KRA was supported by an NSF Graduate Research Fellowship. 16S sequence data are available from the authors upon request.

References

Aiello LC, Wells JCK (2002) Energetics and the evolution of the genus *Homo*. Annu Rev Anthropol 31:323–338

Altmann J, Alberts S (1987) Body mass and growth rates in a wild primate population. Oecologia 72:15–20

Altmann J, Samuels A (1992) Costs of maternal care: infant-carrying in baboons. Behav Ecol Sociobiol 29:391–398

Altmann SA (2009) Fallback foods, eclectic omnivores, and the packaging problem. Am J Phys Anthropol 140:615–629

Amato KR (2013) Black howler monkey (Alouatta pigra) nutrition: integrating the study of behavior, feeding ecology, and the gut microbial community. Ph.D. dissertation, Program in Ecology, Evolution and Conservation Biology, University of Illinois, Urbana

Amato KR, Yeoman CJ, Kent A, Carbonero F, Righini N, Estrada AE, Gaskins HR, Stumpf RM, Yildirim S, Torralba M, Gillis M, Wilson BA, Nelson KE, White BA, Leigh SR (2013) Habitat degradation impacts primate gastrointestinal microbiomes. ISME J 7: 1344-1353

Amato KR, Yeoman CJ, Kent A, Righini N, Estrada AE, Stumpf RM, Yildirim S, Torralba M, Gillis M, Wilson BA, Nelson KE, White BA, Leigh SR (in review) Habitat degradation impacts primate gastrointestinal microbiomes.

Armougom F, Henry M, Vialettes B, Raccah D, Raoult D (2009) Monitoring bacterial community of human gut microbiota reveals an increase in *Lactobacillus* in obese patients and *Methanogens* in aneroxic patients. PLoS One 4:e7125

Arroyo-Rodriguez V, Asensio N, Cristobal-Azkarate J (2008) Demography, life history and migrations in a Mexican mantled howler group in a rainforest fragment. Am J Primatol 70:114–118

Arroyo-Rodriguez V, Dias PAD (2010) Effects of habitat fragmentation and disturbance on howler monkeys: a review. Am J Primatol 72:1–16

Bailey M, Coe CL (1999) Maternal separation disrupts the integrity of the intestinal microflora in infant rhesus monkeys. Dev Psychobiol 35:146–155

Barboza PS, Parker KL, Hume ID (2009) Integrative wildlife nutrition. Springer, Berlin

Bauer E, Williams BA, Smidt H, Verstegen MW, Mosenthin R (2006) Influence of the gastrointestinal microbiota on development of the immune system in young animals. Curr Issues Intest Microbiol 7:35–51

Behie AM, Pavelka MS (2005) The short-term effects of a hurricane on the diet and activity of black howlers (*Alouatta pigra*) in Monkey River, Belize. Folia Primatol 76:1–9

Benson AK, Kelly SA, Legge R, Ma F, Low SJ, Kim J, Zhang M, Oh PL, Nehrenberg D, Hua K, Kachman SD, Moriyama EN, Walter J, Peterson DA, Pomp D (2010) Individuality in gut microbiota composition is a complex polygenic trait shaped by multiple environmental and host genetic factors. Proc Natl Acad Sci U S A 107:18933–18938

Benveniste J, Lespinats G, Adam C, Salomon JC (1971a) Immunoglobulins in intact, immunized, and contaminated axenic mice: study of serum IgA. J Immunol 107:1647–1655

Benveniste J, Lespinats G, Salomon JC (1971b) Serum and secretory IgA in axenic and holoxenic mice. J Immunol 108:1656–1662

Bicca-Marques JC (2003) How do howler monkeys cope with habitat fragmentation? In: Marsh LK (ed) Primates in Fragments: Ecology and Conservation Kluwer Academic, New York

Bjorkholm B, Bok CM, Lundin A, Rafter J, Hibberd ML, Pettersson S (2009) Intestinal microbiota regulate xenobiotic metabolism in the liver. PLoS One 4:e6958

Bo X, Zun-Xi H, Xiao-Yan W, Run-Chi G, Xiang-Hua T, Yue-Lin M, Yun-Juan Y, Hui S, Li-Da Z (2010) Phylogenetic analysis of the fecal flora of the wild pygmy loris. Am J Primatol 72:699–706

Bonilla-Sanchez YM, Serio-Silva JC, Pozo-Montuy G, Chapman CA (2012) Howlers are able to survive in *Eucalyptus* plantations where remnant and regenerating vegetation is available. Int J Primatol 33:233–245

Bovee-Oudenhoven IM, Wissink ML, Wouters JT, Van der Meer R (1999) Dietary calcium phosphate stimulates intestinal *Lactobacilli* and decreases the severity of a *Salmonella* infection in rats. J Nutr 129:607–612

Bradley BJ, Stiller M, Doran-Sheehy DM, Harris T, Chapman CA, Vigilant L, Poinar H (2007) Plant DNA sequences from feces: potential means for assessing diets of wild primates. Am J Primatol 69:699–705

Brourton MR, Perrin MR (1991) Comparative gut morphometrics of Vervet (*Cercopithecus aethiops*) and Samango (*C. mitis erytharchus*) monkeys. Z Saugetierkunde 56:65–71

Buhnik-Rosenblau K, Danin-Poleg Y, Kashi Y (2011) Host genetics and gut microbiota. In: Rosenberg E, Gophna U (eds) Beneficial Microorganisms in Multicellular Life Forms. Springer, Berlin

Carbonero F, Benefiel AC, Gaskins HR (2012) Contributions of the microbial hydrogen economy on colonic homeostasis. Nat Rev Gasteroenterol Hepatol 9:504–518

Cavedon K, Leschine SB, Canale-Parola E (1990) Cellulase system of a free-living mesophilic *Clostridium* (strain C7). J Bacteriol 172:4222–4230

Chaney AL, Marbach EP (1962) Modified reagents for the determination of urea and ammonia. Clin Chem 8:130–132

Chilvers BL, Wilkinson IS (2009) Diverse foraging strategies in lactating New Zealand sea lions. Marine Ecol 378:299–308

Chivers DJ, Hladik CM (1980) Morphology of the gastrointestinal tract in primates: comparisons with other mammals in relation to diet. J Morphol 166:337–386

Chivers DJ, Langer P (1994) Gut form and function: variations and terminology. In: Chivers DJ, Langer P (eds) The Digestive System in Mammals: Food, Form and Function. Cambridge University, Cambridge

Clark KR, Gorley RN. (2006). PRIMER v6: User Manual/Tutorial. PRIMER-E, Plymouth

Clayton JB, Kim HB, Glander K, Isaacson RE, Johnson TJ (2012) Fecal bacterial diversity of wild mantled howling monkeys (Alouatta palliata). Am J Phys Anthropol 147:116

Costello EK, Stagaman K, Dethlefsen L, Bohannan BJ, Relman DA (2012) An application of ecological theory toward an understanding of the human microbiome. Science 336:1255–1262

Cristobal-Azkarate J, Arroyo-Rodriguez V (2007) Diet and activity pattern of howler monkeys (*Alouatta palliata*) in Los Tuxtlas, Mexico: effects of habitat fragmentation and implications for conservation. Am J Primatol 69:1013–1029

David LA, Maurice CF, Carmody RN, Gootenberg DB, Button JE, Wolfe BE, Ling AV, Devlin AS, Varma Y, Fischbach MA, Biddinger SB, Dutton RJ, Turnbaugh PJ (2014) Diet rapidly and reproducibly alters the human gut microbiome. Nat 505:559–566

Deschner T, Kratzsch J, Hohmann G (2008) Urinary c-peptide as a method for monitoring body mass changes in captive bonobos (*Pan paniscus*). Horm Behav 54:620–626

Dethlefsen L, McFall-Ngai M, Relman DA (2007) An ecological and evolutionary perspective on human-microbe mutualism and disease. Nature 449:811–818

Di Fiore A, Link A, Campbell C (2011) The atelines: behavioral and socioecological diversity in a New World radiation. In: Campbell C, Fuentes A, MacKinnon KC, Panger M, Bearder SK (eds) Primates in Perspective, vol 2. Oxford University, Oxford

Dias PAD, Rangel-Negrin A, Canales-Espinosa D (2011) Effects of lactation on the time-budgets and foraging patterns of female black howlers (*Alouatta pigra*). Am J Phys Anthropol 145:137–146

Donnet-Hughes A, Perez PF, Dore J, Leclerc M, Levenez F, Benyacoub J, Serrant P, Segura-Roggero I, Schiffrin EJ (2010) Potential role of the intestinal microbiota of the mother in neonatal immune education. Proc Nutr Soc 69:407–415

Dunn JC, Cristobal-Azkarate J, Vea JJ (2009) Differences in diet and activity pattern between two groups of *Alouatta palliata* associated with the availability of big trees and fruit of top food taxa. Am J Primatol 71:654–662

Eckert KA, Hahn NE, Genz AK, Kitchen DM, Stuart MD, Averbeck GA, Stromberg BE, Markowitz H (2006) Coprological surveys of *Alouatta pigra* at two sites in Belize. Int J Primatol 27:227–238

Edwards MS, Ullrey DE (1999) Effect of dietary fiber concentration on apparent digestibility and digesta passage in non-human primates. II. Hindgut- and foregut-fermenting folivores. Zoo Biol 18:537–549

Erwin ES, Marco GJ, Emery EM (1961) Volatile fatty acid analysis of blood and rumen fluid by gas chromatography. J Dairy Sci 44:1768–1771

Fedigan LM, Rose LM (1995) Interbirth interval variation in three sympatric species of neotropical monkey. Am J Primatol 37:9–24

Fischer SG, Lerman LS (1979) Length-independent separation of DNA restriction fragments in two-dimensional gel electrophoresis. Cell 16:191–200

Fisher MM, Triplett EW (1999) Automated approach for ribosomal intergenic spacer analysis of microbial diversity and its application to freshwater bacterial communities. Appl Environ Microbiol 65:4630–4636

Flint HJ, Bayer EA (2008) Plant cell wall breakdown by anaerobic microorganisms from the mammalian digestive tract. Ann N Y Acad Sci 1125:280–288

Flint HJ, Duncan SH, Louis P (2011) Impact of intestinal microbial communities upon health. In: Rosenberg E, Gophna U (eds) Beneficial Microorganims in Multicellular Life Forms. Springer, Berlin

Fons M, Gomez A, Karjalainen T (2000) Mechanisms of colonisation resistance of the digestive tract. Part 2: bacteria/bacteria interactions. Microb Ecol Health Dis 12:240–246

Foster JA, McVey Neufeld KA (2013) Gut-brain axis: How the microbiome influences anxiety and depression. Cell 36:305–312

Forsythe P, Sudo N, Dinan T, Taylor VH, Bienenstock J (2010) Mood and gut feelings. Brain Behav Immun 24:9–16

Frey JC, Rothman JM, Pell AN, Nizeyi JB, Cranfield MR, Angert ER (2006) Fecal bacterial diversity in a wild gorilla. Appl Environ Microbiol 72:3788–3792

Friswell MK, Gika H, Stratford IJ, Theodoridis G, Telfer B, Wilson ID, McBain AJ (2010) Site and strain-specific variation in gut microbiota profiles and metabolism in experimental mice. PLoS One 5:e8584

Fujita S, Kageyama T (2007) Polymerase chain reaction detection of Clostridium perfringens in feces from captive and wild chimpanzees, *Pan troglodytes*. J Med Primatol 36:25–32

Girard-Buttoz C, Higham JP, Heistermann M, Wedegartner S, Maestripieri D, Engelhardt A (2011) Urinary c-peptide measurement as a marker of nutritional status in macaques. PLoS One 6:e18042

Grueter CC, Li DY, Ren BP, Wei FW, Xiang ZF, Van Schaik CP (2009) Fallback foods of temperate-living primates: a case study on snub-nosed monkeys. Am J Phys Anthropol 140:700–715

Hammond KA, Kristan DM (2000) Responses to lactation and cold exposure by deer mice (*Peromyscus maniculatus*). Physiol Biochem Zool 73:547–556

Harris TR, Chapman CA, Monfort SL (2009) Small folivorous primate groups exhibit behavioral and physiological effects of food scarcity. Behav Ecol 21:46–56

Havel PJ (1998) Leptin production and action: relevance to energy balance in humans. Am J Clin Nutr 67:355–356

Hill MJ (1997) Intestinal flora and endogenous vitamin synthesis. Eur J Cancer Prev 6:S43–S45

Hooper LV, Littman DR, Macpherson AJ (2012) Interactions between the microbiota and the immune system. Science 336:1268–1273

Jaroszewska M, Wilczynska B (2006) Dimensions of surface area of alimentary canal of pregnant and lactating female common shrews. J Mammal 87:589–597

Kau AL, Abern PP, Griffin NW, Goodman AL, Gordon JI (2011) Human nutrition, the gut microbiome and the immune system. Nature 474:327–336

Kay RNB, Davies AG (1994) Digestive physiology. In: Davies AG, Oates JF (eds) Digestive physiology. Cambridge University, Cambridge

Kelaita M, Dias PAD, Aguilar Cucurachi MS, Canales-Espinosa D, Cortés-Ortiz L (2011) Impact of intrasexual selection on sexual dimorphism and testes size in the Mexican howler monkeys *Alouatta palliata* and *A. pigra*. Am J Phys Anthropol 146:179–187

Kelly D, Campbell JI, King TP, Grant G, Jansson EA, Coutts AGP, Pettersson S, Conway S (2003) Commensal anaerobic gut bacteria attenuate inflammation by regulating nuclear-cytoplasmic shuttling of PPAR-g and RelA. Nat Immun 5:104–112

Kisidayova S, Varadyova Z, Pristas P, Piknova M, Nigutova K, Petrzelkova KJ, Profousova I, Schovancova K, Kamler J, Modry D (2009) Effects of high- and low-fiber diets on fecal fermentation and fecal microbial populations of captive chimpanzees. Am J Primatol 71:548–557

Kleiber M (1975) The fire of life: an introduction to animal energetics. Krieger, Huntington

Kruszynskia YT, Home PD, Hanning I, Alberti K (1987) Basal and 24-h c-peptide and insulin secretion rate in normal man. Diabetology 30:16–21

Lambert JE (1998) Primate digestion: interactions among anatomy, physiology, and feeding ecology. Evol Anthropol 7:8–20

Lambert JE (2011) Primate nutritional ecology: feeding biology and diet at ecological and evolutionary scales. In: Campbell C, Fuentes A, MacKinnon KC, Panger M, Bearder SK (eds) Primates in Perspective, 2nd edn. Oxford University, New York

Lambert JE, Fellner V (2012) In vitro fermentation of dietary carbohydrate consumed by african apes and monkeys: preliminary results for interpreting microbial and digestive strategy. Int J Primatol 33:263–281

Lantz EL, Santymire RM, Murray CM, Heintz M, Lipende I, Travis DA, Lonsdorf EV (2011) Characterization of immunocompetence via immunoglobulin A in wild chimpanzees (Pan troglodytes schweinfurthii) at Gombe Stream National Park, Tanzania. Am J Primatol 73:52.

Larimer SC, Fritzsche P, Song Z, Johnston J, Neumann K, Gattermann R, McPhee ME, Johnston RE (2011) Foraging behavior of golden hamsters (*Mesocricetus auratus*) in the wild. J Ethology 29:275–283

Ley RE, Backhed F, Turnbaugh PJ, Lozupone C, Knight R, Gordon JI (2005) Obesity alters gut microbial ecology. Proc Natl Acad Sci U S A 102:11070–11075

Ley RE, Hamady M, Lozupone C, Turnbaugh PJ, Ramey RR, Bircher JS, Schlegel ML, Tucker TA, Schrenzel MD, Knight R, Gordon JI (2008a) Evolution of mammals and their gut microbes. Science 320:1647–1651

Ley RE, Lozupone C, Hamady M, Knight R, Gordon HA (2008b) Worlds within worlds: evolution of the vertebrate gut microbiota. Nature 6:776–788

Ley RE, Turnbaugh PJ, Klein S, Gordon JI (2006) Human gut microbes associated with obesity. Nature 444:1022–1023

Louis P, Scott KP, Duncan P, Flint HJ (2007) Understanding the effects of diet on bacterial metabolism in the large intestine. J Appl Microbiol 102:1197–1208

Mackie RI (2002) Mutualistic fermentative digestion in the gastrointestinal tract: diversity and evolution. Integr Comp Biol 42:319–326

Mackie RI, Gilchrist FMC, Robberts AM, Hannah PE, Schwartz HM (1978) Microbiological and chemical changes in the rumen during the stepwise adaptation of sheep to high concentrate diets. J Agric Sci 90:241–254

Mackie RI, Sghir A, Gaskins HR (1999) Developmental microbial ecology of the neonatal gastrointestinal tract. Am J Clin Nutr 69:1035S–1045S

Macpherson AJ, McCoy KD, Johansen FE, Brandtzaeg P (2008) The immune geography of IgA induction and function. Nat Rev 1:11–22

Marques S, Ramos JL (1993) Transcriptional control of the *Pseudomonas putida* TOL plasmid catabolic pathways. Mol Microbiol 9:923–929

Marshall AJ, Boyko CM, Feilen KL, Boyko RH, Leighton M (2009) Defining fallback foods and assessing their importance in primate ecology and evolution. Am J Phys Anthropol 140:603–614

Martinez-Mota R, Valdespino C, Sanchez-Ramos MA, Serio-Silva JC (2007) Effects of forest fragmention on the physiological stress of black howler monkeys. Anim Cons 10:374–379

McNab BK (2002) The physiological ecology of vertebrates: a view from energetics. Cornell University, Ithaca

McCord AI, Chapman CA, Weny G, Tumukunde A, Hyeroba D, Klotz K, Koblings AS, Mbora DNM, Cregger M, White BA, Leigh SR, Goldberg TL (2013) Fecal microbiomes of non-human primates in western Uganda reveal species-specific communities largely resistant to habitat perturbation. Am J Primatol

Medani M, Collins D, Docherty NG, Baird AW, O'Connell PR, Winter DC (2011) Emerging role of hydrogen sulfide in colonic physiology and pathophysiology. Inflamm Bowel Dis 17:1620–1625

Mellado M, Rodriguez A, Villareal JA, Olvera A (2005) The effect of pregnancy and lactation on diet composition and dietary preference of goats in a desert rangeland. Small Ruminant Res 58:79–85

Meserve PL et al (2003) Thirteen years of shifting top-down and bottom-up control. Bioscience 53:633–646

Messaoudi M, Lalonde R, Violle N, Javelot H, Desor D, Nejdi A, Bisson JF, Tougeot C, Pichelin M, Cazaubiel M, Cazaubiel JM (2010) Assessment of psychotropic-like properties of a probiotic formulation (*Lactobacillus helveticus* R0052 and *Bifidobacterium longum* R0175) in rats and human subjects. Br J Nutr 105:755–764

Milton K (1980) The foraging strategy of howler monkeys. Columbia University, New York

Milton K (1984) The role of food-processing factors in primate food choice. In: Rodman PS, Cant JGH (eds) Adaptations for foraging in nonhuman primates: Contribution to an organismal biology of prosimians, monkeys, and apes. Columbia University, New York

Milton K, McBee RH (1983) Rates of fermentative digestion in the howler monkey, *Alouatta Palliata* (Primates: Ceboidea). Comp Biochem Physiol 74A:29–31

Milton K, Van Soest P, Robertson J (1980) Digestive efficiencies of wild howler monkeys. Physiol Zool 53:402–409

Moeller AH, Peeters M, Ndjango JB, Li Y, Hahn BH, Ochman H (2013) Sympatric chimpanzees and gorillas harbor convergent gut microbial communities. Genome Res 23: 1715–1720

Moreau MC, Ducluzeau R, Guy-Grand D, Muller MC (1978) Increase in the population of duodenal immunoglobulin a plasmocytes in axenic mice associated with different living or dead bacterial strain of intestinal origin. Infect Immun 21:532–539

Nagy KA, Milton K (1979) Energy metabolism and food consumption by wild howler monkeys (*Alouatta palliata*). Ecology 60:475–480

Nakamura N, Amato KR, Garber PA, Estrada AE, Mackie RI, Gaskins HR (2011) Analysis of the hydrogenotrophic microbiota of wild and captive black howler monkeys (*Alouatta pigra*) in Palenque National Park, Mexico. Am J Primatol 73:909–919

Nakamura N, Leigh SR, Mackie RI, Gaskins HR (2009) Microbial community analysis of rectal methanogens and sulfate reducing bacteria in two non-human primate species. J Med Primatol 38:360–370

Nakata PA, McConn MM (2007) Calcium oxalate content affects the nutritional availability of calcium from *Medicago truncatula*. Plant Sci 172:958–961

Neish AS (2009) Microbes in gastrointestinal health and disease. Gastroenterology 136:65–80

Norconk MA, Wright BW, Conklin-Brittain NL, Vinyard CJ (2009) Mechanical and nutritional properties of food as factors in platyrrhine dietary adaptations. In: Garber PA, Bicca-Marques JC, Estrada AE, Heymann EW, Strier KB (eds) South American Primates, Developments in Primatology: Progress and Prospects. Springer, New York

O'Brien TG, Kinnaird M, Dierenfeld ES, Conklin-Brittain NL, Wrangham RW, Silver SC (1998) What's so special about figs? Nature 392:668

Ochman H, Worobey M, Kuo CH, Ndjango JBN, Peeters M, Hahn BH, Hugenholtz P (2010) Evolutionary relationships of wild hominids recapitulated by gut microbial communities. PLoS Biol 8:e1000546

Oftedal OT, Whiten A, Southgate DAT, Van Soest P (1991) The nutritional consequences of foraging in primates: the relationship of nutrient intakes to nutrient requirements. Philos Trans R Soc Lond B Biol Sci 334:161–170

Ohara H, Karita S, Kimura T, Sakka K, Ohmiya K (2000) Characterization of the cellulolytic complex (cellulosome) from *Ruminococcus albus*. Biosci Biotechnol Biochem 64:254–260

Osborn AM, Morre RB, Timmis KN (2000) An evaluation of terminal-restriction fragment length polymorphism (T-RFLP) analysis for the study of microbial community structure and dynamics. Environ Microbiol 2:39–50

Ostfeld RS, Keesing F (2000) Pulsed resources and community dynamics of consumers in terrestrial ecosystems. Trends Ecol Evol 15:232–237

Overdorff DJ, Strait SG, Telo A (1997) Seasonal variation in activity and diet in a small-bodied folivorous primate, *Hapalemur griseus*, in southeastern Madagascar. Am J Primatol 43:211–223

Pavelka MSM, Knopff KH (2004) Diet and activity in black howler monkeys (*Alouatta pigra*) in southern Belize: does degree of frugivory influence activity level? Primates 45:105–111

Park AJ, Collins J, Blennerhassett P, Ghia JE, Verdu EF, Bercik P, Collins SM (2013) Altered colonic function and microbiota profile in a mouse model of chronic depression. Neurogastroenterology and Motility 25: 733–e575

Peles JD, Barrett GW (2008) The golden mouse: a model of energetic efficiency. In: Barrett GW, Feldhamer GA (eds) The Golden Mouse: Ecology and Conservation. Springer, New York

Phillips KA, Abercrombie CL (2003) Distribution and conservation status of the primates of Trinidad. Primate Conserv 19:19–22

Pozo-Montuy G, Serio-Silva JC, Bonilla-Sanchez YM (2011) Influence of the landscape matrix on the abundance of arboreal primates in fragmented landscapes. Primates 52:139–147

Raguet-Schofield ML (2009) The ontogeny of feeding behavior of Nicaraguan mantled howler monkeys (Alouatta palliata). Ph.D. dissertation, Department of Anthropology University of Illinois, Urbana

Redford KH, Segre JA, Salafsky N, Martinez del Rio C, McAloose D (2012) Conservation and the microbiome. Conserv Biol 26:195–197

Righini N (2014) Primate nutritional ecology: the role of food selection, energy intake, and nutrient balancing in Mexican black howler monkey (Alouatta pigra) foraging strategies. Ph.D. dissertation, Department of Anthropology, University of Illinois at Urbana-Champaign, Urbana

Ronaghi M, Uhlen M, Nyren P (1998) A sequencing method based on real-time pyrophosphate. Science 281:363

Rothman JM, Dierenfeld ES, Molina DO, Shaw AV, Hintz HF, Pell AN (2006) Nutritional chemistry of foods eaten by gorillas in Bwindi Impenetrable National Park, Uganda. Am J Primatol 68:675–691

Rothman JM, Plumptre AJ, Dierenfeld ES, Pell AN (2007) Nutritional composition of the diet of the gorilla (*Gorilla beringei*): a comparison between two montane habitats. J Trop Ecol 23:673–682

Santacruz A, Collado MC, Garcia-Valdes L, Segura MT, Martin-Lagos JA, Anjos T, Marti-Romero M, Lopez RM, Florido J, Campoy C, Sanz Y (2010) Gut microbiota composition is associated with body weight, weight gain and biochemical parameters in pregnant women. Br J Nutr 104:83–92

Schoeninger MJ, Iwaniec UT, Glander K (1997) Stable isotope ratios indicate diet and habitat use in New World monkeys. Am J Phys Anthropol 103:69–83

Sekirov I, Russel SI, Antunes CM, Finlay BB (2010) Gut microbiota in health and disease. Physiol Rev 90:859–904

Serio-Silva JC, Hernandez-Salazar LT, Rico-Gray V (1999) Nutritional composition of the diet of *Alouatta palliata mexicana* females in different reproductive states. Zoo Biol 18:507–513

Servin AL (2004) Antagonistic activities of lactobacilli and bifidobacteria against microbial pathogens. FEMS Microbiol Rev 28:405–440

Sherry DS, Ellison PT (2007) Potential applications of urinary c-peptide of insulin for comparative energetics research. Am J Phys Anthropol 133:771–778

Spor A, Koren O, Ley RE (2011) Unravelling the effects of the environment and host genotype on the gut microbiome. Nat Rev 9:279–290

Stevens CE, Hume ID (1995) Comparative physiology of the vertebrate digestive system. Cambridge University, New York

Stoner KE, Gonzalez Di Pierro A (2006) Intestinal parasitic infections in *Alouatta pigra* in tropical rainforest in Lacandona, Chiapas, Mexico: implications for behavioral ecology and conservation. In: Estrada AE, Garber PA, Pavelka MS, Luecke L (eds) New perspectives in the study of Mesoamerican primates. Springer, New York

Strachan DP (1989) Hay fever, hygiene, and household size. BMJ 299:1259–1260

Strier KB (1992) Atelinae adaptations: behavioral strategies and ecological constraints. Am J Phys Anthropol 88:515–524

Stuart CS, Duncan SH, Cave DR (2004) Oxalobacter formigenes and its role in oxalate metabolism in the human gut. FEMS Microbiol Lett 230:1–7

Sudo N (2006) Stress and gut microbiota: does postnatal microbial colonization programs the hypothalamic-pituitary-adrenal system for stress response? Int Cong Ser 1287:350–354

Suzuki K, Harasawa R, Yoshitake Y, Mitsuoka T (1983) Effect of crowding and heat stress on intestinal flora, body weight gain, and feed efficiency of growing rats and chicks. Nippon Juigaku Zasshi 45:331–338

Szekely BA, Singh J, Marsh TL, Hagedorn C, Werre SR, Kaur T (2010) Fecal bacterial diversity of human-habituated wild chimpanzees (*Pan troglodytes schweinfurthii*) at Mahale Mountains National Park, western Tanzania. Am J Primatol 72:566–574

Talham GL, Jiang HQ, Bos NA, Cebra JJ (1999) Segmented filamentous bacteria are potent stimuli of a physiologically normal state of the murine gut mucosal immune system. Infect Immun 67:1992–2000

Thompson ME, Knott CD (2008) Urinary c-peptide of insulin as a non-invasive marker of energy balance in wild orangutans. Horm Behav 53:526–535

Thompson ME, Muller MN, Wrangham RW, Lwanga JS, Potts KB (2008) Urinary c-peptide tracks seasonal and individual variation in energy balance in wild chimpanzees. Horm Behav 55:299–305

Trejo-Macias G, Estrada AE, Mosqueda Cabrera MA (2007) Survey of helminth parasites in populations of *Alouatta palliata mexicana* and *A. pigra* in continuous and in fragmented habitat in Southern Mexico. Int J Primatol 28:931–945

Turnbaugh PJ, Ley RE, Mahowald MA, Magrini V, Mardis ER, Gordon JI (2006) An obesity-associated gut microbiome with increased capacity for energy harvest. Nature 444:1027–1031

Turnbaugh PJ, Ridaura VK, Faith JJ, Rey FE, Knight R, Gordon HA (2009) The effect of diet on the human gut microbiome: a metagenomic analysis in humanized gnotobiotic mice. Sci Transl Med 1:6ra14. doi: 10.1126/scitranslmed.3000322.

Uenishi G, Fujita S, Ohashi G, Kato A, Yamauchi S, Matsuzawa T, Ushida K (2007) Molecular analyses of the intestinal microbiota of chimpanzees in the wild and in captivity. Am J Primatol 69:367–376

Vitazkova SK, Wade SE (2007) Effects of ecology on the gastrointestinal parasites of *Alouatta pigra*. Int J Primatol 28:1327–1343

Wu GD, Chen J, Hoffmann C, Bittinger K, Chen YY, Keilbaugh SA, Bewtra M, Knights D, Walters WA, Knight R, Sinha R, Gilroy E, Gupta K, Baldassano R, Nessel L, Li H, Bushman FD, Lewis JD (2011) Linking long-term dietary patterns with gut microbial enterotypes. Science 334:105–108

Xu B, Xu W, Yang F, Li J, Yang Y, Tang X, Mu Y, Zhou J, Huang Z (2013) Metagenomic analysis of the pygmy loris fecal microbiome reveals unique functional capacity related to metabolism of aromatic compounds. PLoS One 8:e56565

Yannarell AC, Triplett EW (2005) Geographic and environmental sources of variation in lake bacterial community composition. Appl Environ Microbiol 71:227–239

Yeoman CJ, Chia N, Yildirim S, Berg Miller ME, Kent A, Stumpf RM, Leigh SR, Nelson KE, White BA, Wilson BA (2011) Towards an evolutionary model of animal-associated microbiomes. Entropy 13:570–594

Yildirim S, Yeoman CJ, Sipos M, Torralba M, Wilson BA, Goldberg TL, Stumpf RM, Leigh SR, White BA, Nelson KE (2010) Characterization of the fecal microbiome from non-human wild primates reveals species specific microbial communities. PLoS One 5:e13963

Zoetendal EG, Akkermans ADL, Akkermans-va Vliet WM, de Visser JAGM, De Vos WM (2001) The host genotype affects the bacterial community in the human gastrointestinal tract. Microb Ecol Health Dis 13:129–134

Zunino GE, Kowalewski MM, Oklander LI, Gonzalez V (2007) Habitat fragmentation and population size of the black and gold howler monkey (*Alouatta caraya*) in a semideciduous forest in Northern Argentina. Am J Primatol 69:966–975

Chapter 10
Ecological Determinants of Parasitism in Howler Monkeys

Rodolfo Martínez-Mota, Martín M. Kowalewski, and Thomas R. Gillespie

Abstract Infectious diseases caused by pathogens are now recognized as one of the most important threats to primate conservation. The fact that howler monkeys (*Alouatta* spp.) are widely distributed from Southern Mexico to Northern Argentina, inhabit a diverse array of habitats, and are considered "pioneers," particularly adapted to exploit marginal habitats, provides an opportunity to explore general trends of parasitism and evaluate the dynamics of infectious diseases in this genus. We take a meta-analysis approach to examine the effect of ecological and environmental variables on parasitic infection using data from 7 howler monkey species at more than 35 sites throughout their distribution. We found that different factors including precipitation, latitude, altitude, and human proximity may influence parasite infection depending on the parasite type. We also found that parasites infecting howler monkeys followed a right-skewed distribution, suggesting that only a few individuals harbor infections. This result highlights the importance of collecting large sample sizes when developing these kinds of studies. We suggest that future studies should focus on obtaining fine-grained measurements of ecological and microclimate changes to provide better insights into the proximate factors that promote parasitism.

Resumen Las enfermedades infecciosas causadas por patógenos son reconocidas en la actualidad como una de las principales amenazas para la conservación de primates. Los monos aulladores (*Alouatta* spp.) son los primates con mayor distribución en Las Américas, desde el sur de México hasta el noreste de la Argentina. Además, habitan una gran variedad de hábitats y son considerados "pioneros." al encontrarse frecuentemente en áreas marginales. Esto los convierte en modelos

---

R. Martínez-Mota (✉)
Department of Anthropology, University of Illinois at Urbana-Champaign, Urbana, IL, USA
e-mail: rmarti39@illinois.edu

M.M. Kowalewski
Estación Biológica Corrientes, Museo Argentino de Ciencias Naturales, Consejo Nacional de Investigaciones Científicas y Técnicas (CONICET), Buenos Aires, Argentina
e-mail: martinkow@gmail.com

T.R. Gillespie
Departments of Environmental Sciences and Environmental Health, Emory University, Atlanta, GA 30322, USA
e-mail: thomas.gillespie@emory.edu

ideales para explorar tendencias generales de parasitismo y evaluar la dinámica de enfermedades infecciosas. Se realizó un meta-análisis para examinar los efectos de variables ecológicas y ambientales sobre infecciones por parásitos utilizando datos de siete especies de monos aulladores distribuidos en más de 35 sitios a lo largo de su distribución. Se encontró que factores tales como precipitación, latitud, altitud y la proximidad a asentamientos humanos afectan en diferentes grados a la infección parasitaria según el tipo de parásito considerado. También se encontró que los parásitos de monos aulladores siguen una distribución sesgada, indicando que pocos individuos dentro de una población muestran infecciones por parásitos. Esto sugiere la importancia de colectar un número de muestras apropiado. Se recomienda que los estudios futuros se enfoquen en obtener estimaciones detalladas de cambios ecológicos y microclimáticos. Esto permitirá identificar en forma más precisa cuáles son los factores próximos que promueven el parasitismo.

Keywords Disease ecology • Prevalence • Richness • Latitude • Precipitation • Habitat disturbance

10.1 Introduction

Infectious diseases caused by pathogens are now recognized as one of the most important threats for wildlife and primate conservation (Daszak et al. 2000; Leendertz et al. 2006; Gillespie et al. 2008). Several studies have documented that pathogens are capable of reducing wildlife populations (e.g., amphibians (Daszak et al. 1999); Ethiopians wolves (Laurenson et al. 1998)). In primates, the most dramatic cases come from studies of apes impacted by respiratory pathogens or the Ebola hemorrhagic fever (Bermejo et al. 2006; Köndgen et al. 2008; Palacios et al. 2011). Yellow fever outbreaks have impacted populations of mantled (*Alouatta palliata*), brown (*A. guariba*), and black-and-gold (*A. caraya*) howler monkeys (Rifakis et al. 2006; Milton et al. 2009; Holzmann et al. 2010; de Almeida et al. 2012). These studies have demonstrated the vulnerability of primates to infectious diseases and have highlighted the importance of health monitoring to detect primate populations at risk due to pathogenic infection (Leendertz et al. 2006).

Howler monkeys (genus *Alouatta*) have a wide distribution from Southern Mexico to Northern Argentina and inhabit diverse habitats including tropical rain forests, dry deciduous forests, mountain forests, lowland forests, and mangroves, due to their dietary flexibility and ability to exploit difficult-to-digest food items, such as mature leaves and unripe fruits (Di Fiore et al. 2011). Howlers have been studied extensively, including aspects of their behavior (e.g., male and female reproductive behavior (Van Belle et al. 2009; Kowalewski and Garber 2010)), demography (e.g., population change (Clarke et al. 2002; Rudran and Fernandez-Duque 2003)), ecology (e.g., feeding ecology (Milton 1980; Silver et al. 1998)), and parasitism (Table 10.1). More than 60 % of the studies reported in Table 10.1 have focused on gastrointestinal parasites voided in feces, given that fecal samples can be

Table 10.1 Studies of parasitic infection in wild howler monkeys (genus *Alouatta*)

Species	Study site	Latitude	Habitat type	Altitude (m)	Rainfall (mm/year)	Human proximity	Sample type	Sample size	# of individuals	# of groups	Source
A. arctoidea	Hato El Frio, Venezuela	7° 30′ N	MF	60	1,424	Rural	Collected specimens	Unk	38	6	1
A. belzebul	Rio Tocantins, Tucurui, Brazil	3° 40′ S	MF, C	75	2,740	Rural	Feces	212	Unk	50	2
A. caraya	Parana River, Parana, Brazil	22° 46′ S	MF	252	1,700	Rural	Blood	17	17	Unk	3
	Nova Querencia, Mato Grosso do Sul, Brazil	20° 43′ S	LF	450	1,379	Rural	Feces	59	6	1	4
	Tocantins River, Goias, Brazil	13° 49′ S	SF	460	1,750	Rural	Blood	42	42	Unk	5
	Porto Primavera, Sao Paulo-Mato Grosso do Sul, Brazil	21° 15′ S	SF	302	1,500	Rural	Blood	590	590	Unk	5
	Bella Vista, Corrientes, Argentina	28° 30′ S	F*	60	1,200	Remote	Collected specimens	302	302	Unk	6
	Parana River, Chaco, Argentina	27° 20′ S	SF	60	1,200	Remote	Feces	28	28	Unk	7
	Rio Riachuelo, Corrientes, Argentina	27° 30′ S	SF	55	1,200	Rural	Blood	30	30	Unk	8
		27° 30′ S	SF	55	1,200	Rural	Feces	256	16	2	9
		27° 30′ S	SF	55	1,200	Rural	Collected specimens	110	110	Unk	10
	Las Lomas, Corrientes, Argentina	27° 23′ S	MF	67	1,200	Remote	Feces	60	20	2	11
	Isla Brasilera, Corrientes, Argentina	27° 20′ S	MF	54	1,200	Remote	Blood/feces/fur	12	14	Unk	12
		27° 20′ S	MF	54	1,200	Remote	Feces	30	30	Unk	13
	Yacireta, Corrientes, Argentina	27° 28′ S	SF	65	1,200	Rural	Blood/feces/fur	9	9	Unk	12
	Estación Biológica, Corrientes, Argentina	27° 30′ S	SF	59	1,200	Rural	Feces	30	30	Unk	13
	San Cayetano, Corrientes, Argentina	27° 34′ S	SF	60	1,200	Urban	Blood/feces/fur	21	21	Unk	12
		27° 34′ S	SF	60	1,200	Urban	Feces	30	30	Unk	13
A. guariba	Morro Sao Pedro, Porto Alegre, Brazil	30° 01′ S	F*	230	1,324	Rural	Feces	53	Unk	Unk	14
	Reserva Biológica Lami, Porto Alegre, Brazil	30° 15′ S	F*	200	1,324	Urban	Feces	114	Unk	Unk	14
	Mata de Ribeirão Cachoeira, Brazil	22° 50′ S	MF	650	1,049	Remote	Feces	112	Unk	Unk	15

(continued)

Table 10.1 (continued)

Species	Study site	Latitude	Habitat type	Altitude (m)	Rainfall (mm/year)	Human proximity	Sample type	Sample size	# of individuals	# of groups	Source
A. macconnelli	Sinnamary River, Petit Saut Dam, French Guiana	5° 04′ N	C	45	3,000	Remote	Blood	117	117	Unk	16
		5° 04′ N	C	45	3,000	Remote	Blood	81	81	Unk	17
		5° 04′ N	C	45	3,000	Remote	Blood	50	50	Unk	18
	Biological Dynamics of Forest Fragments Project, Manaus, Brazil	2° 30′ S	C	38	2,606	Rural	Feces	35	24	3	19
	Balbina, Uatuma River, Amazonas State, BRA	1° 55′ S	C	34	2,262	Rural	Blood	31	31	Unk	20
A. palliata	Los Tuxtlas Biosphere Reserve, Mexico	18° 34′ N	C	300	4,900	Remote	Feces	38	Unk	Unk	21
	Los Tuxtlas Biosphere Reserve, Mexico	18° 34′ N	SF	300	4,900	Rural	Feces	63	Unk	Unk	21
		18° 34′ N	F*	100	4,900	Rural	Darted individuals[a]	6	6	5	22
		18° 38′ N	MF, SF	100	4,900	Rural	Feces	288	12	3	23
		18° 18′ N	SF	180	4,900	Rural	Feces	278	43	5	24
	Carlos Green, Mexico	17° 41′ N	F*	15	4,014	Rural	Blood	1	1	Unk	25
	Pochitocal, Mexico	18° 15′ N	F*	5	4,014	Rural	Blood	19	19	Unk	25
	Macuspana, Mexico	17° 38′ N	SF	15	3,186	Rural	Feces	Unk	27	2	26
	El Zapotal, Mexico	16° 43′ N	SF	700	950	Urban	Feces	67	15	1	27
	Barro Colorado Island, Panama	9° 10′ N	LF	120	2,612	Remote	Darted individuals[a]	Unk	Unk	Unk	28
	Chomes, Costa Rica	10° 02′ N	MF	9	1,950	Rural	Feces	9	Unk	Unk	29
	Parque Nacional Palo Verde, Costa Rica	10° 22′ N	LF	15	1,950	Remote	Feces	20	Unk	Unk	29
	Parque Nacional Cahuita, Costa Rica	9° 43′ N	LF	20	3,000	Remote	Feces	29	29	Unk	29
A. palliata	San Ramón, Costa Rica	10° 05′ N	SF	1,020	1,500	Urban	Feces	7	5	Unk	29
	Chira, Costa Rica	10° 06′ N	MF	8	2,000	Rural	Feces	5	5	Unk	29
	Gran Nicoya, Costa Rica	9° 59′ N	MF	100	1,950	Rural	Feces	18	Unk	Unk	29
	Playa Potrero, Costa Rica	10° 27′ N	SF	20	1,500	Urban	Feces	8	Unk	Unk	29
	Parque Nacional Manuel Antonio, Costa Rica	9° 24′ N	LF	45	4,000	Urban	Feces	6	Unk	Unk	29
	La Pacifica, Costa Rica	10° 28′ N	MF	45	1,553	Rural	Feces	200	108	13	30
	La Selva Biological Reserve, Costa Rica	10° 26′ N	LF	35	3,962	Rural	Feces	84	13	2	31

A. pigra	Reforma Agraria, Mexico	16° 15′ N	LF	181	3,000	Rural	Feces	17	Unk	Unk	21
	Calakmul Biosphere Reserve, Mexico	18° 06′ N	C	50	820	Remote	Feces	29	Unk	Unk	21
		18° 06′ N	C	50	820	Remote	Feces	Unk	4	2	32
	Palenque Biosphere Reserve, Mexico	17° 27′ N	LF	500	2,200	Remote	Feces	29	Unk	Unk	21
		17° 27′ N	LF	500	2,200	Remote	Feces	Unk	3	2	32
	Community Baboon Sanctuary, Belize	17° 33′ N	MF	28	1,995	Rural	Feces	Unk	13	3	32
	Cockscomb Basin Wildlife Sanctuary, Belize	16° 49′ N	C	200	2,700	Remote	Feces	Unk	5	2	32
	Montes Azules Biosphere Reserve, Mexico	16° 07′ N	C	181	3,000	Remote	Feces	23	Unk	Unk	21
		16° 07′ N	C	181	3,000	Remote	Feces	151	15	3	33
	El Tormento, Mexico	18° 36′ N	LF	90	1,380	Rural	Feces	39	Unk	Unk	21
		18° 36′ N	LF	90	1,380	Rural	Feces	81	69	17	34
		18° 36′ N	LF	90	1,380	Rural	Blood	12	12	3	25
	Carlos Green, Mexico	17° 41′ N	F*	15	4,014	Rural	Blood	2	2	Unk	25
	Pochitocal, Mexico	18° 15′ N	F*	5	4,014	Rural	Blood	6	6	Unk	25
	Lamanai Archaeological Reserve, Belize	17° 46′ N	MF	16	1,200	Rural	Feces	99	Unk	22	35
	Catazaja, Mexico	17° 35′ N	SF	20	1,400	Rural	Feces	218	43	6	36
	Punta Laguna, Mexico	20° 38′ N	MF	14	1,500	Remote	Feces	3	3	2	37
	Petcacab, Mexico	19° 17′ N	C	31	1,200	Remote	Feces	8	8	4	37
	Macuspana, Mexico	17° 38′ N	SF	15	3,186	Rural	Feces	Unk	7	2	26
	Monkey River, Belize	16° 21′ N	LF	12	2,500	Rural	Feces	315	18	4	38
A. sara	Tambopata National Reserve, Peru	13° 8′ S	C	250	2,400	Rural	Feces	16	16	4	39

Habitat type: *C* continuous forest, *SF* small-size fragments, *MF* medium-size fragments, *LF* large-size fragments, *F** indicates a fragmented forest but the size cannot be obtained from the literature. *Unk* unknown. We wrote "unknown" when ambiguous information regarding sample size, number of individuals, or number of groups was presented in the studies

Source: [1]Braza (1980), [2]Martins et al. (2008), [3]Garcia et al. (2005), [4]Godoy et al. (2004); [5]Duarte et al. (2006), [6]Pope (1966), [7]Venturini et al. (2003), [8]Travi et al. (1986), [9]Delgado (2006), [10]Coppo et al. (1979), [11]Milozzi et al. (2012), [12]Santa Cruz et al. (2000), [13]Kowalewski et al. (2011), [14]Cabral et al. (2005), [15]Santos et al. (2005), [16]Fandeur et al. (2000), [17]Volney et al. (2002), [18]Carme et al. (2002), [19]Gilbert (1994), [20]Lourenco de Oliveira and Deane (1995), [21]Trejo-Macías et al. (2007), [22]Cristóbal-Azkarate et al. (2010), [23]Cristóbal-Azkarate et al. (2010), [24]Valdespino et al. (2010), [25]Rovirosa-Hernández et al. (2013), [26]González-Hernández et al. (2011), [27]Castillejos (1993), [28]Milton (1996), [29]Chinchilla Carmona et al. (2005), [30]Stuart et al. (1990), [31]Stoner (1996), [32]Vitazkova and Wade (2007), [33]Stoner and González Di Pierro (2006), [34]Martínez-Mota unpublished data, [35]Eckert et al. (2006), [36]Alvarado-Villalobos (2010), [37]Bonilla-Moheno (2002), [38]Kowalzik et al. (2010), [39]Phillips et al. (2004)

[a]Individuals were darted for clinical evaluation and collection of botfly parasites

collected noninvasively without disturbing individuals (Gillespie 2006). The content of these papers ranges from descriptions (e.g., reports of parasites infecting *A. pigra* (Vitazkova and Wade 2006)) to studies relating parasitic infection to demographic (e.g., group size (Stoner and González Di Pierro 2006)) or ecological variables (e.g., forest fragmentation (Valdespino et al. 2010); contact with domesticated animals (Kowalewski et al. 2011)). In this chapter, we take a meta-analysis approach to examine the effect of ecological and environmental variables on parasitic infection of howler monkeys. First, we review variables that affect patterns of parasite infection, and second, we test whether different predictors such as forest fragmentation, human proximity, and climatic factors influence parasitism in howler monkeys.

10.2 Background

10.2.1 Habitat Disturbance, Forest Fragmentation, and Parasitic Infection

Habitat disturbance associated with anthropogenic activities, such as extensive logging, agriculture, cattle ranching, and ecotourism, has been added to the set of factors that promote the spread of parasites and increase the probability of pathogen exchange (Patz et al. 2000; Smith et al. 2009). These environmental changes favor the dispersal, establishment, and abundance of parasites that were previously rare (Wilcox and Ellis 2006). Evidence suggests that transformation of primate habitats alters parasite–host dynamics, affecting the potential for parasite transmission among primate hosts. For example, Gillespie et al. (2005) found that redtail guenons (*Cercopithecus ascanius*) inhabiting logged forest showed an approximate 85 % increased prevalence of the gastrointestinal parasite *Oesophagostomum* spp., compared to individuals living in an undisturbed forest. Similarly, Goldberg et al. (2008) found that humans harbored bacteria that were genetically more similar to those hosted by redtail guenons that inhabited fragments located near their settlements, compared to bacteria from guenons living in an undisturbed forest, suggesting that bacterial transmission between humans and primates had occurred. Primates inhabiting fragmented forests may be at greater risk of infectious diseases, in particular those living in proximity to human populations, due to increased exposure to pathogens that proliferate in anthropogenically disturbed habitats (Gillespie and Chapman 2006, 2008). Other studies, however, have not found clear differences in measures of parasitic infection when comparing populations of primates inhabiting forests with different degrees of disturbance (sifakas, *P. edwardsi* (Wright et al. 2009); mangabeys, *Cercocebus galeritus galeritus* (Mbora and McPeek 2009)). Clearly teasing out generalities and site-specific variation in how habitat disturbance affects the transmission of parasites in different primate species will be a major area of research in coming years.

10.2.2 Effects of Climate on Parasites

Studies of parasites hosted by wild primates should also take into consideration other factors that may play an interactive role in parasite–host dynamics. For example, climatic conditions, such as the amount of rainfall or moisture, have been identified as important variables for the proliferation of parasite vectors (Altizer et al. 2006; Vittor et al. 2006). In the case of malaria, a vector-borne disease caused by the protozoan *Plasmodium* spp., changes in patterns of precipitation were followed by malaria outbreaks in several African human populations (Zhou et al. 2004; Pascual et al. 2008). In this regard, Odongo-Aginya et al. (2005) reported that density of malaria parasites found in blood samples of human patients fluctuated with mean monthly rainfall during a year in the Entebbe Municipality, Uganda. Parasite vectors, such as mosquitos (e.g., *Anopheles* spp.), benefit from changes in rainfall patterns, given that these conditions increase humidity and availability of water sources, which provide more breeding sites, speed vector development, and increase vector abundance, potentially spreading a disease more efficiently (Patz et al. 2000; Vittor et al. 2006).

Rainfall also has been associated with an increase in protozoan infections such as cryptosporidiosis and giardiasis in human populations (Jagai et al. 2009) as well as in nonhuman primates (chimpanzees (Gonzalez-Moreno et al. 2013); black howler monkeys (Vitazkova and Wade 2006)). This might be the result of a high concentration of oocytes and cysts in water sources that tend to accumulate after heavy rainfall (Muchiri et al. 2009). In addition, precipitation plays an important role in the survival, development, and transmission of soil-transmitted helminths including hookworms (e.g., *Necator americanus*), whipworms (e.g., *Trichuris trichiura*), pinworms (e.g., *Enterobius vermicularis*), or roundworms (e.g., *Ascaris lumbricoides*) which are gastrointestinal parasites of public health concern (Bethony et al. 2006) reported to infect several nonhuman primates such as howler monkeys, orangutans, red langurs, gibbons, and chimpanzees (Vitazkova 2009; Gillespie et al. 2010, 2013; Hilser 2011). Moisture favors the survival and development of different parasite stages that are otherwise compromised by desiccation during dry periods (Gillespie 2006). Thus, we would expect that precipitation also affects patterns of parasite infection in howler monkeys.

Temperature is one of the critical climate factors affecting pathogen survival, distribution, and transmission (Harvell et al. 2002; Poulin 2006). For example, climate variability (e.g., short-term fluctuations around mean temperature) has been found to be a driver of malaria epidemics in African human populations (Lindblade et al. 2000; Zhou et al. 2004). This is most likely due to changes in land use and habitat modification that have led to an increase in temperature that in turn has altered vector distribution and parasite infection patterns (Lindblade et al. 2000; Harvell et al. 2002; Zamora-Vilchis et al. 2012). In parasite studies, altitude has been used as a proxy of temperature, implying that temperature decreases as elevation increases. In fact, a negative relationship between blood parasite prevalence and altitude has been described in birds (Zamora-Vilchis et al. 2012). Patterns of parasitism in primates also may vary according to an altitudinal gradient; for instance, Appleton and Henzi (1993) found that diversity of gastrointestinal parasites was lower in chacma baboons (*Papio cynocephalus ursinus*) that ranged at a high altitude (1,835–2,250 m), where temperature changes can be extreme representing a hostile environment for parasites, than in baboons ranging at 100–200 m altitude in Natal, South Africa. Given that parasites can be sensitive to temperature and be affected by an altitudinal gradient, it may be expected that at higher altitudes howler hosts present lower parasitic infection compared to howlers ranging at a low altitude. Since howler monkeys may inhabit forests both at sea level and at high altitude, this feature allows us to explore whether parasitic infection in howler monkeys follows an altitudinal gradient.

10.2.3 Host Distribution

A latitudinal gradient may affect patterns of parasitic infection, given that abundance and diversity of species increase in tropical areas at lower latitudes (Guernier et al. 2004; Hillebrand 2004). In general, it is acknowledged that geographic zones close to the equator may encompass a large variety of habitats and are characterized by high-energy productivity and favorable climatic conditions (Pianka 1966; Rohde 1992; Luo et al. 2012), which, in turn, may allow the establishment and proliferation of a diverse array of vertebrate hosts compared to temperate zones (Hawkins et al. 2003). This availability and diversity of hosts might favor transmission rates among generalist parasites (Nunn et al. 2005). Parasite species also may follow this latitudinal gradient; for example, species richness of pathogens responsible for infectious diseases in humans was found to be higher in tropical areas at lower latitudes (Guernier et al. 2004). In a meta-analysis of 119 primate host species, Nunn et al. (2005) found that species richness of protozoan parasites, but not helminths and viruses, increased towards the equator. According to this, howler monkeys that range in tropical areas close to the equator are expected to harbor more parasite species compared to howlers found at higher latitudes.

10.3 Goals and Expectations

Existing published data on parasites harbored by different species of howlers creates an opportunity to explore general trends of parasitism in these New World primates. Therefore, the main goal of this chapter is to examine the effect of multiple variables on measures of parasitic infection reported for several species of howler monkeys. We predict that:

1. Howler geographic distribution will have an effect on parasitic infection. We expect that parasite prevalence and species richness as measures of parasitic infection will be higher in howlers living close to the equator compared to howlers living at higher latitudes.

2. Given that humidity and rainfall may favor the development of parasites at different stages, we expect that measures of parasitic infections will be positively correlated with precipitation in howler monkeys. Furthermore, we expect that howlers living at lower altitudes show higher parasitic infection than howlers living at higher altitudes.
3. Habitat disturbance and forest fragmentation have been recognized as factors that modify parasitic infection dynamics; in this regard, we expect that howlers living in fragmented/disturbed habitats show higher parasite prevalence and richness than howlers inhabiting undisturbed forests. In addition, in anthropogenically disturbed habitats, the likelihood of contact between human and non-human primates is higher compared to remote areas, increasing the probabilities of pathogen exchange (Gillespie et al. 2008; Rwego et al. 2008). Thus, we expect that howlers inhabiting areas close to human settlements show an increase in measures of parasitic infection.

10.4 Methods

10.4.1 Data Collection

We conducted a literature review and analyzed published material including scientific articles, brief reports, and dissertation theses that reported parasitic infection in howler monkeys including mantled howlers (*Alouatta palliata*), black howlers (*A. pigra*), red howlers (*A. macconnelli* and *A. sara*), red-handed howlers (*A. belzebul*), brown howlers (*A. guariba*), and black-and-gold howler monkeys (*A. caraya*). We also searched any record of published material in the Global Mammal Parasite Database (www.mammalparasites.org, Nunn and Altizer 2005). We obtained parasite prevalence data reported for each species of parasite and recorded the number of parasite species reported in each study case. For each study site, we obtained ecological/environmental data including latitude, altitude (meters), and annual precipitation (millimeters) from primary literature (i.e., when reported in the study) or from websites such as WorldClim and Google Earth.

We categorized the howler monkey habitats as fragmented or continuous based on forest size (Marsh 2003; Kowalewski and Gillespie 2009). We assigned the category of small forest fragments to those with 1–100 ha forest cover. Fragments ranging in size from 100 to 1,000 ha were considered medium-size fragments, and those ranging from 1,000 to 10,000 ha of forest cover were assigned to the large-fragment category. Continuous habitats were those characterized by having $\geq$10,000 ha of forest area. Moreover, howler habitats were divided in three categories according to their proximity of human settlements, following Kowalewski and Gillespie (2009): (1) we considered an area as "remote" when the site was almost or totally isolated from human settlements. (2) We assigned the category of "rural" area to howler habitats that were close to rural populations, fishing camps, and/or were regularly visited by people. This applies mostly to forest fragments located nearby human settlements, where locals possibly carry out activities such as selective logging, cattle ranching, or hunting, showing a constant presence in howler habitats. (3) An "urban" site was considered when howler habitats were in close proximity to or immersed within human settlements characterized by dense human populations.

10.4.2 Data Analysis

We divided prevalence data into two broad categories, helminth and protozoan parasites: (1) We divided the helminth parasite data set into nematodes (82 records), trematodes (38 records), and cestodes (13 records) and also analyzed the effect of predictor variables on prevalence of *Trypanoxyuris* parasites, given that this was a well-represented genus in 4 out of 7 howler species (exception were *A. guariba*, *A. macconnelli*, and *A. sara*). (2) We separately analyzed prevalence data of protozoan parasites: we first divided this data set in a general category named amoebae parasites (34 records), which included the genera *Entamoeba*, *Endolimax*, *Iodamoeba*, and unknown reported amoebae. Thereafter, we analyzed *Giardia* prevalence (21 records) separately since these parasites were represented in 5 of 7 howler species (*A. belzebul*, *A. caraya*, *A. guariba*, *A. palliata*, and *A. pigra*) in our database. Finally, we analyzed data on *Plasmodium* prevalence (17 records). *Plasmodium* data were only available for 2 South American howler species (*A. caraya* and *A. macconnelli*); however, given that malaria infection is frequently associated with ecological changes (Zhou et al. 2004), we decided to explore the effect of ecological/environmental variables on the prevalence of this genus.

Parasite prevalence usually follows an aggregated distribution (e.g., negative binomial (Wilson et al. 2002)), thus we log-transformed helminth and protozoan prevalence and analyzed these data using generalized linear models with an identity link function in the R software (MASS library, version 2.15.1) (Crawley 2007). We considered the following predictor variables: forest type as a categorical variable, which includes fragments of different size and continuous forests. Similarly, human proximity was included as a categorical variable with three levels (1 = remote, 2 = rural, 3 = urban). Latitude, annual precipitation (millimeters), and altitude (meters) were included as continuous variables. We ran each model taking into account all predictor variables and selected the best model using the Akaike information criterion. Thereafter we ran a deviance test to assess model adequacy.

We also tested the effects of forest type, latitude, altitude, and precipitation and the effect of human proximity on parasite species richness (i.e., number of parasite species reported per howler population). We analyzed these data with a generalized linear model with a negative binomial link function (Wilson and Grenfell 1997; Crawley 2007) using the glm.nb procedure of the MASS library in the R software (version 2.15.1).

10.5 Results

10.5.1 Helminth Analysis

Nematodes: We found that precipitation was a predictor of nematode prevalence ($\chi^2 = 13.53$, $p = 0.003$) in howler monkeys. Figure 10.1 shows that nematode prevalence increases with precipitation. Other terms included in the model, such as forest type, latitude, or altitude, did not have an effect on the response variable. Similarly, human proximity did not have any effect on nematode prevalence.

Trematodes and Cestodes: We did not find any significant effect of forest type, latitude, precipitation, altitude, or human proximity on the prevalence of trematodes and cestodes hosted by howler monkeys. However, we found a trend of cestode prevalence being higher in howlers from remote forests compared to howlers inhabiting rural areas (Fig. 10.2). We did not find any record of cestode parasites at the "urban" level in the "human proximity" categorical variable in our data set; thus, this level was not considered in the analysis.

Trypanoxyuris: Prevalence of *Trypanoxyuris* parasites was not predicted by any of our predictor variables; however, we found a trend in which prevalence was higher at lower altitudes and decreased at higher altitudes (Fig. 10.3).

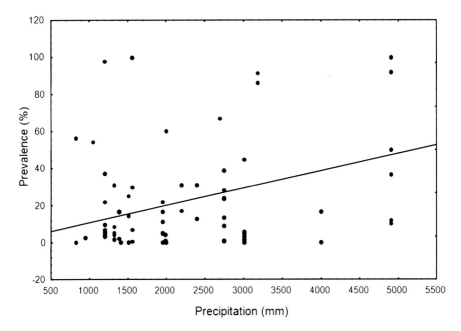

Fig. 10.1 Relationship between nematode prevalence hosted by howler monkeys and precipitation

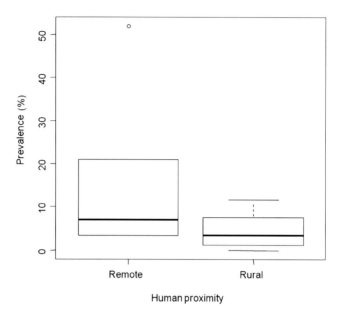

Fig. 10.2 Effects of human proximity on cestode prevalence (%) hosted by howler monkeys. Human proximity categories included in the analysis were remote and rural (see methods for description). Box and whisker plot shows the median, percentiles (25 and 75 %), and the minimum and maximum value

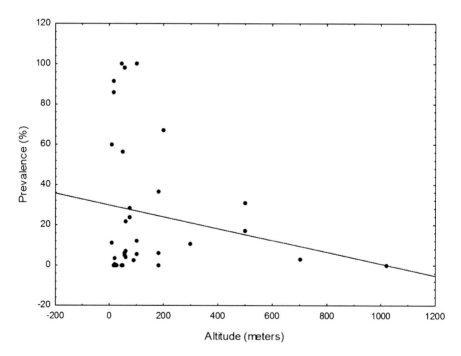

Fig. 10.3 Scatterplot showing a negative relationship between altitude (m) and prevalence (%) of *Trypanoxyuris* spp. reported to infect different howler monkeys

10.5.2 Protozoan Analysis

Amoebae Parasites: We found that the interaction between latitude and precipitation had an effect on the prevalence of amoeba parasites ($\chi^2 = 9.08$, $p < 0.001$). Amoebae prevalence increased close to the equator and at sites where precipitation was high (Fig. 10.4). Other predictors, such as forest type, altitude, or human proximity had no affect on overall amoebae prevalence.

Giardia: Precipitation predicted *Giardia* prevalence ($\chi^2 = 8.6$, $p < 0.05$), producing a negative (exponential) relationship between precipitation and *Giardia* prevalence (Fig. 10.5). Other predictors were not significant.

Plasmodium: *Plasmodium* prevalence was not predicted by any of our independent variables.

10.5.3 Parasite Richness Analysis

We did not find any effect of forest type, latitude, altitude, precipitation, or the degree of human proximity on parasite species richness.

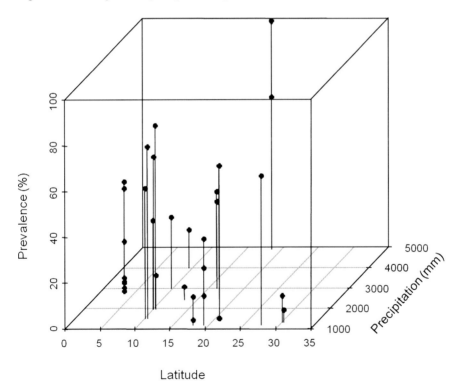

Fig. 10.4 Relationship among amoebae prevalence (%), latitude, and precipitation (mm) in howler monkeys (Genus *Alouatta*)

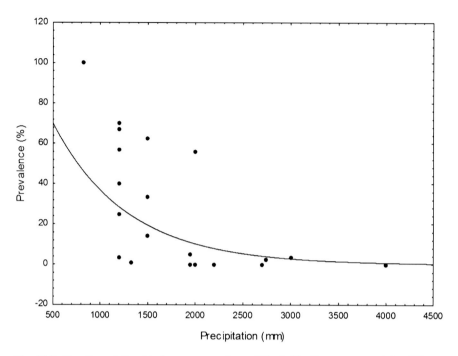

Fig. 10.5 Negative relationship between prevalence (%) of *Giardia* spp. reported for different species of howler monkeys and precipitation (mm)

10.6 Discussion

10.6.1 Effects of Climatic Factors

In this review, we found that different factors including precipitation, latitude, altitude, and human proximity may influence parasite infection in howler monkeys. However, the effect of each of these predictor variables varies depending on the parasite category (Table 10.2). Table 10.2 summarizes general trends found in our analysis. For example, in the case of helminth parasites, precipitation positively predicted nematode, but not trematode and cestode prevalence. Moreover, altitude only affected prevalence of the nematode *Trypanoxyuris*. Humidity and rainfall are critical climatic factors for the survival and spread of parasites, especially soil-transmitted helminths (e.g., *Ascaris* spp.) that are sensitive to desiccation (Patz et al. 2000). It is possible that the encounter rate with nematodes that proliferate in forests characterized by high precipitation is higher for howlers inhabiting these sites than for howlers living in drier environments. To our surprise, trematode prevalence was not predicted by precipitation, despite the majority of these parasites requiring intermediate hosts dependent on water sources (e.g., mollusks such as snails) during

Table 10.2 Effects of precipitation, latitude, and altitude on parasite prevalence and richness in howler monkeys (genus *Alouatta*)

	Nematodes	Trematodes	Cestodes	*Trypanoxyiuris*[a]	Amoeba	*Giardia*	*Plasmodium*	Parasite richness
Precipitation	+	0	0	0	+	–	0	0
Latitude	0	0	0	0	–	0	0	0
Altitude	0	0	0	–	0	0	0	0

+ = positive relationship; – = negative relationship; 0 = no significant effect
[a]Indicates that although not significant there exists a trend

their life cycles. The lack of connection between precipitation and trematode prevalence in howlers may be the result of spatial variability in the intermediate host distribution (Wilson et al. 2002), which may limit the probability of contact between trematode-infective stages and howler monkeys as definitive hosts. Alternately, trematodes that proliferate in howler habitats may be using vertebrates other than howlers as definitive hosts. It is also possible that lower trematode prevalence and richness at some sites simply are the result of using different procedures varying in efficiency to isolate trematode eggs from feces (e.g., flotation and sedimentation techniques), which makes comparing the results of studies difficult (Gillespie 2006).

On the other hand, prevalence of protozoan parasites such as amoebae was affected by the interaction between rainfall and latitude. Howler monkeys living in sites characterized by high amount of annual precipitation and close to the equator have higher prevalence of amoeba parasites compared to howler hosts at higher latitudes living in areas with lower rainfall. Amoebae are waterborne protozoan parasites transmitted via fecal–oral route, and while some species like *Entamoeba coli* are not pathogenic, others, such as *E. histolytica* and *Endolimax nana*, may cause dysentery and diarrheic events, respectively, in human populations (Graczyk et al. 2005). Howler monkeys inhabiting tropical areas characterized by heavy rainfall may be infected by amoebae while drinking water accumulated in tree holes following rainfall events (*A. caraya* (Giudice and Mudry 2000); *A. pigra* (Martinez-Mota, unpubl. data)). However, in our experience, howler monkeys rarely show diarrheic episodes or clinical signs of enteric disease. In fact, in this review, only 11.7 % of our amoebae records were of the diarrheic-causing protozoa *E. nana*, while the majority were *Entamoeba* spp. (20.6 %) and unknown amoebae (29.4 %). Further studies using molecular tools (e.g., PCR) should be used to determine whether amoebae parasites infecting howlers are of pathogenic potential.

In contrast with the pattern found in the amoebae analysis, we found that howlers inhabiting areas with lower annual precipitation have higher *Giardia* prevalence. Giardiasis is a waterborne reemerging infectious disease widely distributed in the tropics. Transmission of *Giardia* occurs by the fecal–oral route, usually when a host ingests cyst-contaminated water and food. Typical symptoms may involve diarrhea, abdominal pain, and weight loss (Thompson 2000; Fayer et al. 2004). Given the zoonotic potential of these protozoa found infecting wildlife, livestock, and humans, giardiasis has become a disease of human health concern (Thompson 2000; Volotao et al. 2008). We found in our meta-analysis that *Giardia* spp. was reported to infect *A. belzebul*, *A. caraya*, *A. guariba*, *A. palliata*, and *A. pigra*. Although prevalence of this parasite has been associated with heavy rainfall and water sources (Hunter 2003; Fayer et al. 2004), our results indicate the opposite trend. Kowalewski et al. (2011) suggested that due to the interplay of additional factors associated with anthropogenic disturbance, such as presence of infected cattle and the common use of small water reservoirs, howler habitat use and stress levels, together with human presence may drive *Giardia* infection patterns in howler monkeys.

In human patients, *Plasmodium* infection correlates negatively with altitude and increases in parallel with precipitation (Drakeley et al. 2005; Odongo-Aginya et al. 2005); however, in our study, none of our predictor variables had a significant effect on *Plasmodium* prevalence. New World primates are potential hosts for *Plasmodium*, and prevalence of this pathogen increases with primate group size (Nunn and Heymann 2005). The fact that we failed to detect any trend in *Plasmodium* infection may be associated to the small number of reported cases in our data base ($n=17$). The small number of reports of *Plasmodium* infection in howlers probably reflects that only few studies have carried out health monitoring initiatives. We suggest that as part of complete health monitoring (or during translocation and/or capture procedures for marking purposes), howlers should be tested for *Plasmodium*. With this new information we will be able to determine if howlers have been exposed to this parasite along their entire distribution.

10.6.2 Parasite Species Richness

Although parasite species richness in primates increases towards the equator (Nunn et al. 2005), we failed to find this relationship in howler monkeys. Our results differ from those of Nunn et al. (2005), who found that latitude negatively predicts protozoan parasite diversity in primates. Nunn et al. used a large database (119 primate taxa) including species with distinct life histories and ecological features (e.g., arboreal and terrestrial, insectivores, folivores, and frugivores), which may explain variation in diversity of parasites hosted by primates. Despite the fact that the genus *Alouatta* is widely distributed from Mexico to South America, with species inhabiting different forest types and ecosystems, all howler species share similar life histories and behavioral ecology, and this might be the reason for the lack of variation in parasite species richness along a latitudinal gradient. Our results suggest that other factors, which in some instances covary with latitude, must be responsible for changes in parasite species diversity within primate hosts.

Parasite species richness has been considered an important disease risk indicator (Nunn and Altizer 2006). Poulin and Morand (2000) suggest that the observed parasite diversity within a host is the result of coevolutionary processes between parasites and host and may reflect the susceptibility of hosts to be colonized by parasites. Furthermore, parasite colonization process and diversity are driven to some extent by host ecological traits (Poulin and Morand 2000). In our analysis, we found that the number of parasite species reported to infect howler monkeys is rather low (average: 5.2 ± 2.3 per population, range: 2–12). This might be the result of howler monkey ecological traits such as arboreality, which may prevent monkeys from contacting infective stages of some parasite species that are more commonly found on the ground. Gillespie et al. (2005) reported that in logged forests the arboreal black-and-white colobus (*Colobus guereza*) showed lower parasite diversity compared to redtail guenons (*Cercopithecus ascanius*), which frequently feed on insects in the lower strata of the canopy (Rode et al. 2006). Such feeding habits may expose primates to parasites that use invertebrates as intermediate hosts. Howler monkeys do not actively feed on insects; moreover, the ingestion of substantial amounts of leaves during certain seasons may contribute to their resistance to parasites, since leaves of species such as *Ficus* spp. may act as natural antiparasitic agents due to their secondary compound content (Huffman 1997; Stoner and González Di Pierro 2006).

We cannot discard the possibility that howler monkeys are intrinsically prone to host few parasite species. Due to their high dispersal and colonizing ability (Ford 2006), and their ecological flexibility, howlers are considered "pioneers," specially adapted to exploit marginal habitats (Rosenberger et al. 2011). The latter probably contributed to their higher resistance to pathogen infections. Studies examining the immune function in howlers may shed light on this possibility.

10.6.3 Human Proximity and Habitat Disturbance

Although we did not find any significant effect of human proximity on parasite prevalence (nematodes, trematodes, amoebae parasites, and the specific genera *Trypanoxyuris*, *Giardia*, and *Plasmodium*), we found a trend in which cestode prevalence was slightly higher in howlers inhabiting remote and less disturbed areas compared to howlers from rural sites that are characterized by a constant presence of people. Surprisingly, we did not find any effect of human proximity on species richness. Howler monkeys inhabiting more conserved and remote areas may interact with a diverse array of fauna, which could increase the probability of parasite transmission, especially generalist parasites that infect different host species. In contrast, it is possible that howlers that inhabit forests located in rural areas do not come into close contact with other vertebrates such as small mammals. In rural areas, hunting is a common activity practiced by local people and decreases the abundance of vertebrates (Peres 2001) serving as potential hosts. In addition, howler habitats located near rural areas are often characterized by anthropogenic impact, such as slash-and-burn agriculture, which involves burning a piece of land before cultivation. Bloemers et al. (1997) found that forest fragments that have been impacted by slash-and-burn agriculture had lower nematode diversity. It is possible that fire associated with this practice, as well as changes in microclimatic conditions, such as decreased humidity and increased desiccation associated with edge effects in forest fragments (Laurance 2000), negatively affect the survival of infective stages of parasites.

Although habitat disturbance has been related to increases in parasitic infection and clearly modifies parasite–host dynamics in primates (Gillespie et al. 2005; Gillespie and Chapman 2006, 2008), mechanisms for such change may be highly influenced by the nature and magnitude of the disturbance experienced. Our analysis failed to detect an effect of the habitat type on parasite prevalence and richness of howler monkeys. Our results paralleled those of Kowalewski and Gillespie (2009) who found that habitat disturbance did not predict parasitism in South American howler monkeys and agree with recent findings which show that primates inhabiting disturbed forests do not have higher parasitic infections than primates living in conserved habitats (Young et al. 2013). This lack of effect of forest type on parasitism may be related to our classification scheme of continuous or fragmented forest (e.g., small, medium, and large fragments). These are artificial categories that do not take into consideration other interacting variables affecting parasitism. Consequently, we suggest that future parasite studies in howler monkeys and other primates avoid the continuous-fragmented forest dichotomy since this categorical variable does not add any explanation power to results. Instead, we encourage primatologists to collect and quantify ecological and environmental data in order to provide better explanations of parasitic infection patterns. We believe that including a general quantitative assessment of habitat disturbance such as an index of logging extraction (Gillespie and Chapman 2006, 2008), size and shape of howler habitats (Valdespino et al. 2010), exposure rates of individuals to a matrix of human-transformed habitat (Zommers et al. 2012), human and domestic animal proximity (Rwego et al. 2008), together with quantitative data of microclimatic variation (e.g., humidity, temperature, rainfall) will improve the quality of explanatory variables and give us better insights into the proximate factors that affect parasitism in howler monkeys.

10.7 Final Remarks

One characteristic of parasites is that they are not evenly distributed in a host population (Wilson et al. 2002). According to the data we analyzed from 31 studies, parasites infecting howler monkeys followed an aggregated (right-skewed) distribution (Fig. 10.6a, b) in which only few individuals in the population harbor parasites. For example, Fig. 10.6 shows that the proportion of infected howlers (i.e., prevalence) with helminth (A) and protozoan (B) parasites is rather low, suggesting that only few sampled individuals per study presented evidence of parasitic infection. This characteristic has significant implications for detecting parasitic infection in a specific population and calls attention to the importance of gathering a large sample size (number of individuals sampled and number of samples collected per individual; Gillespie 2006). This is particularly important given that many studies reporting parasites in howlers are based only on brief surveys and small sample sizes.

More than 60 % of the studies analyzed in this chapter (Table 10.1) used fecal material to recover gastrointestinal parasite eggs, cysts, oocysts, and larvae. Egg counts have been used as a proxy of parasite intensity or load in many studies in nonhuman primates including red colobus monkeys (*Procolobus rufomitratus* (Chapman et al. 2009)), olive baboons (*Papio anubis* (Weyher et al. 2006)), and howler monkeys (mantled howlers, *A. palliata* (Stoner 1996); black howlers, *A. pigra* (Stoner and González Di Pierro 2006; González-Hernández et al. 2011)). Intensity is defined as the number of adult individuals of a specific parasite species within a host (Bush et al. 1997). Because intensity of adult individuals can induce morbidity, this measure provides an important index of disease risk (Bethony et al. 2006); however, helminth parasite egg production does not correlate with the number of adult parasites infecting a single host (Anderson and Schad 1985), which makes egg counts a limited measure of parasite intensity. Despite this being reiterated in the primate literature (Gillespie 2006), primatologists continue using this measure as an index of intensity. This is an incorrect procedure that should be avoided in howler parasite studies. First of all, egg output rate is characterized by day-to-day variability

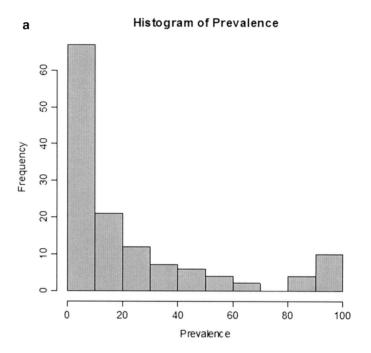

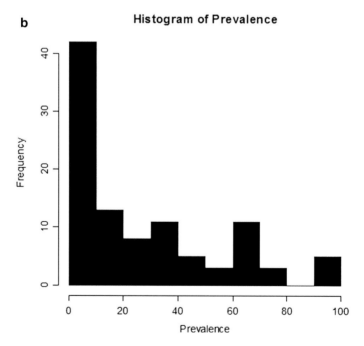

Fig. 10.6 Frequency of helminth (**a**) and protozoan (**b**) prevalence reported in howler monkey studies

within and between individual hosts (Anderson and Schad 1985; Wilson et al. 2002), which may lead to incorrect conclusions, based on false-negative results, such as claiming that a howler monkey population is not infected by certain parasite species. Second, number of eggs shed in feces is not constant over time and does not indicate degree of infection (Ezenwa 2003). Actually, parasite egg output in humans has been found to decrease when worm burden increases due to a density-dependent effect on parasite fecundity (Anderson and Schad 1985). Because of this, helminth egg counts do not provide an accurate measure of parasite intensity.

Our goal is not to minimize the damaging effects that pathogens may have on howler monkeys, but rather to draw attention to the fact that parasitic infection in howler monkeys is driven by complex interactions among environmental and ecological factors, which vary according to parasite type. There is strong evidence that infectious diseases have the potential to increase mortality in howler populations (Holzmann et al. 2010; de Almeida et al. 2012). Unfortunately, there is a disconnection between such sporadic evidence of pathogenic threats to howlers and the ubiquitous data typically collected in the study of howler parasites. Howler parasite studies are generally focused on relating parasitic infections to seasonal periods (e.g., wet vs. dry), forest type (e.g., disturbed vs. undisturbed), or sex (e.g., male vs. female), and although these are important variables to be taken into account, fine-grained estimations of ecological and microclimate change will provide better insights into the proximate factors that promote parasitism in howler monkeys. Finally, we want to point out that there is a gap in primate gastrointestinal parasite taxonomy, which highlights the need to collaborate with molecular parasitologists to correctly identify parasite taxa hosted by howler monkeys. In this way, we will be able to accurately determine parasites with pathogenic potential and then assess disease risk.

Acknowledgments We are very grateful to two anonymous reviewers for their valuable input. We thank Nicoletta Righini for helpful comments during the development of this manuscript.

References

Altizer S, Dobson A, Hosseini P, Hudson P, Pascual M, Rohani P (2006) Seasonality and the dynamics of infectious diseases. Ecol Lett 9:467–484

Alvarado-Villalobos MA (2010) Prevalencia e intensidad de parásitos intestinales de *Alouatta pigra* en fragmentos de selva en Playas de Catazajá. Chiapas. Bachelors thesis, Universidad Autónoma de Ciudad Juárez, Chihuahua, México

Anderson RM, Schad GA (1985) Hookworm burdens and faecal egg counts: an analysis of the biological basis of variation. Trans R Soc Trop Med Hyg 79:812–825

Appleton CC, Henzi SP (1993) Environmental correlates of gastrointestinal parasitism in montane and lowland baboons in Natal, South Africa. Int J Primatol 14:623–635

Bermejo M, Rodríguez-Teijeiro JD, Illera G, Barroso A, Vilà C, Walsh PD (2006) Ebola outbreak killed 5000 gorillas. Science 314:1564

Bethony J, Brooker S, Albonico M, Geiger SM, Loukas A, Diemert D, Hotez PJ (2006) Soil-transmitted helminth infections: ascariasis, trichuriasis, and hookworms. Lancet 367:1521–1532

Bloemers GF, Hodda M, Lambshead PJD, Lawton JH, Wanless FR (1997) The effects of forest disturbance on diversity of tropical soil nematodes. Oecologia 111:575–582

Bonilla-Moheno M (2002) Prevalencia de parásitos gastroentéricos en primates (*Alouatta pigra* y *Ateles geoffroyi yucatanensis*) localizados en hábitat conservado y fragmentado de Quintana Roo, México. Bachelors thesis, Universidad Nacional Autónoma de México, México DF

Braza F (1980) El araguato rojo (*Alouatta seniculus*). Doñana Acta Vertebr 75:1–175

Bush AO, Lafferty KD, Lotz JM, Shostak AW (1997) Parasitology meets ecology on its own terms: Margolis et al. revisited. J Parasitol 83:575–583

Cabral JNH, Rossato RS, de M Gomes MJT, Araujo FAP, Oliveira F, Praetzel K (2005) Gastrointestinais de bugios-ruivos (*Alouatta guariba clamitans* Cabrera 1940) da regiao extremo-sul de Porto Alegre/RS- Brasil, diagnosticados atraves da coproscopia: implicacoes para a conservacao da especie e seus ha. Congresso Brasileiro de Parasitologia, Porto Alegre, RS, Brasil

Carme B, Aznar C, Motard A, Demar M, De Thoisy B (2002) Serologic survey of *Toxoplasma gondii* in noncarnivorous free-ranging neotropical mammals in French Guiana. Vector Borne Zoonotic Dis 2:11–17

Castillejos M (1993) Identificación de parásitos gastrointestinales en monos aulladores (*Alouatta palliata*), en la reserva "El Zapotal" Chiapas, México. Bachelors thesis, Facultad de Medicina Veterinaria y Zootecnia, Universidad Veracruzana, México

Chapman CA, Rothman JM, Hodder SAM (2009) Can parasite infections be a selective force influencing primate group size? A test with red colobus. In: Huffman MA, Chapman CA (eds) Primate parasite ecology. The dynamics and study of host-parasite relationships. Cambridge University, Cambridge

Chinchilla Carmona M, Guerrero Bermúdez O, Gutiérrez-Espeleta GA, Sánchez Porras R, Rodríguez Ortiz B (2005) Parásitos intestinales en monos congo *Alouatta palliata* (Primates: Cebidae) de Costa Rica. Rev Biol Trop 53:437–445

Clarke MR, Crockett CM, Zucker EL, Zaldivar M (2002) Mantled howler population of Hacienda La Pacifica, Costa Rica, between 1991 and 1998: effects of deforestation. Am J Primatol 56:155–163

Coppo JA, Moreira RA, Lombardero OJ (1979) El parasitismo en los primates del CAPRIM. Acta Zool Lilloana 35:9–12

Crawley MJ (2007) The R book. Wiley, Sussex

Cristóbal-Azkarate J, Hervier B, Vegas-Carrillo S, Osorio-Sarabia D, Rodríguez-Luna E, Veà JJ (2010) Parasitic infections of three Mexican howler monkey groups (*Alouatta palliata mexicana*) living in forest fragments in Mexico. Primates 51:231–239

Cristóbal-Azkarate J, Colwell DD, Kenny D, Solórzano B, Shedden A, Cassaigne I, Rodríguez-Luna E (2012) First report of bot fly (*Cuterebra baeri*) infestation in howler monkeys (*Alouatta palliata*) from Mexico. J Wildl Dis 48:822–825

Daszak P, Berger L, Cunningham AA, Hyatt AD, Green DE, Speare R (1999) Emerging infectious diseases and amphibian population declines. Emerg Infect Dis 5:735–748

Daszak P, Cunningham AA, Hyatt AD (2000) Emerging infectious diseases of wildlife- threats to biodiversity and human health. Science 287:443–449

de Almeida MA, Dos Santos E, da Cruz CJ, da Fonseca DF, Noll CA, Silveira VR, Maeda AY, de Souza RP, Kanamura C, Brasil RA (2012) Yellow fever outbreak affecting *Alouatta* populations in southern Brazil (Rio Grande do Sul State), 2008–2009. Am J Primatol 74:68–76

Delgado A (2006) Estudio de patrones de uso de sitios de defecación y su posible relación con infestaciones parasitarias en dos grupos de monos aulladores negros y dorados (*Alouatta caraya*) en el nordeste argentino. Bachelors thesis, Universidad Nacional de Córdoba, Argentina

Di Fiore A, Link A, Campbell CJ (2011) The Atelines. Behavioral and socioecological diversity in a New World monkey radiation. In: Campbell CJ, Fuentes AF, MacKinnon KC, Bearder SK, Stumpf RM (eds) Primates in perspective, 2nd edn. Oxford University, New York

Drakeley CJ, Carneiro I, Reyburn H, Malima R, Lusingu JPA, Cox J, Theander TG, Nkya WMMM, Lemnge MM, Riley EM (2005) Altitude-dependent and -independent variations in *Plasmodium falciparum* prevalence in northeastern Tanzania. J Infect Dis 191:1589–1598

Duarte AMRC, Porto MAL, Curado I, Malafronte RS, Hoffmann EHE, de Oliveira SG, da Silva AMJ, Kloetzel JK, Gomes AC (2006) Widespread occurrence of antibodies against circumsporozoite protein and against blood forms of *Plasmodium vivax, P falciparum* and *P malariae* in Brazilian wild monkeys. J Med Primatol 35:87–96

Eckert KA, Hahn NE, Genz A, Kitchen DM, Stuart MD, Averbeck GA, Stromberg BE, Markowitz H (2006) Coprological surveys of Alouatta pigra at two sites in Belize. Int J Primatol 27:227–238

Ezenwa VO (2003) Thee effect of time of day on the prevalence of coccidian oocysts in antelope faecal samples. Afr J Ecol 41:192–193

Fandeur T, Volney B, Peneau C, De Thoisy B (2000) Monkeys of the rainforest in French Guiana are natural reservoirs for *P. brasilianum/P. malariae* malaria. Parasitology 120:11–21

Fayer R, Dubey JP, Lindsay DS (2004) Zoonotic protozoa: from land to sea. Trends Parasitol 20:531–536

Ford SM (2006) The biogeographic history of Mesoamerican primates. In: Estrada A, Garber PA, Pavelka MSM, Luecke L (eds) New perspectives in the study of Mesoamerican primates: distribution, ecology, behavior, and conservation. Springer, New York

Garcia JL, Svoboda WK, Chryssafidis AL, de Souza Malanski L, Shiozawa MM, de Moraes Aguiar L, Teixeira GM, Ludwig G, da Silva LR, Hilst C, Navarro IT (2005) Sero-epidemiological survey for toxoplasmosis in wild New World monkeys (*Cebus* spp.; *Alouatta caraya*) at the Paraná river basin, Paraná State, Brazil. Vet Parasitol 133:307–311

Gilbert KA (1994) Endoparasitic infection in red howling monkeys (*Alouatta seniculus*) in the Central Amazonian basis: a cost of sociality? PhD dissertation, The State University of New Jersey at New Brunswick Rutgers

Gillespie TR (2006) Noninvasive assessment of gastrointestinal parasite infections in free-ranging primates. Int J Primatol 27:1129–1143

Gillespie TR, Chapman CA (2006) Prediction of parasite infection dynamics in primate metapopulations based on attributes of forest fragmentation. Conserv Biol 20:441–448

Gillespie TR, Chapman CA (2008) Forest fragmentation, the decline of an endangered primate, and changes in host-parasite interactions relative to an unfragmented forest. Am J Primatol 70:222–230

Gillespie TR, Chapman CA, Granier EC (2005) Effects of logging on gastrointestinal parasite infections and infection risk in African primates. J Appl Ecol 42:699–707

Gillespie TR, Nunn CL, Leendertz FH (2008) Integrative approaches to the study of primate infectious disease: implications for biodiversity conservation and global health. Am J Phys Anthropol 51:53–69

Gillespie TR, Lonsdorf EV, Canfield EP, Meyer DJ, Nadler Y, Raphael J, Pusey AE, Pond J, Pauley J, Mlengeya T, Travis DA (2010) Demographic and ecological effects on patterns of parasitism in eastern chimpanzees (*Pan troglodytes schweinfurthii*) in Gombe National Park, Tanzania. Am J Phys Anthropol 143:534–544

Gillespie TR, Barelli C, Heistermann M (2013) Effects of social status and stress on patterns of gastrointestinal parasitism in wild white-handed gibbons (*Hylobates lar*). Am J Phys Anthropol 150:602–608

Giudice AM, Mudry MD (2000) Drinking behavior in the black howler monkey (*Alouatta caraya*). Zoocriaderos 3:11–19

Godoy KCI, Odalia-Rimoli A, Rimoli J (2004) Infeccao por endoparasitos em um grupo de bugios-pretos (*Alouatta caraya*), em um fragmento florestal no Estado de Mato Grosso do Sul. Neotrop Primates 12:63–68

Goldberg TL, Gillespie TR, Rwego IB, Estoff EL, Chapman CA (2008) Forest fragmentation as cause of bacterial transmission among non-human primates, humans, and livestock, Uganda. Emerg Infect Dis 14:1375–1382

González-Hernández M, Dias PAD, Romero-Salas D, Canales-Espinosa D (2011) Does home range use explain the relationship between group size and parasitism? A test with two sympatric species of howler monkeys. Primates 52:211–216

Gonzalez-Moreno O, Hernandez-Aguilar RA, Piel AK, Stewart FA, Gracenea M, Moore J (2013) Prevalence and climatic associated factors of *Cryptosporidium* sp. infection in savanna chimpanzees from Ugalla, Western Tanzania. Parasitol Res 112:393–399

Graczyk TK, Shiff CK, Tamang L, Munsaka F, Beitin AM, Moss WJ (2005) The association of *Blastocystis hominis* and *Endolimax nana* with diarrheal stools in Zambia school-age children. Parasitol Res 98:38–43

Guernier V, Hochberg ME, Guégan JF (2004) Ecology drives the worldwide distribution of human diseases. PLoS Biol 2:740–746

Harvell CD, Mitchell CE, Ward JR, Altizer S, Dobson AP, Ostfeld RS, Samuel MD (2002) Climate warming and disease risks for terrestrial and marine biota. Science 296:2158–2162

Hawkins BA, Porter EE, Dinis-Filho JAF (2003) Productivity and history as predictors of the latitudinal diversity gradient of terrestrial birds. Ecology 84:1608–1623

Hillebrand H (2004) On the generality of the latitudinal diversity gradient. Am Nat 163:192–211

Hilser H (2011) An assessment of primate health in the Sabangau peat-swamp forest, central Kalimantan, Indonesian Borneo. Masters thesis, Oxford Brookes University, Oxford

Holzmann I, Agostini I, Areta JI, Ferreyra H, Beldomenico P, Di Bitetti MS (2010) Impact of yellow fever outbreaks on two howler monkey species (*Alouatta guariba clamitans* and *A. caraya*) in Misiones, Argentina. Am J Primatol 72:475–480

Huffman MA (1997) Current evidence for self-medication in primates: a multidisciplinary perspective. Yearbook Phys Anthropol 40:171–200

Hunter PR (2003) Climate change and waterborne and vector-borne disease. J Appl Microbiol 94:37S–46S

Jagai JS, Castronovo DA, Monchak J, Naumova EN (2009) Seasonality of cryptosporidiosis: a meta-analysis approach. Environ Res 109:465–478

Köndgen S, Kühl H, N'Goran PK, Walsh PD, Schenk S, Ernst N, Biek R, Formenty P, Mätz-Rensing K, Schweiger B, Junglen S, Ellerbrok H, Nitsche A, Briese T, Lipkin WI, Pauli G, Boesch C, Leendertz FH (2008) Pandemic human viruses cause decline of endangered great apes. Curr Biol 18:260–264

Kowalewski MM, Garber PA (2010) Mating promiscuity and reproductive tactics in female black and gold howler monkeys (*Alouatta caraya*) inhabiting an island on the Parana River, Argentina. Am J Primatol 72:734–748

Kowalewski MM, Gillespie TR (2009) Ecological and anthropogenic influences on patterns of parasitism in free-ranging primates: a meta-analysis of the Genus *Alouatta*. In: Garber PA, Estrada A, Bicca-Marques JC, Heymann EW, Strier KB (eds) South American primates. Comparative perspectives in the study of behavior, ecology, and conservation. Springer, New York

Kowalewski MM, Salzer JS, Deutsch JC, Raño M, Kuhlenschmidt MS, Gillespie TR (2011) Black and gold howler monkeys (*Alouatta caraya*) as sentinels of ecosystem health: patterns of zoonotic protozoa infection relative to degree of human-primate contact. Am J Primatol 73:75–83

Kowalzik BK, Pavelka MSM, Kutz SJ, Behie A (2010) Parasites, primates, and ant-plants: clues to the life cycle of *Controrchis* sp. in black howler monkeys (*Alouatta pigra*) in southern Belize. J Wildl Dis 46:1330–1334

Laurance WF (2000) Do edge effects occur over large spatial scales? Trends Ecol Evol 15:134–135

Laurenson K, Sillero-Zubiri C, Thompson H, Shiferaw F, Thirgood S, Malcolm J (1998) Disease as a threat to endangered species: Ethiopian wolves, domestic dogs and canine pathogens. Anim Conserv 1:273–280

Leendertz FH, Pauli G, Maetz-Rensing K, Boardman W, Nunn C, Ellerbrok H, Jensen SA, Junglen S, Boesch C (2006) Pathogens as drivers of population declines: the importance of systematic monitoring in great apes and other threatened mammals. Biol Conserv 131:325–337

Lindblade KA, Walker ED, Onapa AW, Katungu J, Wilson ML (2000) Land use change alters malaria transmission parameters by modifying temperature in a highland area of Uganda. Trop Med Int Health 5:263–274

Lourenco de Oliveira R, Deane LM (1995) Simian malaria at two sites in the Brazilian Amazon. The infection rates of Plasmodium brasilianum in non-human primates. Mem Inst Oswaldo Cruz 90:331–339

Luo Z, Tang S, Li C, Fang H, Hu H, Yang J, Ding J, Jiang Z (2012) Environmental effects on vertebrate species richness: testing the energy, environmental stability and habitat heterogeneity hypotheses. PLoS One 7(e35514):1–7

Marsh LK (2003) The nature of fragmentation. In: Marsh LK (ed) Primates in fragments: ecology and conservation. Kluwer Academic, New York

Martins SS, Ferrari SF, Silva CS (2008) Gastro-intestinal parasites of free-ranging red-handed howlers (*Alouatta belzebul*) in Eastern Amazonia. In: Ferrari SF, Rímoli J (eds) A Primatologia no Brasil—9 (.) Aracaju, Sociedade Brasileira de Primatologia, Biologia Geral e Experimental—UFS, pp 114–124

Mbora DNM, McPeek MA (2009) Host density and human activities mediate increased parasite prevalence and richness in primates threatened by habitat loss and fragmentation. J Anim Ecol 78:210–218

Milozzi C, Bruno G, Cundom E, Mudry MD, Navone GT (2012) Intestinal parasites of *Alouatta caraya* (Primates, Ceboidea): preliminary study in semi-captivity and in the wild in Argentina. Mastozool Neotrop 19:271–278

Milton K (1980) The foraging strategy of howler monkeys: a study in primate economics. Columbia University, New York

Milton K (1996) Effects of bot fly (*Alouattamyia baeri*) parasitism on free-ranging howler monkey (*Alouatta palliata*) population in Panama. J Zool 239:39–63

Milton K, Lozier JD, Lacey EA (2009) Genetic structure of an isolated population of mantled howler monkeys (*Alouatta palliata*) on Barro Colorado Island, Panama. Conserv Genet 10:347–358

Muchiri JM, Ascolillo L, Mugambi M, Mutwiri T, Ward HD, Naumova EN, Egorov AI, Cohen S, Else JG, Griffiths JK (2009) Seasonality of *Cryptosporidium* oocyst detection in surface waters of Meru, Kenya as determined by two isolation methods followed by PCR. J Water Health 7:67–75

Nunn CL, Altizer SM (2005) The global mammal parasite database: an online resource for infectious disease records in wild primates. Evol Anthropol 14:1–2

Nunn CL, Altizer SM (2006) Infectious diseases in primates: behavior, ecology, and evolution. Oxford University, Oxford

Nunn CL, Heymann EW (2005) Malaria infection and host behavior: a comparative study of Neotropical primates. Behav Ecol Sociobiol 59:30–37

Nunn CL, Altizer SM, Sechrest W, Cunningham AA (2005) Latitudinal gradients of parasite species richness in primates. Divers Distrib 11:249–256

Odongo-Aginya E, Ssegwanyi G, Kategere P, Vuzi PC (2005) Relationships between malaria infection intensity and rainfall pattern in Entebbe peninsula, Uganda. Afr Health Sci 5:238–245

Palacios G, Lowenstine LJ, Cranfield MR, Gilardi KVK, Spelman L, Lukasik-Braum M, Kinaki JF, Mudakikwa A, Nyirakaragire E, Busetti AV, Savji N, Hutchison S, Egholm M, Lipkin WI (2011) Human metapneumovirus infection in wild mountain gorillas, Rwanda. Emerg Infect Dis 17:711–713

Pascual M, Cazelles B, Bouma MJ, Chaves LF, Koelle K (2008) Shifting patterns: malaria dynamics and rainfall variability in African highland. Proc R Soc B 275:123–132

Patz JA, Graczyk TK, Geller N, Vittor AY (2000) Effects of environmental change on emerging parasitic diseases. Int J Parasitol 30:1395–1405

Peres CA (2001) Synergistic effects of subsistence hunting and habitat fragmentation on Amazonian forest vertebrates. Conserv Biol 15:1490–1504

Phillips KA, Haas ME, Grafton BW, Yrivarren M (2004) Survey of the gastrointestinal parasites of the primate community at Tambopata National Reserve, Peru. J Zool 264:149–151

Pianka ER (1966) Latitudinal gradients in species diversity: a review of concepts. Am Nat 100:33–46

Pope BL (1966) Some parasites of the howler monkey of northern Argentina. J Parasitol 52:166–168

Poulin R (2006) Global warming and temperature-mediate increases in cercarial emergence in trematode parasites. Parasitology 132:143–151

Poulin R, Morand S (2000) The diversity of parasites. Q Rev Biol 75:277–293

Rifakis PM, Benítez JA, Pineda JDLP, Rodríguez-Morales AJ (2006) Epizootics of yellow fever in Venezuela (2004-2005). Ann N Y Acad Sci 1081:57–60

Rode KD, Chapman CA, McDowell LR, Stickler C (2006) Nutritional correlates of population density across habitats and logging intensities in redtail monkeys (*Cercopithecus ascanius*). Biotropica 38:625–634

Rohde K (1992) Latitudinal gradients in species diversity: the search for the primary cause. Oikos 65:514–527

Rosenberger AL, Halenar L, Cooke SB (2011) The making of Platyrrhine semifolivores: models for the evolution of folivory in primates. Anat Rec 294:2112–2130

Rovirosa-Hernández MJ, Cortés-Ortiz L, García-Orduña F, Guzmán-Gómez D, López-Monteon A, Caba M, Ramos-Ligonio A (2013) Seroprevalence of *Trypanosoma cruzi* and *Leishmania mexicana* in free-ranging howler monkeys in Southeastern Mexico. Am J Primatol 75:161–169

Rudran R, Fernandez-Duque E (2003) Demographic changes over thirty years in a red howler population in Venezuela. Int J Primatol 24:925–947

Rwego IB, Isabirye-Basuta G, Gillespie TR (2008) Gastrointestinal bacterial transmission among humans, mountain gorillas, and domestic livestock in Bwindi Impenetrable National Park, Uganda. Conserv Biol 22:1600–1607

Santa Cruz ACM, Borda JT, Patiño EM, Gomez L, Zunino GE (2000) Habitat fragmentation and parasitism in howler monkeys (*Alouatta caraya*). Neotrop Primates 8:146–148

Santos MVS, Ueta MT, Setz EZF (2005) Levantamento de helmintos intestinais em bugio- ruivo, *Alouatta guariba* (Primates, atelidae), na mata de ribeir̃ao cachoeira no Distrito de Souzas/ Campinas, SP. Congresso Brasileiro de Parasitologia, Porto Alegre, RS, Brasil

Silver SC, Ostro LET, Yeager CP, Horwich R (1998) Feeding ecology of the black howler monkey (*Alouatta pigra*) in northern Belize. Am J Primatol 45:263–279

Smith KF, Acevedo-Whitehouse K, Pedersen AB (2009) The role of infectious diseases in biological conservation. Anim Conserv 12:1–12

Stoner KE (1996) Prevalence and intensity of intestinal parasites in mantled howling monkeys (*Alouatta palliata*) in Northeastern Costa Rica: implications for conservation biology. Conserv Biol 10:539–546

Stoner KE, González Di Pierro AM (2006) Intestinal parasitic infections in *Alouatta pigra* in tropical rain forest in Lacandona, Chiapas, Mexico: implications for behavioral ecology and conservation. In: Estrada A, Garber PA, Pavelka MSM, Luecke L (eds) New perspectives in the study of Mesoamerican primates: distribution, ecology, behavior, and conservation. Springer, New York

Stuart MD, Greenspan LL, Glander KE, Clarke MR (1990) A coprological survey of parasites of wild mantled howling monkeys, *Alouatta palliata palliata*. J Wildl Dis 26:547–549

Thompson RCA (2000) Giardiasis as a re-emergence infectious disease and its zoonotic potential. Int J Primatol 30:1259–1267

Travi BL, Colillas OJ, Segura EL (1986) Natural trypanosome infection in Neotropical monkeys with special reference to *Saimiri sciureus*. In: Taub DM, King FA (eds) Current Perspectives in primate biology. Van Nostrand Rehinold, New York

Trejo-Macías G, Estrada A, Mosqueda Cabrera MA (2007) Survey of helminth parasites in populations of *Alouatta palliata mexicana* and *A. pigra* in continuous and in fragmented habitat in southern Mexico. Int J Primatol 28:931–945

Valdespino C, Rico-Hernández G, Mandujano S (2010) Gastrointestinal parasites of howler monkeys (*Alouatta palliata*) inhabiting the fragmented landscape of the Santa Marta mountain range, Veracruz, Mexico. Am J Primatol 72:539–548

Van Belle S, Estrada A, Ziegler TE, Strier KB (2009) Sexual behavior across ovarian cycles in wild black howler monkeys (*Alouatta pigra*): male mate guarding and female choice. Am J Primatol 71:153–164

Venturini L, Santa Cruz AM, González JA, Comolli JA, Toccalino PA, Zunino GE (2003) Presencia de *Giardia duodenalis* (Sarcomastigophora, Hexamitidae) en mono aullador (*Alouatta caraya*)

de vida silvestre. Comunicaciones Científicas y Tecnológicas, Universidad Nacional del Nordeste

Vitazkova SK (2009) Overview of parasites infecting howler monkeys, *Alouatta* sp., and potential consequences of human-howler interactions. In: Huffman MA, Chapman CA (eds) Primate parasite ecology. The dynamics and study of host-parasite relationships. Cambridge University, Cambridge

Vitazkova SK, Wade SE (2006) Parasites of free-ranging black howler monkeys (*Alouatta pigra*) from Belize and Mexico. Am J Primatol 68:1089–1097

Vitazkova SK, Wade SE (2007) Effects of ecology on the gastrointestinal parasites of *Alouatta pigra*. Int J Primatol 28:1327–1343

Vittor AY, Gilman RH, Tielsch J, Glass G, Shields T, Sanchez Lozano W, Pinedo-Cancino V, Patz JA (2006) The effect of deforestation on the human-biting rate of *Anopheles darlingi*, the primary vector of falciparum malaria in the Peruvian Amazon. Am J Trop Med Hyg 74:3–11

Volney B, Pouliquen JF, De Thoisy B, Fandeur T (2002) A sero-epidemiological study of malaria in human and monkey populations in French Guiana. Acta Trop 82:11–23

Volotao AC, Junior JC, Grassini C, Peralta JM, Fernandes O (2008) Genotyping of *Giardia duodenalis* from southern brown howler monkeys (*Alouatta clamitans*) from Brazil. Vet Parasitol 158:133–137

Weyher AH, Ross C, Semple S (2006) Gastrointestinal parasites in crop raiding and wild foraging *Papio anubis* in Nigeria. Int J Primatol 27:1519–1534

Wilcox BA, Ellis B (2006) Forest and emerging infectious diseases of humans. Unasylva 57:11–18

Wilson K, Grenfell BT (1997) Generalized linear modelling for parasitologists. Parasitol Today 13:33–38

Wilson K, Bjørnstad ON, Dobson AP, Merler S, Poglayen G, Randolph SE, Read AF, Skorping A (2002) Heterogeneities in macroparasite infections: patterns and processes. In: Hudson PJ, Rizzoli A, Grenfell BT, Heesterbeek H, Dobson AP (eds) The ecology of wildlife disease. Oxford University, Oxford

Wright PC, Arrigo-Nelson SJ, Hogg KL, Bannon B, Morelli TL, Wyatt J, Harivelo AL, Ratelolahy F (2009) Habitat disturbance and seasonal fluctuations of lemur parasites in the rain forest of Ranomafana National Park, Madagascar. In: Huffman MA, Chapman CA (eds) Primate parasite ecology. The dynamics and study of host-parasite relationships. Cambridge University, Cambridge

Young H, Griffin RH, Wood CL, Nunn CL (2013) Does habitat disturbance increase infectious disease risk for primates? Ecol Lett 16:656–663

Zamora-Vilchis I, Williams SE, Johnson CN (2012) Environmental temperature affects prevalence of blood parasites of birds on an elevation gradient: implications for disease in a warming climate. PLoS One 7(e39208):1–8

Zhou G, Minakawa N, Githeko AK, Yan G (2004) Association between climate variability and malaria epidemics in the East African highlands. Proc Natl Acad Sci U S A 101:2375–2380

Zommers Z, Macdonald DW, Johnson PJ, Gillespie TR (2012) Impact of human activity on chimpanzee ground use and parasitism (*Pan troglodytes*). Conserv Lett 6:264–273

Part IV
Ontogeny and Sensory Ecology

Chapter 11
An Ontogenetic Framework for *Alouatta*: Infant Development and Evaluating Models of Life History

Melissa Raguet-Schofield and Romina Pavé

Abstract This review investigates the ontogeny of the genus *Alouatta*, with the goal of determining whether howler monkey development follows a "fast-slow" continuum or whether individual life history features are dissociable from one another. Data indicate that while many aspects of howler life history are relatively accelerated compared to other atelines, a consideration of only the end parameters (e.g., age at weaning) obscures important variation within each trait. Moreover, sexual dimorphism in the pace and timing of *Alouatta* developmental events (e.g., somatic and craniodental maturation) provides support for a framework of life history dissociability. Based on these results, we propose a life history model for *Alouatta* ontogeny that recognizes that within the context of an overall rapid development, dissociabilities occur both among and within individual life history parameters.

Resumen Este trabajo revisa la ontogenia del género *Alouatta*, con el objetivo de determinar si el desarrollo de los monos aulladores procede de una manera continua o si las características individuales de historia de vida son disociables unas de otras. Los datos presentados indican que aunque muchos aspectos de la historia de vida de los monos aulladores son relativamente acelerados comparados con otros atelinos, cuando sólo se consideran los parámetros del final del proceso (por ejemplo, edad al destete) variación importante que existe dentro de cada variable se oscurece. Además, el dimorfismo sexual en el ritmo y tiempo de las distintas etapas de desarrollo de *Alouatta* (por ejemplo, la maduración somática y craniodental) apoya el marco teórico de la disociabilidad de la historia de vida. Con estos resultados proponemos un modo de historia de vida para la ontogenia de *Alouatta* el cual reconoce que dentro del contexto general de un desarrollo rápido, existe disociabilidad entre y dentro los parámetros individuales de historia de vida.

M. Raguet-Schofield (✉)
Department of Anthropology, University of Illinois, Urbana, IL, USA
e-mail: raguet@illinois.edu

R. Pavé
Instituto Nacional de Limnología (INALI), Consejo Nacional de Investigaciones Científicas y Técnicas CONICET, Santa Fe, Argentina
e-mail: rominaepave@yahoo.com.ar

Keywords *Alouatta* • Infants • Ontogeny • Life history • Dissociability

11.1 Introduction

In a historical context, the formulation of r-/K-selection theory (MacArthur and Wilson 1967) may have provided biologists with an initial impetus to conduct research on life history evolution (Reznick et al. 2002). MacArthur and Wilson (1967) and later Pianka (1970) described r-selected species as those that live in unpredictable or variable environments and exhibit small body size, rapid development, high reproductive output, early reproduction, and shorter life spans. K-selected species, on the other hand, are expected to inhabit predictable or constant environments and have larger body size, slower development (and, therefore, greater competitive ability during an extended maturation period), delayed reproduction, and longer life spans (Pianka 1970). The r-/K-selection paradigm predominated biological investigations during the next several decades, and accordingly, emergent life history theory centered around the concept of a fast-slow continuum (Reznick et al. 2002; Leigh and Blomquist 2007). Drawing from the predictions of r/K selection, this model of life history variation viewed developmental and reproductive traits as highly integrated features proceeding along a uniformly fast or uniformly slow trajectory (Ross 1988, 1992, 1998; Sacher and Staffeldt 1974; Hofman 1983; Stearns 1983; Harvey and Clutton-Brock 1985; Harvey et al. 1987; Read and Harvey 1989; Promislow and Harvey 1990; Charnov and Berrigan 1993). Unlike the theoretically derived principles of r/K selection, the fast-slow life history model received empirical support from influential studies such as Harvey and Clutton-Brock (1985) and Harvey et al. (1987) (Promislow and Harvey 1990). This research showed that most primate life history variables (e.g., neonatal body mass, weaning age, and interbirth interval) correlated tightly with one another and were highly associated with both adult brain and body mass.

r/K selection remained prominent in the life history literature until the 1990s, when Stearns (1977, 1983, 1992) largely discredited the concept by demonstrating that it oversimplified natural selection and disregarded ontogeny and organismal design constraints (Reznick et al. 2002; Hawkes 2006). Stearns (1992) instead proposed a demographic theory of life history evolution, emphasizing population age structure and taking into consideration which age groups are most influenced by selection (Reznick et al. 2002). While subsequent research often still aligned with the fast-slow continuum [for example, Charnov (1993) proposed that relationships among life history traits were invariant ratios that scaled to each other at constant values], Leigh and Blomquist (2007) regard Stearns (1992) as helping initiate a shift in perspective that paved the way for future life history studies.

Following the move away from r/K selection, primate researchers increasingly began to recognize the importance of contributions from evolutionary biologists such as Williams (1957, 1966), who discussed the concept of life history trade-offs and differential selection pressures throughout ontogeny (Leigh and Blomquist 2007).

Specifically, Williams (1966) found that nestling birds grew rapidly during developmental stages with high risk of mortality (i.e., while learning to fly), but their growth rate slowed dramatically during low-risk stages (i.e., after becoming capable of flight). Influenced by this ontogenetic framework, Leigh and Blomquist (2007) applied more strict phylogenetic controls to Harvey et al.'s (1987) data and demonstrated that many of the original life history correlations (including adult body mass and age at first reproduction, adult body mass and weaning age, and weaning age and neonatal brain size) were greatly diminished. Additional primatological research provides evidence that within taxa, brain growth, somatic growth, and dental development vary throughout ontogeny and need not be correlated (Garber and Leigh 1997; Godfrey et al. 2003; Pereira and Leigh 2003; Leigh and Bernstein 2006; Blomquist et al. 2009). Moreover, developmental and evolutionary biology support Williams' (1966) findings by demonstrating dissociations among somatic growth, tissue and organ system differentiation, and metabolic processes (Needham 1933; Gould 1977; Raff 1996; Wagner 1996; Bolker 2000; Raff and Raff 2000).

As a result of the problems with the fast-slow continuum, Pereira and Leigh (2003), Leigh and Blomquist (2007), and Blomquist et al. (2009) proposed a dissociability model of life history evolution. This alternative framework regards developmental and reproductive traits as separate, dissociable features that may vary independently from one another throughout an organism's lifetime and may even vary in their relationships to each other across the developmental period (Leigh and Blomquist 2007). Blomquist et al. (2009) suggest that the dissociability of life history traits may be of particular importance for primates, considering that the Primate order has a longer developmental period than most other mammals and, therefore, presumably a more extended time frame to modify a suite of developmental features into an adaptive pattern. Pereira and Leigh (2003), Leigh and Blomquist (2007), and Blomquist et al. (2009) define this type of dissociation as a *life history mode*—a developmental pattern that involves a distinct rate of growth and completion schedules for organs, organ systems, and developmental modules. The observation of life history modes within the Primate order prompted Blomquist et al. (2009: 117) to question the utility of the fast-slow continuum altogether and assert that this perspective "is a heuristic that, unfortunately, inhibits the investigation and understanding of important variation in primate life histories and demography."

Because *Alouatta* females have a smaller body mass, an earlier age at sexual maturation, a shorter gestation length, and a shorter interbirth interval (IBI) than the other atelines (Di Fiore and Campbell 2007; references in Table 11.1), traditional life history studies have placed this genus along the "fast" continuum of life history variation (Harvey et al. 1987; Fedigan and Rose 1995; Ross 1991). Recent research, however, suggests dissociability by providing evidence that howler life history is not uniformly accelerated [e.g., howlers experience delays in attaining locomotor competence relative to Cebines (Bezanson 2005, 2009)] or tightly integrated [weaning lags behind the adoption of an adult-like diet (Raguet-Schofield 2010)]. Moreover, life history differences exist across *Alouatta* species. For example, *A. caraya* and *A. seniculus* appear to have a shorter IBI and an earlier weaning

Table 11.1 Life history parameters in the atelines. Interbirth interval includes only cases in which infants survive the first year of life

Taxon	Body mass (kg)	Age at first birth (years)	Interbirth interval (months)	First solid food intake (months)	Weaning age (months)	Gestation (days)
Alouatta palliata	Range: 4.02–6.6	3.5	22.5	1–6	12–18	186±6
Alouatta caraya	4.33	4–5	14–16	1	10±1.51	187±7
Alouatta seniculus	Range: 4.5–6.02	Median: 5.2 range: 4–7	17	1	10.5–14	184–194
Alouatta belzebul	5.52	–	–	–	–	–
Alouatta guariba	4.35	–	21.2	1–4	–	–
Alouatta pigra	6.4, range: 6.29–6.56	–	-	–	–	184
Ateles	Range: 7.29–9.33	7	31.9–50	–	24–36	226–232
Brachyteles	8.07	8.9–9.25	36.4	–	18–24	216±1.5
Lagothrix	Range: 7.02–7.16	9	36.7	–	N/A	210–225

A. palliata: this study; Domingo-Balcells and Veà Baró (2009); Smith and Jungers (1997); Peres (1994); Glander et al. (1991); Froehlich et al. (1981); Glander (1980); *A. caraya*: this study; Kowalewski and Garber (2010); Rumiz (1990); Pave et al. (In prep.); *A. seniculus*: Smith and Jungers (1997); Crockett and Pope (1993); Crockett and Rudran (1987a); Crockett and Sekulic (1982); *A. belzebul*: Peres (1994); *A. fusca*: Strier et al. (2001); Smith and Jungers (1997); *A. pigra*: Kelaita et al. (2011); van Belle et al. (2009); Ford and Davis (1992); *Ateles*: Di Fiore and Campbell (2007); Smith and Jungers (1997); Glander et al. (1991); Milton (1981); Eisenberg (1973); *Brachyteles*: Martins and Strier (2004); Strier et al. (2006); Strier (1991); *Lagothrix*: Di Fiore and Campbell (2007); Nishimura (2003); Smith and Jungers (1997); Peres (1993)

age (Crockett and Sekulic 1982; Rumiz 1990) than *A. palliata* (Glander 1980; Froehlich et al. 1981), but their gestation lengths and adult female body sizes are within *A. palliata*'s range (Table 11.1). Additionally, *A. caraya* and *A. seniculus* may have a slightly later age at first birth—a life history variable that (according to the fast-slow continuum) would predict an increased IBI, a delayed weaning age, and a longer gestation length than *A. palliata*. This evidence suggests that life history variations within the genus do not follow a clear pattern aligned with the fast-slow continuum and indicates that further research on howler monkey developmental and reproductive tactics is necessary.

To this end, the current review investigates the ontogeny of the genus *Alouatta* and determines whether howler monkey life history ascribes to the fast-slow continuum or the dissociability model. We also investigate whether *Alouatta* species deviate from one another in particular life history variables (e.g., craniodental development, weaning, and locomotor independence). We accomplish these goals by analyzing both previously collected and new data on *Alouatta* growth rate, craniodental development, feeding and foraging behavior, weaning, social maturation, and locomotor/positional proficiency.

11.2 Study Cases

11.2.1 Mantled Howler Monkeys *(*Alouatta palliata*)*

11.2.1.1 Methods

M. Raguet-Schofield conducted field observations on *A. palliata* near the village of Mérida on La Isla de Ometepe, Nicaragua (11.44°N, 85.55°W), from August 2006 through August 2007. The forest at this location is dry and semideciduous, receiving approximately 1,500 mm of rain during the year of study. Rain is concentrated from May through November; during the 4 driest months of the year (December through March), a total of only 70 mm of rain fell.

Two groups of *A. palliata*, inhabiting a 19-hectare anthropogenically disturbed forest patch, were observed. Focal animals included 8 adult males, 20 adult females, and 10 immatures (sex unknown) from approximately 6 to 20 months of age. Following Clarke (1982, 1990), immatures were considered infants from birth through 12 months of age, after which they were considered juveniles until month 20. This classification is based on Clarke's (1990) observation that immatures nursed throughout the infant period but became fully weaned during juvenility. In the present study, all immatures were 6–8 months old at the beginning of research; these individuals were followed throughout the project and were 18–20 months old at the end of the study. For the majority of the analyses, only data from 6 to 12 months (i.e., the infant stage) are presented for immatures.

A total of 1,285 h of data were collected throughout the study. Focal animals were randomly selected each day, and each age/sex class was observed for 2 full days per month in both groups. Two-minute instantaneous focal animal samples were collected during full day follows (dawn to dusk) (Altmann 1974). At each 2-min interval, the focal animal's activity was scored as rest, nurse, forage, feed, travel, play, social, other, or unknown. Bouts were scored as "nurse" when immatures were positioned ventrally on their mothers and appeared to be suckling. It is important to note that suckling time does not necessarily reflect nutrition transfer (Cameron et al. 1999; Cameron 1998; Tanaka 1992), and in this study, it should be regarded primarily as "nipple contact time" rather than a metric of milk consumption. Activities were scored as "forage" when individuals searched for or manipulated possible food items, and animals were considered to be "feeding" while they ingested food items. During feeding, observations on food type (young leaves, mature leaves, unknown leaves, unripe fruits, ripe fruits, flowers, stems, buds, pods) and plant species were noted continuously. Data were analyzed using Kruskal-Wallis tests, with significance levels set at $p<0.05$.

11.2.1.2 Results

Feeding and Foraging Proficiency. Infants (aged 6–12 months) devoted significantly more of their daily activity budget to foraging (3.09 %) than did adults (females: 1.83 %, males: 1.45 %) ($H = 9.27, p < 0.01$). Infants also had a lower return rate for their foraging efforts. Specifically, infants had a significantly higher forage/feed ratio than adults, which indicates that they spent more time foraging relative to the time that they spent feeding ($H = 11.41, p < 0.004$). These foraging discrepancies between adults and infants appear to relate directly to infants' inexperience, strength, or skill, as there were no instances of increased aggression or social exclusion by adults toward infants at feeding sites.

The feeding time that *A. palliata* infants devoted to individual food types was largely similar to that of adults (Table 11.2). Infants and adults consumed the same proportions of mature leaves, ripe and unripe fruits, flowers, buds, stems, and pods. A significant age-class difference exists only in infants' increased time spent consuming young leaves compared to adult males ($H = 7.24, p < 0.04$). Additionally, infants consumed all the same plant species that adults did, and infants did not consume any additional plant species that were not part of the adult diet.

Weaning Process. Nipple contact time differed significantly across months, as individuals aged throughout the study (Table 11.3: $H = 33.45, p < 0.0005$). Nipple contact ceased for most individuals at 15–17 months of age. A peak in nipple contact occurred in December (at which point infants were 9–11 months old), when nipple contact accounted for over 10 % of infants' daily activity budget. The December nipple contact maximum was an increase in daily nursing time compared to the previous months—thus indicating that as immatures aged, they did not steadily decline the time that they spent in nipple contact with their mother.

11.2.2 Black and Gold Howler Monkeys (Alouatta caraya)

11.2.2.1 Methods

R. Pavé conducted field observation of infant and mother behavior of *A. caraya* in San Cayetano, Corrientes Province, Argentina (27° 30'S, 58° 41'W), from September 2008 through November 2010. The study site is subtropical with an average annual temperature of 21.7 °C and an average annual of rainfall of 1,230 mm; rainfall decreases slightly in the winter, from June to August (Zunino et al. 2007). The site is a fragmented forest, and vegetation is characterized by dense, semideciduous upland and riparian forests, open lowland forests with palm trees, and grasslands (Rumiz 1990; Zunino et al. 2007).

Following Clarke (1990) and Rumiz (1990), individuals were considered infants for the first year of life because they generally continued suckling and remained in contact with their mothers throughout this time. Focal animals included 18 infants (9 males, 8 females, and 1 unsexed) from month 0 (1–4 weeks) through month 11

Table 11.2 Proportion of feeding time spent consuming the different dietary categories for adult male, adult female, and infant (6–12 months) *A. palliata* at La Isla de Ometepe, Nicaragua

	Mature leaves	Young leaves	Unknown leaves	Ripe fruit	Unripe fruit	Flowers	Stems	Buds	Pods	Unknown
Infant	35.53	11.85	4.63	14.28	12.53	10.94	4.84	1.59	4.43	2.36
Adult females	29.02	13.38	6.41	18.51	10.57	12.25	4.28	2.80	0.54	2.21
Adult males	38.43	3.95	6.73	13.92	15.66	7.86	4.31	4.31	4.14	1.32

The values are means of all individuals in each age/sex class

Table 11.3 Proportion of daily activity budget devoted to feeding and nursing throughout the study period in *A. palliata* at La Isla de Ometepe, Nicaragua

Months of study	Ages of immatures (months)	Feed	Nurse	Nurse as proportion of feeding time
September	6–8	8.03	5.50	40.65
October	7–9	10.78	4.39	28.95
November	8–10	7.37	5.71	43.65
December	9–11	16.51	10.59	28.56
January	10–12	16.01	4.13	20.51
February	11–13	6.77	1.49	18.07
March	12–14	11.51	0.17	1.47
April	13–15	12.15	0.31	2.50
May	14–16	17.15	0.82	4.58
June	15–17	14.72	0	0
July	16–18	18.14	0	0

The values are means of all individuals in each age/sex class

(weeks 48–52) and 9 adult multiparous females who each gave birth to 1–3 infants during the study. Most infants could not be studied during each month of life. Study subjects belonged to six groups, each of which contained 1–2 adult males and 2–4 adult females. Individuals were recognized based on color patterns, body sizes, and natural or artificial marks like scars or ear tags. A total of 1,244 h (877 h for infants and 367 h for mothers) of data were collected throughout the study. Each infant and mother dyad ($N=18$) was followed for 1 full day (dawn to dusk) per month using focal animal sampling (Altmann 1974).

For infants, activities recorded included: nursing (as in *A. palliata*, when an infant had oral nipple contact with its mother or another lactating adult female or allomother), out-of-sight nursing (OSN: infants appeared to have nipple contact but their faces were not visible), feeding on solid food, movement exploration (infants' movement while they manipulated objects such as leaves and twigs), transport, independent locomotion, resting, exploration, and others. Recorded social activities for infants included grooming, interest (a non-mother sniffs, touches, or embraces the infant), care (an infant attaches to a non-mother during resting or feeding), nursing attempts, rejection by mother (refusing to let infants ride or suckle), and others. For mothers, activity pattern and social activities were recorded. When mothers or infants were feeding, food type [leaves (including shoots), flowers, fruits, stems (including petioles), or others (bark and nectar)] and plant species were recorded. Time spent in each activity was adjusted for the total time that each infant was observed. Kruskal-Wallis and Spearman's rank correlations were conducted with statistical significance set at $p \leq 0.05$.

Table 11.4 Time in which infant *A. caraya* spent in different activities at San Cayetano, Corrientes, Argentina

Infants by age	Feeding on solid food	Nursing	Nursing +OSN	Nursing +OSN as proportion of feeding	Movement exploration	Locomotion	Transport
0 (N=5)	0	2.57	70.26	100	0.47	5.11	12.61
1 (N=6)	0.02	4.20	51.31	99.96	1.19	10.09	9.8
2 (N=6)	0.42	4.06	34.49	98.79	4.01	10.33	10.21
3 (N=6)	4.06	3.71	23	84.99	18.63	11.2	6.46
4 (N=6)	6.19	11.08	23.70	79.29	17.13	13.5	7.20
5 (N=8)	8.36	5.31	18.29	68.63	10.97	12.69	4.02
6 (N=8)	10.56	4.19	10.55	49.97	12.47	15.98	5.56
7 (N=8)	13.06	7.67	14.40	52.45	2.88	19.03	1.60
8 (N=9)	15.63	3.90	16.66	51.59	2.36	20.26	0.56
9 (N=5)	22.11	3.45	7.74	25.94	2.79	18.74	0.16
10 (N=5)	24.18	1.53	15.9	39.67	0.67	15.12	0.01
11 (N=7)	23.17	0.62	3.31	12.5	0	14.70	0.04

Time is expressed as percentage of the total time of activity. The values are means of all infants present per age. Nursing includes allonursing. Transport includes transport by mother and other individuals

11.2.2.2 Results

Feeding Behavior. Infant *A. caraya* began to eat solid food when they were 5 weeks (i.e., 1 month) old. At this age, two female infants (N=6; 5 females and 1 male) ate while they were in contact with their mothers, but in only one case the mother of the infant was concurrently eating. Infants between 1 and 11 months differed statistically in time spent feeding per month ($H=32.86$, $p=0.0001$); for example, infants at 1 and 2 months fed less (0.02–0.42 % of their activity budget) than infants at 8 through 11 months (15.63–23.17 %) (Table 11.4). Mothers devoted 17.8 %±4.07 (range=14.05–25.74; $n=9$) of their activity budget to feeding time. By the time infants were between 8 and 9 months old, their time spent feeding on solid food did not differ from their mothers. The diet of *A. caraya* infants was similar to that of their mothers (Table 11.5). No significant differences existed across age classes in the proportions of leaves, fruits, flowers, or stems consumed. However, three *A. caraya* infants (between 5 and 11 months old) consumed four leaf species that their mothers did not.

Weaning Process. Time spent in nipple contact differed across ages, considering only nipple contact ($H=26.473$, $p=0.0055$) or when analyzing nipple contact plus out-of-sight nipple contact ($H=41.282$, $p<0.0001$) (Table 11.4, Fig. 11.1). Infants were nursed by their mothers from birth until month 12 and by allomothers from 2 to 11 months old. The average age at cessation of suckling (on both mothers and allomothers) was 10±1.51 months (range=7–12 months; $n=15$ infants). Considering only suckling, at 10 months this activity represented 1.53 %±2.78

Table 11.5 Proportion of feeding time spent consuming the different dietary categories for each of the *A. caraya* infant age classes and for mothers at San Cayetano, Corrientes, Argentina

	Leaves	Fruits	Flowers	Stem	Others
3–5 months	53.58	38.24	3.99	2.79	1.40
6–8 months	48.88	45.18	2.98	2.21	0.75
9–11 months	48.69	47.79	1.22	2.30	0
Mothers	46.23	44.79	2.24	5.84	0.85

The values are means of all individuals in each class

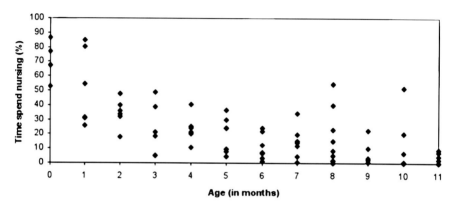

Fig. 11.1 Time spent nursing (including out-of-sight nursing and allonursing) of all the infant *A. caraya* at San Cayetano, Corrientes, Argentina. Each data point represents one infant's suckling time

(range = 0–6.48; *n* = 5 infants) of the activity budget and 0.62 % ± 1.14 (range = 0–3.12; *n* = 7 infants) at 11 months. At 11 months, only 3 (2 females, 1 male) of the 7 infants (2 females, 5 males) maintained nipple contact. Two of these (1 male and 1 female) had nipple contact with their mothers, and the third (a female) had nipple contact with an allomother.

Nipple contact attempts on mothers (SAM) began at birth and continued to month 11; however, time spent in SAM showed no linear relationship with infant age (r_s = 0.433, *n* = 12, *p* = 0.159; Fig. 11.2). Nipple contact attempts peaked in month 4 (0.39 % ± 0.27, range = 0.08–0.72, *n* = 6 infants) and month 7 (0.46 % ± 0.49, range = 0.03–1.49, *n* = 8 infants). At 11 months, 3 (2 males, 1 female) of the 7 infants (2 females, 5 males) attempted to suckle from their mother, but only 2 successfully obtained nipple contact. Three infants (2 females, 1 male) had new siblings at 12–13 months old, and the 2 female infants attempted to access the nipple but were rejected.

Maternal rejection (to suckling and riding) began at month 1 and continued to month 11; however, time spent in rejection showed no linear relationship with infant age (r_s = 0.091, *n* = 12, *p* = 0.778; Fig. 11.2). Rejection peaked in month 4 (0.18 % ± 0.26, range = 0.01–0.71, *n* = 6 infants), coinciding with high values for suckling attempts.

Social Development. Infants began socially interacting with non-mothers during month 0. These social interactions primarily involved non-mothers expressing

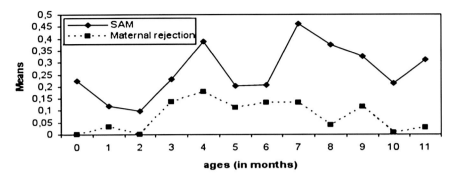

Fig. 11.2 Time spent in suckling attempts on mothers (SAM) of all the infant *A. caraya* at San Cayetano, Corrientes, Argentina. The points represent the mean values of all infants present per age

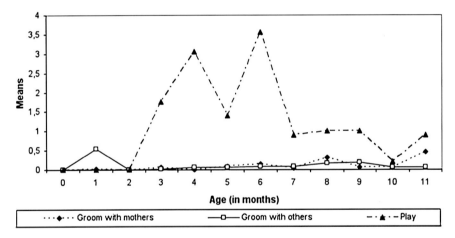

Fig. 11.3 Time spent in social interactions (grooming and play) to all the infant *A. caraya* at San Cayetano, Corrientes, Argentina. The points represent the mean values of all infants present per age

interest and giving care to infants. Grooming (given and received) with mothers and other individuals began at month 1. Grooming with mothers increased with age, peaking at 8 months (0.31 % ± 0.94, range = 0–2.82, $n = 9$ infants) and 11 months (0.46 % ± 1.13, range = 0–3.02, $n = 7$ infants); however, differences among age classes failed to attain statistical significance ($H = 12.101$, $p = 0.3561$; Fig. 11.3). Grooming with non-mothers decreased with age; however, differences among age classes did not reach statistical significance ($H = 9.59$, $p = 0.5672$; Fig. 11.3). Social play began at 3 months and differed across ages ($H = 40.229$, $p < 0.0001$; Fig. 11.3). Infants played with other infants, juveniles, and adults (including their mothers). Play activities included wrestling, biting, pinching, and chasing. The highest values of playing time occurred during months 4 (3.06 % ± 3.67, range = 0.22–9.99, $n = 6$ infants) and 6 (3.56 % ± 3.07, range = 0.73–8.71, $n = 8$ infants).

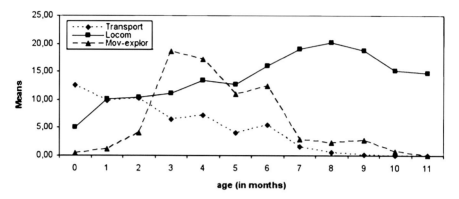

Fig. 11.4 Time spent in movement exploration, independent locomotion, and transport to all the infant *A. caraya* at San Cayetano, Corrientes, Argentina. The points represent the mean values of all infants present per age

Independent Locomotion and Transport. At birth, infants were transported by their mothers 85.1 % of the time and by other individuals 14.9 %. During month 0, infants were transported mainly ventrally (78.2 % ventral, 14.9 % lateral, and 6.9 % dorsal), and independent locomotion occurred either entirely on the mother or on a branch while in contact with her. Transport by mothers continued until month 9. During months 10 and 11, infants were transported only by other individuals (sub-adult and adult females) on rare occasions (Table 11.4). Transport of older infants (8–10 months) occurred mainly during intergroup encounters, when the group moved quickly. A significant negative correlation existed between time spent in transport and infant age ($r_s = -0.972$, $n = 12$, $p < 0.001$; Table 11.4; Fig. 11.4).

Movement exploration differed statistically across ages ($H = 47.768$, $p < 0.0001$; Fig. 11.4). Movement exploration began at month 0 and ceased at month 10 (Table 11.4). The highest values occurred at month 3 (18.63 % ± 11.54, range = 5.01–37.50, $n = 6$ infants) and 4 (17.13 % ± 8.51, range = 1.47–25.01, $n = 6$ infants), when infants began to move away from their mothers and explore their environment. Independent locomotion rose continuously beginning at month 1 (reaching maximum values between months 7 and 9) and was significantly positively correlated with infant age ($r_s = -0.797$, $n = 12$, $p = 0.002$; Table 11.4, Fig. 11.4). Bridging occurred from month 4 to month 9, and both mothers (22.34 %) and non-mothers (77.66 %) bridged for infants during group movement.

11.3 Discussion

11.3.1 Growth Rate

In order to align with the predictions of the fast-slow model, *Alouatta* somatic growth rate (i.e., weight gain) is expected to be accelerated relative to the other atelines as part of a complex of overall rapid development. Data on fetal growth rate (measured in grams per day) across haplorhine primates indicate that both *Alouatta* and *Cebus* (a New World monkey generally regarded as having a "slow" life history) have a higher relative fetal growth rate than *Ateles*, *Lagothrix*, and *Saimiri*; however, these differences fail to attain statistical significance (Ross 1991). Hartwig (1996) likewise found nearly uniform prenatal growth (calculated by neonatal weight divided by gestation length) in *Alouatta*, *Lagothrix*, and *Ateles*. Prenatal growth rate in these taxa therefore appears to involve phylogenetic effects and does not clearly conform to the fast-slow expectation that howlers grow more quickly than other atelines.

Postnatally, sexual dimorphism characterizes *Alouatta* growth patterns (Froehlich et al. 1981; Leigh 1994). For example, in captive *A. caraya* ($n=26$ females, 27 males) and wild *A. palliata* ($n=28$ females, 26 males) males gain weight more rapidly than females during the first year of life (Froehlich et al. 1981; Leigh 1994). Male *A. palliata* weight gain then slows for the following 2.5 years, and at the onset of puberty (around age 4) males exhibit a pronounced growth spurt that lasts until approximately 5 or 6 years (Froehlich et al. 1981). Growth spurts in males are not uncommon throughout the Primate order and characterize many species of cercopithecines, hominoids, and *Cebus apella* (Leigh 1996). *Alouatta* is the most sexually dimorphic genus in the ateline clade (Ford 1994; Plavcan 2001); although there is some suggestion that *A. caraya* displays a higher degree of body mass dimorphism than *A. palliata* (Ford 1994), there does not appear to be a clear distinction between the species' growth patterns. Across the genus, Froehlich et al. (1981) and Leigh (1994) demonstrate that howler males attain their larger adult size than females by both a relative extension and acceleration of growth (particularly during spurts).

Compared to other primates (*Macaca* and *Cercopithecus*), male *A. caraya* exhibit an earlier onset of rapid weight gain, and they attain adult mass at an earlier age (under 5 years in *A. caraya* and over 5 years in the other species). Female *A. caraya* also have accelerated growth trajectories compared to *Macaca* and *Cercopithecus*; however, only negligible growth rate differences exist between female *A. caraya* and *Ateles geoffroyi*. Leigh (1994) shows that both ateline species exhibit similarly rapid growth rates and that *Ateles* females reach larger terminal sizes (and a later age at maturation) as a result of their longer growth duration. Froehlich et al. (1981) and Leigh (1994) suggest that howlers' more rapid attainment of adulthood may be an adaptation to folivory, enabling juveniles to develop digestive morphology necessary to masticate and digest leaves. Similarly, Janson and van Schaik (1993) argue that rapid growth is possible in folivorous juvenile primates because a leaf-based diet reduces risk of starvation due to feeding competition. While the link between rapid growth and folivory is compelling, relating these two variables is problematic. First of all, Leigh's (1994) study shows no significant differences in the growth rate of female *Alouatta* and the more frugivorous *Ateles* (van Roosmalen 1986). Secondly, in some habitats, *Alouatta* spends nearly half of its annual feeding time on fruit (see Garber et al. 2014) [e.g., 49.9 % in *A. palliata* at Los Tuxtlas, Mexico (Estrada 1984), 40.8 % in *A. pigra* at Community Baboon Sanctuary, Belize (Silver et al. 1998)], strongly indicating that this taxon exploits a diet composed of both fruits and leaves. Finally, the opposite trend occurs in strepsirrhines (slow growth in the more folivorous indriids and rapid growth in the more frugivorous lemurids [Godfrey et al. 2003]).

Overall, *Alouatta* somatic growth data provide support for the dissociability model of life history variation (Table 11.6). While postnatal growth rates of both males and females are more rapid than some similarly sized cercopithecoids, female growth rate is indistinguishable in *A. caraya* and *Ateles* (Leigh 1994), thus diminishing the fast-slow contention that howlers grow more quickly than other atelines. An additional striking feature of *Alouatta* growth is the difference between male and female conspecifics. Males exhibit more rapid postnatal growth than females, yet even within the preadult time frame, male growth does not remain uniformly accelerated and instead alternates spurts with periods of slower growth. Leigh (2001: 232) argues that variable growth rates indicate "there are…different ways of 'assembling' ontogenies" and aligns the presence of growth spurts with life history dissociability. Moreover, in this review, we propose that differential growth processes between the sexes (i.e., linear growth in females, spurts alternating with slower growth in males) indicate additional support for a framework of dissociability.

11.3.2 Craniodental Development

Schultz's rule (Schultz 1935), which posits that primates with rapid somatic growth rates will also have accelerated dental eruption, aligns with the wholesale life history accelerations predicted by the fast-slow model (Smith 2000; Godfrey et al. 2001, 2005; Henderson 2007). Early studies on laboratory *A. caraya* ($n=2$) indicated that its dentition (including canines and third molars) was complete at only 18 months of age (Malinow et al. 1968); similarly, Coppo and Resoagli (1978) found that 2-year-old wild *A. caraya* ($n=6$ males, 3 females) had permanent canines and third molars in the eruptive stage. Wild *A. palliata*, however, appears to reach dental maturation at a later age (at 4–5 years, Glander 1980; Froehlich et al. 1981; De Gusta et al. 2003). It is unclear whether these data reflect a difference in the timing of dental maturation between *A. palliata* and *A. caraya* or whether these age discrepancies may be a result of sampling methods. Regardless, in a broad analysis of 40 primate species, Godfrey et al. (2001, 2003) found that both *A. caraya* and *A. palliata* ($n=14$ immatures, 17 adults) exhibit a relative acceleration of dental eruption, as measured by the proportion of postcanine teeth present at 4 and 12 months. *Alouatta*, however, has fewer teeth present at weaning than does *Ateles* (Godfrey et al. 2001, 2003). Godfrey et al. (2003) attribute *Ateles*' more complete dental endowment at weaning to its substantially delayed weaning age and suggest that alternative pathways exist for weanlings to achieve masticatory competence: either accelerate dental development or postpone weaning. In terms of Schultz's (1935) predictions, Godfrey et al. (2001, 2003) find little evidence linking the pace of skeletal development with that of dental development; instead, they demonstrate an accelerated dental development in primates with smaller relative brain masses, a life history association also noted by Smith (1989), and in "folivorous" species. Godfrey et al. (2001, 2003) link this dental precocity in "folivorous" primates with the masticatory challenges posed by a diet containing tough and fibrous leaves.

An additional prediction of Schultz's rule is that primates with shorter and more rapid somatic growth periods will exhibit an earlier eruption of the molars relative to the incisors (Schultz 1935; Henderson 2007). Despite howlers' shorter growth duration, Henderson (2007) finds that *Alouatta* (*A. belzebul*, *A. caraya*, *A. guariba*, *A. palliata*, *A. seniculus*, and *A. pigra*) exhibits an eruption sequence similar to *Lagothrix*, *Brachyteles*, and *Ateles*. Henderson (2007) argues that in New World primates, dental eruption sequences do not correlate with diet, growth rate, or body mass; however, an association between brain mass and dental eruption exists (i.e., taxa with relatively large brains have a delayed emergence of the molars relative to the anterior teeth). Although it is important to note the association of brain mass with dental eruption sequences, Henderson's (2007) results suggest that the emergence of the primate permanent dentition is more strongly associated with phylogeny than with any life history parameter.

Like skeletal development, male and female howlers exhibit sexual dimorphism in their dental and cranial maturation. Whereas female dentition (including canines and third molars) is complete in *A. palliata* by 42–48 months, male dentition continues to develop until as late as 60 months (Glander 1980). *A. seniculus*, *A. palliata*, and *A. caraya* males also attain larger eventual cranial dimensions than females, primarily as a result of an extended period of male cranial growth, and, to a lesser degree, sexual differences in growth rate (Ravosa and Ross 1994; Flores and Casinos 2011). Although Plavcan (2001) finds that sexual dimorphism in *Alouatta* cranial growth is more similar within the genus than it is with members of other genera (e.g., *Macaca*), differences exist across *Alouatta* species. For example, *A. seniculus* exhibits a greater degree of sexual dimorphism at younger developmental stages than *A. palliata*, suggesting that an acceleration of male growth rate plays more of a role in cranial sexual dimorphism (Ravosa and Ross 1994). Additionally, *A. caraya* males exhibit larger mandibular and zygomatic measurements than females in early ontogenetic stages, although these differences become nonsignificant by adulthood (Flores and Casinos 2011). Flores and Casinos (2011: 11) label these features "transitional dimorphisms" and suggest that *A. caraya* juveniles may be more effective competitors with adult males than are juveniles of other howler species, such as *A. palliata*, which do not exhibit cranial dimorphism until adulthood.

Alouatta craniodental development supports the dissociability model of life history by showing that even within the context of rapid dental eruption, howlers have fewer teeth at weaning than spider monkeys and dental eruption occurs along a standard platyrrhine sequence rather than an accelerated schedule matching Schultz's (1935) predictions for primates with shorter growth durations (Table 11.6). Similar to somatic growth, sexual dimorphism exists in the timing of craniodental maturation, with males experiencing delayed completion of the dentition and cranium relative to females. This delay aligns with males' later attainment of adult body size, but it may occur through different processes across howler species. In particular, cranial sexual dimorphism in *A. palliata* arises primarily from extended male growth, while an accelerated male growth rate plays more of a role in *A. seniculus* craniofacial morphology. It is unclear how these differences correspond to distinct socio-ecological pressures; nonetheless, this evidence indicates that variable pathways to sexually dimorphic adult cranial forms exist within the genus.

11.3.3 Feeding and Foraging Proficiency

Multiple researchers have suggested that rapid somatic and craniodental development in "folivorous" species may be an evolutionary tactic to contend with fibrous or toxic leaf resources in the adult diet (Froehlich et al. 1981; Leigh 1994; Godfrey et al. 2001, 2003) or that rapid growth in "folivorous" juveniles results from reduced risk of starvation due to feeding competition with adults (Janson and van Schaik 1993). Few studies, however, have addressed the timing of feeding and foraging proficiency in the genus *Alouatta*.

The current research documents that *A. caraya* begins consuming food during month 1; similarly, initial intake of solid food occurs between months 1 and 4 in *A. guariba* and *A. seniculus* (Mack 1979; Kats and Otta 1991; Miranda et al. 2005; Podgaiski and Assis Jardim 2009). Serio-Silva and Rodriguez-Luna (1994) and Lyall (1996) also observed *A. palliata* to ingest food as early as the first month; however, Clarke (1990) found slightly later ages of food ingestion in *A. palliata*, as well as differences between males and females in this behavior. Based on a 22-month study of 11 *A. palliata* infants (6 female and 5 male), Clarke (1990) reports that females begin to consume plants at 10 weeks whereas males do not do so until 24 weeks. Because the other studies do not discriminate between male and female infants, it is unknown whether the initiation of solid food intake differs between the sexes in other howler species or if these initial differences in *A. palliata* translate to sexual dimorphism in the acquisition of feeding proficiency. Nonetheless, Clarke's (1990) data suggest a decoupling of male *A. palliata*'s rapid somatic growth rate during this time frame and their later initiation of solid food intake compared to females.

Across *Alouatta* species, several studies suggest dietary similarity between immatures and adults (*A. caraya*: Bicca-Marques and Calegaro-Marques 1994; Prates and Bicca-Marques 2008; *A. guariba*: Koch and Bicca-Marques 2007; *A. pigra*: Pavelka and Knopff 2004). However, these reports do not provide exact ages of the immatures in question, and age-class dietary differences may be obscured if very young infants were excluded from the studies. The present analysis remedies this lack of research by providing dietary data on *A. caraya* infants from birth and *A. palliata* infants from 6 months of age to 12 months of age.

In *A. caraya*, feeding time differs significantly across age class, but by 8–9 months, infant feeding time does not differ from that of their (lactating) mothers. By 7 months, infant feeding time does not differ from non-lactating females (Perez Rueda 2010). A limitation of this study is that food categories do not distinguish between mature/young leaves and ripe/unripe fruit, but available data indicate a lack of statistical significance across *A. caraya* age classes in the time spent feeding on major food categories. Additionally, it is noteworthy that *A. caraya* infants consumed four leaf species that were not part of their mothers' diets and that the youngest age class (3–5 months) spent slightly less time feeding on fruits (38 % of feeding budget) than older age classes and mothers (45 % of feeding time). These differences, while nonsignificant, corroborate observations (R. Pavé) that young infants avoided difficult to process, hard-shelled fruits (*Enterolobium contortisiliquum*) and palms (*Acromia* sp. and *Copernicia alba*).

Table 11.6 Feeding and foraging proficiency in relation to reproductive maturation (age at first birth)

Species	Age at feeding and foraging proficiency	Age at female reproductive maturation
Gorilla beringei	3–4 years (Watts 1985)	10–12 years (Watts 1991; Harcourt et al. 1981)
Saimiri sciureus	8–12 months (Stone 2004), approximately 6–8 months (Boinksi and Fragaszy 1989)	3.5–4 years (in captivity, Taub 1980)
Papio cynocephalus	1.4 years (Altmann 1998)	5.5 years (Altmann and Alberts 2003)
Eulemur fulvus	10–12 months (Tarnaud 2004)	2–4 years (Overdorff et al. 1999)
Cebus capucinus	1–2 years (MacKinnon 2005)	5–7 years (Fedigan and Jack 2001)
Alouatta palliata	6–12 months (Raguet-Schofield 2010; this study)	3.5 years (Froehlich et al. 1981; Glander 1980)

In *A. palliata*, infants from 6 to 12 months devoted more of their daily activity budget to foraging and had a lower return rate for foraging effort than adults of either sex, but like *A. caraya* infants, their dietary patterns were statistically identical to adult females. The only gross dietary category difference among *A. palliata* age classes was infants' increased feeding time on young leaves as compared to adult males—a dietary pattern that may relate to infants' relatively increased protein requirements (Raguet-Schofield 2010; however see Garber et al. 2014 for a discussion of the fact that young leaves may not contain significantly higher levels of protein than mature leaves). Unlike *A. caraya*, *A. palliata* immatures were not observed consuming additional leaf species compared to adults.

While the current study documents that *Alouatta* attains adult dietary patterns early in ontogeny, this result may not be remarkable across the Primate order. Gorillas (Harcourt et al. 1981; Watts 1985, 1991), baboons (Altmann 1998; Altmann and Alberts 2003), capuchins (Fedigan and Jack 2001; MacKinnon 2005), squirrel monkeys (Taub 1980; Boinksi and Fragaszy 1989; Stone 2004), and lemurs (Overdorff et al. 1999; Tarnaud 2004) all have been reported to exhibit adult-like feeding behaviors years in advance of reaching reproductive maturation (as evidence by females' age at first birth) (Table 11.6). The early adoption of adult-like dietary proficiency therefore appears to broadly characterize primates rather than being a component of an overall rapid life history trajectory.

The ontogeny of *Alouatta* feeding and foraging behavior supports the dissociability model of life history variation by showing that even within the context of an early adoption of an adult-like diet, uniform acceleration does not characterize the entire process (Table 11.7). For example, mantled howler foraging competence lags behind the attainment of an adult diet, and male *A. palliata* may initiate feeding behaviors later than females (Clarke 1990). Additionally, dietary proficiency appears decoupled from *Alouatta* craniodental maturation—in *A. palliata*, infants consume adult-like diets as early as 6–12 months, but they do not attain adult craniodental form until 4–5 years (see above). These data may call into question the assumption

Table 11.7 *Alouatta* life history variables and the model with which they are associated

Trait	Model	Description
Body mass	Fast	Sexual dimorphism; interspecific variation
Sexual maturation	Fast	Sexual dimorphism; interspecific variation
Gestation length	Fast	No indication of interspecific variation despite possible differences in female body mass
Interbirth interval	Fast	Interspecific variation
Somatic growth	Dissociable, with rapid elements	Sexual dimorphism; periods of spurts and slow growth in males, linear in females; more rapid than some Cercopithecoids, but in females similar to *Ateles*
Craniodental development	Dissociable, with rapid elements	Rapid tooth eruption, but less teeth present at weaning than *Ateles*; correlates with brain size; may be linked to folivory; sexual dimorphism; interspecific variation
Adult feeding/ foraging behaviors	Dissociable, with rapid elements	Corresponds to general primate pattern; foraging inefficiencies remain after attainment of adult diet; decoupled from craniodental maturation
Weaning	Dissociable, with rapid elements	Monthly fluctuations in nipple time; linger in mixed feeding period; interspecific variation
Social maturation	Undetermined	Sexual dimorphism; interspecific variation
Adult locomotion/ positional behavior	Dissociable	Delayed compared to *Cebus*: interspecific variation

that the masticatory challenges of a folivorous diet underlie the pace of primate craniodental development. Furthermore, early dietary proficiency is not unique to howlers or even to other primates regarded as having "fast" life histories. Primates with a broad range of dietary patterns and growth rates can effectively feed like adults well in advance of physical and sexual maturation.

11.3.4 The Weaning Process

Weaning is often an extended process with multiple dissociable phases (Lee 1996; Langer 2003, 2008). Relative to the "milk-only" phase (all nutrition derived from the mother), Langer (2003) predicts that "folivorous" species will have a longer mixed-feeding phase (infants supplement their mothers' milk through independent feeding) in order to compensate for difficulties digesting a leaf-based diet. Limited weaning data in atelines do not allow us to determine the relative length of weaning stages across genera, but existing information indicates that howlers cease nursing at an earlier age compared to other members of the clade (Table 11.1). Early weaning in *Alouatta* likely contributes to a shorter IBI relative to other atelines. Specifically, *Alouatta* weans offspring from 10 to 18 months and has an IBI between

14 and 22 months; *Ateles* and *Brachyteles* wean between 18 and 36 months and have IBIs of 32–50 months (references in Table 11.1). The timing of weaning influences IBI because continued nipple contact (even when milk is not transferred) can reduce a mother's ability to conceive by inhibiting ovulation (Lee 1987; Brown 2001); however, *A. caraya*, *A. palliata*, and *A. seniculus* mothers can all resume ovulation and become pregnant while they are still nursing their previous infant (Crockett and Rudran 1987b; Clarke 1990; Rumiz 1992; Pavé et al. 2010). It is therefore unclear how weaning relates to other life history variables in *Alouatta*, and further investigation is necessary.

The present research documents that *A. palliata* ceases nipple contact at 15–17 months, corroborating Clarke (1990), and on average *A. caraya* is fully weaned at 10 ± 1.51 months. The later weaning age in *A. palliata* corresponds with its slightly longer IBI than both *A. caraya* and *A. seniculus* but is dissociated from its earlier age at first reproduction compared to other *Alouatta* species (see Table 11.1). This study also demonstrates that fluctuations occur in the monthly time that howlers spend nursing/nipple contact. For example, *A. palliata* nursing time decreases from 6 to 9 months of age, but it then increases again from approximately 10–11 months, before dropping off precipitously around 1 year. In *A. caraya* both nursing and nursing + OSN decline for the first 6 months before increasing again in month 7 (and in months 8 and 10 for nursing + OSN). Combined with data on the ontogeny of feeding behavior (see above), weaning data indicate that *A. caraya* and *A. palliata* infants establish adult-like dietary patterns as much as 6–12 months prior to cessation of nipple contact. Raguet-Schofield (2010) found that nipple contact peaks in *A. palliata* coincide with seasonal periods of food scarcity and suggested infant howlers rely on their mothers' milk as a "fall back" food when resources become scarce. This interpretation should be regarded cautiously, however, given that it is unknown how nipple time translates to nutrition transfer in howlers. Other studies have demonstrated that suckling often has a social function (i.e., soothing after a distress situation) rather than nutritive function (Cameron 1998; Baldovino and Di Bitetti 2008). Nonetheless, the current data suggest that in howlers, full weaning is decoupled from the acquisition of feeding competence.

It is also important to consider the implications of the howler weaning process for maternal energy expenditure. Langer (2008) specifies that once infants begin to supplement their mothers' milk through independent feeding, they support some of their own costs of growth and reduce maternal energetic burden—thus enabling mothers to shift their investment from current to future offspring. The faster growth rate of male *Alouatta* (see above), coupled with the possible later initiation of solid food intake in male *A. palliata*, suggests that male offspring may place more of an energetic burden on howler mothers than female offspring, as is the case for primiparous *Macaca mulatta* (Hinde 2007) and red deer (Landete-Castillejos et al. 2005). The latter two species produce richer milk (in terms of fat and protein) for sons as compared to daughters. Red deer mothers accordingly experience increased IBIs following the birth of a male offspring (Clutton-Brock et al. 1983), but no such delays in future reproduction are apparent in howlers: in *A. guariba* and *A. seniculus*,

IBI is the same following the birth of a male or female infant (Crockett and Rudran 1987b; Strier et al. 2001). These results indicate that even if males are more costly to rear during early ontogenetic periods, mothers of male offspring do not experience immediate trade-offs between current and future reproduction.

While the timing of the shift from current to future reproduction remains uncertain across atelines, life history data suggest that howler females may have more flexibility in maternal investment tactics than other members of the clade. Considering howlers' age at first reproduction, IBI, weaning age, and reproductive lifespan, Raguet-Schofield (2010) calculated that *A. palliata* females could produce 6.62–7.48 offspring during their lifetimes. Clarke and Glander (1984) similarly estimated 8 infants per *A. palliata* female and Pavé (personal observation) found 7–8 offspring per female in *A. caraya*. With regard to the other atelines, Raguet-Schofield (2010) calculated that *Ateles* and *Lagothrix* females could produce 5.64 and 3.59 infants, respectively, and that *Brachyteles* may approach the reproductive output of howlers, with a potential maximum of 6.83–7.0 offspring. Because howlers (and perhaps *Brachyteles*) have a higher potential lifetime reproductive output than *Ateles* and *Lagothrix*, *Alouatta* females may be able to modulate investment in current offspring depending on whether the environmental conditions warrant withholding such investment and shifting reserves to a future reproductive event. Clarke and Glander (1984: 123) similarly suggest that female *A. palliata* "should vary [their] reproductive tactics" according to ecological and social conditions that influence infant survivability. Pavé et al. (2012) support this viewpoint by demonstrating a spike in infant mortality during a 2-month period of flooding that drastically reduced food availability (8 of 40 infants [mean age: 4 months] died). These authors suggest that mothers stopped investing in offspring when the infants' chances of survival became low (Pavé et al. 2012). *Ateles* and *Lagothrix* females may not have the same opportunity to shift their investment: because they produce fewer offspring in their lifetimes, their reproductive success may depend more on the survival of the current offspring, even if it diminishes their future reproductive efforts (Raguet-Schofield 2010).

An investigation of the weaning process in howlers provides support for the dissociability model of life history variation (Table 11.7). While howlers cease nursing at a younger age compared to other atelines, differences in weaning age and IBI within the genus do not necessarily correspond with the predictions of the fast-slow continuum (Table 11.1). The current study documents that monthly fluctuations occur in the amount of time that *A. caraya* and *A. palliata* spend nursing during infancy, suggesting that weaning does not proceed along a tightly integrated schedule of rapid decline. Moreover, these data indicate that whether infants nurse for nutritive or social purposes, the end of the nursing period is decoupled from the attainment of adult-like dietary patterns. Finally, this investigation of the phases of weaning suggests that howler mothers may have more flexibility in their level of maternal investment than other atelines. Continued research on ateline ontogeny will shed light on how the trade-off between current and future reproduction impacts the evolution of life history and maternal investment strategies in this clade.

11.3.5 Social Development

Interactions between individuals facilitate socialization within the group and therefore are crucial to primate infant development (Altmann 1980). This study documents that *A. caraya* infants begin socially interacting with non-mothers during the first 4 weeks of life, and by month 1, infants begin to receive grooming. Social play in *A. caraya* begins during month 3. Similarly, *A. palliata* and *A. guariba* infants initiate social play between 2 and 4 months (Clarke 1990; Kats and Otta 1991; Serio-Silva and Rodriguez-Luna 1994; Lyall 1996; Miranda et al. 2005). Affiliative interactions between kin and non-kin comprise the majority of *A. caraya* social behaviors in this study; however, Clarke et al. (1998) find that agonistic or competitive interactions characterize *A. palliata* infant socialization. Clarke (1990) also documents that male and female *A. palliata* infants proceed along different trajectories of social development—corresponding with the types of sex-specific social interactions they will engage in as adults. For example, females are more social than males along all points of development as well as during adulthood.

Baldwin and Baldwin (1978) observe that *A. palliata* social play increases as infants' motor skills develop and then decreases steadily after motor proficiency is attained. Clarke (1990) found that female *A. palliata* increased play time at 10 weeks, with peaks at approximately 4 and 7 months; playing reached very low levels at 12 months. This study documents a similar trajectory in *A. caraya*: social playing peaks just prior to 7 months; afterwards all infant social interactions with non-mothers decrease. The reduction in infant social activity corresponds with an increase in feeding time, an observation that corroborates Baldwin and Baldwin's (1978) argument that the time and energy necessary to digest plant food affects play in howlers.

Lack of comparative data on socialization across atelines prevents us from clearly placing social development within the life history models. The present study indicates that howlers begin to engage in social play during the first months of life, peak around 7–10 months, and decline after 1 year; however, it is unknown whether this trajectory is accelerated relative to other atelines. Interspecific differences in *Alouatta* socialization may exist: whereas antagonistic interactions characterize *A. palliata* socialization (Clarke et al. 1998), affiliation appears more common in *A. caraya*. Finally, Clarke (1990) reports sexual dimorphism in social behavior throughout *A. palliata* infancy, which corresponds to differential male–female behaviors in adults.

11.3.6 Locomotor and Positional Proficiency

The attainment of adult-like locomotor and positional behaviors in howlers is expected to occur rapidly under the fast-slow model, whereas within a framework of dissociability, these features may be decoupled from other aspects of development. Considering that howlers adopt adult-like diets early in their ontogeny, a correspondingly rapid postural maturation may enable them to attain this dietary proficiency.

In *A. caraya*, infants begin independent locomotion at month 0 (this study) and in *A. palliata* at month 1 (Lyall 1996) or 2 (Clarke 1990). By 11 weeks of age (2.75 months), Lyall (1996) reports that *A. palliata* infants spend 34.4 % of their daily time budget engaged in independent locomotion and environmental exploration; *A. caraya* infants spend slightly less time devoted to these combined categories (14.34 % at month 2 and 29.83 % at month 3). By 6 months of age, *A. palliata* are no longer transported by adults (Clarke 1990). In *A. caraya*, however, infants are transported by their mothers until month 9 and by other group members until month 11 but only during dangerous situations. These interspecific differences may indicate that *A. palliata* reaches locomotor independence more rapidly than *A. caraya*. If so, this trait appears to be dissociable from *A. palliata*'s relatively delayed weaning and increased IBI but aligns with this taxon's earlier age at first reproduction.

Compared to *Cebus capucinus*, Bezanson (2005, 2009) shows that *A. palliata* is delayed in attaining adult-like locomotor and positional behavior. This finding is contrary to expectations of the fast-slow life history model: *Cebus* takes longer to reach adult body mass and limb proportions than *Alouatta* and is therefore expected to attain locomotor proficiency later in life (Bezanson 2009). Instead, Bezanson (2009, 2005) finds that by 6 months, infant capuchins' locomotor and positional behaviors are statistically identical to those of adults, while the more quickly growing *A. palliata* continues to exhibit distinct locomotor patterns until at least 24 months of age. Specifically, young howlers leap more frequently and bridge tree crown gaps less frequently than adults. Prates and Bicca-Marques (2008) also found differences in locomotion among infant, juvenile, and adult *A. caraya*. They show that *A. caraya* infants and juveniles (exact ages undefined) leap and climb more frequently than adults, but unlike Bezanson (2009, 2005), the young *A. caraya* bridged significantly more than adults.

Alouatta locomotor and postural ontogenies provide evidence of dissociability among life history parameters. Although direct comparisons with other atelines are not possible, data from the more slowly maturing *Cebus* suggest that howlers are relatively delayed in their attainment of adult locomotor and positional behaviors. Moreover, infant and juvenile howlers retain distinct locomotor and postural patterns from adults even after their diets are the same. These data demonstrate that howler positional development is dissociable from other, more rapid, aspects of howler life history. In addition, differences in the onset of independent transport in *A. palliata* and *A. caraya* suggest that interspecific variation occurs within the genus.

11.4 Conclusion

While the fast-slow continuum may retain a broad explanatory power for life history variation across organisms, adhering too closely to this paradigm obscures important variation that occurs within and across Primate genera and even between males and females of the same species. As a result, we propose that the dissociability model offers a preferable explanation for *Alouatta* life history. More precisely, the howler mode of development involves a standard ateline prenatal growth rate, followed by a period of sexually dimorphic postnatal growth (i.e., weight gain). Female growth is more linear (and similar to that of female *Ateles*), but male growth evidences dissociability with its repeated periods of spurts and stasis. Tooth eruption sequences appear phylogenetically mediated, and while the teeth of *Alouatta* erupt at younger ages compared to other atelines, craniodental maturation is sexually dimorphic and variable across *Alouatta* species. Attainment of adult diet is decoupled from craniodental maturation and follows the general primate pattern of early proficiency, yet howler infants remain inefficient foragers and continue to nurse for as much as a year after their diet matches that of adults. Finally, howlers are delayed in attaining adult-like locomotor and postural proficiency, and the process involves variation across howler species.

The data compiled in this review further support the dissociability model by indicating that life history variation within the genus *Alouatta* does not conform to the predictions of the fast-slow continuum. For example, *A. palliata* has a slightly younger age at first reproduction than *A. caraya* and *A. seniculus*, but it has a longer IBI and later age at weaning. *A. palliata* may also be delayed relative to *A. caraya* in locomotor independence—a characteristic that corresponds with *A. caraya*'s more rapid IBI and weaning but conflicts with *A. palliata*'s earlier age at first reproduction. Within individual *Alouatta* species, sexual dimorphism also exists in several aspects of development (e.g., postnatal growth rate, craniodental maturation, and initiation of solid food intake). In this review, we align the presence of sexually dimorphic growth patterns with the dissociability model of life history evolution. Ultimately, the howler mode of development may enable *Alouatta* females to more rapidly shift their resources from current to future offspring compared to other atelines, considering that howler females bear young earlier and over shorter gestation periods, have shorter IBIs, and wean the offspring at younger ages. While further research on ateline life history modes can elucidate the way this clade negotiates life history trade-offs, the current review advises taking a dissociability perspective of life history evolution in order to yield a more complete picture of primate growth, development, and reproduction.

Acknowledgments Melissa Raguet-Schofield thanks Drs. Steve Leigh, Paul Garber, Martín Kowalewski, and E.A. Quinn, all research assistants, field station staff, Eduardo Gonzalez, and Rob and William Raguet-Schofield. MR's project was funded by the National Science Foundation (DDIG 0622411), the Graduate College at University of Illinois, a Beckman Institute CS/AI Fellowship, Idea Wild, and Sigma Xi (Grant-in-Aid of Research). Romina Pavé thanks Drs. Martin Kowalewski and Gabriel Zunino and all field assistants, especially M. Amparo Perez Rueda, Dana Banega, and Nancy Lopez. RP's project was funded by the American Society of Mammalogists, Idea Wild, Zoo of Barcelona, and the National Council of Research and Technology of Argentina (CONICET).

References

Altmann J (1974) Observational study of behavior: sampling methods. Behaviour 49:227–266
Altmann J (1980) Baboon mothers and infants. University of Chicago Press, Chicago
Altmann SA (1998) Foraging for survival: yearling baboons in Africa. University of Chicago Press, Chicago
Altmann J, Alberts SC (2003) Variability in reproductive success viewed from a life-history perspective in baboons. Am J Hum Biol 15:401–409

Baldovino MC, Di Bitetti MS (2008) Allonursing in tufted capuchin monkeys (*Cebus nigritus*): milk or pacifier? Int J Primatol 79:79–92

Baldwin JD, Baldwin JI (1978) Exploration and play in howler monkeys (*Alouatta palliata*). Primates 19:411–422

Bezanson MF (2005) Leap, bridge, or ride? Ontogenetic influences on positional behavior in *Cebus* and *Alouatta*. In: Estrada A, Garber PA, Pavelka MSM, Leucke L (eds) New perspectives in the study of Mesoamerican primates: distribution, ecology, behavior, and conservation, 1st edn. Springer, New York

Bezanson M (2009) Life history and locomotion in *Cebus capucinus* and *Alouatta palliata*. Am J Phys Anthropol 140:508–517

Bicca-Marques J, Calegaro-Marques C (1994) Activity budget and diet of *Alouatta caraya*: an age-sex analysis. Folia Primatol 63:216–220

Blomquist GE, Kowalewski MM, Leigh SR (2009) Demographic and morphological perspectives on life history evolution and conservation of new world monkeys. In: Garber PA, Estrada A, Bicca-Marques JC, Heymann EW, Strier KB (eds) South American primates: comparative perspectives in the study of behavior, ecology, and conservation. Springer, New York

Boinksi S, Fragaszy DM (1989) The ontogeny of foraging in squirrel monkeys, *Saimiri oerstedi*. Anim Behav 37:415–428

Bolker JA (2000) Modularity in development and why it matters to evo-devo. Am Zool 40:770–776

Brown GR (2001) Sex-biased investment in nonhuman primates: can Trivers & Willard's theory be tested? Anim Behav 61:683–694

Cameron EZ (1998) Is suckling behaviour a useful predictor of milk intake? A review. Anim Behav 56:521–532

Cameron EZ, Stafford KJ, Linklater WL, Veltman CJ (1999) Suckling behaviour does not measure milk intake in horses, *Equus caballus*. Anim Behav 57:673–678

Charnov EL (1993) Life history invariants. Oxford University Press, Oxford

Charnov EL, Berrigan D (1993) Why do female primates have such long lifespans and so few babies? Or life in the slow lane. Evol Anthropol 1:191–194

Clarke M (1982) Socialization, infant mortality, and infant-nonmother interactions in howling monkeys (*Alouatta palliata*) in Costa Rica. PhD dissertation, University of California, Davis

Clarke MR (1990) Behavioral development and socialization of infants in a free-ranging group of howling monkeys (*Alouatta palliata*). Folia Primatol 54:1–15

Clarke MR, Glander KE (1984) Female reproductive success in a group of free-ranging howling monkeys (Alouatta palliata) in Costa Rica. In: Small MF (ed) Female primates: studies by women primatologists. Alan R. Liss, New York, pp 111–118

Clarke MR, Glander KE, Zucker E (1998) Infant-nonmother interactions of free-ranging mantled howlers (*Alouatta palliata*) in Costa Rica. Int J Primatol 19:451–472

Clutton-Brock TH, Guinness FE, Albon SD (1983) The costs of reproduction to red deer hinds. J Anim Ecol 52:367–383

Coppo J, Resoagli E (1978) Etapas de crecimiento en monos caraya. Facena 2:29–39

Crockett CM, Pope TR (1993) Consequences of sex differences in dispersal for juvenile red howler monkeys. In: Pereira ME, Fairbanks LA (eds) Juvenile primates: life history, development and behavior. Oxford University Press, New York

Crockett CM, Rudran R (1987a) Red howler monkey birth data I: seasonal variation. Am J Primatol 13:347–368

Crockett CM, Rudran R (1987b) Red howler monkey birth data II: interannual, habitat, and sex comparisons. Am J Primatol 13:369–384

Crockett CM, Sekulic R (1982) Gestation length in red howler monkeys. Am J Primatol 3:291–294

De Gusta D, Everett MA, Milton K (2003) Natural selection on molar size in a wild population of howler monkeys (*Alouatta palliata*). Proc R Soc Lond B Biol 270:515–517

Di Fiore A, Campbell C (2007) The atelines: variation in ecology, behavior, and social organization. In: Campbell C, Fuentes A, MacKinnon KC, Panger M, Bearder SK (eds) Primates in Perspective. Oxford University Press, New York

Domingo-Balcells C, Veà Baró J (2009) Developmental stages in the howler monkey, subspecies *Alouatta palliata mexicana*: a new classification using age-sex categories. Neotrop Primates 16:1–8

Eisenberg JF (1973) Reproduction in two species of spider monkeys, *Ateles fusciceps* and *Ateles geoffroyi*. J Mammal 54:955–957

Estrada A (1984) Resource use by howler monkeys (*Alouatta palliata*) in the rain-forest of Los Tuxtlas, Veracruz, Mexico. Int J Primatol 5:105–131

Fedigan LM, Jack K (2001) Neotropical primates in a regenerating Costa Rican dry forest: a comparison of howler and capuchin population patterns. Int J Primatol 22:689–713

Fedigan LM, Rose LM (1995) Interbirth interval variation in three sympatric species of neotropical monkey. Am J Primatol 37:9–24

Flores D, Casinos A (2011) Cranial ontogeny and sexual dimorphism in two New World monkeys: *Alouatta caraya* (Atelidae) and *Cebus apella* (Cebidae). J Morphol 272:744–757

Ford SM (1994) Evolution of sexual dimorphism in body weight in platyrrhines. Am J Primatol 34:221–244

Ford SM, Davis LC (1992) Systematics and body size: implications for feeding adaptations in new world monkeys. Am J Phys Anthropol 88:415–468

Froehlich JW, Thorington RW, Otis JS (1981) The demography of howler monkeys (*Alouatta palliata*) on Barro Colorado Island, Panamá. Int J Primatol 2:207–236

Garber P, Leigh SR (1997) Ontogenetic variation in small-bodied New World primates: implications for patterns of reproduction and infant care. Folia Primatol 68:1–22

Garber P, Righini N, Kowalewski M (2014) Evidence of alternative dietary syndromes and nutritional goals in the genus *Alouatta*. In: Kowalewski M, Garber P, Cortés-Ortiz L, Urbani B, Youlatos D (eds) Howler monkeys: behavior, ecology and conservation. Springer, New York

Glander KE (1980) Reproduction and population growth in free-ranging mantled howling monkeys. Am J Phys Anthropol 53:25–36

Glander KE, Fedigan LM, Fedigan L, Chapman C (1991) Field methods for capture and measurement of three monkey species in Costa Rica. Folia Primatol (Basel) 57:70–82

Godfrey LR, Samonds KE, Jungers WL, Sutherland MR (2001) Teeth, brains, and primate life histories. Am J Phys Anthropol 114:192–214

Godfrey LR, Samonds KE, Jungers WL, Sutherland MR (2003) Dental development and primate life histories. In: Kappeler PM, Pereira ME (eds) Primate life histories and socioecology, 1st edn. University of Chicago Press, Chicago

Godfrey LR, Samonds KE, Wright PC, King SJ (2005) Schultz's unruly rule: dental developmental sequences and schedules in small-bodied, folivorous lemurs. Folia Primatol 76:77–99

Gould SJ (1977) Ontogeny and phylogeny. Belknap, Boston

Harcourt A, Stewart K, Fossey D (1981) Gorilla reproduction in the wild. In: Graham CE (ed) Reproductive biology of the great apes. Academic, New York

Hartwig WC (1996) Perinatal life history traits in New World monkeys. Am J Primatol 40:99–130

Harvey PH, Clutton-Brock TH (1985) Life history variation in primates. Evolution 39:559–581

Harvey RH, Martin RD, Clutton-Brock TH (1987) Life histories in comparative perspective. In: Smuts BB, Cheney DL, Seyfarth RM, Wrangham RW, Struhsaker TT (eds) Primate societies, 1st edn. University of Chicago Press, Chicago

Hawkes K (2006) Life history theory and human evolution: a chronicle of theory and ideas. In: Hawkes K, Pine RR (eds) The evolution of human life history. School of American research, Santa Fe

Henderson E (2007) Platyrrhine dental eruption sequences. Am J Phys Anthropol 134:226–239

Hinde K (2007) First-time macaque mothers bias milk composition in favor of sons. Curr Biol 17:R958–R959

Hofman MA (1983) Energy metabolism, brain size and longevity in mammals. Q Rev Biol 58:495–512

Janson CH, van Schaik CP (1993) Ecological risk aversion in juvenile primates: slow and steady wins the race. In: Pereira ME, Fairbanks LA (eds) Juvenile primates. Oxford University Press, Oxford

Kats B, Otta E (1991) Comportamento ludico do bugio (*Alouatta fusca clamitans*, Cabrera, 1940) (Primates: Cebidae: Alouattinae). Biotemas 4:61–82

Kelaita M, Dias PA, Aguilar-Cucurachi Mdel S, Canales-Espinosa D, Cortés-Ortiz L (2011) Impact of intrasexual selection on sexual dimorphism and testes size in the Mexican howler monkeys *Alouatta palliata* and *A. pigra*. Am J Phys Anthropol 146:179–187

Koch F, Bicca-Marques JC (2007) Padrão de atividades e dieta de *Alouatta guariba clamitans* Cabrera, 1940: uma análise sexo-etária. In: Bicca-Marques JC (ed) A Primatologia no Brasil–10. Sociedade Brasileira de Primatologia, Porto Alegre

Kowalewski MM, Garber PA (2010) Mating promiscuity and reproductive tactics in female black and gold howler monkeys (*Alouatta caraya*) inhabiting an island on the Parana River, Argentina. Am J Primatol 72:734–748

Landete-Castillejos T, García A, López-Serrano FR, Gallego L (2005) Maternal quality and differences in milk production and composition for male and female Iberian red deer calves (*Cervus elaphus hispanicus*). Behav Ecol Sociobiol 57:267–274

Langer P (2003) Lactation, weaning period, food quality, and digestive tract differentiations in eutheria. Evolution 57:1196–1215

Langer P (2008) The phases of maternal investment in eutherian mammals. Zoology 111: 148–162

Lee PC (1987) Nutrition, fertility, and maternal investment in primates. J Zool 213:409–422

Lee PC (1996) The meanings of weaning: growth, lactation, and life history. Evol Anthropol 5:87–98

Leigh SR (1994) Ontogenetic correlates of diet in anthropoid primates. Am J Phys Anthropol 94:499–522

Leigh SR (1996) Evolution of human growth spurts. Am J Phys Anthropol 101:455–474

Leigh SR (2001) Evolution of human growth. Evol Anthropol 10:223–236

Leigh SR, Bernstein RM (2006) Ontogeny, life history, and maternal reproductive strategies in baboons. In: Leigh SR, Swedell L (eds) Life history, reproductive strategies, and fitness in baboons. Kluwer Academic/Plenum, New York

Leigh SR, Blomquist GE (2007) Life history. In: Campbell CJ, Fuentes A, MacKinnon KC, Panger M, Bearder SK (eds) Primates in perspective. Oxford University Press, New York

Lyall Z (1996) The early development of behavior and independence in howler monkeys, *Alouatta palliata mexicana*. Neotrop Primates 4:4–8

MacArthur RH, Wilson EO (1967) The theory of island biogeography. Princeton University Press, Princeton

Mack D (1979) Growth and development of infant red howling monkeys (*Alouatta seniculus*) in a free-ranging population. In: Eisenberg J (ed) Vertebrate ecology in the northern neotropics. Smithsonian Institution Press, Washington, DC

MacKinnon KC (2005) Food choice by juvenile capuchin monkeys (*Cebus capucinus*) in a tropical dry forest. In: Estrada A, Garber PA, Pavelka MSM, Leucke L (eds) New perspectives in the study of Mesoamerican primates: distribution, ecology, behavior, and conservation, 1st edn. Springer, New York

Malinow MR, Locker Pope B, Depaoli JR, Kats S (1968) Laboratory observations on living howlers. In: Malinow MR (ed) Biology of the howler monkey, Bibliotheca Primatologica. Karger, New York

Martins WP, Strier KB (2004) Age at first reproduction in philopatric female muriquis (*Brachyteles arachnoides hypoxanthus*). Primates 45:63–67

Milton K (1981) Estimates of reproductive parameters for free-ranging *Ateles geoffroyi*. Primates 22:574–579

Miranda JMD, Aguiar LM, Ludwig G, Moro-Rios RF, Passos FC (2005) The first seven months on an infant of *Alouatta guariba* (Humboldt) (Primates, Atelidae): interactions and the development of behavioral patterns. Rev Bras Zool 22:1191–1195

Needham J (1933) On the dissociability of the fundamental processes in ontogenesis. Biol Rev 8:180–223

Nishimura A (2003) Reproductive parameters of wild female *Lagothrix lagotricha*. Int J Primatol 24:707–722

Overdorff DJ, Merenlender AM, Talata P, Telo A, Forward ZA (1999) Life history of *Eulemur fulvus rufus* from 1988–1998 in southeastern Madagascar. Am J Phys Anthropol 108:295–310

Pavé R, Kowalewski MM, Peker SM, Zunino GE (2010) Preliminary study of mother-offspring conflict in *Alouatta caraya*. Primates 51:221–226

Pavé R, Kowalewski M, Garber P, Zunino G, Fernandez V, Peker S (2012) Infant mortality in black-and-gold howlers (*Alouatta caraya*) living in a flooded forest in northeastern Argentina. Int J Primatol 33:937–957

Pavelka M, Knopff K (2004) Diet and activity in black howler monkeys (*Alouatta pigra*) in southern Belize: does degree of frugivory influence activity level? Primates 45:105–111

Pereira ME, Leigh SR (2003) Modes of primate development. In: Kappeler PM, Pereira ME (eds) Primate life histories and socioecology. University of Chicago Press, Chicago

Peres CA (1993) Structure and spatial organization of an Amazonian terra firme forest primate community. J Trop Ecol 9:259–276

Peres CA (1994) Which are the largest New World monkeys? J Hum Evol 26:245–249

Perez Rueda, M (2010) Patrones de alimentación en hembras de *Alouatta caraya* (Primates: Atelidae) durante diferentes estados reproductivos. Bachelors Thesis, Corrientes, Argentina: Facultad de Ciencias Exactas y Naturales y Agrimensura, Universidad Nacional del Nordeste

Pianka ER (1970) On r- and K-selection. Am Nat 104:592–597

Plavcan MJ (2001) Sexual dimorphism in primate evolution. Am J Phys Anthropol 116:25–53

Podgaiski L, Assis Jardim M (2009) Early behavioral development of a free-ranging howler monkey infant (*Alouatta guariba clamitans*) in southern Brazil. Neotrop Primates 16:27–31

Prates HM, Bicca-Marques JC (2008) Age-sex analysis of activity budget, diet, and positional behavior in *Alouatta caraya* in an orchard Forest. Int J Primatol 29:703–715

Promislow DEL, Harvey PH (1990) Living fast and dying young: a comparative analysis of life history variation among mammals. J Zool 220:417–437

Raff R (1996) The shape of life: genes, development, and the evolution of animal form. University of Chicago Press, Chicago

Raff EC, Raff RA (2000) Dissociability, modularity, evolvability. Evol Dev 2:235–237

Raguet-Schofield ML (2010) The ontogeny of feeding behavior of Nicaraguan mantled howler monkeys (*Alouatta palliata*). PhD dissertation, University of Illinois, Urbana-Champaign

Ravosa MJ, Ross CF (1994) Craniodental allometry and heterochrony in two howler monkeys: *Alouatta seniculus* and *A. palliata*. Am J Primatol 33:277–299

Read A, Harvey PH (1989) Life history differences among the eutherian radiations. J Zool 219:329–353

Reznick D, Bryant MJ, Bashey F (2002) r- and K-selection revisited: the role of population regulation in life-history evolution. Ecology 83:1509–1520

Ross C (1988) The intrinsic rate of natural increase and reproductive effort in primates. J Zool 214:199–219

Ross C (1991) Life history patterns of New World monkeys. Int J Primatol 12:481–502

Ross C (1992) Environmental correlates of the intrinsic rate of natural increase in primates. Oecologia 90:383–390

Ross C (1998) Primate life histories. Evol Anthropol 6:54–63

Rumiz DI (1990) *Alouatta caraya*: population density and demography in northern Argentina. Am J Primatol 21:279–294

Rumiz DI (1992) Effects of demography, kinship, and ecology on the behavior of the red howler monkey, *Alouatta seniculus*. PhD dissertation, University of Florida, Gainesville

Sacher GA, Staffeldt EF (1974) Relation of gestation time to brain weight for placental mammals: implications for the theory of vertebrate growth. Am Nat 108:593–615

Schultz AH (1935) Eruption and decay of the permanent teeth in primates. Am J Phys Anthropol 19:489–581

Serio-Silva J, Rodriguez-Luna E (1994) Howler monkey (*Alouatta palliata*) behavior during the first weeks of life. Masters thesis, Universidad Veracruzana, Veracruz, Mexico

Silver SC, Ostro LET, Yeager CP, Horwich R (1998) Feeding ecology of the black howler monkey (*Alouatta pigra*) in northern Belize. Am J Primatol 45:263–279

Smith BH (1989) Dental development as a measure of life history in primates. Evolution 43:683–688

Smith BH (2000) "Schultz's Rule" and the evolution of tooth emergence and replacement patterns in primates and ungulates. In: Teaford MF, Smith MM, Ferguson MWJ (eds) Development, function and evolution of teeth. Cambridge University Press, Cambridge

Smith RJ, Jungers WL (1997) Body mass in comparative primatology. J Hum Evol 32:523–559

Stearns SC (1977) The evolution of life history traits: a critique of the theory and a review of the data. Annu Rev Ecol Evol Syst 8:145–171

Stearns SC (1983) The influence of size and phylogeny on patterns of covariation among life-history traits in the mammals. Oikos 41:173–187

Stearns SC (1992) The evolution of life histories. Oxford University Press, Oxford

Stone A (2004) Juvenile feeding ecology and life history in a neotropical primate the squirrel monkey (*Saimiri sciureus*). PhD dissertation, University of Illinois, Urbana-Champaign

Strier KB (1991) Demography and conservation of an endangered primate, *Brachyteles arachnoides*. Conserv Biol 5:214–218

Strier KB (1996) Reproductive ecology of female muriquis. In: Norconk MA, Rosenberger AL, Garber PA (eds) Adaptive radiations of New World primates. Plenum, New York

Strier KB, Boubli JP, Possamai, CB, Mendes SL (2006) Population demography of Northern muriquis (Brachyteles hypoxanthus) at the Estação Biológica de Caratinga/Reserva particular do Patrimônio Natural-Felìciano Miguel Abdala, Minas Gerais, Brazil. Am J Phys Anthropol 130:227–237

Strier K, Mendes S, Santos R (2001) Timing of births in sympatric brown howler monkeys (*Alouatta fusca clamitans*) and northern muriquis (*Brachyteles arachnoides hypoxanthus*). Am J Primatol 55:87–100

Tanaka I (1992) Three phases of lactation in free-ranging Japanese macaques. Anim Behav 44:129–139

Tarnaud L (2004) Ontogeny of feeding behavior of *Eulemur fulvus* in the dry forest of Mayotte. Int J Primatol 25:803–824

Taub DM (1980) Age at first pregnancy and reproductive outcome among colony-born squirrel monkeys (*Saimiri sciureus*, Brazilian). Folia Primatol 33:262–272

van Belle S, Estrada A, Ziegler TE, Strier KB (2009) Sexual behavior across ovarian cycles in wild black howler monkeys (*Alouatta pigra*): male mate guarding and female mate choice. Am J Primatol 71:153–164

van Roosmalen MGM (1986) Habitat preferences, diet, feeding behavior, and social organization of the black spider monkey, *Ateles paniscus paniscus*, in Surinam. Acta Amazon 15:1–238

Wagner GP (1996) Homologues, natural kinds and the evolution of modularity. Am Zool 36:36–43

Watts DP (1985) Observations on the ontogeny of feeding behavior in mountain gorillas (*Gorilla gorilla beringei*). Am J Primatol 8:1–10

Watts DP (1991) Mountain gorilla reproduction and sexual behavior. Am J Primatol 24:211–225

Williams GC (1957) Pleiotropy, natural selection, and the evolution of senescence. Evolution 11:398–411

Williams GC (1966) Adaptation and natural selection: a critique of some current evolutionary thought. Princeton University Press, Princeton

Zunino GE, Kowalewski MM, Oklander LI, Gonzalez V (2007) Habitat fragmentation and population trends of the black and gold howler monkey (*Alouatta caraya*) in a semideciduous forest in northern Argentina. Am J Primatol 69:966–975

Chapter 12
The Sensory Systems of *Alouatta*: Evolution with an Eye to Ecology

Laura T. Hernández Salazar, Nathaniel J. Dominy, and Matthias Laska

Abstract Our knowledge about the perceptual world of howler monkeys is unevenly distributed between the five senses. Whereas there is abundant knowledge about the sense of vision in the genus *Alouatta*, only limited data on the senses of hearing, smell, taste, and touch are available. The discovery that howler monkeys are the only genus among the New World primates to possess routine trichromacy has important implications for the evolution of color vision and therefore has been studied intensively. Detailed information about the genetic mechanisms and physiological processes underlying color vision in howler monkeys are available. Although the sound production, vocal repertoire, and acoustic communication in the genus *Alouatta* have been well documented, basic physiological measures of hearing performance such as audiograms are missing. Similarly, despite an increasing number of observational studies on olfactory communication in howler monkeys, there is a complete lack of physiological studies on the efficiency of their sense of smell. Information about the senses of taste and touch is even scarcer and mainly restricted to a description of their anatomical basis. A goal of this chapter is to summarize our current knowledge of the anatomy, physiology, genetics, and behavioral relevance of the different senses in howler monkeys in comparison to other platyrrhines.

Resumen El conocimiento que tenemos de las percepciones en los monos aulladores no es igual para los cinco sentidos. Mientras que para el género *Alouatta* existe un vasto conocimiento acerca de la visión, existen datos limitados para el resto de sus sentidos, oído, olfato, gusto y tacto. El hallazgo de que los monos aulladores son el único género entre los primates del Nuevo Mundo que poseen una visión tricrómata, tiene una importante implicación para la evolución de la visión a color y por ello, ha sido ampliamente estudiada. Existe información detallada acerca

---

L.T.H. Salazar (✉)
Instituto de Neuroetologia, Universidad Veracruzana, Xalapa, Veracruz 91000, Mexico
e-mail: terehernandez@uv.mx

N.J. Dominy
Department of Anthropology, Dartmouth College, Hanover, NH 03755, USA

M. Laska
IFM Biology, Section of Zoology, Linköping University, Linköping 581 83, Sweden

de los mecanismos genéticos y procesos fisiológicos subyacentes a la visión del color en los aulladores, así como para la producción de sonido, el repertorio vocal y la comunicación acústica, aunque falta información de las medidas fisiológicas básicas de la audición, como son los audiogramas. De manera similar, a pesar del creciente número de estudios observacionales sobre la comunicación olfativa en los monos aulladores, no se cuentan con estudios fisiológicos sobre la eficiencia de su sentido del olfato. La información sobre los sentidos del gusto y el tacto es aún más limitada y, es restringida a su base anatómica. Este capítulo tiene como objetivo resumir el conocimiento actual acerca de la anatomía, fisiología, genética y la relevancia a nivel conductual del uso de los diferentes sentidos en los monos aulladores en comparación con otros platirrinos.

Keywords Sensory modalities • Communication • Neotropical primates

12.1 Introduction

Primates developed sensory capacities that enabled them to survive in remarkably diverse habitats. The senses can be considered as the window to the outside world which allows primates to make informed decisions about food, sexual selection, reproduction, and social life, inter alia. Thus, the study of primate sensory systems allows us to understand more about their perceptual world and their behavioral responses. Within the group of New World monkeys, the members of the genus *Alouatta* are characterized by being highly selective with regard to food and sexual partners (Carpenter 1934; Milton 1980; Calegaro-Marques and Bicca-Marques 1993; Kitchen 2004; Espinosa-Gómez et al. 2013). This should lead to an array of evolutionary adaptations both in their sensory capabilities and their behavior. However, the number of studies concerning the sensory systems in howler monkeys is relatively low.

A literature search in the Web of Knowledge^{SM} performed in March 2012 demonstrates that scientific publications on the sensory systems of *Alouatta* are generally scarce and unevenly distributed among the five senses. The search words *Alouatta* or howl* in combination with vision*/visual* yielded 126 original papers published between 1975 and 2012. The corresponding numbers for hear*/acoust* (10), olfact*/smell* (2), tast*/gustat* (4), and touch*/somatosens* (1) are dramatically fewer. A similar search performed on the genus *Saimiri*, another member of the New World primates, found 675, 160, 85, 40, and 314 publications, respectively. These differences may relate to the common use of squirrel monkeys in captive research for several decades (Abee 1989; Brady 2000). There are two reasons for the comparatively low number and uneven distribution of scientific publications on the sensory systems of howler monkeys: first, howlers are rarely used as laboratory animals; and second, they are difficult to maintain in captivity due to their selective feeding habits and susceptibility to stress. Thus, studies of their sensory systems are mainly restricted to anatomical and observational methods, whereas physiological studies are almost nonexistent. Second, the discovery of uniform trichromatic vision in *Alouatta caraya* and *A. seniculus* (Jacobs et al. 1996) has fueled a disproportionate focus on the howler visual system.

12.2 The Sense of Vision

Uniform or routine trichromatic vision depends on the presence of two X-linked opsin genes that encode middle (M)- and long (L)-wavelength-sensitive visual pigments. This character state exists in all catarrhine primates and the genus *Alouatta* among platyrrhines (Jacobs et al. 1996). The routine trichromatic vision of *Alouatta* distinguishes it from other platyrrhine monkeys, which have one X-linked opsin gene, although the locus is polymorphic with 2–5 alleles depending on the genus (Jacobs and Deegan 2005; Talebi et al. 2006). As a result of this polymorphism, nonhowler platyrrhine males possess dichromatic vision, whereas approximately 33 % of females are dichromatic and 66 % of females are heterozygous and possess trichromatic vision. This condition, termed allelic or polymorphic trichromatic vision, is the probable ancestral state that gave rise to the routine trichromatic vision of *Alouatta* (Boissinot et al. 1997, 1998). Moreover, the nucleotide divergence between the M- and L-opsin genes of *Alouatta* (2.7 %) is lower than the average divergence within catarrhine primates (6.1 %), indicating a relatively recent duplication event (Hunt et al. 1998). Thus, routine trichromatic color vision has almost certainly evolved independently and convergently in catarrhines and the lineage that gave rise to *Alouatta* (Kainz et al. 1998; Dulai et al. 1999).

Since the discovery of routine trichromatic vision in *A. caraya* and *A. seniculus* (Jacobs et al. 1996), a finding that has since been verified using electrophysiological, molecular, and behavioral methods (Boissinot et al. 1997; Dulai et al. 1999; Silveira et al. 2007; Araújo et al. 2008) and reported in another species, *A. palliata* (Joganic et al. 2009), considerable interest has been focused on how and why routine trichromatic vision emerged at least twice during primate evolution (Dulai et al. 1999; Lucas et al. 2003). And recently, Matsushita et al. (2013) reported that some individuals of *A. palliata* and *A. pigra* possess hybrid M- and L-opsin genes, resulting in anomalous trichromatic vision. Importantly, *Alouatta* also differs from other platyrrhine monkeys in having an unusually high density of cone photoreceptors in the fovea of the central retina (Franco et al. 2000; Finlay et al. 2008). Thus, *Alouatta* also enjoys enhanced visual acuity, indicating an increased ability to discriminate not only color but also fine detail.

For primates, the adaptive advantages of high-acuity routine trichromatic vision are uncertain. Debate has tended to focus on the adaptive advantages of detecting food targets against a background of mature foliage; for instance, ripe fruit (Regan et al. 1998, 2001), young leaves (Lucas et al. 1998; Dominy and Lucas 2001, 2004), or both (Sumner and Mollon 2000). A central problem with much of this debate is that most captive- and field-based studies have focused on the foraging advantages of female platyrrhines with relatively low-acuity trichromatic vision (e.g., Caine and Mundy 2000; Dominy et al. 2003a; Smith et al. 2003; Osorio et al. 2004; Vogel et al. 2007; Hiramatsu et al. 2008). A frequent assumption in this work is that the selective advantages that favored polymorphic trichromatic vision are similar to those that favored the high-acuity routine trichromatic vision of *Alouatta*. This view has been questioned in recent years (Dominy et al. 2003b), and most current research is focused on how and why polymorphic color vision has been maintained by balancing selection (Melin et al. 2007; Hiwatashi et al. 2010; Caine et al. 2010; Smith et al. 2012).

As a result, relatively little comparative research has been focused specifically on the visual ecology of *Alouatta*. Field studies of two species, *A. seniculus* and *A. palliata*, have called attention to the benefits of trichromatic color vision for detecting yellow, orange, and red fruit against a background of mature foliage (Regan et al. 1998, 2001; Urbani 2002; Lucas et al. 2003; Stoner et al. 2005) (Fig. 12.1). Yet, for *Alouatta*, there was no obvious foraging advantage over sympatric monkeys, such as *Ateles*, which rely on such fruit to a greater extent despite a lower proportion of trichromatic individuals (Regan et al. 2001; Lucas et al. 2003; Stoner et al. 2005). Such findings have cast doubt on the adaptive importance of high-acuity routine trichromatic vision for detecting and discriminating ripe fruit. Based on field observations of *A. palliata*, Lucas et al. (2003) found that routine trichromatic color vision contributed significantly to the detection and discrimination of young leaves and that *A. palliata* consumed leaves with redder chromaticities than did a sympatric

Fig. 12.1 The high-acuity trichromatic color vision of *Alouatta* facilitates the chromatic discrimination of certain hues (yellows, oranges, and reds) from a background of mature foliage. For howlers, the functional ecology of trichromatic vision is debated, but it likely improves foraging efficiency. Here a mantled howler monkey (*A. palliata*) consumes the ripe fruits of *Astrocaryum standleyanum* on Barro Colorado Island, Panama (photograph by Greg Willis, reproduced with permission)

polymorphic species, *Ateles geoffroyi*. This pattern unites *Alouatta* with catarrhine species (Lucas et al. 1998; Sumner and Mollon 2000; Dominy and Lucas 2001, 2004) and suggests that a diet dependent on young leaves, which differ texturally and chromatically from mature leaves, was the primary factor favoring the independent evolution of high-acuity routine trichromatic vision in primates.

12.3 The Sense of Hearing

The genus *Alouatta* is most renowned for its howling, and much attention has been focused on the production and acoustic structure of this signature vocalization, particularly in *A. palliata* (Kelemen and Sade 1960; Chivers 1969; Schön 1970, 1971; Baldwin and Baldwin 1976; Schön Ybarra 1986, 1988; Whitehead 1996; Boscarol et al. 2004; Piazza et al. 2004; Bustos et al. 2008; De Boer 2009), *A. pigra* (Kitchen et al. 2004; Kitchen 2004, 2006), and *A. caraya* (Garber and Kowalewski 2012) (Fig. 12.2). Perhaps not surprisingly, the adaptive significance of the howl or roar

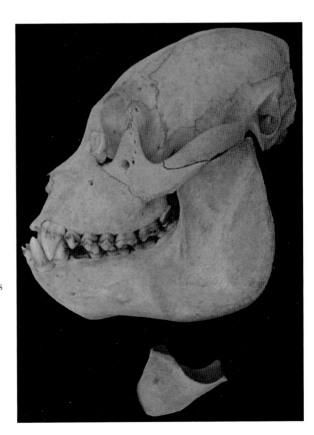

Fig. 12.2 Skull and hyoid bone of a mantled howler monkey (*A. palliata*) from the El Zota Biological Field Station, Costa Rica (photograph by N.J. Dominy). The cavernous hyoid is an outstanding anatomical characteristic of *Alouatta*. The primary acoustic function of the enlarged hyoid is uncertain, but it appears to amplify the peak amplitude of howls. Hyoid bones in males are larger than those of females (see Youlatos et al. 2014)

Fig. 12.3 The eponymous vocalization of howlers is produced with enlarged hyolaryngeal structures, which are evident beneath the skin surface (photograph by David Tipling, reproduced with permission). The adaptive significance of howling is debated, but it is widely interpreted as a costly and honest signal of male quality or coalition size that functions to deter conflict with rival males

has also been a topic of long-standing interest (Sekulic 1982a, 1983; Sekulic and Chivers 1986; Chiarello 1995; Cornick and Markowitz 2002; Kitchen et al. 2004; Da Cunha and Byrne 2006; Delgado 2006; Da Cunha and Jalles-Filho 2007). For example, the playback experiments of Kitchen (2004, 2006) suggest that howling is a costly and honest signal of male quality or that coalition size functions to deter conflict with rival males (Fig. 12.3). However, the effectiveness of any vocal signal depends in part on the auditory system of the intended receiver, and the hearing sense of howlers is practically unstudied.

On the surface, *Alouatta* is presumed to have ordinary hearing abilities. The anatomy of the outer ear and middle ear structures is comparable to those of other New World monkeys (Coleman and Ross 2004). Further, the hearing sensitivities of New World monkeys are broadly comparable to those of catarrhine primates (Heffner 2004; Coleman 2009). Accordingly, it appears reasonable to assume that *Alouatta* and its human observers hear the low fundamental frequencies (300–1,000 Hz) of howls similarly well. It is perhaps significant that the recorded or estimated amplitudes of howls (*A. palliata*: 91 dB at 5 m, Whitehead 1989; *A. seniculus*: 90 dB at 5 m, Sekulic 1983) must be greater for the group members in closer proximity to callers. For example, sound levels could be as high as 120 dB for as long as 60 min when the males of *A. seniculus* assemble, or even embrace, during intense bouts of unified howling (Sekulic 1982b; estimate based on a 6-dB decrease per doubling of distance). Among humans, a 120-dB sound can induce permanent hearing loss from 15 min of exposure per day (Daniel 2007). Thus, the potential for hearing loss in *Alouatta* is perhaps underappreciated in discussions on the adaptive function of howling, the costs of which are already presumed to be high, both energetically and through increased exposure to predators (Kitchen 2004).

Recently, Ramsier et al. (2012) described the audiogram of four captive adult *A. palliata*. The mean audiogram was dual peaked (w shaped) with two regions of enhanced sensitivity separated by a mid-frequency "dip" of decreased sensitivity. The lower-frequency area of enhanced sensitivity was centered at 0.7–1 kHz and the upper-frequency region of enhanced (best) sensitivity was centered at 11.3 kHz. These two frequency regions corresponded, respectively, with the dominant frequencies of male howling and an infant distress call, the *wrah-ha* (Baldwin and Baldwin 1976; Ramsier et al. 2009), which contained harmonics at very high frequencies (10 and 21 kHz). For a genus in which some species are reported to frequently engage in infanticide (Crockett 2003; Van Belle and Bicca-Marques 2014), it is possible that natural selection has biased the adult auditory system toward these high frequencies in order to favor the emitter's response showing the energetic and psychological state of the signaler. This concept, termed receiver bias (Endler and Basolo 1998), is hypothesized to constrain auditory sensitivity at lower frequencies. Thus, the extreme loudness of howling is possibly a compensatory adaptation for increasing the propagation distance of a call that is constrained by the auditory sensitivities of the intended conspecific receivers.

No experimental data are available on the spatial sound localization ability in howler monkeys. However, behavioral observations on the orientation responses of howler monkeys toward conspecific (Whitehead 1987, 1989, 1994; Drubbel and Gatier 1993; Kitchen 2004, 2006; Kitchen et al. 2004) as well as toward aerial predator vocalizations (Gil-da-Costa et al. 2003) suggest that the ability to localize a behaviorally meaningful source of sound is comparable to other monkeys.

12.4 The Sense of Smell

Olfactory communication appears to be an important part of social and sexual behavior in howler monkeys. Anatomical data on both odor-perceiving as well as odor-producing structures and at least some behavioral studies on the sense of smell have been published, whereas physiological studies on the efficiency of the olfactory system are completely lacking.

The anatomy of the main olfactory system in the genus *Alouatta* has been described as similar to that in other New World primate species (Smith and Rossie 2006). The size of the main olfactory bulbs, the first relay station of the olfactory pathway, has been found to be 45.2 mm^3 in *A. seniculus* and is thus markedly smaller than that reported in another atelid species of comparable body size, the spider monkey *Ateles geoffroyi*, whose main olfactory bulbs have a size of 90.4 mm^3 (Stephan et al. 1988). Nevertheless, the relative size of the main olfactory bulbs, expressed as proportion of total brain size, is similar in both species (0.8‰ in *Alouatta* and 0.9‰ in *Ateles*) and thus typical of New World primates in general. Howler monkeys also have been found to possess a well-developed vomeronasal organ (Smith et al. 2002), a feature that is typical of New World primates. It should be mentioned, however, that physiological evidence confirming the vomeronasal organ to be functional in howler monkeys is missing. This also is true for many other species of New World primates. Nevertheless, behavioral studies reporting flehmen-like behavior in the mantled howler monkey *A. palliata* (Glander 1980) and the black howler monkey *A. pigra* (Horwich 1983) suggest that the vomeronasal organ is functional. Additionally, tongue-flicking behavior, which is also interpreted as being indicative of a functional vomeronasal organ, has been observed in several species of howler monkeys (Van Belle et al. 2009). It is interesting to note that tongue flicking was also described in Carpenter's monograph (1934). However, the author interpreted tongue flicking as a visual display or "gesture" rather than as the sampling of chemical volatiles in a reproductive context. The size of the accessory olfactory bulbs, the first relay station of the vomeronasal system, has been found to be 0.595 mm^3 in *A. seniculus* and is markedly smaller than the 2.189 mm^3 reported in spider monkeys (Stephan et al. 1982).

With regard to odor-producing structures, specialized skin glands on the throat, sternum, and the anogenital region have been described (Epple and Lorenz 1967; Machida and Giacometti 1968; Hirano et al. 2003) in several members of the genus *Alouatta*, a pattern that is consistent with data available for the majority of New World primates studied (Epple and Lorenz 1967).

Observational studies on olfactory-related behavior in howler monkeys have focused generally on single behavioral patterns. There are reports of throat rubbing (Sekulic and Eisenberg 1983), chest rubbing (Young 1982), and rubbing of the anogenital region (Hirano et al. 2008) against substrates such as branches or tree trunks in several species of howler monkeys. Urine washing, the deposition of urine onto the palms of the hands or the soles of the feet with subsequent rubbing of the wetted palms or soles against other body parts or against branches or tree trunks, seemingly with the aim to impregnate the fur or the substrate with urine odor, also has been observed repeatedly in members of the genus *Alouatta* (Milton 1975; Jones 2003; Hirano et al. 2008). Some of these studies also report that the sites where body odor or scent gland secretions or urine were deposited onto a substrate were subsequently inspected by conspecifics and sometimes induced behaviors such as flehmen, tongue flicking, vocalization, or orientation responses, suggesting that these deposits may serve as scent marks in the context of social communication (Jones and Van Cantfort 2007). A chemical analysis of the urine of a male *A. caraya* using gas chromatography–mass spectrometry found a mixture of compounds that is similar to that reported in other species of New World primates (Jones 2002).

The only observational study to date that attempted to establish a complete ethogram of olfactory-related behaviors in a species of howler monkeys reported nine types of scent-marking behaviors (urine washing, anogenital rubbing, face rubbing, throat rubbing, lateral neck rubbing, chest rubbing, back rubbing, urination, and defecation) and 11 types of odor-evoked behaviors (urine sniffing, anogenital sniffing, face sniffing, body sniffing, hand sniffing, place sniffing, sniffing at anogenital scent mark, sniffing at other types of scent mark, flehmen, licking of scent mark, licking of anogenital region) in a group of free-ranging mantled howler monkeys, *A. palliata* (Baltisberger 2003). Further, this study found that significant sex differences in the frequency of their occurrence were reported for only three of these 20 olfactory-related behaviors (anogenital rubbing and back rubbing: females>males; anogenital sniffing: males>females). Olfactory-related behaviors directed toward a particular conspecific occurred about four times more often in the direction male → female than vice versa. Adult females directed their olfactory-related behaviors about twice as often toward juveniles than toward adult males (Baltisberger et al. 2003). Interestingly, in 488 recorded bouts of food choice and consumption, not a single instance of olfactory-related behavior such as sniffing at food was observed. This result suggests that olfaction might play a more important role in social communication in *A. palliata*, particularly in the context of reproduction, than in the context of food selection.

A recent genetic study reported that the proportion of genes coding for olfactory receptors (OR) that are pseudogenized, i.e., nonfunctional, is markedly higher in *A. caraya* compared to *Aotus azarai*, *Ateles fusciceps*, *Callithrix jacchus*, *Cebus apella*, *Lagothrix lagotricha*, and *Saimiri sciureus* (Gilad et al. 2004). This deterioration in the repertoire of olfactory receptor types, which is shared by catarrhine primates, was interpreted as a sensory trade-off associated with the evolution of routine trichromacy vision. However, this conclusion has been partly retracted (Gilad et al. 2007) and a subsequent study comparing *C. jacchus* with catarrhine primates found comparable numbers of OR pseudogenes (Matsui et al. 2010). Physiological studies on the sense of smell in the howler monkey are completely missing. Thus, we have no data on olfactory sensitivity in terms of detection thresholds or discrimination capabilities with odor stimuli (food odors and social odors) under controlled conditions. However, behavioral observations suggest that howler monkeys are able to discriminate between conspecific and heterospecific odors, as well as between the odors of males and females, and possibly also between different reproductive states and individuals (Baltisberger 2003).

12.5 The Sense of Taste

Anatomical data on the structures perceiving taste stimuli and behavioral observations of food selection behavior in *Alouatta* have been published, whereas physiological studies on the efficiency of the gustatory system are lacking. The mean number of fungiform papillae on the surface of the tongue (Fig. 12.4) in the mantled howler monkey *A. palliata* (24.6) has been found to be markedly lower than that in the spider monkey *Ateles geoffroyi* (64.5) (Alport 2007). In contrast to humans, no differences in the number of fungiform papillae were found between male and female howler monkeys (Alport 2008). Similarly, the mean number of taste buds on the *Plica sublingualis*, an anatomical structure in the oral cavity below the proper tongue, has been found to be lower in howler monkeys (636) than in spider monkeys (1,765) (Hofer et al. 1979). An accumulation of taste buds on this sublingual structure can be found close to the openings of the sublingual salivary glands suggesting that they may serve the perception of fresh saliva (Hofer 1977).

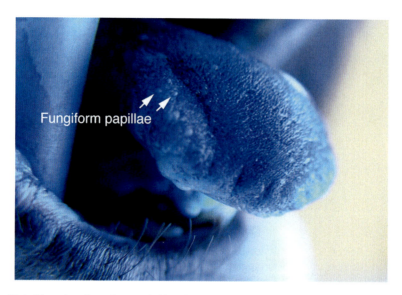

Fig. 12.4 Lingual surface of a mantled howler monkey (*A. palliata*) from Hacienda La Pacifica, Costa Rica (photograph by N.J. Dominy). The relatively small number of fungiform papillae among howlers has been interpreted as an adaptation to their folivorous diet. The blue color is due to the application 0.5 % methylene blue biological stain, which adheres to all papillae except the fungiform type. The lighter contrast facilitates visual quantification under field conditions

It is interesting to note, however, that the occurrence of taste buds on the *Plica sublingualis* is restricted to New World primates and not found in prosimians and Old World primates (Rommel 1981), irrespective of dietary habits. The presence of other types of taste papillae found in primates (filiform, foliate, and circumvallate papillae) has been confirmed in howler monkeys without reporting quantitative data (Machida et al. 1967). The central processing of taste sensation and associated neuroanatomical structures have been well studied in other primate species including *Macaca mulatta* and *M. nemestrina* (Scott and Plata-Salaman 1999; Rolls 2007), but no information on this topic has been published in *Alouatta*.

Physiological studies on the sense of taste in the howler monkey are completely missing. Thus, no data on sensitivity in terms of taste detection thresholds for gustatory stimuli or on the ability to discriminate between different qualities or intensities of tastants are available. The taste function has been studied intensively in several nonhuman primates using a simple psychophysical method, the two-bottle preference test (e.g., Glaser 1979; Laska et al. 2003, 2009; Laska and Hernández-Salazar 2004). This test assesses gustatory responsiveness to any tastant by giving an animal the option to choose between two bottles containing either tap water or a defined concentration of a tastant dissolved in tap water. This allows the researcher to evaluate differences in consumption between the two stimuli and determine taste preference thresholds (Richter and Campbell 1940). To the best of our knowledge, there are no published studies that have performed this test with howler monkeys.

One possible explanation for not using howler monkeys in two-bottle preference tests is their apparent lack of curiosity, which makes it difficult to engage them in licking or sucking at the spouts of water bottles (Hernández-Salazar and Laska, unpublished data). This cannot be attributed to the fact that howler monkeys usually meet their water requirements by consuming juicy plant material (Bicca-Marques 1992) because the same is true for spider monkeys and squirrel monkeys, which readily learn the concept of the two-bottle preference test and eagerly cooperate as long as at least one alternative is savory.

The diet composition of howler monkeys as well as their feeding behavior has been studied in detail (Glander 1977, 1982; Milton 1980; Estrada 1984; Chiarello 1994; Galetti et al. 1994; Guillotin et al. 1994; Pinto and Setz 2004; Simmen and Sabatier 1996; Dias and Rangel-Negrín, 2014). Chemical analyses of the diet of *A. palliata* and *A. pigra* suggest that their food selection can be affected by the presence of plant secondary compounds which appear to be avoided when present at elevated concentrations (Milton 1979; Bilgener 1995). The vast majority of plant secondary compounds such as alkaloids and glycosides taste bitter to humans, and tannins are perceived as both bitter and astringent by humans (Hladik and Simmen 1996). Whether the rejection of certain types of food observed in howler monkeys is based on taste perception or on post-ingestive factors and thus on a gradual learning process, based on the experience of negative physiological consequences of ingestion, or perhaps based on a combination of both remains unknown.

12.6 The Sense of Touch

The sense of touch appears to play an important role in behavioral contexts as diverse as locomotion, posture, social interaction, food selection, and thermoregulation in howler monkeys. Here, again, limited anatomical data on the structures mediating somatosensory perception (Kaas 2004) and only a limited number of behavioral observations on the sense of touch have been published. Physiological studies on the efficiency of the somatosensory system in the genus *Alouatta* are completely lacking.

The skin of howler monkeys has been studied histologically and seems to contain all the types of mechanoreceptors (Pacinian corpuscles, Meissner's corpuscles, Ruffini's endings, and Merkel's disks) as well as temperature and pain receptors (free nerve endings) that are typical of primates (Machida and Giacometti 1968; Perkins 1975; Darian-Smith 1982). The skin of the genital area in the black and gold howler monkey, *A. caraya*, has been reported to have a higher density of touch receptors relative to adjacent skin areas (Machida and Giacometti 1967), and this feature is typical of primates. Unfortunately, no information on the density of somatosensory receptors in other areas of glabrous skin such as the palms of the hands, the soles of the feet, or the distal ventral part of the prehensile tail is available for any member of the genus *Alouatta*. Similarly, no neuroanatomical data concerning the central processing of somatosensory information in howler monkeys have been published, whereas there is abundant information on this topic from another atelid, the spider monkey (Pubols 1968; Pubols and Pubols 1969, 1971).

Physiological studies on the sense of touch in the howler monkey are completely missing. Thus, no data on sensitivity in terms of detection thresholds for the perception of touch, pressure, point localization, two-point discrimination, temperature, or pain are available. Similarly, no data on the ability to discriminate between somatosensory stimuli have been published in the genus *Alouatta*. However, laboratory studies suggest that touch perception is important for howler monkeys in the context of positional behavior (Schmitt and Larson 1995). The postural behavior and locomotor behavior of wild howlers have been analyzed in considerable detail (Mendel 1976; Cant 1986; Bicca-Marques and Calegaro-Marques 1995; Bezanson 2009; Youlatos and Gasc 2011; Youlatos and Guillot 2014, this volume). These studies indicate that both adult and juvenile howlers engage in a variety of suspensory climbing and above-branch quadrupedal walking behaviors which appear to be sensitive somatosensory or proprioceptive mechanisms to maintain balance in both a small and large branch setting.

Although howler monkeys spend a relatively small proportion of their time budget on allogrooming (Jones 1979; Sanchez-Villagra et al. 1998; Chiarello 2005), the sense of touch appears to play an important role in social interaction. Hand-holding has been observed in adult black howler monkeys, *A. pigra*, and interpreted to serve a communicatory function in the context of reproduction (Brockett et al. 2005).

The food selection behavior of howler monkeys also suggests that tactile information may be used to assess the hardness and/or texture of potential food items (Teaford et al. 2006; Raguet-Schofield 2010). Mantled howler monkeys *A. palliata*, have been reported to prefer using their mouth rather than their hands for food retrieval and manipulation, which may be attributed more to their limited manual dexterity than to differences in somatosensory capabilities between these two areas of glabrous skin (Carpenter 1934). Howler monkeys also appear to be able to perceive temperature information and to respond appropriately by performing thermoregulatory behavior. At temperatures below thermal neutrality, howler monkeys have been described to adopt a curled posture, with their arms and legs kept flexed and close to the body, while the back is bent and the head is near the tail, giving the body a spherical shape in which the belly or ventral surface of the body is covered (Paterson 1981). Such a posture serves to reduce the body surface area to volume ratio, which in turn reduces the conductive heat loss of the animal to its surrounding environment (Bicca-Marques and Calegaro-Marques 1998). Additionally, mantled howler monkeys have been reported to wrap their tails around their bodies as a means of thermoregulation at low ambient temperatures (Bopp 1954; Laska and Tutsch 2000), a behavior that is widespread among mammals (Hickman 1979). Heat-dissipating postures such as sprawling have been described when howler monkeys are exposed to high ambient temperatures (Paterson 1981; Bicca-Marques and Calegaro-Marques 1998) which also suggest a well-developed ability to perceive and respond to temperature cues.

12.7 General Conclusion and Future Research

Whereas the anatomical structures of howler monkeys involved in vision, hearing, olfaction, taste, and touch appear to be generally similar to those reported in other atelines and platyrrhines, some differences are present. With regard to color vision, for example, howler monkeys have developed adaptations that make them more similar to the Old World monkeys, apes, and humans than to other platyrrhines. It has been suggested that a folivorous diet might favor the independent evolution of routine trichromatic vision in howlers. Here the question that is still to be answered is: what would be the behavioral advantage for howlers to have routine trichromacy compared to other platyrrhines having dichromacy? Perhaps the answer to this question requires not only to compare foraging behavior between species but also to explore some other behavioral contexts, such as visual social signals, spatial memory, and predator avoidance, paying special attention to the biogeography and ecological scenarios. *Alouatta* is distinguished by its eponymous howling, and most of the studies on this topic so far have focused on the production and acoustic structure of the vocalizations or the socioecological function of howling. Some studies propose that the loud vocalizations made by males are honest signals of male quality and are used as deterrent signal in the competition for resources. It has also been suggested that the extreme loudness of howling might be a compensatory strategy in order to favor the propagation of the call (Fig. 12.3). However, in order to test each of these suggestions, it would be necessary to study the hearing capabilities and limitations of howlers in much more detail. Recent studies that established audiograms in captive mantled howlers (*A. palliata*) revealed interesting results suggesting that the hearing sensitivity is tuned to the dominant frequency of infant calls and of the male howling. However, it will be important to include physiological measurements of hearing in future research and to relate these to the behavior of the howlers.

The vocal repertoires of howlers and their behavioral role for group cohesion are an important and understudied aspect of the acoustic communication in this genus. A theoretical study on several nonhuman primates species (including *A. palliata*) showed that there is a positive correlation between the vocal repertoires and group size as well as grooming time (as a measure of social interaction) (McComb and Semple 2005). Assessing similar correlations within the genus, considering the richness of their vocal repertoires and their social structure, would contribute to our understanding of the evolutionary processes that promoted this behavior.

The sense of smell in primates is traditionally considered to be of minor importance and capacity. In howler monkeys, the majority of studies have so far focused on anatomical descriptions of the odor-perceiving and odor-producing structures. Although these studies suggest howlers to have a functional vomeronasal organ, very little is known about the use of social odors and chemical communication in the genus *Alouatta* and the performance of its main olfactory system. It would therefore be important to assess their olfactory sensitivity and discrimination capabilities using techniques based on operant conditioning procedures, which have been successfully applied with other platyrrhines (*Ateles geoffroyi*, *Saimiri sciureus*, *Callithrix jacchus*).

The sense of taste in platyrrhines has been related to dietary specialization. Studies in *Ateles geoffroyi* and *Saimiri sciureus*, for example, have shown that compounds that are relevant for their diets (such as sugars, salts, or secondary plant compounds) are detected with high sensitivity and accuracy. Given that howler monkeys adopted a nutritional niche that clearly differs from that of the aforementioned platyrrhines, it would be important to study the gustatory sensitivity and discrimination performance in different species of the genus *Alouatta* and assess possible correlations between these physiological measures and evolutionary adaptations to their dietary specialization.

Finally, the sense of touch in platyrrhines has been linked to both social interaction and food choice. Both aspects of behavior have hardly been studied in howlers so far. Future studies should therefore pay special attention to the assessment of possible correlations between the importance of touch for social interaction and group size and composition. Additionally, studies should assess the relevance of hardness as a basis of food choice in howlers.

Besides the behavioral correlates to senses, we still lack major information on the genetic bases of the sensory systems, as well as developmental cues that would be important in order to understand the selective pressures that promoted the evolution of specific senses in howlers compared to those of other platyrrhines. Therefore, studies that attempt to integrate the sensory capabilities and limitations of howler monkeys and the function of sensory perception in response to specific ecological, social, sexual, and communicative challenges will bring forth the importance of senses in the evolutionary steps that shaped this unique genus.

Acknowledgments We thank the editors of this volume for their invitation to contribute to the present volume. Many thanks go to Greg Willis for providing photographs of *A. palliata*. This chapter benefited from the comments of three anonymous referees whose input shaped its final form. We are greatly indebted to them and their effort.

References

Abee CR (1989) The squirrel monkey in biomedical research. ILAR J 31:11–20
Alport L (2007) A comparative analysis of lingual fungiform papillae and diet in primates. Am J Phys Anthropol 132:62
Alport L (2008) Intraspecific sex differences among primates in the density of lingual fungiform papillae. Am J Phys Anthropol 135:59
Araújo AC Jr, Didonet JJ, Araújo CS, Saletti PG, Borges TRJ, Pessoa VF (2008) Color vision in the black howler monkey (*Alouatta caraya*). Vis Neurosci 25:243–248
Baldwin JD, Baldwin JI (1976) Vocalizations of howler monkeys (*Alouatta palliata*) in Southwestern Panama. Folia Primatol 26:81–108
Baltisberger C (2003) Untersuchungen zur geruchlichen Kommunikation bei Brüllaffen (*Alouatta palliata*)—Eine Freilandstudie. Diploma Thesis, Biology, University of Munich, Germany
Baltisberger C, Laska M, Rodríguez-Luna E (2003) Olfactory communication in mantled howler monkeys (*Alouatta palliata*)—a field study. Folia Primatol 74:183–184

Bezanson M (2009) Life history and locomotion in *Cebus capucinus* and *Alouatta palliata*. Am J Phys Anthropol 140:508–517

Bicca-Marques JC (1992) Drinking behavior in the black howler monkey (*Alouatta caraya*). Folia Primatol 58:107–111

Bicca-Marques JC, Calegaro-Marques C (1995) Locomotion of black howlers in a habitat with discontinuous canopy. Folia Primatol 64:55–61

Bicca-Marques JC, Calegaro-Marques C (1998) Behavioral thermoregulation in a sexually and developmentally dichromatic neotropical primate, the black-and-gold howling monkey (*Alouatta caraya*). Am J Phys Anthropol 196:533–546

Bilgener M (1995) Chemical factors influencing food choice of howler monkeys (*Alouatta palliata*). Turk J Zool 19:291–303

Boissinot S, Zhou Y-H, Qiu L, Dulai KS, Neiswanger K, Schneider H, Sampaio I, Hunt DM, Hewett-Emmett D, Li W-H (1997) Origin and molecular evolution of the X-linked duplicate color vision genes in howler monkeys. Zool Stud 36:360–369

Boissinot S, Tan Y, Shyue S-K, Schneider H, Sampaio I, Neiswanger K, Hewett-Emmett D, Li W-H (1998) Origins and antiquity of X-linked triallelic color vision systems in New World monkeys. Proc Natl Acad Sci U S A 95:13749–13754

Bopp P (1954) Schwanzfunktionen bei Wirbeltieren. Rev Suisse Zool 61:83–100

Boscarol G, Falcone I, Fiore R, Gamba M, Giacoma C, Martinoli L, Piazza S, Pizzigalli C, Luna R (2004) Vocal repertoire of *Alouatta palliata* mexicana in Agaltepec. Folia Primatol 75:410–411

Brady AG (2000) Research techniques for the squirrel monkey (*Saimiri* sp). ILAR J 41:10–18

Brockett RC, Horwich RH, Jones CB (2005) Hand-holding by Belizean black howler monkeys: intentional communication in a neotropical primate. Folia Primatol 76:227–230

Bustos CA, Corkum LD, Slater KY, Mennill DJ (2008) Acoustic characteristics of the vocalizations of mantled howler monkeys (*Alouatta palliata*) in the fragmented low-land forests of Honduras. Am J Primatol 70:68

Caine NG, Mundy NI (2000) Demonstration of a foraging advantage for trichromatic marmosets (*Callithrix geoffroyi*) dependent on food colour. Proc R Soc Lond B 267:439–444

Caine NG, Osorio D, Mundy NI (2010) A foraging advantage for dichromatic marmosets (*Callithrix geoffroyi*) at low light intensity. Biol Lett 6:36–38

Calegaro-Marques C, Bicca-Marques JC (1993) Reprodução de *Alouatta caraya* Humboldt, 1812 (Primates, Cebidar). A Primatologia Brasil 4:51–66

Cant JGH (1986) Locomotion and feeding postures of spider and howling monkeys. Field study and evolutionary interpretation. Folia Primatol 46:1–14

Carpenter CR (1934) A field study of the behavior and social relations of howling monkeys. Comp Psychol Monogr 10:1–168

Chiarello AG (1994) Diet of the brown howler monkey, *Alouatta fusca*, in a semi-deciduous forest fragment of southeastern Brazil. Primates 35:25–34

Chiarello AG (1995) Role of loud calls in brown howlers, *Alouatta fusca*. Am J Primatol 36:213–222

Chiarello AG (2005) Grooming in brown howler monkeys, *Alouatta fusca*. Am J Primatol 35:73–81

Chivers DJ (1969) On the daily behavior and spacing of howling monkey groups. Folia Primatol 10:48–102

Coleman MN (2009) What do primates hear? A meta-analysis of all known nonhuman primate behavioral audiograms. Int J Primatol 30:55–91

Coleman MN, Ross CF (2004) Primate auditory diversity and its influence on hearing performance. Anat Rec 281A:1123–1137

Cornick LA, Markowitz H (2002) Diurnal vocal patterns of the black howler monkey (*Alouatta pigra*) at Lamanai, Belize. J Mammal 83:159–166

Crockett CM (2003) Re evaluating the sexual selection hypothesis for infanticide by *Alouatta* males. In: Jones CB (ed) Sexual selection and reproductive competition in primates: new perspectives and directions, vol 3, Special topics in primatology. American Society of Primatology, Norman, pp 327–365

Da Cunha RGT, Byrne RW (2006) Roars of black howler monkeys (*Alouatta caraya*): evidence for a function in inter-group spacing. Behaviour 143:1169–1199

Da Cunha RGT, Jalles-Filho E (2007) The roaring of southern brown howler monkeys (*Alouatta guariba clamitans*) as a mechanism of active defense of borders. Folia Primatol 78:259–271

Daniel E (2007) Noise and hearing loss: a review. J Sch Health 77:225–231

Darian-Smith I (1982) Touch in primates. Annu Rev Psychol 33:155–194

De Boer B (2009) Acoustic analysis of primate air sacs and their effect on vocalization. J Acoust Soc Am 126:3329–3343

Delgado RA (2006) Sexual selection in the loud calls of male primates: signal content and function. Int J Primatol 27:5–25

Dias PAD, Rangel-Negrín A (2014) Diets in Howler Monkeys. In: Kowalewski M, Garber P, Cortes-Ortíz L, Urbani B, Youlatos D (eds) Howler monkeys: behavior, ecology and conservation. Springer, New York

Dominy NJ, Lucas PW (2001) Ecological importance of trichromatic vision to primates. Nature 410:363–366

Dominy NJ, Lucas PW (2004) Significance of color, calories, and climate to the visual ecology of catarrhines. Am J Primatol 62:189–207

Dominy NJ, Garber PA, Bicca-Marques JC, de Azevedo-Lopes MA (2003a) Do female tamarins use visual cues to detect fruit rewards more successfully than do males? Anim Behav 66:829–837

Dominy NJ, Svenning JC, Li WH (2003b) Historical contingency in the evolution of primate color vision. J Hum Evol 44:25–45

Drubbel RV, Gatier JP (1993) On the occurrence of nocturnal and diurnal loud calls, differing in structure and duration, in red howlers (*Alouatta seniculus*) of French Guyana. Folia Primatol 60:195–209

Dulai KS, von Dornum M, Mollon JD, Hunt DM (1999) The evolution of trichromatic color vision by opsin gene duplication in New World and Old World primates. Genome Res 9:629–638

Endler JA, Basolo AL (1998) Sensory ecology, receiver biases and sexual selection. Trends Ecol Evol 13:415–420

Epple G, Lorenz R (1967) Vorkommen, Morphologie und Funktion der Sternaldrüse bei den Platyrrhini. Folia Primatol 7:98–126

Espinosa-Gómez FC, Gómez-Rosales S, Wallis IR, Canales-Espinosa D, Hernández-Salazar LT (2013) Digestive strategies and food choice in mantled howler monkeys *Alouatta palliata mexicana*: bases of their dietary flexibility. J Comp Physiol B. doi:10.1007/s00360-013-0769-9

Estrada A (1984) Resource use by howler monkeys (*Alouatta palliata*) in the rain forest of Los Tuxtlas, Veracruz, Mexico. Int J Primatol 5:105–131

Finlay BL, Franco ECS, Yamada ES, Crowley JC, Parsons M, Muniz JAPC, Silveira LCL (2008) Number and topography of cones, rods and optic nerve axons in New and Old World primates. Vis Neurosci 25:289–299

Franco ECS, Finlay BL, Silveira LCL, Yamada ES, Crowley JC (2000) Conservation of absolute foveal area in New World monkeys: a constraint on eye size and conformation. Brain Behav Evol 56:276–286

Galetti M, Pedroni F, Morellato LP (1994) Diet of the brown howler monkey *Alouatta fusca* in a forest fragment in southeastern Brazil. Mammalia 57:111–118

Garber PA, Kowalewski M (2012) Collective action and male affiliation in howler monkeys (*Alouatta caraya*). In: Sussman RW, Cloninger CR (eds) Origins of altruism and cooperation, developments in primatology: progress and prospects. Springer, New York

Gilad Y, Wiebe V, Przeworski M, Lancet D, Pääbo S (2004) Loss of olfactory receptor genes coincides with the acquisition of full trichromatic vision in primates. PLoS Biol 2:120–125

Gilad Y, Wiebe V, Przeworski M, Lancet D, Pääbo S (2007) Correction: Loss of olfactory receptor genes coincides with the acquisition of full trichromatic vision in primates. PLoS Biol 5:e148

Gil-da-Costa R, Palleroni A, Hauser MD, Touchton J, Kelley P (2003) Rapid acquisition of an alarm response by a neotropical primate to a newly introduced avian predator. Proc R Soc Lond B 270:605–610

Glander KE (1977) Poison in a monkey's Garden of Eden. Nat Hist 86(3):35–41
Glander KE (1980) Reproduction and population growth in free-ranging mantled howling monkeys. Am J Phys Anthropol 53:25–36
Glander KE (1982) The impact of plant secondary compounds on primate feeding behavior. Yearb Phys Anthropol 25:1–18
Glaser D (1979) Gustatory preference behavior in primates. In: Kroeze JHA (ed) Preference behavior and chemoreception. IRL Press, London
Guillotin M, Dubost G, Sabatier D (1994) Food choice and food composition among the three major primate species of French Guiana. J Zool 233:551–579
Heffner RS (2004) Primate hearing from a mammalian perspective. Anat Rec 281A:1111–1122
Hickman GC (1979) The mammalian tail: a review of functions. Mammal Rev 9:143–157
Hiramatsu C, Melin AD, Aureli F, Schaffner CM, Vorobyev M, Matsumoto Y, Kawamura S (2008) Importance of achromatic contrast in short-range fruit foraging of primates. PLoS One 3:e3356
Hirano ZMB, Tramonte R, Silva ARM, Braga RR, dos Santos WF (2003) Morphology of epidermal glands responsible for the release of colored secretions in *Alouatta guariba clamitans*. Lab Primate Newslett 42:4–7
Hirano ZMB, Correa IC, de Oliveira DAG (2008) Contexts of rubbing behavior in *Alouatta guariba clamitans*: a scent-marking role? Am J Primatol 70:575–583
Hiwatashi T, Okabe Y, Tsutsui T, Hiramatsu C, Melin AD, Oota H, Schaffner CM, Aureli F, Fedigan LM, Innan H, Kawamura S (2010) An explicit signature of balancing selection for color-vision variation in New World monkeys. Mol Biol Evol 27:453–464
Hladik CM, Simmen B (1996) Taste perception and feeding behavior in nonhuman primates and human populations. Evol Anthropol 2:58–71
Hofer H (1977) Taste areas of the Plica sublingualis of *Alouatta* and *Aotus* (Primates, Platyrrhini). Cell Tissue Res 177:415–429
Hofer H, Meinel W, Rommel C (1979) Taste buds in the epithelium of the Plica sublingualis of New World monkeys. Anat Anz 145:17–31
Horwich RH (1983) Breeding behaviors in the black howler monkey (*Alouatta pigra*) of Belize. Primates 24:222–230
Hunt DM, Dulai KS, Cowing JA, Julliot C, Mollon JD, Bowmaker JK, Li W-H, Hewett-Emmett D (1998) Molecular evolution of trichromacy in primates. Vision Res 38:3299–3306
Jacobs GH, Deegan JF (2005) Polymorphic New World monkeys with more than three M/L cone types. J Opt Soc Am A 22:2072–2080
Jacobs GH, Neitz M, Deegan JF, Neitz J (1996) Trichromatic colour vision in New World monkeys. Nature 382:156–158
Joganic JL, Perry GH, Cunningham AJ, Dominy NJ, Verrelli BC (2009) Molecular evolution of color vision genes in howling monkeys (*Alouatta*). Am J Phys Anthropol 48(suppl):159–160
Jones CB (1979) Grooming in the mantled howler monkey, *Alouatta palliata*. Primates 20:289–292
Jones CB (2002) How important are urinary signals in *Alouatta*? Lab Primate Newslett 41:15–17
Jones CB (2003) Urine-washing behaviors as condition-dependent signals of quality by adult mantled howler monkeys (*Alouatta palliata*). Lab Primate Newslett 42:12–14
Jones CB, van Cantfort TE (2007) Multimodal communication by male mantled howler monkeys (*Alouatta palliata*) in sexual contexts: a descriptive analysis. Folia Primatol 78:166–185
Kaas JH (2004) Evolution of somatosensory and motor cortex in primates. Anat Rec 281A:1148–1156
Kainz PM, Neitz J, Neitz M (1998) Recent evolution of uniform trichromacy in a New World monkey. Vision Res 38:3315–3320
Kelemen G, Sade J (1960) The vocal organ of the howling monkey (*Alouatta palliata*). J Morphol 107:123–140
Kitchen DM (2004) Alpha male black howler monkeys responses to loud calls: effect of numeric odds, male companion behaviour and reproductive investment. Anim Behav 67:125–139
Kitchen DM (2006) Experimental test of female black howler monkey (*Alouatta pigra*) responses to loud calls from potentially infanticidal males: effects of numeric odds, vulnerable offspring, and companion behavior. Am J Phys Anthropol 131:73–83

Kitchen DM, Horwich RH, James RA (2004) Subordinate male black howler monkey (*Alouatta pigra*) responses to loud calls: experimental evidence for the effects of intra-group male relationships and age. Behaviour 141:703–723

Laska M, Hernández-Salazar LT (2004) Gustatory responsiveness to monosodium glutamate and sodium chloride in four species of nonhuman primates. J Exp Zool 301A:898–905

Laska M, Tutsch M (2000) Laterality of tail resting posture in three species of New World primates. Neuropsychologia 38:1040–1046

Laska M, Scheuber HP, Hernández-Salazar LT, Rodriguez-Luna E (2003) Sour-taste tolerance in four species of nonhuman primates. J Chem Ecol 29:2637–2649

Laska M, Rivas-Bautista RM, Hernández-Salazar LT (2009) Gustatory responsiveness to six bitter tastants in three species of nonhuman primates. J Chem Ecol 35:560–571

Lucas PW, Darvell BW, Lee PKD, Yuen TDB, Choong MF (1998) Colour cues for leaf food selection by long-tailed macaques (*Macaca fascicularis*) with a new suggestion for the evolution of trichromatic colour vision. Folia Primatol 69:139–154

Lucas PW, Dominy NJ, Riba-Hernandez P, Stoner KE, Yamashita N, Lorí-Calderön E, Petersen-Pereira W, Rojas-Durán Y, Salas-Pena R, Solis-Madrigal S, Osorio D, Darvell WB (2003) Evolution and function of routine trichromatic vision in primates. Evolution 57:2636–2643

Machida H, Giacometti L (1967) The anatomical and histochemical properties of the skin of the external genitalia of the primates. Folia Primatol 6:48–69

Machida H, Giacometti L (1968) The skin. In: Malinow MR (ed) Biology of the howler monkey. Karger, New York, pp 126–140

Machida H, Perkins E, Giacometti L (1967) The anatomical and histochemical properties of the tongue of primates. Folia Primatol 5:264–279

Matsui A, Go Y, Niimura Y (2010) Degeneration of olfactory receptor gene repertoires in primates: no direct link to full trichromatic vision. Mol Biol Evol 27:1192–1200

Matsushita Y, Oota H, Welker BJ, Pavelka MS, Kawamura S (2013) Color vision variation as evidenced by hybrid L/M opsin genes in wild populations of trichromatic Alouatta New World monkeys. Int J Primatol. doi:10.1007/s10764-013-9705-9

McComb K, Semple S (2005) Coevolution of vocal communication and sociality in primates. Biol Lett 1:381–385

Melin AD, Fedigan LM, Hiramatsu C, Sendall CL, Kawamura S (2007) Effects of colour vision phenotype on insect capture by a free-ranging population of white-faced capuchins, *Cebus capucinus*. Anim Behav 73:205–214

Mendel F (1976) Postural and locomotor behavior of *Alouatta palliata* on various substrates. Folia Primatol 26:36–53

Milton K (1975) Urine-rubbing behavior in the mantled howler monkey, *Alouatta palliata*. Folia Primatol 23:105–112

Milton K (1979) Factors influencing leaf choice by howler monkeys: a test of some hypotheses of food selection by generalist herbivores. Am Nat 114:362–378

Milton K (1980) The foraging strategy of howler monkeys. Columbia University Press, New York

Osorio D, Smith AC, Vorobyev M, Buchanan-Smith HM (2004) Detection of fruit and the selection of primate visual pigments for color vision. Am Nat 164:696–708

Paterson JD (1981) Postural-positional thermoregulatory behavior and ecological factors in primates. Can Rev Phys Anthropol 3:3–11

Perkins EM (1975) Phylogenetic significance of the skin of New World monkeys (order Primates, infraorder Platyrrhini). Am J Phys Anthropol 42:395–424

Piazza S, Boscarol G, Falcone I, Fiore R, Gamba M, Giacoma C, Martinoli L, Pizzigalli C (2004) Acoustic features of loud calls in a group of *Alouatta palliata mexicana*: individual differences and connections with age. Folia Primatol 75:405–406

Pinto L, Setz E (2004) Diet of *Alouatta belzebul discolor* in an Amazonian rain forest of Northern Mato Grosso State, Brazil. Int J Primatol 25:1197–1211

Pubols LM (1968) Somatic sensory representation in the thalamic ventrobasal complex of the spider monkey (*Ateles*). Brain Behav Evol 1:305–323

Pubols BH, Pubols LM (1969) Forelimb, hindlimb, and tail dermatomes in the spider monkey (*Ateles*). Brain Behav Evol 2:132–159

Pubols BH, Pubols LM (1971) Somatotopic organization of spider monkey somatic sensory cortex. J Comp Neurol 141:63–76

Raguet-Schofield M (2010) The role of dietary toughness in the ontogeny of Nicaraguan mantled howler monkeys (*Alouatta palliata*) feeding behavior. Am J Phys Anthropol 141:193

Ramsier MA, Glander KE, Finneran JJ, Cunningham AJ, Dominy NJ (2009) Hearing, howling, and Hollywood: auditory sensitivity in *Alouatta palliata* is attuned to high-frequency infant distress calls. Am J Primatol 71:69

Ramsier M, Cunningham AJ, Patiño M, Villanea FA, Spoor F, Demes B, Larson S, Glander KE, Talebi M, Dominy NJ (2012) Hearing sensitivity and the evolution of acoustic communication in platyrrhine monkeys. Am J Phys Anthropol 147[suppl]:243

Regan BC, Julliot C, Simmen B, Viénot F, Charles-Dominique P, Mollon JD (1998) Frugivory and colour vision in *Alouatta seniculus*, a trichromatic platyrrhine monkey. Vision Res 38:3321–3327

Regan BC, Julliot C, Simmen B, Viénot F, Charles-Dominique P, Mollon JD (2001) Fruits, foliage and the evolution of primate colour vision. Philos Trans R Soc Lond B 356:229–283

Richter CP, Campbell KH (1940) Taste thresholds and taste preferences of rats for five common sugars. J Nutr 20:31–46

Rolls ET (2007) Sensory processing in the brain related to the control of food intake. Proc Nutr Soc 66:96–112

Rommel C (1981) Die sublingualen Strukturen der Primaten. Teil 1: Prosimiae, Platyrrhini und Cercophitecinae. Gegenbaurs Morph Jb 127:153–175

Sanchez-Villagra MR, Pope TR, Salas V (1998) Relation of intergroup variation in allogrooming to group social structure and ectoparasite loads in red howlers (*Alouatta seniculus*). Int J Primatol 19:473–491

Schmitt D, Larson SG (1995) Heel contact as a function of substrate type and speed in primates. Am J Phys Anthropol 96:39–50

Schön MA (1970) On the mechanism of modulating the volume of the voice in howling monkeys. Acta Oto Laryngol 70:443–447

Schön MA (1971) The anatomy of the resonating mechanism in howling monkeys. Folia Primatol 15:117–132

Schön Ybarra MA (1986) Loud calls of adult male red howling monkeys (*Alouatta seniculus*). Folia Primatol 47:204–216

Schön Ybarra MA (1988) Morphological adaptation for loud phonations in the vocal organ of howling monkeys. Primate Rep 22:19–24

Scott TR, Plata-Salaman CR (1999) Taste in the monkey cortex. Physiol Behav 67:489–511

Sekulic R (1982a) Daily and seasonal patterns of roaring and spacing in four red howler (*Alouatta seniculus*) troops. Folia Primatol 39:22–48

Sekulic R (1982b) The function of howling in red howler monkeys (*Alouatta seniculus*). Behavior 81:38–54

Sekulic R (1983) The effect of female call on male howling in red howler monkeys (*Alouatta seniculus*). Int J Primatol 4:291–305

Sekulic R, Chivers JR (1986) The significance of call duration in howler monkeys. Int J Primatol 7:183–190

Sekulic R, Eisenberg JF (1983) Throat-rubbing in red howler monkeys (*Alouatta seniculus*). In: Müller-Schwarze D, Silverstein M (eds) Chemical signals in vertebrates III. Plenum, New York, pp 347–350

Silveira LCL, Saito CA, Da Silva Filho M, Bowmaker JK, Kremers J, Lee BB (2007) Physiological properties of photoreceptors and retinal ganglion cells from a trichromatic platyrrhine: the howler monkey, *Alouatta* sp. In: Silveira LCL, Ventura DF, Lee BB (eds) 19th Symposium of the international colour vision society abstracts book EDUFPA, Belém, Brazil, pp 69–70

Simmen B, Sabatier D (1996) Diets of some French Guianan primates: food composition and food choices. Int J Primatol 17:661–693

Smith TD, Rossie JB (2006) Primate olfaction: anatomy and evolution. In: Brewer W, Castle D, Pantelis C (eds) Olfaction and the brain: window to the mind. Cambridge University Press, Cambridge, pp 135–166

Smith TD, Bhatnagar KP, Shimp KL, Kinzinger JH, Bonar CJ, Burrows AM, Mooney MP, Siegel MI (2002) Histological definition of the vomeronasal organ in human and chimpanzees, with a comparison to other primates. Anat Rec 267:166–176

Smith AC, Buchanan-Smith HM, Surridge AK, Osorio D, Mundy NI (2003) The effect of colour vision status on the detection and selection of fruits by tamarins (*Saguinus* spp.). J Exp Biol 206:3159–3165

Smith AC, Surridge AK, Prescott MJ, Osorio D, Mundy NI, Buchanan-Smith HM (2012) Effect of colour vision status on insect prey capture efficiency of captive and wild tamarins (*Saguinus* spp.). Anim Behav 83:479–486

Stephan H, Baron G, Frahm HD (1982) Comparison of brain structure volumes in insectivora and primates. II. Accessory olfactory bulb (AOB). J Hirnforsch 23:575–591

Stephan H, Baron G, Frahm HD (1988) Comparative size of brains and brain components. In: Steklis H, Erwin J (eds) Comparative primate biology, vol 4. Alan R. Liss, New York

Stoner KE, Riba-Hernández P, Lucas PW (2005) Comparative use of color vision for frugivory by sympatric species of platyrrhines. Am J Primatol 67:399–409

Sumner P, Mollon JD (2000) Catarrhine photopigments are optimized for detecting targets against a foliage background. J Exp Biol 203:1963–1986

Talebi MG, Pope TR, Vogel ER, Neitz M, Dominy NJ (2006) Polymorphism of visual pigment genes in the muriqui (Primates, Atelidae). Mol Ecol 15:551–558

Teaford MF, Lucas PW, Ungar PS, Glander KE (2006) Mechanical defenses in leaves eaten by Costa Rican howling monkeys (*Alouatta palliata*). Am J Phys Anthropol 129:99–104

Urbani B (2002) A field observation on color selection by new world sympatric primates, *Pithecia pithecia* and *Alouatta seniculus*. Primates 43:95–101

Van Belle S, Bicca-Marques JC (2014) Insights into Reproductive Strategies and Sexual Selection in Howler Monkeys. In: Kowalewski M, Garber P, Cortes-Ortiz L, Urbani B, Youlatos D (eds) Howler monkeys: Behavior, ecology and conservation. Springer, New York

Van Belle S, Estrada A, Ziegler TE, Strier KB (2009) Sexual behavior across ovarian cycles in wild black howler monkeys (*Alouatta pigra*): male mate guarding and female mate choice. Am J Primatol 71:153–164

Vogel ER, Neitz M, Dominy NJ (2007) Effect of color vision phenotype on the foraging of wild white-faced capuchins, *Cebus capucinus*. Behav Ecol 18:292–297

Whitehead JM (1987) Vocally mediated reciprocity between neighbouring groups of mantled howling monkeys, *Alouatta palliata palliata*. Anim Behav 35:1615–1627

Whitehead JM (1989) The effect of the location of a simulated intruder on responses to long-distance vocalizations of mantled howling monkeys, *Alouatta palliata*. Behaviour 108: 73–103

Whitehead JM (1994) Acoustic correlates of internal states in free-ranging primates: the example of the mantled howling monkey, *Alouatta palliata*. In: Thierry B, Roeder JJ, Anderson JR, Herrenschmidt N (eds) Current primatology, vol 2. University Louis Pasteur, Strasbourg, pp 221–226

Whitehead JM (1996) Vox Alouattinae: a preliminary survey of the acoustic characteristics of long-distance calls of howling monkeys. Int J Primatol 16:121–144

Youlatos D, Gasc JP (2011) Gait and kinematics of arboreal quadrupedal walk of free-ranging red howler in French Guiana. In: D'Août K, Vereecke EE (eds) Primate locomotion. Springer, New York

Youlatos D, Kowalewski M, Garber PA, Cortés-Ortiz L (2014) New challenges in the study of howler monkey anatomy, physiology, sensory ecology, and evolution: where we are and where we need to go? In: Kowalewski MM, Garber PA, Cortés-Ortiz L, Urbani B, Youlatos D (eds) Howler monkeys: adaptive radiation, systematics, and morphology. Springer, New York

Young OP (1982) Tree-rubbing behavior of a solitary male howler monkey. Primates 23:303–306

Chapter 13
Production of Loud and Quiet Calls in Howler Monkeys

Rogério Grassetto Teixeira da Cunha, Dilmar Alberto Gonçalves de Oliveira, Ingrid Holzmann, and Dawn M. Kitchen

Abstract One of the most striking features of howler monkeys' natural history is their loud call, which gives the genus *Alouatta* its common name in English. However, the disproportionate focus on functional aspects of those calls has driven attention away from other relevant issues related to their vocal behavior. In this chapter, we review the studies of acoustic structure conducted so far on these peculiar calls, highlighting the variation among and within the species of this genus. The variation we uncover runs against the notion of uniformity among howler monkeys, but we do find that the relationship between loud call structure and phylogeny compliments genetic work in this genus. We also show how the anatomy of howler monkey's vocal organs can explain the unusual features of their loud calls and possibly the variation found between species, while also pointing to the various gaps that exist in our knowledge regarding the role of the several components of their highly specialized vocal apparatus. Additionally, we review some basic concepts about sound propagation and geographic variation in long-distance communication. Unlike loud calls, we know relatively little about the low-amplitude calls of howler monkeys.

R.G.T. da Cunha (✉)
Instituto de Ciências da Natureza, Universidade Federal de Alfenas,
Alfenas-MG 37130-000, Brazil
e-mail: rogcunha@hotmail.com

D.A.G. de Oliveira
Departamento de Fauna/CBRN, Centro de Manejo de Fauna Silvestre,
Secretaria do Meio Ambiente do Estado de São Paulo, São Paulo, Brazil
e-mail: dilmar_oliveira@yahoo.com.br

I. Holzmann
CONICET (Consejo Nacional de Investigaciones Científicas y Técnicas), IBS (Instituto de Biología Subtropical), CeIBA (Centro de Investigaciones del Bosque Atlántico), Buenos Aires, Argentina
e-mail: holzmanningrid@yahoo.com.ar

D.M. Kitchen
Department of Anthropology, The Ohio State University, Columbus, OH 43210, USA

Department of Anthropology, The Ohio State University-Mansfield, Mansfield, OH 44906, USA
e-mail: kitchen.79@osu.edu

Such calls have received a great deal of attention in the literature, particularly in Old World monkeys, because they can offer insights into the social lives of these animals. Because few comparable studies have been conducted on howler monkeys, we propose some lines of future research that we deemed potentially interesting. We conclude with some methodological approaches to recording howler monkey calls in the field and for sharing vocalizations with other researchers.

Resumen Una de las características más llamativas de la historia natural de los monos aulladores son sus vocalizaciones de larga distancia, las cuales son responsables del nombre popular en inglés, y algunos nombres en español, para el género *Alouatta*. Sin embargo, el enfoque desproporcionado que ha recibido la funcionalidad de estas vocalizaciones, ha desviado la atención de otros aspectos relevantes del comportamiento vocal de los monos aulladores. En este capítulo revisamos los estudios llevados a cabo hasta el momento, sobre la estructura acústica de estas peculiares voces, remarcando la variación de las mismas entre y dentro de las diferentes especies del género. Las variaciones que aquí dejamos al descubierto desafían la noción de uniformidad en estos primates y muestran una relación entre la estructura vocal y las relaciones filogenéticas que complementan estudios genéticos recientes realizados en este género. También mostramos cómo la anatomía de los órganos vocales de los monos aulladores puede explicar tanto las características inusuales de sus vocalizaciones de larga distancia, como posiblemente la variación entre las diferentes especies, y señalamos los vacíos existentes en el conocimiento acerca del papel que poseen diversos componentes –altamente especializados- de los aparatos vocales de estos primates. Adicionalmente, revisamos conceptos básicos sobre la propagación del sonido y la variación geográfica en la comunicación a grandes distancias. Sonidos de baja amplitud producidos en otros grupos taxonómicos, particularmente en monos del Viejo Mundo han recibido gran atención en la literatura debido a que ofrecen una mirada interna a la vida social de estos animales. Debido a que pocos estudios comparables se han llevado a cabo en monos aulladores, proponemos algunas futuras investigaciones que consideramos potencialmente interesantes. Finalmente concluimos con aproximaciones metodológicas para grabar voces de monos aulladores en el campo y para compartir las grabaciones obtenidas con otros investigadores.

Keywords Structure of loud calls • Morphology of vocal apparatus • Call design • Sound propagation • Geographic variation • Soft calls • Recording methods

Abbreviations

dB	Decibels
e.g.,	For example
Hz	Hertz
i.e.,	In other words

kHz	Kilohertz
m	Meters
min	Minutes
ms	Milliseconds
pers. comm.	Personal communication
pers. obs.	Personal observation
s	Seconds
SPL	Sound pressure level
unpubl. data	Unpublished data

13.1 Introduction

When it comes to loud calling, howler monkeys do not stand alone in the primate world (Mitani and Stuht 1998). In fact, even in the broader mammalian world, many species produce loud calls—lions (*Panthera leo* (McComb et al. 1994)), wolves (*Canis lupus* (Mech 1966)), and African elephants (*Loxodonta africana* (Leighty et al. 2008)), to name just a few. On the other hand, howler monkeys do stand out in this noisy crowd. They utter the most powerful primate vocalization in the Neotropics, probably rivaled only by jaguars (*Panthera onca*) and bellbirds (*Procnias* spp.). In fact, if we consider both call duration and amplitude per body size, then competitors, even worldwide, lag far behind. Such a striking feature of their natural history gives the genus *Alouatta* its common name in several languages.

As one would expect, a modified and specialized anatomy of the vocal apparatus, with the most noteworthy component being the greatly enlarged hyoid bone (Schön 1971; Schön Ybarra 1988), is associated with the production of these calls. It has even been suggested that this anatomical commitment might affect other aspects of the howler monkeys' lives, such as positional behavior (Schön Ybarra 1984). Such an anatomical suite of characters, coupled with the time and presumably the energy invested in loud calling, contrasting with their otherwise phlegmatic lifestyle suggests an important role for these calls in the lives of howler monkeys. As in other species, howler monkey loud calls probably play a vital role in fitness in that they are involved in intergroup competition, mate attraction, or defense, and predator avoidance. Such functional aspects of the loud calls will be dealt with in Kitchen et al. (2014, this volume), where the issue will be analyzed at different explanatory levels.

In this current chapter, we consider loud calls from a proximate, structural perspective, including acoustic features, the specialized anatomy of the vocal apparatus, long-range propagation issues, and a consideration of geographical variation on call structure. We also give attention to the neglected female loud calls and the quieter/soft calls in the repertoire. We conclude with a brief rough guide to howler monkey vocal research, in which we address various methodological issues.

In both this chapter and in Kitchen et al. (2014), we attempt to critically review studies conducted on these peculiar vocalizations in order to highlight the variation present among the different howler monkey species. In addition to varying body mass, degree of sexual dimorphism, coat color variation, and other aspects of their behavior and ecology (see chapters throughout these volumes), we argue that the structure and putative functions of howler monkey loud calls may also vary widely across different *Alouatta* species.

13.2 Structure of Male Loud Calls

Since Carpenter's (1934) work with *A. palliata*, most authors have described two main categories of howler monkey loud calls: barks and roars. The presence of both has been confirmed for every species of howler monkey studied so far. Altmann (1959), followed by Baldwin and Baldwin (1976), named the male forms of these calls the A series (roars) and C series (barks or woofs). Categorizing these calls in series stresses the high degree of variation found in each type of howler monkey loud call, reflecting high levels of gradation from "incipient" (low amplitude: Baldwin and Baldwin 1976) to very loud emissions of each form (Neville et al. 1988; Drubbel and Gautier 1993; Oliveira 2002). These low-frequency, harsh/atonal sounds also have a common structure, with marked peaks of amplitude at stable frequency bands (Schön Ybarra 1986; Drubbel and Gautier 1993; Whitehead 1995; Oliveira 2002; da Cunha 2004). Female forms of these vocalizations were labeled the B (roars or "roar accompaniments") and D (barks) series, and they are clearly distinct from the male forms (Baldwin and Baldwin 1976). Because most studies have focused on male calling behavior, there are scarce bioacoustic analyses of female repertoires (see Sect. 13.3).

We believe it is important to stress at the onset that in addition to diversity among all species, we will reveal clear distinctions in roar types used and temporal patterns of loud calling between two groups: the two Central American species, *A. pigra* and *A. palliata*, compared to the remaining South American species. Given that Cortés-Ortiz and colleagues (2003) found closer phylogenetic relationships within than between these two clades, the acoustic and structural trends we describe below do not conflict with this taxonomic hypothesis.

13.2.1 Incipient Forms of Roaring and Barking

Because they can be only perceived at short range, the "incipient" forms of barks and roars cannot officially be viewed as loud calls. However, they usually have a clear structural relationship with the louder forms and are often emitted during loud calling bouts. The "incipient roar" (Figs. 13.1a, e) is made up of very short pulses ("strings of short subunits" (Drubbel and Gautier 1993); "a gruff, popping" noise

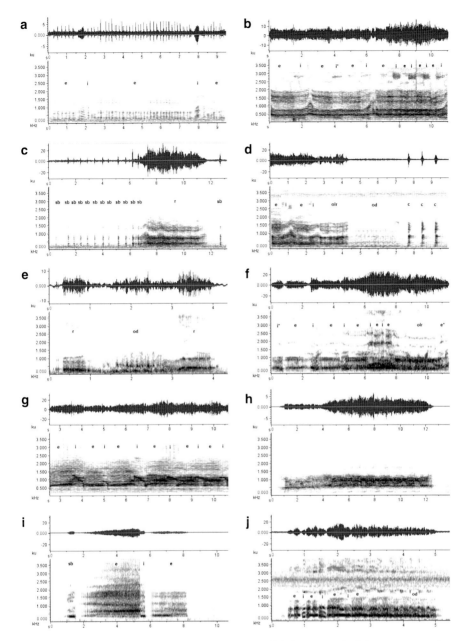

Fig. 13.1 Roars. (**a**) *A. guariba*, incipient roars with two inhalatory sounds (i) between successive exhalations (e); (**b**) *A. guariba*, loud roars, with inhalatory (i) and exhalatory (e) phases and faster respiratory cycles at the climax in amplitude (i*: an apparent inhalation, but without frequency modulation); (**c**) *A. guariba*, brief roar (r), preceded and followed by single-pulsed barks (sb); (**d**) *A. guariba*, roar ending, with two normal cycles indicated by their inhalatory and exhalatory phases, followed by an oodle-like roar (olr), an oodle (od), and three coughs (c); (**e**) *A. caraya*, incipient roars, with two emissions (r) and an oodle (od) between them; (**f**) *A. caraya*, loud roars, with inhalatory (i) and exhalatory (e) phases and faster respiratory cycles at the climax in amplitude, followed by oodle-like roaring (olr) (i*, an apparent inhalation; e*, an apparent exhalation); (**g**) *A. belzebul*, roar, with exhalatory (e) and inhalatory (i) phases; (**h**) *A. belzebul*, brief roar; (**i**) *A. pigra*, a roar with exhalation (e) and inhalation (i), preceded by a single-pulsed bark (sb) and followed by a faint exhalation; (**j**) *A. palliata*, single roar, with inhalatory (i) and exhalatory (e) phases and ending in an oodle (od)

(Altmann 1959)). At least for some South American species (*A. caraya* (da Cunha RGT unpubl. data); *A. guariba*[1] and *A. belzebul* (Oliveira 2002); *A. macconnelli* (formerly *A. seniculus* (Drubbel and Gautier 1993)), a series of incipient roars (e.g., this phase lasts 24–114 s in *A. macconnelli* (Drubbel and Gautier 1993)) often precede full roaring bouts, as a kind of warming-up phase where pulses gradually become louder and uttered at shorter intervals in the transition to "full roars."

Although Baldwin and Baldwin (1976) describe incipient roars in *A. palliata*, these calls are apparently not as frequently produced by the Central American species (*A. palliata* and *A. pigra*) and are instead heard as a short burst of popping or an "aw" or "er" sound (Baldwin and Baldwin 1976) typically at the beginning of roars (Kitchen DM unpubl. data). Schön Ybarra (1986) also gave a similar description of this use of incipient roars at the onset of "brief roars" (defined below) in *A. arctoidea* (formerly *A. seniculus*).

A. palliata (Baldwin and Baldwin 1976) and *A. guariba* (Oliveira 2002) have also been described as producing "incipient barks" (Fig. 13.2a)—short-range, simple pulses usually emitted with a closed mouth. Similar muffled sounds have been observed in *A. caraya*, both before barking and on their own (da Cunha pers. obs.). In the Central American species, Baldwin and Baldwin (1976) describe this in *A. palliata* as a muffled "unf unf unf" sound, and these calls often occur before the onset of loud calling in *A. pigra* (also referred to as "grunting" (Kitchen pers. obs.)).

13.2.2 Full Roars

Common features of howler monkey roars or "howls" are their high amplitude (up to 90 dB sound pressure level (SPL) at 5 m of distance (Whitehead 1995)), low frequency, and harshness. In the South American species analyzed so far, full roars are composed of two sections: a longer exhalatory phase and a shorter inhalatory one, with higher frequencies of the dominant band occurring in the inhaling periods (Whitehead 1995). For example, the pattern in *A. guariba* (Fig. 13.1b) is that short inhalatory sounds, varying in structure from tonal with low fundamental frequency (90–150 Hz) to harsh and usually with an ascendant modulation, can occur intercalated with incipient roars (Oliveira 2002). These inhalations acquire a harsh structure and merge with the exhalatory pulses (derived from the popping incipient roar but with an ascending modulation of the lower dominant band) to produce full roars (Oliveira 2002). During a roaring bout, these respiratory cycles become faster and louder until they reach a climax in amplitude (Oliveira 2002). This alternating pattern allows most South American howler monkey species to utter continuous emissions of roars lasting up to several minutes. For example, *A. caraya* long roars

---

[1] All studies on *A. guariba* (formerly *A. fusca*) loud calls are restricted to the southern subspecies, *A. g. clamitans*. The authors found no reference to studies on the more restricted and lesser known northern subspecies, *A. g. guariba*.

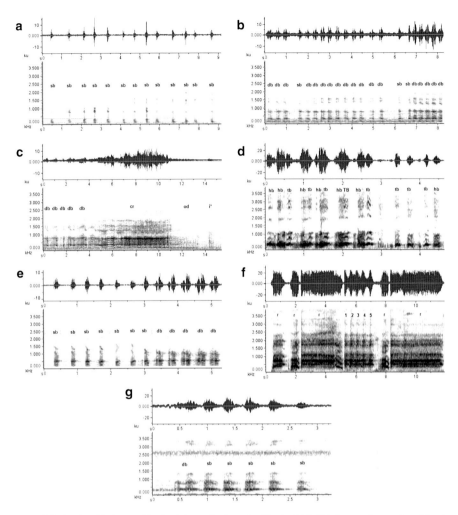

Fig. 13.2 Barks. (**a**) *A. guariba*, incipient, single-pulsed barks (sb); (**b**) *A. guariba*, series of loud barks, including both single-pulsed (sb) and double-pulsed (db) calls; (**c**) *A. guariba*, five longer barks, followed by a composite roar (cr) and an oodle (od) (i*: sigh-like sound, perhaps an inhalatory sound); (**d**) *A. caraya*, multiple callers, notice more tonal voice (tb) in one caller (probably a female), while remaining barks are typically harsh calls (hb); (**e**) *A. belzebul*, single-pulsed (sb) and double-pulsed (db) barks; (**f**) *A. pigra*, five barks (1–5) or a five-pulsed single bark, interspersed with roars (r); (**g**) *A. palliata*, double-pulsed bark (db) followed by single-pulsed barks (sb)

(Fig. 13.1f: da Cunha 2004) last up to 1 min 43 s (Whitehead 1995), in fact much more, da Cunha pers. obs.) and the long roars of *A. macconnelli* have a median duration of 3 min 28 s (range, 1–10 min (Drubbel and Gautier 1993)). The respiratory cycles of *A. belzebul* roars are marked by a higher degree of frequency modulation than found in other species (Fig. 13.1g) and have the longest periods of uninterrupted calling, lasting up to 12 min (Oliveira 2002).

South American species that produce these long, continuous roars also sometimes emit short-duration "brief roars" (*A. caraya*: *A. belzebul* (Oliveira 2002; da Cunha unpubl. data; Fig. 13.1h); *A. guariba* (Oliveira 2002, Fig. 13.1c); *A. macconnelli* (Drubbel and Gautier 1993)). Oliveira (2002) found a range of 2–18 s for brief roars in *A. belzebul* and 2–8 s in *A. guariba*. Some studies discuss only brief forms of roaring in *A. arctoidea* (up to 8 s (Schön Ybarra 1986); median value of 19 s (Sekulic and Chivers 1986)), but Drubbel and Gautier (1993) confirm the presence of both brief ("short calls," average duration of 11 s; range 1–40 s) and continuous ("long calls" more than 60 s) forms of roaring in *A. macconnelli*. Bouts consisting only of brief roars can last up to 20 min (*A. guariba* (Oliveira D unpubl. data); *A. macconnelli* (Drubbel and Gautier 1993)).

The Central American species, *A. palliata* (Sekulic and Chivers 1986; Whitehead 1995; Fig. 13.1j) and *A. pigra*, are the exception in that they emit only brief roars lasting a few seconds. While Whitehead (1995) describes *A. pigra* as a species capable of continuous roaring, they actually produce clear pauses between consecutive "brief" roars (Kitchen unpubl. data). The false impression by Whitehead is likely because this species can emit loud calls in quick succession and their loud calling bouts overall last much longer than in the southern species.

Although superficially different, the roars of *A. pigra* are similar to *A. palliata*, except the syllables are much longer, and far fewer syllables are produced per roar in the former species (Kitchen DM, Bergman TJ, Cortés-Ortiz L unpubl. data). Individual roars by *A. pigra* consist of a single long exhalatory emission (lasting 2.2 s on average: Kitchen 2000), sometimes preceded by a short inhalation, followed by a shorter low-amplitude inhalatory sound (Fig. 13.1i and 13.2f). For *A. palliata*, Baldwin and Baldwin (1976) describe solo male roars as a series of 1–4 respiratory cycles ("exhaled separated by shorter inhaled syllables"), while roars emitted in choruses, the most frequent form, are usually longer and more variable with 2–14 cycles per emission. Whitehead (1987, 1989) describes the roars of *A. palliata* in a similar way, with the typical roar consisting of a legato series of cycles ("notes"), increasing in duration and intensity, followed by a single note (probably a single exhalation phase) with maximum duration and intensity, and ending usually with a diminuendo of progressively shorter notes (similar to an "oodle"; see below). Sekulic and Chivers (1986) report that the series of notes that make up a roar in *A. palliata* lasts an average of 3.5 s.

We want to emphasize one important message in this section so far: brief roars are rare in several South American species, whereas they are the only roars produced by the Central American species. However, despite the difference in how frequently they are produced, there may be overall similarities in the brief roars in the two clades. For example, the description of the brief roars of *A. arctoidea* provided by Schön Ybarra (1986) is similar to what Whitehead (1987, 1989) characterized as normal roars in *A. palliata*—a crescendo, followed by a climax and a short, low-intensity coda (diminuendo). Perhaps these brief roars are the ancestral form of roaring (Oliveira and Ades 2004) given that they most closely resemble the loud vocalizations found in other primate species in terms of the duration of elements

Table 13.1 Dominant frequencies for the exhaling phase of howler monkey roars

Species	Frequency (Hz) First band	Second band	Third band[a]	Source
A. guariba	300–400	450–600	700–1,000	1 (chorus)[b]
	556			1 (solo)[c]
	300–450	650–800		2
A. belzebul	504	612		1
	740			2[d]
A. caraya	302	498		1
	200–450	600–1,000		3
A. pigra	408	694		1
A. palliata	420–480	700–840		1
	327	646		4
A. arctoidea	450–500	900	1,900–2,000	5
	555	690		1

Sources: (1) Whitehead 1995; (2) Oliveira 2002; (3) da Cunha 2004; (4) Eisenberg 1976; (5) Schön Ybarra 1986
[a]A third peak of amplitude is not described for most studies
[b]Range values for recordings of adult male choruses
[c]Average value of the lower dominant frequency (solo emissions)
[d]A second peak of amplitude was not found in this study

(colobus monkeys (Teichroeb and Sicotte 2010); Mentawai macaques, langurs, leaf monkeys, and gibbons (Schneider et al. 2008); gibbons (Geissmann 2002); guenons (Gautier 1989)).

Besides differences in structural pattern, there is also some interspecific variation in the acoustic structure of the howls. In Table 13.1, the analysis is focused on the exhaling phase of roars, since the inhaling phase is usually shorter and modulated in variable patterns (Drubbel and Gautier 1993; Whitehead 1995; Oliveira 2002), making its description less precise. Specifically, although all species have their lowest emphasized bands in the 200–700 Hz range, *A. caraya* and *A. palliata* produce some of the lowest peak frequencies. *Alouatta belzebul* and *A. pigra* produce some of the highest peak frequencies, and they are the only two species who have a peak that is higher in frequency than their second most emphasized frequency (Table 13.1; Whitehead 1995). Whitehead (1995) also found that the peak frequencies for *A. belzebul* and *A. pigra* were higher than the second most emphasized frequency during the inhalation phase, whereas the reverse pattern was seen in the other species (see also Drubbel and Gautier 1993; Oliveira 2002)—the sole exception was *A. palliata*, which always emphasized their lowest frequency band in both phases. In the power spectra of roars of all species studied to date, there is a confounding factor caused by the lack of distinction between more precise amplitude peaks (dominant frequencies) and wider bands ("frequency clusters": Drubbel and Gautier 1993). The wider bands (usually two) cover hundreds of Hz and contain one or more amplitude peaks each.

The most striking pattern was found in the roars of *A. belzebul* of the Atlantic rainforest of northeastern Brazil (Oliveira 2002), whose high-pitched roars presented a single dominant peak with the widest variation observed among howler monkeys: values ranging from 550 to 1,100 Hz (average values: 740 Hz exhaling phase; 920 Hz inhaling phase). However, data for the same species in Brazilian Amazon region show the typical pattern of other South American species, with two dominant peaks per phase (Table 13.1; average values inhaling phase: 732 and 823 Hz (Whitehead 1995)). The fact that both *A. belzebul* populations are regarded as the same subspecies (Cortés-Ortiz et al. 2003) makes this difference in pattern intriguing.

13.2.3 Barks

As in roars, barks or "woofs" (Baldwin and Baldwin 1976; Neville et al. 1988) have a large degree of gradation, ranging from shorter, single pulses of low amplitude (incipient forms, Fig. 13.2a) to double pulses of increasing duration, rate, and amplitude (Fig. 13.2b–g). In *A. guariba* (Oliveira 2002; Fig. 13.2b, c), the duration of double-pulsed barks ranges from 100 to 800 ms, and sonograms available from other species fall within this range (Baldwin and Baldwin 1976; Schön Ybarra 1986; da Cunha 2004). However, even the shortest double-pulsed barks have a longer duration (>100 ms) than the single pulses of incipient roars (usually <70 ms (Schön Ybarra 1986; Oliveira 2002)).

There are of course variations on this pattern. For example, Schön Ybarra (1986) describes the presence of triple pulses of barks for *A. arctoidea*. In *A. pigra*, barks can have multiple pulses in a sonographically continuous emission that, however, sounds like distinct emissions given that the amplitude variation is observed, with the lower-amplitude periods being quiet enough to be possibly misconstrued as silent "breaks" (Kitchen unpubl. data; Fig. 13.2f). In *A. caraya*, da Cunha (2004) describes male barks as having double or single pulses of a similar frequency structure to that found in roars (Fig. 13.2d) and that bouts of barking by dominant males usually include a roar climax-like vocalization, similar to the "composite roars" in *A. guariba* (described below). Regardless of the nature of the pulses themselves, the usual emission pattern is one made up of a long to a very long string of pulses.

Although only scarcely described acoustically, available data on barks shows that their dominant frequencies are similar to those found in the roars of the same species (Eisenberg 1976; Baldwin and Baldwin 1976; Schön Ybarra 1986; Whitehead 1995; Oliveira 2002). The barks of Central American howler monkeys, however, have even greater structural resemblance to their roars (*A. pigra* (Kitchen 2000); *A. palliata* (Baldwin and Baldwin 1976)) than those in South American species, perhaps reflecting a lower degree of functional divergence between call types (see also Sect. 13.2.5).

Since the calls are akin in their frequency spectra and high amplitude, barking is likely generated through similar processes as roaring (see Sect. 13.4). However, barks are not produced continuously as roaring can be in South American species, rendering the complex respiratory maneuvers found in sustained roaring unnecessary. Schön Ybarra (1986) noticed that most *A. arctoidea* barks appeared to be uttered in exhalation, but the incorporation of inhalation phases could explain the merging of longer barks, which coalesce into the loud, composite roars described below (Oliveira 2002).

13.2.4 Oodles and Roar Variants

There are many loud calls in howler monkey repertoires that do not seem to fall exactly, or sometimes at all, into either graded series of barks or roars. For example, Drubbel and Gautier (1993) describe "oodles" in *A. macconnelli* as "blowing sounds," occurring as short-range sounds after the coda (ending phase) of a long-lasting roaring period. In *A. caraya*, oodle calls seem to be unvoiced (not generated by vibrating vocal folds or other anatomical structures, such as in whisper), since they have a muffled nature (da Cunha RGT unpubl. data; see also Sect. 13.4). They are heard at the end of sessions or before a brief pause that is followed by the resumption of the continuous roaring session (Fig. 13.1e). In *A. guariba*, oodles are found in the ending of long, continuous roars or in pauses between them (Oliveira 2002; Fig. 13.1d). Additionally, *A. guariba* produces a loud, roar-like call with an oodle quality at its ending, typically emerging as a fusion of very loud and long barks, usually heard in intense barking bouts (Oliveira 2002; Fig. 13.2c). We will refer to these calls as "composite roars," as their characteristics are intermediary between regular roars, barks, and oodles. This call also resembles a brief roar but has faster cycles that sound muffled during the ending phase, just like the oodles that usually follow them.

Central American species also produce similar oodles (Kitchen unpubl. data; Fig. 13.1j) during pauses between roars. Baldwin and Baldwin (1976) discuss a "roar terminus" in *A. palliata* as a series of fast cycles of usually declining pitch that frequently occurs at the end of normal roars, sometimes grading into oodles. This is likely similar to the harsher and louder form of oodle that is often described as occurring at the end of a roar in *A. palliata* (Altmann 1959; Whitehead 1987, 1989) and *A. pigra* (Kitchen 2000) and is part of the complex gradation found in the loud call repertoire of these two species. Another example of this complexity is the "roar variant" in *A. palliata*, characterized by a start as a sudden intense note that is followed by a trailing off, without the oodle-like ending (Whitehead 1987, 1989).

13.2.5 Pattern of Loud Calling Bouts

Among the South American species we have been able to analyze, there seems to be a distinction between roaring and barking bouts. Pauses (defined as <1 min by Oliveira 2002) followed by either a gradual or a sudden return to full, continuous roaring can occur in roaring bouts, but otherwise (not including the "warm up phase") males emit full roars the entire time (*A. caraya* (da Cunha unpubl. data); *A. guariba* (Oliveira 2002); *A. macconnelli* (Drubbel and Gautier 1993)). In contrast to these fairly ritualized roaring bouts, the barking bouts of *A. guariba* (one of the few South American species where this data is available) are more variable, with frequent diminuendos and crescendos in amplitude, duration, and rate of bark pulses (Oliveira 2002). However, the scarce evidence found in the literature indicates that the barking bouts of other South American species may be more stable, with uniformity in the bark pulses emitted, at least during some periods (*A. belzebul* (Oliveira 2002); *A. arctoidea* (Schön Ybarra 1986)). Barking bouts can also have a much longer duration than a roaring bout and, although a composite roar or some kind of roar-like call can sometimes constitute a climax of amplitude in these bouts, barking bouts typically do not contain full or brief roars. For example, in *A. caraya*, a barking bout can be sustained for around 40 min, and during some periods the calls are stable, interspersed with something like roar climaxes, and then going back through diminuendo/crescendo phases (da Cunha unpubl. data).

Although Central American species produce some bouts with only barks, roaring bouts always include at least some barks and variants of both roar and bark vocalizations (*A. pigra* (Kitchen 2000); *A. palliata* (Baldwin and Baldwin 1976)). As we said above, this mixed pattern may be occasionally observed but is apparently not typical of any South American species (Schön Ybarra 1986; Oliveira 2002). Thus, the patterns of these mixed roar/bark bouts of Central American howler monkeys are much more variable than the stereotyped roaring bouts of the South American species.

A few trends are common to both *A. pigra* and *A. palliata*—bouts are often preceded by a quieter build-up phase (e.g., incipient barks/grunts) followed by "loud calling periods" (defined by Kitchen (2000) as including any loud calls and short "breaks" of <1 s). Roars become less frequent and pauses (<1 min as defined by Kitchen (2000)) between loud calling periods get longer toward the end of a bout. Additionally, loud calling periods/roars occur at a faster rate in bouts when another group is nearby. Entire bouts (including loud calling periods and silent periods) can last over an hour in both species (Kitchen DM, Bergman TJ, Cortés-Ortiz L unpubl. data).

13.2.6 Male Loud Calls: Concluding Remarks

There is wide variation in acoustic properties of calls, the temporal patterning of calling bouts, and the nature and duration of such bouts in the howler monkey species studied so far. Perhaps the clearest trend is a division between Central and South American species in that features of their loud calls parallel the two identified phylogenetic clades of the genus *Alouatta* (Cortés-Ortiz et al. 2003; Villalobos et al. 2004). Both *A. palliata* and *A. pigra* produce only simple, short-duration roars (a few seconds each), their barks are essentially just shorter syllables of their species-typical roars, and both barks and roars usually occur in the same bout of loud calling. However, although the individual vocalizations are shorter than in South American species, they are produced during bouts that last much longer than in the southern species, with pauses between calls. On the other hand, barks and roars are much more easily distinguished in South American species and the two call types are not typically combined in the same bout. These species produce roar vocalizations in both brief and long-lasting forms, and bouts of the latter consist of continuous emissions (up to several minutes) of inhalatory and exhalatory phases. Such respiratory cycles can also be noticed on roars of Central American species, but the inhalatory phase has a much lower amplitude compared to the South American species and may not play a role in long-distance communication (Kitchen DM pers. obs.).

We found that one major difficulty in making comparisons across species is due to the fact that authors vary widely both in how they define call types and in what is considered a "bout." Some researchers define a bout from a functional perspective; that is, sessions close in time but apparently related to the same triggering stimulus are considered part of the same bout. Others choose some arbitrary period of silence as the criteria to define a new bout. Therefore, a determinant future step in the study of howler monkey vocalizations is to unify criteria and establish a nomenclature valid for all species based on clear and objective criteria. We believe this review is a first step in that direction.

Despite decades of research on howler monkeys, their most salient vocal feature – loud calling – remains undescribed in some species and awaits more detailed acoustic data for almost all species. For example, although the calls of *A. belzebul* and of some species of the *A. seniculus* group (*A. arctoidea* and *A. macconnelli*) have been described, those taxa have wide distributions with several discrete populations (Cortés-Ortiz et al. 2003, 2014; Gregorin 2006; Rylands and Mittermeier 2009). Given some of the distinctiveness among populations (e.g., the populations of *A. belzebul* described above), the study of the vocal repertoire of these taxa, as well as the study of hybrid vocalizations (see Sect. 13.5), may shed light on their taxonomic relationships.

13.3 The Structural Features of Female Loud Calls

We have dealt so far with an issue we believe is crucial in understanding howler monkeys' loud calls: variation. Another source of variation, a quite neglected one in fact, lies between the sexes. Howler monkeys are fairly unusual among nonmonogamous primates given that both males and females produce loud calls. Actually, it might be more accurate to say that females often utter a moan-like call, albeit a call that is clearly related to male roars in structural terms ("roar accompaniment": Baldwin and Baldwin 1976). Females can "roar," together with the alpha male, or they can remain silent. Furthermore, male and female calls are commonly emitted (but not always) at the same time during group sessions. This duet-like pattern is normally found in monogamous species that jointly defend a border, such as the titi monkeys (*Callicebus moloch* (Robinson 1979)) and the hylobatids (Geissmann 2002), but it is otherwise rare in nonhuman primates.

The paucity of studies dealing with either structural or functional aspects of female calls probably relates to the difficulty in isolating female calls—they are much lower in amplitude than male calls and are nearly always masked by the overlapping sounds of males during a chorus. In fact, because of the extent of vocal overlap in some species, many authors that have worked with howler monkeys are unable to differentiate among any of the participants in a given chorus.

Still, the structure of female "roars" has been described for *A. palliata* (Baldwin and Baldwin 1976), *A. arctoidea* (Sekulic 1982), and *A. guariba* (Oliveira 2002). Female roars and barks are generally higher pitched than male loud calls (Baldwin and Baldwin 1976; Eisenberg 1976; Sekulic 1982). This is not surprising, given the sexual dimorphism in body size and hyoid volume in howler monkeys (Hershkovitz 1949; Gregorin 2006). Female roars are also reported to be more intense when uttered in roar choruses as an accompaniment to male roars (Baldwin and Baldwin 1976; Sekulic 1982; Oliveira 2002). Analyzing isolated emissions of *A. guariba* female roars, Oliveira (2002) demonstrated variation from tonal to harsh structure, with intense and irregularly oscillating frequency modulation in the tonal sections of these vocalizations (Fig. 13.3). However, besides from the previous example, spectrograms of female calls are absent or of medium or poor quality in the literature.

Female barks are usually simple pulses of lower intensity than male barks in *A. guariba* (Oliveira 2002), although female *A. caraya* sometimes produce more intense forms with greater frequency modulation than male barks (da Cunha pers. obs.). Incipient barks (simple pulses usually emitted with closed mouth) are described for *A. palliata* (Baldwin and Baldwin 1976) and *A. guariba* (Oliveira 2002) and are frequently produced by females and juveniles.

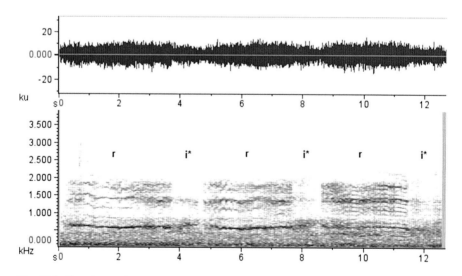

Fig. 13.3 Female roar in *A. guariba*: three successive emissions (r), with apparent inhalatory phases (i*) obscured by a noisy background

We need better recordings and descriptions of female calls before we can further advance the study of their structure. A possible solution to circumvent the drawbacks of their softer calls that are obscured during choruses could be the use of a small microphone attached to a collar, so as to capture the sound more directly. However, despite the lack of data, one can still speculate about structural issues of female vocalizations. For example, given the existence of interspecific differences in hyoid size and shape (Gregorin 2006), one could predict there will be variation in the structure of female calls similar to that observed in males, especially in the formant frequencies (see Sect. 13.2). Of particular interest would be to investigate if differences among species in female vocalizations merely mirror interspecific male differences or if female differences follow a different pattern. In the first case, females' hyoids may simply be species-typical but smaller versions of the male ones, and their calls might accordingly be simply softer and higher-pitched versions of the male calls, with more widely spaced formant frequencies. However, given the many socioecological and behavioral differences between males and females, we predict that female interspecific differences in vocalizations may not simply mirror those of males but may follow a distinct pattern. For example, the differences between males of two species could reflect the fact that one species has stronger intrasexual selection than another, whereas interspecies differences between females could instead reflect the fact that there is infanticide risk in one species but not in the other and females might be either quieter or more aggressive when facing such a risk. These questions remain open for further studies.

13.4 Morphology and Vocal Production

Although the peculiar anatomical features of the howler monkey's vocal apparatus clearly shape their unusual sounds, the phonation mechanisms underlying these calls are complex and have been poorly studied. The hyoid bone is a large, inflated, and hollow structure (the "hyoid bulla"), accommodated within the large, expanded mandibula (Fig. 13.4a, b) and positioned below the tongue (Hershkovitz 1949; Schön 1970; Fig. 13.4c). A pair of lateral air sacs borders the bulla (Kelemen and Sade 1960; Schön 1970). The "tentorium" is a subchamber of the hyoid bulla formed by a folding at the upper border of the hyoid opening (Fig. 13.4b–d). This structure is absent in *A. palliata*, rudimentary in *A. caraya*, and variably developed and shaped in the remaining South American species (Hershkovitz 1949). The most developed tentorium is present in the *A. seniculus* group, in which individuals have large hyoids and inflated tentorium chambers containing bony lateral partitions or trabeculae (Hershkovitz 1949; Gregorin 2006).

Kelemen and Sade (1960) attributed the loudness of howler monkey calls to the presence of rigid cavities formed by the hyoid bulla (Fig. 13.4c) and nonrigid lateral air sacs (Fig. 14.3c shows lateral aperture probably leading to an air sac in *A. guariba*). Since then, most phonation studies address the role of the hyoid as a Helmholtz resonator, amplifying the glottal source (Schön Ybarra 1986, 1988; Riede et al. 2008; de Boer 2009).

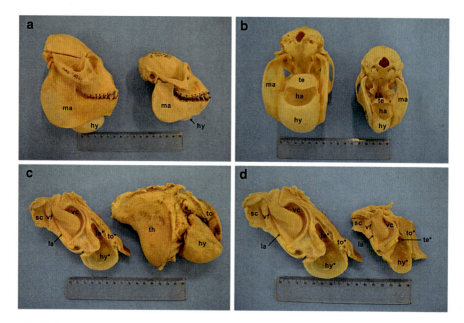

Fig. 13.4 Vocal anatomy of *A. guariba*. (**a**) lateral view of adult male (*left*) and adult female (*right*) skulls, both showing an enlarged mandible (ma) that houses an inflated hyoid bulla (hy) but also remarkable sexual dimorphism; (**b**) same structures in ventral view, notice the hyoid aperture (ha) and the upper tentorium (te) subchamber; (**c**) longitudinal view of adult male vocal apparatus, the inside view (*left*) shows the subglottal chamber (sc), a large vocal fold (vf), the lateral aperture (la) that probably leads to a lateral air sac (not confirmed for the species), the contorted supraglottal vocal tract (vc), the hyoid chamber (hy*), with the tentorium subchamber (te*) and the sectioned tongue (to*), while the outside view (*right*) shows the large thyroid cartilage (th), hyoid bulla, and tongue (to) after removal of layers of muscle and connective tissue; (**d**) the inside view of the same adult male vocal apparatus (*left*) compared to the same structure from an adult female (*right*). The ruler in the images shows scale in centimeters. All photos by Júlio César de Souza Júnior

The large glottis (Fig. 13.4c, showing an enlarged vocal fold) can produce loud, low-frequency sounds that are further amplified by the resonators (hyoid, air sacs) and the constrictions in the post-glottal structures (Fig. 13.4c depicts narrow and curved supraglottal vocal tract), features that reduce the velocity of the air flow, elevating its pressure and, consequently, raising its volume (Schön Ybarra 1988, 1995). Recent modeling studies (Riede et al. 2008; de Boer 2009) have also indicated that the hyoid is largely responsible for the low frequency of the first formant in howler monkey vocalizations and allows a greater efficiency in the generation of loud sounds.

The sound produced at the larynx encounters a contorted pathway before reaching the mouth, given the enlargement of several structures (hyoid, cartilages, vocal folds—Fig. 13.4c, enlarged subglottic chamber). Forced air passage would also result in the generation of irregular, noisy vibrations—at least partially responsible for the harshness found in roars and barks (Schön Ybarra 1986, 1995). Whitehead (1995) suggested that the acoustic features of howler monkey loud calls were derived both from hyoid involvement and sub- and supraglottal maneuvers.

As an example, he mentions the generation of the broadband (noisy) bursts of loud calling by an increase in subglottal pressure and a coupling of the extra-laryngeal structures (hyoid bulla, lateral air sacs) with the supraglottal air tract, leading to wide frequency fluctuations. We have found no mention in the literature to subglottal mechanisms, but a possible way to generate a sound so high at the laryngeal source is the production of large abdominal pressures. Anecdotally, one of us (RGTC) observed the eversion of tissue in the anal region during the exhalatory phase of *A. caraya* roars, a likely indication of extremely high abdominal pressure.

Kelemen and Sade (1960) argued that the rigidity of the laryngeal organ, containing large ossified cartilages (see thyroid cartilage in Fig. 13.4c) restricted the modulatory capacity in howler monkeys when compared to human and ape larynges. However, Schön Ybarra (1986, 1988) argued that howler monkeys could show some vocal plasticity through changes in the width and length of the mouth chamber and that even the hyoid position could be changed by the action of some muscles (Schön 1964). As another form of modulation, Riede and colleagues (2008) suggested that the hyoid creates interactions between the vocal cords and the vocal tract that could explain the dynamic changes usually found in roar pitch.

Few studies have focused on the role of the elastic, inflatable air sacs (Kelemen and Sade 1960; Schön Ybarra 1988). Drubbel and Gautier (1993) interpreted the oodles ("blowing sounds"), usually occurring at the end or pauses of continuous roaring in South American species, as a product of the emptying of air sacs. "Coughs" are also reported at these times (Schön Ybarra 1986) and may be a kind of choking sound caused by swallowing saliva (Oliveira D pers. obs.), which often dribbles from an individual's mouth during the bouts (Schön Ybarra 1986). Whether these phenomena occur in the Central American species is unknown.

In sum, our present knowledge of the mechanisms underlying loud call production in howler monkeys is still very limited. Although recent modeling approaches are promising (Riede et al. 2008; de Boer 2009), conceiving a way of examining phonation in living animals would be valuable as it would allow closer investigation of the dynamic processes involved in call modulation. Additionally, the few studies on morphology and phonation published to date have centered on just *A. palliata* (Kelemen and Sade 1960) and on the *A. seniculus* group (Schön 1970, 1971; Schön Ybarra 1988). The high degree of interspecific variation found in the morphology of the vocal apparatus and the structure, duration, and temporal patterning of calls highlights the need to investigate vocal production in other species.

The large variation in hyoid size and shape among different howler monkey species has implications for systematic arrangements (Hershkovitz 1949; Gregorin 2006) and since Ihering (1914) has been used as a taxonomic character (e.g., Lönnberg 1941). Hershkovitz (1949) regarded the smaller hyoid found in *A. palliata* as an ancestral state, from which the larger and complex hyoids found in other *Alouatta* species diverged; however, genetic evidence now places *A. palliata* as part of a clade with the other Central American species, *A. pigra*, being no longer considered basal for the genus (Cortés-Ortiz et al. 2003). *Alouatta palliata* hyoids are also less sexually dimorphic than other species (Hershkovitz 1949; Gregorin 2006; Fig. 13.4b, d), including *A. pigra* (Cortés-Ortiz L. pers. comm.).

Sekulic and Chivers (1986) proposed that loud calls in *A. palliata* are shorter in duration than in *A. arctoidea* because of the presence of a smaller hyoid with a smaller air reservoir. However, this explanation is unlikely given the hyoid's rigid structure and the fact that roars, produced during the whole respiratory cycle, do not need an air reservoir. Although both Central American species, particularly *A. palliata*, have shorter, simpler roars than other howler monkeys, there are traits unique to only *A. palliata* such as the absence of significant energy above 1,000 Hz (other species have spectral energy to 2,000 Hz: Whitehead 1995). Although Thorington and colleagues (1984) suggested that large hyoid sizes meant lower frequencies, this hypothesis has not been supported (e.g., *A. palliata* produce low-frequency calls as compared to *A. belzebul*, a species with a large hyoid that produces some of the highest frequency calls in the genus (Gregorin 2006)). Thus, it remains unclear whether and how the atypical calls produced by *A. palliata* are linked to their distinctive hyoid morphology.

13.5 Sound Propagation and Geographic Variation

Howler monkeys, like many other primate species, produce loud calls to communicate over long distances. The acoustic structure of any sound can be altered and degraded as it travels, due to physical phenomena such as attenuation (intensity of acoustic signals generally decreases 6 dB each time the distance from the source is doubled, due to factors such as atmospheric absorption and sound scattering: Wiley and Richards 1978; Brenowitz 1982) and reverberation (when sound is reflected and scattered by stationary objects during propagation: Naguib and Wiley 2001). However, sounds with most of their energy concentrated at low frequencies (a common feature of primate loud calls) are less degradable by attenuation than are higher-frequency sounds (frequency-dependent attenuation (Waser and Waser 1977; Mitani and Stuht 1998; Naguib and Wiley 2001)). One exception is that the ground may cause relatively large attenuation effects, particularly in low frequencies (<1 kHz), but this effect becomes negligible above 1 m from the ground (Roberts et al. 1977; Mitani and Stuht 1998; Nelson 2003; Maciej et al. 2011).

How the different types of habitat influence the sound over distances (e.g., due to vegetation absorption and/or reverberation) is debated and the evidence is mixed (Date and Lemon 1993; Naguib 1996; Daniel and Blumstein 1998; Blumenrath and Dabelsteen 2004; Schneider et al. 2008). Contrary to intuitive expectations, sound is less scattered and travel farther distances (at almost every frequency) in closed than in open habitats (Wiley and Richards 1982; Waser and Brown 1986). In contrast, reverberation is stronger in closed habitats and constrains long-range communication (Waser and Brown 1986). However, calls of certain frequencies, given from particular heights and/or at specific times of the day, can transmit over long distances in closed habitats almost free of attenuation (Morton 1975; Marten et al. 1977; Waser and Brown 1986; Brown and Handford 2000). A sound window (frequency range that attenuates less and propagates farther in a given habitat

(Morton 1975; Waser and Brown 1986)) of between 100 and 400 Hz exists in rainforests for sounds produced around 7–8 m above the ground. Howler monkey roars, with their high amplitudes, relatively low emphasized frequencies (between 300 and 1,000 Hz, well within the forest sound window), and harshness (noisy, atonal sound structure), are among the primate vocalizations capable of propagating the greatest distances (at least 1 km (Baldwin and Baldwin 1976; Schön Ybarra 1986; Whitehead 1989; Whitehead 1987; Whitehead 1995)). However, future howler monkey research might focus on how acoustic degradation in different habitats and under different conditions affects their vocalizations.

Many forest primate species seem to concentrate their long-distance calls around dawn, and howler monkeys are no exception (Sekulic 1982; Whitehead 1995; da Cunha and Byrne 2006). A commonly cited reason is that sound propagation is better during this "time window" (Gautier and Gautier 1977; Horwich and Gebhard 1983; Waser and Brown 1986; Brown and Handford 2000; Cornick and Markowitz 2002), despite increased background noise (Wiley and Richards 1982). However, there may also be other proximate explanations for such a temporal pattern; for example, research on birds found that calling at daybreak allowed animals to avoid heat stress (Ricklefs and Hainsworth 1968; see also Sekulic 1982). On the other hand, *A. pigra* (Horwich and Gebhard 1983; Cornick and Markowitz 2002) and perhaps other species (e.g., *A. arctoidea*: Braza et al. 1981) seem to have a bimodal pattern with a secondary peak at afternoon/sunset and with reductions at midday. Sekulic (1982) also reported a reduction in the midday calling activity in *A. arctoidea* in Venezuela, possibly the time of the day with the worst environmental conditions for sound propagation (Wiley and Richards 1982). Conversely, Drubbel and Gautier (1983) reported that *A. macconnelli* in Guyana frequently produce long roaring choruses at night (also heard frequently in *A. pigra* (Kitchen pers. obs.)), when temperature gradients are favorable and wind turbulence is scarce, helping sound propagation (Wiley and Richards 1982). A fourth pattern is a notable absence of a dawn chorus in *A. guariba* at several sites (Chiarello 1995; Oliveira 2002; Steinmetz 2005; da Cunha and Jalles-Filho 2007; Holzmann et al. 2012). Whether the lack of a dawn chorus in this species, or the lack of a secondary afternoon peak in species with a dawn chorus, is the result of varying environmental conditions or other factors, such as population densities, requires further investigation.

To evaluate how well howler monkey long-distance calls are adapted to local conditions, both in their structure and in their timing, it will be necessary to explore geographic variation between populations of the same species. Because different degradation processes act differently in diverse types of habitats or under different conditions, we might expect interpopulation variation due to selective pressures such as (1) vegetation structure of local environment (e.g., closed vs. open habitats (Wiley and Richards 1978)), (2) social factors such as population density (Delgado 2006), and (3) other environmental sound characteristics mostly based on local biota and local conditions (like wind and rain) that provoke sound interference (Martin 1981; Brenowitz 1982; Sorjonen 1986; Waser and Brown 1986; de la Torre and Snowdon 2002). Given that many *Alouatta* species concentrate calling at dawn and dusk, a noisy time in tropical forests, the frequency window is likely the most important mechanism to cope with interspecific acoustic competition, although this possibility has yet to be tested.

Rather than arising due to selection for particular call features, geographic variation in vocalizations could also arise indirectly due to differences between populations in anatomy (e.g., body size: Bowman 1979), genetics (as a result of reproductive isolation between populations of the same species: Wich et al. 2008; Thinh et al. 2011), or flexible adjustments to local conditions (e.g., increasing amplitude in a noisy habitat: Lombard 1911). A howler monkey species with a wide distribution range, present in different types of habitats (e.g., *A. caraya, A. arctoidea*, or *A. palliata*), would be an ideal model to test these different hypotheses related to geographical variation in long-distance calls.

Since howler monkey roars have been proposed to function in intergroup spacing, judging the distance from a caller can be very important (Whitehead 1987, 1989). Out of a set of sound degradation phenomena that potentially provide receivers with distance information, reverberation is the only one that might apply to howler monkeys, given the characteristics of their calls and habitats (following Wiley and Richards 1978). By manipulating this parameter in a series of playback experiments, Whitehead (1987) demonstrated that howler monkeys were able to perceive approaches and withdrawals based on barks alone (see Sect. 13.6). However, Naguib and Wiley (2001) proposed that longer barks could simulate the reverberation of shorter pulses, providing the basis for potential deceptive communication of distance in howler monkeys. To date, no one has explored a possible test between these somewhat opposing hypotheses about honesty and deception.

In summary, there are many interesting questions that remain unexplored in relation to sound propagation in howler monkeys. For example, little has been done to explore inter- and intraspecific variation in long-distance calls based on aspects things such as habitat differences. One potential confounding effect in such studies is that structural variation can also occur within a population based on individual variation. More studies should focus on uncovering the existence of individual variation between same sex individuals. For example, based on spectrographic analysis of roars, researchers found evidence for individuality in the acoustic features from two different populations of *A. pigra* in Belize (Bocian et al. 1999; Kitchen 2000). Additionally, we have not yet scratched the surface in understanding the ultimate and proximate factors governing the timing features of howler monkey calls. Of particular interest would be to investigate deviations from the most common timing patterns in parallel with the function of loud calls. For example, is the absence of dawn chorus in some species (such as *A. guariba*) related to functional, habitat, or call structure differences? Are there intraspecific differences in timing? If so, what causes them?

13.6 Going Soft: The Neglected Calls

Howler monkeys prodigious loud calls, as impressive and theoretically interesting as they are, have a downside. They have drawn attention away from the rich repertoire of more subtle calls. Yes, howler monkeys can and do call quietly. What is more, they have a broad repertoire of such calls, and some species are actually highly vocal in this category (*A. caraya* (da Cunha pers. obs., Holzmann 2012)). In this section, we will review the available work conducted on this topic and point to some lines of research we believe could be particularly fruitful. In our review, we mainly discuss studies that focused either on the entire repertoire or just soft calls. An attempt to survey all published works to uncover sources where soft calls were mentioned en passant was not feasible.

In Table 13.2, we summarize the scarce information on soft calls. The few classic published studies that have dealt with low-amplitude calls are restricted in scope, mainly descriptive, conducted only on *A. palliata* and *A. caraya* (but see *A. guariba* (Holzmann 2012)), with no or poor spectrograms and with functional interpretations that are not solidly grounded (Carpenter 1934; Altmann 1959; Baldwin and Baldwin 1976; Calegaro-Marques and Bicca-Marques 1997). These limitations impose serious restrictions on comparative work.

Although a number of these calls might provide interesting research projects, we chose to discuss three categories of soft calls whose study in howler monkeys we believe could be particularly fruitful. These are some of the most commonly produced call types. They have been discussed in at least some previous literature, and they pose interesting theoretical issues of potentially broader relevance: contact calls, immature calls, and alarm calls.

13.6.1 Contact Calls

In primates, one of the most ubiquitous categories of calls is that used to promote or retain spatial cohesion, particularly when group members become spread out or separated (see da Cunha and Byrne (2009) for a review on Neotropical primates). A variety of specific functions have been proposed for these calls, commonly labeled as contact, isolation, or "lost" calls: maintaining contact at close, visual range (Epple 1968; Pook 1977) or at longer ranges in situations likely to lead to separation, such as rapid travel or dispersed foraging, regaining contact (Daschbach et al. 1981; Byrne 1981; Palombit 1992; Harcourt et al. 1993; Halloy and Kleiman 1994), monitoring the position of others (Caine and Stevens 1990), initiating and directing or coordinating group travel (Boinski 1991, 1993), and attracting others in particular circumstances (Dittus 1988; Mitani and Nishida 1993). Before proceeding, a cautionary note: although conventional, terms like "contact call" and "alarm call" are functional labels and, as such, not adequate until appropriate studies have been conducted (Martin and Bateson 2007).

In the case of howler monkeys, contact calls have only been partially studied in a few species. Given their ubiquity, it is surprising that references to these calls are so scant in the literature (see also Kitchen et al. 2014, regarding loud contact calls), even more so if we exclude those calls performed by infants when separated from their mothers (more properly included within the subsection on immature calls

Table 13.2 Descriptions of low-amplitude calls in two species of howler monkeys

Species	Call type	Age and sex of caller	Context/situation of use	Function/other remarks
A. palliata[1]	Deep metallic cluck	Adult male	Before and during group travel	Initiation and control of group travel; coordination of animals in the clan
A. palliata[1]	Gurgling grunts and crackling sounds	Adult male	"Mildly disturbing and 'apprehensive' situation"	Preparation and coordination of defensive action by males
A. palliata[1]	Wail ending with a grunt and a groan	Adult females with infants	When a young fell on the ground	Aid in location and retrieval of fallen infants
A. palliata[1]	Cries	Infants	When they fell on the ground; facing large gaps	Location and retrieval of fallen infants
A. palliata[1]	Purr	Infants/juveniles	"Contact with mother"	Facilitation of mother-infant relationship
A. palliata[1]	Little chirping squeal	Infants/juveniles	During invitation to play	Facilitation and coordination of play activity
A. palliata[1]	Grunting sound	Adult male	When immatures are play-fighting and an individual is crying	Control of activity of young by an adult male
A. palliata[1]	Grunting similar to who! who![1]	Males	Presence of strange or new stimuli	Directing behavior of the group. Altmann (1959) considers this call and male barks to be the same
A. palliata[2]	Metallic cackling notes	Adult females	"Aggressive interaction among females"	
A. palliata[3]	Eh	Infants		"A locator signal and perhaps an indication that the infant is close to the mother and not in distress"
A. palliata[4]			When exploring the surroundings and when reapproaching mother	In addition to the above, also involved in eliciting embracing. Suggested to reflect mild alarm/anxiety
A. palliata[3]	Whimper	Immatures		"in frustrating situations" (e.g., prevention from suckling and prevention from making a crossing)
A. palliata[4]	Basic *whimper* calls (1), male emphatic *whimpers* (2), and chirping *whimper* (3)[3]	(1) All classes; (2) adult males; (3) juveniles and adult females	During group progression (1, 2, 3); when spotting a troop with which there was no mixing (1, 2, 3); juveniles "when mildly startled, surprised, threatened or pestered" (1); in active play; in adult greeting; and during mother-infant interactions (after falls, when infant was lost)	

13 Production of Loud and Quiet Calls in Howler Monkeys

A. palliata[3]	Grunt	Adult male	During group travel	May be the same call as the "deep metallic cluck" of Carpenter
A. palliata[3]	Cackle laugh			"Unknown"; It may be the same as the metallic cackling notes mentioned by Collias and Southwick (1952)
A. palliata[4]	Cackle laugh or heh	All	During agonistic interactions, particularly the threatened animal	
A. palliata[4]	Caw	Infant	Separation from the mother and when the mother refuses contact	
A. palliata[4]	Wrah-has	Mothers of dependent infants	When they were astray	
A. palliata[4]	Yelps/screeches	Infants, juveniles, adult females	"Sudden fright situations"; when "a large animal startled, lunged at or swatted at"	
A. palliata[4]	Hiccup	Adult males	During pauses in roar sessions	
A. palliata[4]	Squeaks	Infants	Mild distress situations (mother leaving and when attempting to leave rough play)	
A. palliata[4]	Barks	Infants	Rough play, when startled, when access to the nipple was hindered	
A. palliata[5]	Whine	Adult female	"Correlated with contact situations"	
A. palliata[5]	Growl	Adult male	"Correlated with aggressive arousal," also during grappling (sort of play behavior)	
A. palliata[5]	Shriek hiss	Adult female	"in aggressive encounters with the male"	
A. palliata[5]	E-uh-uh (chuck)	Adult female	"during grappling"	
A. palliata[5]	Ó-eh ó-eh	Unclear	"often given by the male when alert and in low stages of autonomic arousal"	"It can serve as a group warning signal"
A. palliata[5]	Soft chirp	Adults	"in ambivalent contexts"	
A. palliata[6]	Low-, medium-, and high-intensity guttural barks	Presumably adults	Used in displacement	

(continued)

Table 13.2 (continued)

Species	Call type	Age and sex of caller	Context/situation of use	Function/other remarks
A. palliata[6]	Guttural barks, intention howls, growls, paedomorphic whines	Adults	In sexual contexts (presentation or postcopulatory)	
A. palliata[6]	Paedomorphic vocalization	Adults	Used in displacement	
A. palliata[7]	Throat rumbling and clucking	Adult males	Ritualized greeting	
A. caraya[8]	Stress screams	Young and subadult females and infants	During aggressive interactions, play, and attempts of allomaternal care of newborns	
A. caraya[8]	Snarl	All except juvenile male	Stress contexts, play	
A. caraya[8]	Cry	Infants/juveniles	During stress situations (e.g., separation from group)	
A. caraya[8]	Quack	Subadult male	Aggressive behaviors of the adult male and a subadult female that occurred during and after copulation attempts by the subadult	
A. caraya[8]	Mutter	All	During feeding and attempts of allomaternal care	
A. caraya[8]	Lullaby	Female infant		
A. caraya[8]	Cough	Adult male	Precopulatory behavior	

[1]Carpenter (1934); [2]Collias and Southwick (1952); [3]Altmann (1959); [4]Baldwin and Baldwin (1976); [5]Eisenberg (1976); [6]Jones (1980, 2007a, b); [7]Wang and Milton (2003); [8]Calegaro-Marques and Bicca-Marques (1995)—this work was published in Portuguese, so the names of the calls and the information were translated by one of us (RGTC)

below). The best examples we found included a report that *A. palliata* individuals emit whimpers in a variety of situations, including group progressions (Baldwin and Baldwin 1976). Also, in their brief report on *A. caraya* calls, Calegaro-Marques and Bicca-Marques (1997) mention a vocalization ("cry") emitted in stressful situations including circumstances in which the caller was away from the group. Finally, based on a 19-month fieldwork study on the vocal behavior of a wild *A. caraya* group, da Cunha and Byrne (2013) suggested that a low-amplitude vocalization, the "moo" call, served a contact function. Based on ad libitum and anecdotal information, these authors (Byrne 2000; da Cunha and Byrne 2009) propose that "moo" calls among *A. caraya* individuals represent a genuine call-and-answer system, albeit one based on first-order intentionality (i.e., without comprehension of others' mental states (Dennett 1978)). The hypothesis that "moos" are produced in antiphonal exchanges awaits rigorous testing (e.g., baboons (Cheney et al. 1996)). Besides, we call the attention that primate contact calls are ideal for studying intentionality in animal communication.

Thus, it is clear that there is a fundamental need for detailed repertoire studies, as the foundation of more advanced studies. Just with regard to contact calls, there are many interesting basic questions to focus on, for example, do other howler monkey species produce contact calls? Are contact calls structurally similar between different howler monkey species? Are there acoustic differences between contact calls produced in slightly different contexts (e.g., by isolated animals vs. those maintaining regular contact during minor spread)?

13.6.2 Immature Calls

Another ubiquitous kind of primate vocalization category is those calls emitted by infants and juveniles in stressful or care-related situations, usually labeled as "distress calls," "cries," "tantrum calls," or just "infant calls" (see Newman (1995) for a review). Once again, information on immature howler monkey calls is scarce and concentrated mostly on *A. palliata* (see Table 13.2). However, in *A. caraya*, it was possible to identify a group of structurally related calls that perform some role related to infant distress situations (da Cunha 2004; Holzmann 2012). Similar calls were reported in *A. guariba* infants (Holzmann 2012; Oliveira unpubl. data). Nevertheless, such calls are so variable and graded that it is difficult to categorize them in a precise way.

13.6.3 Alarm Calls

Notwithstanding the undeniable importance of alarm calls in the primate bioacoustics literature, evidence for such calls in howler monkeys is even scarcer than for the two previous types of vocalizations. As seen in Table 13.2, several quiet vocalizations are produced in a variety of alert or alarm situations; however, no call types have been rigorously described, so once again more recordings in a range of contexts will be necessary in order to uncover consistent patterns. One promising example among the quiet calls is the low-amplitude "incipient barks" that Oliveira (2002) reported were emitted, usually by *A. guariba* females and juveniles, in mild-alarm contexts induced by the close proximity of a human observer. This author reports that these calls are also frequently emitted by females during group choruses of loud barks, possibly functioning to incite male barking, and this might be viewed as a similar context to alarm, given that they are potentially stressful situations.

In Kitchen et al. (2014), we also address the possibility that howler monkeys use loud calls as alarms. Regardless of whether, from the signaler's perspective, quiet or loud calls produced in such contexts are affective responses to stress, referential, or both, these calls may function to alert receivers about danger (e.g., Seyfarth and Cheney 2003; but see Owren et al. 2010), and there may even be different calls for different predators/situations, as is true in other primate species (Seyfarth et al. 1980; Zuberbuhler 2000, 2001; Arnold and Zuberbuhler 2004; Casar et al. 2012). Playback experiments will ultimately be necessary to test these questions.

13.7 The Rough Guide to Recording and Sharing Vocalizations

Regardless of interest levels for researching the soft calls, howler monkey loud calls will certainly keep attracting abundant attention from scientists. Not only are these calls fascinating because they are peculiar in their production and stand out in the jungle soundtrack but also because their functional significance remains unsettled. Thus, we felt we could contribute to the advancement of research on this topic by briefly proposing some guidelines for their study. By doing so, we do not want to imply that this is the only or the best way to tackle the issue. We simply felt others could profit from some of our tips to avoid common mistakes.

First and foremost, authors should make clear which call type they are referring to and do so using the nomenclature already employed in the literature. One important aspect of this is that there is so much variation among species that it becomes difficult for a researcher familiar with, for example, the calls of the Central American species to understand the written description of calls from a South American species. After working together on this chapter, the four of us authors have firsthand experience with this issue. The basis for the classification should be variables extracted from good spectrograms, and, thus, more high-quality spectrograms need to be provided in the literature. Several kinds of free software are capable of producing high-quality images (e.g., Raven Lite: Bioacoustic Research Program 2011, Praat: Boersma and Weenink 2012). Another crucial point is to clearly define what authors consider a bout and, even more importantly, to show data (in the most possible raw form) on both call durations and inter-call intervals, so that others can examine these and make comparisons.

When journals offer such an option, authors should also take advantage of using online supplemental materials to upload audio examples of calls. Such examples should include both isolated individual call types and short sections of longer bouts, in order to demonstrate patterns. Although multi-animal choruses are interesting, solo calling bouts are even more useful (but rare to capture in many species). Sample recordings should also be shared with archives such as that offered by the Macaulay Library from Cornell University (see http://macaulaylibrary.org) or the sound archive of the British Library (http://sounds.bl.uk). Within the first website, it is possible to browse and use their collection for research or education, as long as proper citations are used.

The above websites also provide tips for purchasing equipment, for making proper field recordings, and for documenting information about the caller (see also Geismann and Parsons 2011). In the tropics, researchers need to consider the use of durable recorders that record in high quality without being susceptible to humidity, dust, and the occasional falls during a forest trek. Although it is common sense for most field workers to make recordings using systematic methods and professional equipment, there are common mistakes made both by people recording vocalizations for first time and by those with years of experience (including ourselves). Many important steps can be forgotten during the excitement of recording an intense calling bout. For example, not using headphones when recording causes observers to miss some of the noise that is picked up by strong directional microphones made by their own body movements, leaves under their feet when they adjust their position, and colleagues talking (even at a distance). Additionally, headphones assist the recordist in monitoring the input level (along with level meters on most recorders)—because different calls within a howler monkey bout can range so extensively in amplitude, a common mistake is to record too loudly and this causes clipping and distortion. When using headphones, we recommend in some situations the recordist keeps one ear free in order to locate individual callers—otherwise, directional microphones can be disorienting when both ears are covered. When recording, only practice helps to avoid talking over recordings while also recording information in real time about the identity of the caller, so that individuals can be compared later. This is especially important if observers want to try to capture isolated calls from individuals during group choruses.

Finally, we urge researchers who are not focusing on vocalizations in their projects to still consider carrying recording equipment with them into the field. Once familiar with recording protocols, the real-time recording ability of modern equipment can help with a variety of data collection beyond just vocalizations. And by increasing the number of recordists in the field, we may ultimately be better able to compare the repertoires of different species and to ask questions about the context of infrequently produced calls.

References

Altmann SA (1959) Field observations on a howling monkey society. J Mammal 40:317–330

Baldwin JD, Baldwin JI (1976) Vocalizations of howler monkeys (*Alouatta palliata*) in Southwestern Panama. Folia Primatol 26:81–108

Bioacoustics Research Program (2011) Raven Pro: interactive sound analysis software, version 1.4 [Computer software]. Cornell Lab of Ornithology, Ithaca. http://www.birds.cornell.edu/raven. Accessed 11 April 2012

Blumenrath SH, Dabelsteen T (2004) Degradation of great tit (*Parus major*) song before and after foliation: implication for vocal communication in a deciduous Forest. Behaviour 141:935–958

Bocian D, Aday C, Gavazzi A, Markowitz H, Baptista L (1999) Can spectrographic analysis of roars be used to identify individual male black howler monkeys (*Alouatta pigra*)? Am J Primatol 49:36–37

Boersma P, Weenink D (2012) Praat: doing phonetics by computer, version 5.3.13 [Computer software]. University of Amsterdam. http://www.praat.org/. Accessed 11 April 2012

Boinski S (1991) The coordination of spatial position: a field study of the vocal behaviour of adult female squirrel monkeys. Anim Behav 41:89–102

Boinski S (1993) Vocal coordination of troop movement among white-faced capuchin monkeys, *Cebus capucinus*. Am J Primatol 30:85–100

Bowman RI (1979) Adaptive morphology of song dialects in Darwin's finches. J Ornithol 120:353–389

Braza F, Alvarez F, Azcarate T (1981) Behaviour of the red howler monkey (*Alouatta seniculus*) in the llanos of Venezuela. Primates 22:459–473

Brenowitz EA (1982) The active space of red-winged blackbird song. J Comp Physiol 147:511–522

Brown TJ, Handford P (2000) Sound design for vocalizations: quality in the woods, consistency in the fields. Condor 102:81–92

Byrne RW (1981) Distance vocalisations of Guinea baboons (*Papio papio*) in Senegal: an analysis of function. Behaviour 78:283–312

Byrne RW (2000) How monkeys find their way: leadership, coordination, and cognitive maps of African baboons. In: Boinski S, Garber PA (eds) On the move: how and why animals travel in groups. University of Chicago Press, Chicago

Caine NG, Stevens C (1990) Evidence for a "monitoring call" in red-bellied tamarins. Am J Primatol 22:251–262

Calegaro-Marques C, Bicca-Marques JC (1997) Vocalizações de *Alouatta caraya* (Primates, Cebidae). In: Ferrari SF, Schneider H (eds) A primatologia no Brasil, vol 5. SBPr/UFPA, Belém

Carpenter CR (1934) A field study of the behavior and social relations of howling monkeys *Alouatta palliata*. Comp Psychol Monogr 10:1–168

Casar C, Byrne R, Young RJ, Zuberbuehler K (2012) The alarm call system of wild black-fronted titi monkeys, Callicebus nigrifrons. Behav Ecol Sociobiol 66:653–667

Cheney DL, Seyfarth RM, Palombit RA (1996) The function and mechanisms underlying baboon 'contact' barks. Anim Behav 52:507–518

Chiarello AG (1995) Role of loud calls in brown howlers, *Alouatta fusca*. Am J Primatol 36:213–222

Cornick LA, Markowitz H (2002) Diurnal vocal patterns of the black howler monkey (*Alouatta pigra*) at Lamanai, Belize. J Mammal 83:159–166

Cortés-Ortiz L, Bermingham E, Rico C, Rodríguez-Luna E, Sampaio I, Ruiz-García M (2003) Molecular systematics and biogeography of the Neotropical monkey genus, *Alouatta*. Mol Phylogenet Evol 26:64–81

Cortés-Ortiz L, Rylands AB, Mittermeier R (2014) The taxonomy of howler monkeys: integrating old and new knowledge from morphological and genetic studies. In: Kowalewski M, Garber P, Cortés-Ortiz L, Urbani B, Youlatos D (eds) Howler monkeys: adaptive radiation, systematics, and morphology. Springer, New York

da Cunha RGT (2004) A functional analysis of vocalisations of black howler monkeys (*Alouatta caraya*). PhD Dissertation, University of St. Andrews, St. Andrews

da Cunha RGT, Byrne RW (2006) Roars of black howling monkeys (*Alouatta caraya*): evidence for a function in inter-group spacing. Behaviour 143:1169–1199

da Cunha RGT, Byrne RW (2009) The use of vocal communication in keeping the spatial cohesion of groups: intentionality and specific functions. In: Garber PA, Estrada A, Bicca-Marques JC, Heymann EW, Strier KB (eds) South American primates. Comparative perspectives in the study of behavior, ecology, and conservation. Springer, New York

da Cunha RGT, Byrne RW (2013) Age-related differences in the use of the "moo" call in black howlers (*Alouatta caraya*). Int J Primatol 34:1105–1121

da Cunha RGT, Jalles-Filho E (2007) The roaring of southern brown howler monkeys (*Alouatta guariba clamitans*) as a mechanism of active defense of borders. Folia Primatol 78:259–271

Daniel JC, Blumstein DT (1998) A test of acoustic adaptation hypothesis in four species of marmots. Anim Behav 56:1517–1528

Daschbach NJ, Schein MW, Haines DE (1981) Vocalizations of the slow loris, *Nycticebus coucang* (Primates, Lorisidae). Int J Primatol 2:71–80

Date EM, Lemon RE (1993) Sound transmission: a basis on dialects in birdsongs? Behaviour 124:291–312

de Boer B (2009) Acoustic analysis of primate air sacs and their effect on vocalization. J Acoust Soc Am 126:3329–3343

de la Torre S, Snowdon CT (2002) Environmental correlates of vocal communication of wild pygmy marmosets, *Cebuella pygmaea*. Anim Behav 63:847–856

Delgado RA (2006) Geographic variation in the long calls of male Orangutans (*Pongo* spp.). Ethology 113:487–498

Dennett DC (1978) Beliefs about beliefs. Behav Brain Sci 1:568–569

Dittus W (1988) An analysis of toque macaque cohesion calls from an ecological perspective. In: Todt D, Goedeking P, Symmes D (eds) Primate vocal communication. Springer, Berlin

Drubbel RV, Gautier JP (1993) On the occurrence of nocturnal and diurnal loud calls, differing in structure and duration, in red howlers (*Alouatta seniculus*) of French Guyana. Folia Primatol 60:195–209

Eisenberg JF (1976) Communication mechanisms and social integration in the black spider monkey, *Ateles fusciceps robustus*, and related species. Smithson Contrib Zool 213:1–108

Epple G (1968) Comparative studies on vocalization in marmoset monkeys (Hapalidae). Folia Primatol 8:1–40

Gautier JP (1989) A redrawn phylogeny of guenons based upon their calls: biogeographical implications. Bioacoustics 2:11–21

Gautier JP, Gautier A (1977) Communication in Old World monkeys. In: Sebeok T (ed) How animals communicate. Indiana University Press, Bloomington

Geismann T, Parsons S (2011) Recording primate vocalizations. In: Setchell JM, Curtis DJ (eds) Field and laboratory methods in primatology: a practical guide, 2nd edn. Cambridge University Press, Cambridge

Geissmann T (2002) Duet-splitting and the evolution of gibbon songs. Biol Rev 77:57–76

Gregorin R (2006) Taxonomy and geographic variation of species of the genus *Alouatta* Lacépède (Primates, Atelidae) in Brazil. Rev Bras Zool 23:64–144

Halloy M, Kleiman DG (1994) Acoustic structure of long calls in free-ranging groups of golden lion tamarins, *Leontopithecus rosalia*. Am J Primatol 32:303–310

Harcourt AH, Stewart KJ, Hauser MD (1993) Functions of wild gorilla "close" calls. I. Repertoire, context, and interspecific comparison. Behaviour 124:89–122

Hershkovitz P (1949) Mammals of northern Colombia. Preliminary report no. 4: monkeys (Primates), with taxonomic revisions of some forms. Proc U S Nat Mus 98:323–327

Holzmann I (2012) Distribución geográfica potencial y comportamiento vocal de dos especies de mono aullador (*Alouatta guariba clamitans* y *A. caraya*). PhD Dissertation, Universidad Nacional de La Plata.

Holzmann I, Agostini I, Di Bitetti M (2012) Roaring behavior of two syntopic howler species (*A. caraya* and *A. guariba clamitans*): evidence supports the mate defense hypothesis. Int J Primatol 33:338–355

Horwich RH, Gebhard K (1983) Roaring rhythms in black howler monkeys (*Alouatta pigra*) of Belize. Primates 24:290–296

Ihering HV (1914) Os bugios do gênero *Alouatta*. Rev Mus Paulista 9:231–280

Kelemen G, Sade J (1960) The vocal organ of the howling monkey, *Alouatta palliata*. J Morphol 107:123–140

Kitchen DM (2000) Aggression and assessment among social groups of Belizean black howler monkeys (*Alouatta pigra*). PhD Dissertation, University of Minnesota, Minneapolis

Kitchen DM, da Cunha RGT, Holzmann I, Oliveira DAG (2014) Function of loud calls in howler monkeys. In: Kowalewski M, Garber P, Cortés-Ortiz L, Urbani B, Youlatos D (eds) Howler monkeys: adaptive radiation, systematics, and morphology. Springer, New York

Leighty KA, Soltis J, Wesolek CM, Savage A (2008) Rumble vocalizations mediate interpartner distance in African elephants, *Loxodonta africana*. Anim Behav 67:125–139

Lombard E (1911) Le signe de l'élévation de la voix. Annales Des Maladies de l'Oreille et Du Larynx 37:101–119

Lönnberg E (1941) Notes on members of the genera *Alouatta* and *Aotus*. Ark Zool 33A:1–44

Maciej P, Fischer J, Hammerschmidt K (2011) Transmission characteristics of primate vocalizations: implications for acoustic analyses. PLoS One 6:e23015. doi:10.1371/journal.pone.0023015

Marten K, Quine D, Marler P (1977) Sound transmission and its significance for animal vocalization. Behav Ecol Sociobiol 2:291–302

Martin GR (1981) Avian vocalizations and the sound interference model of Robert's et al. Anim Behav 29:632–633

Martin P, Bateson P (2007) Measuring behaviour: an introductory guide. Cambridge University Press, Cambridge

McComb K, Packer C, Pusey A (1994) Roaring and numerical assessment in contests between groups of female lions, *Panthera leo*. Anim Behav 47:379–387

Mech LD (1966) The wolves of Isle. Royale Fauna Series 7. U.S. Department Printing Office, Washington, DC

Mitani JC, Nishida T (1993) Contexts and social correlates of long-distance calling by male chimpanzees. Anim Behav 45:735–746

Mitani JC, Stuht J (1998) The evolution of nonhuman primate loud calls: acoustic adaptation for long-distance transmission. Primates 39:171–182

Morton ES (1975) Ecological sources of selection on avian sounds. Am Nat 109:17–34

Naguib M (1996) Ranging by song in Carolina wrens *Thryothorus ludovicianus*: effects of environmental acoustics and strength of song degradation. Behaviour 133:541–559

Naguib M, Wiley RH (2001) Estimating the distance to a source of sound: mechanism and adaptations for long-range communication. Anim Behav 62:825–837

Nelson BS (2003) Reliability of sound attenuation in Florida scrub habitat and behavioral habitat implications. J Acoust Soc Am 113:2901–2911

Neville MK, Glander KE, Braza F, Rylands AB (1988) The howling monkeys, genus *Alouatta*. In: Mittermeier RA, Rylands AB, Coimbra-Filho A, Fonseca GAB (eds) Ecology and behavior of neotropical primates, vol 2. World Wildlife Fund, Washington, DC

Newman JD (1995) Vocal ontogeny in macaques and marmosets: convergent and divergent lines of development. In: Zimmermann E, Newman JD, Jürgens U (eds) Current topics in primate vocal communication. Plenum, New York

Oliveira DAG (2002) Vocalizações de longo alcance de *Alouatta fusca clamitans* e *Alouatta belzebul belzeul*: estrutura e contexto. PhD Dissertation, Universidade de São Paulo, São Paulo

Oliveira DAG, Ades C (2004) Long-distance calls in Neotropical primates. An Acad Bras Cienc 76:393–398

Owren MJ, Rendall D, Ryan MJ (2010) Redefining animal signaling: influence versus information in communication. Biol Philos 25:755–780

Palombit RA (1992) A preliminary study of vocal communication in wild long-tailed macaques (*Macaca fascicularis*): II. Potential of calls to regulate intragroup spacing. Int J Primatol 13:183–207

Pook AG (1977) A comparative study of the use of contact calls in *Saguinus fuscicollis* and *Callithrix jacchus*. In: Kleiman DG (ed) The biology and conservation of the Callitrichidae. Smithsonian Institution Press, Washington, DC

Ricklefs RE, Hainsworth FR (1968) Temperature dependent behavior of the cactus wren. Ecology 49:227–233

Riede T, Tokuda IT, Munger JB, Thomson SL (2008) Mammalian laryngeal air sacs add variability to the vocal tract impedance: physical and computational modeling. J Acoust Soc Am 124:634–647

Roberts J, Kacelnik A, Hunter ML Jr (1977) A model of sound interference in relation to acoustic communication. Anim Behav 27:1271–1272

Robinson JG (1979) Vocal regulation of use of space by groups of titi monkeys *Callicebus moloch*. Behav Ecol Sociobiol 5:1–15

Rylands AB, Mittermeier RA (2009) The diversity of the new world primates (Platyrrhini). In: Garber PA, Estrada A, Bicca-Marques JC, Heymann EW, Strier KB (eds) South American primates: comparative perspectives in the study of behavior, ecology, and conservation. Springer, New York

Schneider C, Hodges K, Fischer J, Hammerschmidt K (2008) Acoustic niches of Siberut primates. Int J Primatol 29:601–613

Schön MA (1964) Possible function of some pharyngeal and lingual muscles of the howling monkey (*Alouatta seniculus*). Acta Anat 58:271–283

Schön MA (1970) On the mechanism of modulating the volume of the voice in howling monkeys. Acta Oto Laryngol 70:443–447

Schön MA (1971) The anatomy of the resonating mechanism in howling monkeys. Folia Primatol 15:117–132

Schön Ybarra MA (1984) Locomotion and postures of red howlers in a deciduous forest-savanna interface. Am J Phys Anthropol 63:65–76

Schön Ybarra MA (1986) Loud calls of adult male red howling monkeys (*Alouatta seniculus*). Folia Primatol 47:204–216

Schön Ybarra MA (1988) Morphological adaptations for loud phonations in the vocal organ of howling monkeys. Primate Rep 22:19–24

Schön Ybarra MA (1995) A comparative approach to the non-human primate vocal tract: implications for sound production. In: Zimmermann E, Newman JD, Jürgens U (eds) Current topics in primate vocal communication. Plenum, New York

Sekulic R (1982) Daily and seasonal patterns of roaring and spacing in four red howler *Alouatta seniculus* troops. Folia Primatol 39:22–48

Sekulic R, Chivers DJ (1986) The significance of call duration in howler monkeys. Int J Primatol 7:183–190

Seyfarth RM, Cheney DL (2003) Signalers and receivers in animal communication. Annu Rev Psychol 54:145–173

Seyfarth RM, Cheney DL (2011) Meaning, reference, and intentionality in the natural vocalizations of monkeys. In: Maran T, Martinelli D, Turovski A (eds) Readings in zoosemiotics. Walter de Gruyter GmbH, Berlin

Seyfarth RM, Cheney DL, Marler P (1980) Vervet monkey alarm calls: semantic communication in a free-ranging primate. Anim Behav 28:1070–1094

Sorjonen J (1986) Factors affecting the structure of song and singing behaviour of some Northern European passerine birds. Behaviour 98:286–304

Steinmetz S (2005) Vocalizações de longo alcance como comunicação intra-grupal nos bugios (*Alouatta guariba*). Neotrop Primates 13:11–15

Teichroeb JA, Sicotte P (2010) The function of male agonistic displays in ursine colobus monkeys (*Colobus vellerosus*): male competition, female mate choice or sexual coercion? Ethology 116:366–380

Thinh VN, Hallam C, Roos C, Hammerschmidt K (2011) Concordance between vocal and genetic diversity in gibbons. BMC Evol Biol 11:36

Thorington RW, Ruiz JC, Eisenberg JF (1984) A study of black howler monkeys (*Alouatta caraya*) populations in northern Argentina. Am J Primatol 6:357–366

Villalobos F, Valerio AA, Retana AP (2004) A phylogeny of howler monkeys (Cebidae: *Alouatta*) based on mitochondrial, chromosomal and morphological data. Rev Biol Trop 52:665–677

Waser PM, Brown CH (1986) Habitat acoustics and primate communication. Am J Primatol 10:135–154

Waser PM, Waser MS (1977) Experimental studies of primates vocalization: specializations for long-distance propagation. Z Tierpsychol 43:239–263

Whitehead JM (1987) Vocally mediated reciprocity between neighbouring groups of mantled howling monkeys, *Alouatta palliata palliata*. Anim Behav 35:1615–1627

Whitehead JM (1989) The effect of the location of a simulated intruder on responses to long-distance vocalizations of mantled howling monkeys, *Alouatta palliata palliata*. Behaviour 108:73–103

Whitehead JM (1995) Vox Alouattinae: a preliminary survey of the acoustic characteristics of long distance calls of howling monkeys. Int J Primatol 16:121–144

Wich SA, Schel AM, de Vries H (2008) Geographic variation in Thomas langur (*Presbytis thomasi*) loud calls. Am J Primatol 10:566–574

Wiley RH, Richards DG (1978) Physical constraints on acoustic communication in the atmosphere: implications for the evolution of animal vocalisations. Behav Ecol Sociobiol 3:69–94

Wiley RH, Richards DG (1982) Adaptations for acoustic communication in birds: sound transmission and signal detection. In: Kroodsma DE, Miller EH, Oullet H (eds) Acoustic communication in birds, vol 2. Academic, New York

Zuberbuhler K (2000) Referential labelling in Diana monkeys. Anim Behaviour 59:917–927

Zuberbuhler K (2001) Predator-specific alarm calls in Campbell's guenons. Behav Ecol Sociobiol 50:414–422

Chapter 14
Function of Loud Calls in Howler Monkeys

Dawn M. Kitchen, Rogério Grassetto Teixeira da Cunha, Ingrid Holzmann, and Dilmar Alberto Gonçalves de Oliveira

Abstract Beyond the unique sound of howler monkey vocalizations, their vigorous loud calling displays are perplexing given the otherwise sedentary lifestyle of these primates. Here we provide potential explanations for this energetic investment by reviewing all available functional studies conducted to date. We highlight the variation among and even within species when we explore whether male loud calls are used in group cohesion, predator avoidance, attraction of females, or competition with other males or other groups over resources. In the competition scenario, we examine strategies of avoidance versus direct competition and whether contests are focused on defense of space, food, mates, or infants. We suggest that much of the debate surrounding the function of loud calls stems from methodological differences among researchers and from the varied levels of analyses used, although we also demonstrate that studies of form and function can be intertwined. We emphasize the need to examine different call types separately and discuss the role of howling in intragroup male relationships. Finally, we address the understudied role of female loud calling and the potential use of hybrid populations to examine the evolution of species-typical loud calls. We conclude with some practical hints for designing field tests to uncover functional significance.

D.M. Kitchen (✉)
Department of Anthropology, The Ohio State University, Columbus, OH 43210, USA

Department of Anthropology, The Ohio State University-Mansfield, Mansfield, OH 44906, USA
e-mail: kitchen.79@osu.edu

R.G.T. da Cunha
Instituto de Ciências da Natureza, Universidade Federal de Alfenas,
Alfenas-MG 37130-000, Brazil
e-mail: rogcunha@hotmail.com

I. Holzmann
CONICET (Consejo Nacional de Investigaciones Científicas y Técnicas), IBS (Instituto de Biología Subtropical), CeIBA (Centro de Investigaciones del Bosque Atlántico),
Buenos Aires, Argentina
e-mail: holzmanningrid@yahoo.com.ar

D.A.G. de Oliveira
Departamento de Fauna/CBRN, Centro de Manejo de Fauna Silvestre,
Secretaria do Meio Ambiente do Estado de São Paulo, Brazil
e-mail: dilmar_oliveira@yahoo.com.br

© Springer Science+Business Media New York 2015
M.M. Kowalewski et al. (eds.), *Howler Monkeys*, Developments in Primatology:
Progress and Prospects, DOI 10.1007/978-1-4939-1957-4_14

Resumen Más allá del sonido único de las vocalizaciones de los monos aulladores, estos vigorosos despliegues nos dejan perplejos, dado el sedentario estilo de vida de estos primates. En este capítulo damos explicaciones potenciales a esta inversión de energía, a través de una revisión de los estudios funcionales llevados a cabo hasta la fecha. Remarcamos la variación entre y dentro de las especies cuando exploramos si los aullidos de los machos son utilizados en la cohesión de grupo, evasión de predadores, atracción de hembras o competencia. Sobre este último escenario, examinamos las estrategias de evasión versus la competencia directa y exploramos si la competencia se focaliza en la defensa del espacio, la comida, las parejas o los infantes. Sugerimos que gran parte del debate sobre la función de las vocalizaciones de larga distancia radica en diferencias metodológicas entre investigadores, así como en la variedad de niveles de análisis utilizados, aunque también demostramos que los estudios de forma y función pueden estar entrelazados. Enfatizamos la necesidad de examinar diferentes tipos de llamados separadamente y discutimos el papel de los aullidos en las relaciones intragrupales entre machos. Finalmente, abarcamos el escasamente estudiado papel de las vocalizaciones de larga distancia emitidas por las hembras y la potencial utilización de poblaciones de híbridos para examinar la evolución de las vocalizaciones de larga distancia, típicos de cada especie. Concluimos con consejos prácticos para el diseño de estudios en el campo que permitan descubrir significados funcionales.

Keywords Bark • Mate defense • Infanticide • Resource defense • Resource holding potential • Roar

Abbreviations

%	Percent
>	Greater than
A.	Alouatta
e.g.	For example
i.e.	In other words
kHz	Kilohertz
MA	Massachusetts
Min	Minutes
NY	New York
P.	Pan
pers. obs.	Personal observation
RHP	Resource holding potential
TFT	Tit-for-Tat
UK	United Kingdom
unpubl. data	Unpublished data

14.1 Introduction: Why Howl?

Howler monkeys are unique among the platyrrhines in their complex, loud, long, low-frequency calls (Moynihan 1967; Snowdon 1989). In da Cunha et al. (2015, this volume), we reviewed studies that highlight the acoustic and morphological features that make howler monkeys and their calls unique and the environmental influences on propagation of their sounds. But, a question remains: why should an animal that allots most of its activity budget to inactivity (likely due to the lack of ready energy available from its largely folivorous diet: Milton 1980) invest so much time and effort into loud calling?

Loud calls are ubiquitous in the animal kingdom—occurring in species as distinct as frogs (e.g., Gerhardt 1974; Bee et al. 2000) and whales (e.g., Širović et al. 2007)—and they have always generated an amount of interest proportional to their volume. For all species studied to date, the list of functions can be narrowed down to a few broad categories: (a) maintaining group cohesion (e.g., Cheney et al. 1996), (b) reducing predation risk (e.g., reviewed in Cäsar and Zuberbühler 2012), (c) attracting and bonding with mates (e.g., Blair 1958), and (d) competing with other individuals/groups to protect food/space (reviewed by Fashing 2001), mates (e.g., Steenbeek and Assink 1998; but see Wich and Nunn 2002), or vulnerable offspring (e.g., Steenbeek et al. 1999; Wich et al. 2002). These categories are not mutually exclusive, and howler monkey loud calls may have evolved under a variety of selective pressures.

Here, we critically review all studies conducted so far that have focused on the biological meaning of these peculiar calls, including analyses at different explanatory levels. Because of different fitness limitations on the sexes (Trivers 1972; Emlen and Oring 1977), we discuss the possible functions of male and female loud calls separately. We also discuss the opportunity for evolutionary insights from studies in sympatric zones, particularly those with hybridizing animals. Throughout, we continue to stress the variation among the different howler monkey populations that we highlighted in da Cunha et al. (2015). In our conclusion, we address various methodological issues and provide directions for future research.

14.2 Loud Calls and Group Cohesion

No studies have directly tested whether or not male howler monkey loud calls function in group cohesion (i.e., contact calls during travel or when separated). However, Whitehead (1989) reported that male loud calling (both roars and barks) in *A. palliata* preceded 33 % of all major group travel events, Steinmetz (2005) reported that 14 of 37 (38 %) male *A. guariba* howling bouts were produced during separation of the group (see also Oliveira 2002), and Sekulic (1982b) described cases of males roaring on reunion in *A. arctoidea* (formerly *A. seniculus*). The relatively quieter calls in the repertoire are also good candidates for contact calls (da Cunha et al. 2015).

14.3 Loud Calls and Predators

Several authors describe barks and, less often, roars emitted during encounters with potential predators (including dogs and humans) and following nonthreatening disturbances such as vultures, planes, vehicles, and thunder (e.g., *A. palliata*: Carpenter 1934; Baldwin and Baldwin 1976; Whitehead 1989; *A. arctoidea*: Sekulic 1982c, 1983; *A. pigra*: Horwich and Lyon 1990; *A. guariba*: Oliveira 2002). Although uncommon, human observers sometimes witness predator attacks on howler monkeys. For example, McKinney (2009) observed a male *A. palliata* howling briefly during an attack on the group by northern crested caracaras (*Caracara cheriway*), and Julliot (1994) reported that *A. macconnelli* (formerly *A. seniculus*) gathered together and roared in proximity of crested eagles (*Morphnus guianensis*).

However, most examples in the literature fail to provide specific information on the use of loud calls (e.g., harpy eagle, *Harpia harpyja*, attacks: Eason 1989; Peres 1990). For example, during a playback study, *A. palliata* that had only 1-year experience with introduced harpy eagles responded appropriately to the threat of attack, but no details were given on the call type used or duration of alarm calls produced by the monkeys (Gil-da-Costa et al. 2003). Although Camargo and Ferrari (2007) report that an adult male *A. belzebul* gave "typical" ru-ru-ru alarm calls during an attack on an infant by two tayras (*Eira barbara*), no spectrograms were included. Individuals in a captive group of *A. guariba* responded with barks to the presentation of two taxidermized mammals: an ocelot (*Leopardus pardalis*) and a capybara (*Hydrochoerus hydrochaeris*) (Oliveira et al. unpubl. data). Interestingly, the naive monkeys showed no ability to distinguish predator from non-predator. Finally, it seems that predator presence does not always elicit loud vocal responses; for example, da Cunha and Byrne (2006) reported that four natural encounters with ocelots did not produce any loud vocal response from a group of *A. caraya* nor did a pilot playback study of various predator vocalizations (da Cunha RGT unpubl. data). It is unclear if silence is part of an escape response for howler monkeys (see also silence following black hawk-eagle, *Spizaetus tyrannus*, encounter: Miranda et al. 2006). Given the rarity of predator encounters observed by humans, we suggest future studies increase the use experimental techniques such as acoustic and visual predator models in order to identify differences between loud calls produced in various contexts.

14.4 Loud Calls as Sexually Selected Signals

The exaggerated nature of loud calling displays suggests a role for sexual selection (e.g., Zahavi 1977; but see FitzGibbon and Fanshawe 1988 for exaggerated signaling in predator deterrence). Snowdon (2004) proposed that to be sexually selected, signals must be (1) sexually dimorphic, (2) variable among males, (3) discriminated among individuals, (4) preferred or avoided in context of reproductive access, and (5) related to increased reproductive fitness. Although the first two have been clearly demonstrated in howler monkeys (da Cunha et al. 2015), and criteria 3 and 4 have been established during intrasexual competition (e.g., Kitchen 2000, 2004), the last criterion is difficult to measure in any primate. Indirect measures such as "winning" an encounter or relative access to cycling females are often used to approximate fitness.

Assuming howler monkey loud calls do function in sexual selection, the potential intersexual component has been largely ignored. This is not unique to howler monkeys—female choice is a challenging topic to test on any wild animal, especially when it has to be disentangled from strong male-male competition and male-female sexual coercion (including infanticide). In most howler monkey species with bisexual dispersal patterns, females can join established groups (reviewed in Crockett and Eisenberg 1987; Di Fiore and Campbell 2007) and immigrating females may target groups based on the qualities of a male expressed through his loud calls. Females might choose males based on direct benefits if, for example, aspects of his loud call correlate with his ability to defend a resource or an infant (see Wiley and Poston 1996) or on indirect benefits if call features correlate with "good genes" (Zahavi 1977).

There is evidence that females have preferences among males. In multi-male groups, for example, females frequently keep close proximity to one male (the "central male" following Van Belle et al. 2008) over another during howling bouts or intergroup encounters (e.g., *A. pigra*: Kitchen 2000; Van Belle et al. 2008, 2009a; *A. palliata*: Zucker and Clarke 1986; *A. guariba*: Oliveira et al. unpubl. data). Although not causal evidence for female choice, there also appears to be a relationship between male calling and female reproduction; captive male *A. caraya* with higher calling rates had higher reproductive rates than quieter males, and females in this population were more likely to conceive if they heard male conspecifics calling (Farmer et al. 2011). However, whether females base their preferences on specific acoustic features has not been tested in howler monkeys or, with a few notable exceptions (humans, *Homo sapiens*, red deer, *Cervus elaphus*, and koalas, *Phascolarctos cinereus*: reviewed in Charlton et al. 2012), in any mammalian species.

Taken together, these studies suggest that female choice might be an important influence on the production of loud calls by males (e.g., *A. arctoidea*: Sekulic 1982b). How to construct a female choice study in howler monkeys in light of the relatively low rate of sociosexual behaviors, the lack of external signs of estrus, the potentially confounding effects of male competition, and the threat of infanticide remains problematic. One option is to monitor female dispersal patterns. Although anecdotal evidence exists (e.g., a solitary female moving preferentially toward a calling male in *A. palliata*: Whitehead 1989), only long-term studies would be able to adequately address this question. Additionally, as da Cunha and Jalles-Filho (2007) point out, howling happens daily at some sites, yet female immigration events are rare. It is possible that long-term memory of male howling bouts—either within a multi-male group or over an entire area—eventually affects female mating or dispersal decisions. However, even if females do base their choices on male quality as expressed through loud calls, intersexual selection may not be the sole pressure shaping the evolution of these vocalizations.

14.5 The Competitive Nature of Howler Monkeys

Ever since Carpenter's pioneering work in 1934, pioneering work, most studies on howler monkeys have proposed, in some way or another, a function related to regulating space use between groups (reviewed in da Cunha and Jalles-Filho 2007). Explanations for how and why this spacing is maintained have nevertheless differed widely. Some explanations are based on real population differences, but some, we suspect, are due to the varying perceptions or approaches of different authors. On one end of the spectrum, some researchers have advocated for a "territorial" function of howling displays (e.g., Collias and Southwick 1952; Altmann 1959; Bernstein 1964; Horwich and Gebhard 1983). But, according to Mitani and Rodman (1979), howler monkeys are not territorial because group home ranges overlap too substantially, at least in some species or in high-density populations (e.g., 32–63 % in *A. arctoidea*: Sekulic 1982a; 14–63 % in *A. palliata*: Whitehead 1989; but see Agostini et al. 2010b), and they typically have daily path lengths that are too short to theoretically patrol boundaries (reviewed in Crockett and Eisenberg 1987). In fact, as our knowledge of different *Alouatta* species expands, Milton's (1980) original description seems to hold true: likely due to energetic constraints, howler monkeys appear to be "travel minimizers." Still, although howler monkeys do not patrol the borders of their home range, there is ample indication that at least some populations aggressively defend their group or their space (see below).

At the other extreme, some have described howler monkey spacing in fairly cooperative terms, with individuals apparently calling to indicate where in their range they are so that other groups do not approach. For example, several studies of *A. palliata* have described evidence for "mutual avoidance" between groups (e.g., Carpenter 1934; Southwick 1962; Baldwin and Baldwin 1976). Chivers (1969) found that when two *A. palliata* groups slept close together (<220 m on average), they generally moved away from each other following the dawn chorus, and Whitehead (1987) found that groups of *A. palliata* met each other less frequently than would be expected by chance based on a model of random movement.

Of course, it would be unlikely for such avoidance to evolve as a purely cooperative strategy (defined in West et al. 2007). In a world of cooperators, individuals who opted to cheat and thereby exploit this information would have an advantage (e.g., Maynard Smith and Price 1973). For example, if *group X* announces that it is in *location Y*, then *group A* can exploit *X*'s unguarded fig tree at *location Z*. As Sekulic (1982a) pointed out, although "informing neighbors may reduce energy expended in interaction one day, it could also reduce the resources available at the other side of the home range for the following day." A similar conclusion was drawn by da Cunha and Byrne (2006), who found that the calls of an *A. caraya* group were disproportionately distributed in the exclusive core area and not along the borders. Although they suggested that this regular advertisement of occupancy allows regulation of the space use in this species, they suggested that this was because it is a competitive strategy of assessment for settling disputes without chases and fights.

Instead of some cooperative social contract, avoidance is probably most often explained as a by-product mutualism, where the group's collective behavior (insofar as individual interests overlap) "maximizes its own immediate fitness and any positive effects on the fitness of other individuals are coincidental" (Clutton-Brock 2002). Avoidance between groups is, in fact, one of the outcomes predicted by evolutionary game theory (e.g., Maynard Smith 1974; Maynard Smith and Parker 1976). If groups are avoiding the potential costs, to both winners and losers, of escalating a contest, then only when two individuals (or two groups) are similarly matched should rivals approach one another and compete for some resource. It is then possible that the "mutual avoidance" scenarios proposed in *A. palliata* occur because animals avoid escalating contests they would likely lose. Howler monkeys may use aspects of loud calling as a means to monitor their opponents' relative resource holding potential (RHP: Parker 1974) and avoid one another if the outcome is clear (see RHP discussion below). Other asymmetries could also exist that might be assessed through motivational cues. For example, one group/individual might be less willing to back down if they have more at stake (e.g., an investment in females, vulnerable offspring, or a rich food source), a territory holder may have more to lose than an intruder (i.e., ownership games: Maynard Smith and Parker 1976), or losing a fight to a stranger might be more costly than losing to a familiar neighbor (i.e., "dear enemies": Ydenberg et al. 1988).

Rather than a by-product mutualism, Whitehead (1987) suggested that the mechanism producing "mutual avoidance" was in fact a reciprocation of movements, which took the shape of a Tit-for-Tat (TFT) or reciprocity strategy. TFT is theoretically a stable solution to a problem that mimics a Prisoner's Dilemma (see Axelrod and Hamilton 1981)—where avoiding one another has benefits for both contestants, yet being exploited by a cheating rival has high costs. If interactions are iterated indefinitely, a TFT strategy, unlike a purely cooperative strategy, is successful because it mirrors the response of a rival and thereby avoids exploitation while still being readily "forgiving," so to speak. Whitehead (1987) played calls that mimicked both retreating and approaching neighbors to *A. palliata* individuals, with subjects retreating from rivals in the first case and approaching them in the latter. These are exciting results because few empirical studies have found support for the existence of natural reciprocity strategies (Stevens and Hauser 2004). However, although a few other cases have been documented, (e.g., Seyfarth and Cheney 1984), the psychological and cognitive constraints might make TFT strategies beyond the abilities of animals like howler monkeys (Stevens et al. 2011).

Whether intergroup avoidance in howler monkeys is termed cooperative, competitive, reciprocal, or mutually beneficial in the literature may purely be a semantic issue when researchers use different terms to describe the same type of event (see West et al. 2007 for description of term usage in the literature). Alternatively, it is possible that the difference between these terms has real biological significance if there are tangible differences in the strategies used by animals in different populations. For example, as we described above, overt aggression seems to be a rare phenomenon among groups of *A. palliata*. In contrast, intergroup encounters—rather than avoidance between groups—are common in other species. For example, in Chiarello's (1995) study of *A. guariba*, 93 % of calling bouts were directed at nearby groups and 35 % escalated to chases. Such a striking difference between reports of *A. palliata* compared to other species is a dichotomy that seems to be a theme in this volume (e.g., da Cunha et al. 2015).

However, a third possibility is that species and populations are not actually that different but simply need to be studied under similar population densities and time periods to see similarities. For example, although Chivers (1969) found evidence for mutual avoidance in *A. palliata*, he also contemplated a role for intergroup dominance (based on variability in the amount that some groups roared compared to others), and he noticed that 15 % of the males in his study had fresh wounds or scars on their face. Chivers assumed these wounds were the result of intragroup conflict, but they could easily be the result of intergroup conflict. For example, DeGusta and Milton (1998) analyzed *A. palliata* skeletons from Barro Colorado Island (BCI) and reported: "We attribute the trauma primarily to fighting, and its frequency (16.4 % of adult males) contradicts previous assertions that BCI howlers are nonaggressive." Similarly, in another population of *A. palliata*, 38 % of all males were wounded and the majority of injuries were attributed to takeover attempts (Cristóbal-Azkarate et al. 2004). Thus, although *A. palliata* might be less aggressive than other species, they may actively compete when necessary, particularly at high densities.

If so, rather than uncovering real differences in the aggressive nature of different species, it is possible that sites differ in important ways (e.g., population density, habitat quality, and extent of home range overlap). Such differences could affect the costs and benefits of contest escalation, pointing to an interesting possibility of facultative use of loud calls in this genus (e.g., Lichtenberg et al. 2012). Therefore, populations—rather than species—could fall on a continuum from mutual avoidance to advertisement of occupancy to active defense of space without needing to invoke an explanation that focuses on the cooperative nature of a species. Innovative playbacks, such as those used by Whitehead (1987), and comparative studies are promising ways to test such hypotheses.

At the intraspecific level, Sekulic (1982b) further speculated that there might be a difference among types of calling bouts. She suggested that dawn choruses could function in intergroup spacing/avoidance, whereas bouts produced during the day might serve more directly competitive functions (but see Waser 1977). For example, in *A. palliata*, Chivers (1969) found evidence that the dawn chorus functioned to allow animals to assess their location relative to other groups (described above). Conversely, Chivers noted that when daytime encounters occurred, they were likely to escalate to approaches and vocal battles, with one or both groups eventually retreating (i.e., resulting in a win vs. a draw). Alternatively, we suggest that it might not be necessary to invoke wholly different functional explanations. If howler monkey intergroup encounters follow game theoretical predictions, then dawn versus daytime calling might simply represent different levels of sequential or cumulative assessment (e.g., Payne 1998). Given the variation in calling across the genus (e.g., some populations only call at dawn and others have no dawn chorus: see da Cunha et al. 2015), a comparative study of cost-benefit factors might be particularly fruitful.

14.6 What Is Defended?

Beyond potential differences in the competitive nature of males, it is also likely that species (and even populations) vary in the currency defended—females, vulnerable offspring, or food and other resources in a home range. Once again, answering these questions has proved logistically difficult, a problem compounded by variation within and between species.

14.6.1 Space/Food/Resource Defense

Despite their largely folivorous diet, there is indication that food can be a limiting factor for *Alouatta* (Jones 1980). Howler monkeys are more selective feeders than we once thought (Glander 1978), in part because they have few adaptations to deal with the secondary compounds in leaves (Milton 1980) and there appears to be food competition that limits the optimal group size in at least *A. arctoidea* (Crockett 1984) and *A. pigra* (van Belle et al. 2008).

If howling bouts are related to defense of these resources, we expect a spatial and/or temporal pattern to emerge. Temporally, animals might be expected to refrain from calling if costs become too high such as during food-limited times of the year or when climactic conditions impose a physiological burden. Alternatively, calling might increase at food-limited times, when losing access to valuable resources would be most costly. The empirical findings in howler monkeys are inconclusive. Although some howler monkey populations have demonstrated no seasonal variation in howling patterns, others have shown an increase in calling during the dry season, when fruits and new leaves are least abundant (e.g., *A. macconnelli*: Drubbel and Gautier 1993; *A. arctoidea*: Sekulic 1982b; *A. pigra*: Horwich and Gebhard 1983; *A. guariba*: Chiarello 1995, but see Holzmann et al. 2012).

Many studies have uncovered spatial patterns to calling, though scales range from sites to quadrants to entire areas. For example, Sekulic (1982b) found that during the dry season, most (>70 %) *A. arctoidea* intergroup interactions occurred near patchy distributions of fig trees (*Ficus* spp.). Similarly, Chiarello (1995) found that a disproportionate number of *A. guariba* intergroup encounters (19 of 42) occurred in just two of the 67 delineated home range quadrants and always near large emergent guapinol trees (*Hymenaea courbaril*), which provide important feeding and sleeping sites in this population. Whitehead (1989) also found that simulated *A. palliata* intruders heard from high use areas typically prompted howling and approaches toward the speakers, whereas similar calls heard from low use areas did not and, in fact, typically resulted in movement away from the speakers. Whitehead suggests that aggressive defense is therefore site dependent in *A. palliata*.

Differential behavior in the border versus center of a home range can also be indicative of space or resource defense. As such, da Cunha and Byrne (2006) found evidence that *A. caraya* both used and called more frequently from the center of their home range; whereas the border area overlapped with other groups, the study group had almost exclusive use of the center. Furthermore, these authors found that the group was more likely to call at and approach playbacks simulating intruders in the center of the range than from the border area.

In contrast to the *A. caraya* finding, other populations of howler monkeys tend to concentrate their calls along the boundary of their home range (e.g., *A. pigra*: Horwich and Gebhard 1983; Kitchen DM unpubl. data). For example, in both a dry season (Bernstein 1964) and a wet season (Altmann 1959) study of *A. palliata* on Barro Colorado Island, all non-dawn chorus vocal bouts were directed at another group and occurred at the edge/border of the callers home range. Similarly, Drubbel and Gautier (1993) advocate that *A. macconnelli* acoustically mark their home range borders with their loud calls. Likewise, da Cunha and Jalles-Filho (2007) found that *A. guariba* calls occurred disproportionately on the borders of their range, despite no indication that they used the border more intensely. It appears that calling in this population served to reinforce borders, particularly in areas susceptible to invasion by other groups. Because males defending females or vulnerable offspring should not have a site-specific pattern in their calling, da Cunha and Jalles-Filho instead suggest that the group was defending their entire home range. In another study of *A. guariba*, Oliveira (2002) also argued that defense of space or specific food sources was the cause of most intergroup conflicts, although he did not discard the mating defense hypothesis in some circumstances or at other sites. However, in a third population of *A. guariba* where they live in contact with *A. caraya*, Holzmann et al. (2012) found no relationship between howling frequency and location (exclusive areas, boundary areas, or important feeding sites) or seasonality (despite food availability changing markedly over the study period), although some groups tended to howl in areas of their home range that were closest or overlapped with conspecific groups (but not heterospecific groups: see below). However, as the authors point out, there was a lower roaring than rate other study sites of this species, likely due to a lower population density. Thus, perhaps food competition is not as strong as at other sites.

In sum, many studies have found evidence that males defend aspects of their group's home range, be it an important site, a well-used quadrant, an area, or a boundary. Any variation in how/where space is defended is probably dependent on factors such as population density, habitat quality, and the extent of home range overlap among groups. Whether this defense protects food resources, sleeping sites, or merely space is not clear and may also vary among populations.

14.6.2 Female Defense

Few studies have found evidence for mate defense in *Alouatta* (see also Wich and Nunn 2002), a hypothesis originally proposed by Sekulic (Sekulic 1982b; Sekulic and Chivers 1986). The strongest argument so far for mate defense comes out of a contact zone between *A. guariba* and *A. caraya*. In this population, there is ample heterospecific but almost no conspecific home range overlap (Agostini et al. 2010b), and

Holzmann et al. (2012) reported that subjects called more at conspecifics than heterospecifics (see below). Given both species have nearly perfect overlap in their feeding niche (Agostini et al. 2010a), the howling patterns seen are not consistent with defense of food/space. Instead, these findings are suggestive that howling is, at least in part, used in defense of females from potentially transferring male conspecifics.

14.6.3 Infant Defense

Given infanticide has an obviously strong impact on reproductive fitness and has been documented in at least eight populations of five howler monkey species (reviewed in Van Belle et al. 2010), it seems reasonable that calling could be used to defend vulnerable offspring. Kitchen (2004) conducted a playback study on *A. pigra* where she presented 12 central males with the recordings (both barks and roars) of unfamiliar, and therefore potentially infanticidal, males. Central males had an overall stronger howling response to playbacks if they had offspring in their group that were younger than 9 months old (the age at which they remain vulnerable to infanticide: see Crockett and Sekulic 1984). In fact, the only time males called in trials when the simulated group outnumbered their own was when there was a small offspring in the group. Still, 94 natural interactions between neighboring, and thus familiar, groups in this population revealed no effect of small offspring presence on contest outcome (Kitchen 2000). Although male transfers and takeover events are relatively common at this site (e.g., Horwich et al. 2000) and infanticide has been observed (Brockett et al. 1999), these events are relatively uncommon between neighboring groups. Assuming howler monkeys can discriminate among individuals based on their calls, then the playback study (Kitchen 2004) was more likely than the observational study to simulate an actual infanticidal threat.

Holzmann et al. (2012) observed 79 natural howling bouts produced by four different male *A. caraya* and *A. guariba* and found no pattern related to the presence of small offspring (see also *A. guariba*: da Cunha and Jalles-Filho 2007). However, neighbors were unlikely to pose an infanticidal threat at this site; in fact, to date, no immigration events or infanticide has been observed at this site (Holzmann pers. obs.). More studies employing experimental playback studies (as in Wich et al. 2002) will be necessary to rule out infant defense and to disentangle it from mate and food defense.

14.7 Loud Calls and Within-Group Male Cooperation and Competition

Loud calling bouts may also function in male-male competition within groups. For example, Fialho and Setz (2007) report howling during an event where one resident male permanently ousted another in *A. guariba*. During a year-long study, Sekulic (1982b) reported 20 intragroup aggressive interactions among males in one of her four study groups, six (30 %) of which resulted in short roaring bouts of less than 1 min.

When one group is in a contest with another, the group males might have similar interests and should therefore join each other in a vocal display directed at extra-group competitors; yet, multi-male social groups do not always call together. For example, in a group of *A. guariba*, the central male initiated almost all loud calling bouts, whereas the subordinate male, a newcomer in the group, participated in less than 50 % of episodes (Oliveira et al. unpubl. data). Similarly, the central male initiated all howling bouts in a group of *A. caraya*, and although the subordinate adult male joined him during 87 % of these bouts, the two subadult males only joined during an average of 28 % of bouts (da Cunha 2004). Dias et al. (2010) reported that in two three-male groups of *A. palliata*, one or more noncentral males joined the central males only 65–70 % of the time. Similarly, in *A. pigra*, Kitchen (2000) found that noncentral males in nine groups joined the central male in howling during only 59 % of 112 natural intergroup encounters.

If howling helps defend some aspect of the group (space, food, females, infants), an interesting question is what might lead noncentral males in multi-male groups to participate with the central male as opposed to "free-riding" (see reviews in Nunn 2000; Nunn and Lewis 2001; Kitchen and Beehner 2007). For example, in *A. arctoidea*, Sekulic (1982b, 1983) suggested that males in strong alliances had longer roaring bouts than males in weak or antagonistic male-male relationships. Similarly, Dias et al. (2010) found that a coalitionary dyad of *A. palliata* howled together more often than either male called with the usurped male (who remained a resident in the group following the takeover). Additionally, these three males howled together only half as often as another three-male group that had been in a stable relationship for many years. Likewise, in Belizean *A. pigra*, five noncentral males in long-term relationships with the central male in their group had much stronger responses (i.e., called for longer, were quicker to approach and got closer to the speaker) to playbacks simulating intruders than five noncentral males in short-term relationships with their central male (Kitchen et al. 2004). On the other hand, in an observational study of two multi-male groups of Mexican *A. pigra* (van Belle et al. 2008), coalitionary males and long-term residents were not more likely to have more affiliative or fewer aggressive interactions than other dyads; in other words, males in this population howled together regardless of relationship duration. However, the authors point out that this group was studied during a socially unstable time period with frequent male membership changes.

Future studies should attempt to include measures of relationship status (preferably genetic evidence of relatedness) and of reproductive skew among males. Playback experiments are useful to increase sample size of intergroup encounters and also to simulate unfamiliar and thus potentially more threatening rivals (e.g., Ydenberg et al. 1988; Kitchen 2000).

14.8 Different Types of Loud Calls

Although loud call types are distinctive (see description and spectrograms in da Cunha et al. 2015), Whitehead (1985) is one of only a few contemporary authors who has actually tested for functional variation among call types. He found that *A. palliata* had site-dependent responses to roaring but not to barking.

Responses to barks were instead dependent on acoustic features that mimicked approach or retreat. In an observational study of *A. palliata*, Baldwin and Baldwin (1976) found that whether loud calling bouts included roars depended on the intensity of the eliciting stimulus. Likewise, in *A. pigra*, loud call bouts are more likely to include roars during close than distant interactions, although there may be some individual variation among males in this tendency (Kitchen 2000 pers. obs.). This contextual difference may be even clearer in South American species because they are less likely to combine barking and roaring in the same bout. In *A. guariba*, for example, Chiarello (1995) found that 39 of 43 close encounters elicited loud calling bouts made up of only roars, whereas only four of these encounters elicited barks only or barks plus roars.

There may be even more subtle differences in the graded calls within a call type. For example, Drubbel and Gautier (1993) categorized two types of roars in *A. macconnelli* and found that "long roars" (>1 min) were typically produced during nighttime choruses (58/62 cases), whereas "short roars" were frequently produced during short-range interactions (36/62). Whitehead (1987) also found a difference among roar types in *A. palliata*—he reported that "roar variants" (see da Cunha et al. 2015) were typically produced in dawn choruses or during mild/distant interactions, whereas "full roars" were more likely to be associated with close encounters.

However, a cross-species comparison is premature because it hinges on resolving nomenclature issues that exist in the literature (see da Cunha et al. 2015). The problem is compounded by the different nature of loud calling in the two taxonomic clades of *Alouatta* (Cortés-Ortiz et al. 2003); for example, the howling bouts of Central American species typically include both roars and barks, whereas the two call types are usually produced in separate bouts in South American species. Virtually all studies on South American species have focused on roars; thus, barks remain a largely unexplored vocalization.

14.9 Mechanisms of Competition: Form Meets Function

Since authors do not always state at which level they are working, the bioacoustics literature is full of seemingly different "functions" for the same call; however, we suggest that many researchers are simply approaching the same phenomenon from different angles. Here we consider another level of analysis that we have not considered in detail so far—a more proximate approach. Although there is of course an ultimate function of competition, we focus here on the mechanisms by which males announce "intent," convey individual or group-level RHP, or signal deceptively.

14.9.1 Motivational State

Calls such as those produced by howler monkeys have been interpreted as aggressive signals, part of the motivational-structural rules proposed by (Morton 1977; Owings and Morton 1998). In this theoretical model, aggressive calls are harsh, low-frequency sounds that mimic the larger body size of a more dominant and dangerous animal, because larger individuals have larger vocal folds that vibrate at lower frequencies and in a more unstable pattern than smaller individuals, generating lower-pitched, atonal sounds (e.g., August and Anderson 1987; Hauser 1993). Morton (1977) claimed that a dominant frequency around 1,500 kHz would be effective in long-distance propagation in forests, yet howler monkeys have roars with much lower frequencies (<1,000 kHz, da Cunha et al. 2015); therefore, Morton viewed the unnecessarily low-frequency roars as an aggressive long-range signal that originally evolved from a short-range signal. Aggressive signals are also predicted to be intense (Bradbury and Vehrencamp 1998), and the amplitude of howler monkey loud calls, usually interpreted as necessary for long-range signaling, is also intense in close range confrontations between groups.

In support for Morton's theory, Oliveira (2002) observed that, relative to spontaneous choruses, intergroup encounters in *A. guariba* were associated with longer and louder loud calling bouts, perhaps relating to a greater aggressive motivation in such contexts, and we have noticed this phenomenon in other species as well (*A. caraya*: da Cunha pers. obs.; *A. pigra* and *A. palliata*: Kitchen pers. obs.). Similarly, Whitehead (1994) observed that the first formant in roars of *A. palliata* uttered in response to playbacks of roars from unfamiliar males had lower frequencies than those emitted naturally. He also found that males produced even lower frequency calls when playbacks elicited an approach than when they elicited a retreat, a possible effect of the motivational state of the caller in the production of these sounds, which would parallel the trends predicted by Morton. Although vocal frequency can be correlated with a male's fighting ability or condition, when the same male flexibly changes frequencies during different contexts, it may possibly reflect some sort of affect, motivation, or deception.

Game theoretical predictions suggest that when signaling intentions is low in cost, then individuals should always "lie" about intentions, and these cues will become meaningless to rivals. However, if there is a threat of retaliation, then high cost signals (e.g., Poole 1989) could be used as honest indicators of intention (Zahavi 1977). This is an interesting avenue for further research in howler monkeys.

14.9.2 Resource Holding Potential: RHP

Because they are so salient and so clearly tied to male-male competition, many researchers (e.g., Sekulic 1982b; Chiarello 1995; Kitchen 2000; Oliveira 2002) have suggested that howler monkeys assess one another's RHP using reliable features of the loud calling bouts. The source-filter approach (Owren and Linker 1995; Fitch and Hauser 1995; Frey and Gebler 2010) proposes that the vocal tract can provide cues to the body size of a vocalizing animal: the resonances present in a vocal tract are dependent on its extension, which has a direct relationship with the size of the caller. Harsh sounds or those with a low fundamental frequency and several harmonics should accurately reflect these resonances and could therefore be honest signals of body size. Because howler monkey loud calls have wide frequency bands, they should clearly show these resonances, or formants, in their structure. As we discuss in elsewhere (da Cunha et al. 2015), marked and usually stable frequency peaks are found in the roars of most *Alouatta* species indicating the presence of such formants. Kitchen (2000) found a relationship between the number of formants per roar, some formant frequencies, and the width of some bands with age and size in male *A. pigra*. In a set of playback experiments on this same population, subject responses were strongly correlated with some of these same acoustic features.

However, Fitch and Hauser (1995) also remarked that other cues, like call duration and emission rate, may be more reliable signals of RHP than static body size ones because they provide better indication of the present energetic condition of the caller (e.g., baboons, *Papio* spp.: Kitchen et al. 2003; Fischer et al. 2004; red deer: Clutton-Brock and Albon 1979). If aspects of a loud call are energetically costly to produce, a more fit animal should be able to vocalize louder, longer, and more frequently than a weaker opponent, leading to dynamic processes of evaluation between opponents (e.g., Zahavi 1977; Payne 1998; Frey and Gebler 2010).

In *A. pigra*, Kitchen (2000) reports that higher roaring rates per bout, longer periods of continuous loud calling per bout (including roars, barks, and pauses of less than 1 s), and lower proportion of silent periods per bout were correlated not only to age and body size but also to which group won a natural contest. When she experimentally manipulated the proportion of loud calling per bout, subjects had the strongest howl and move response to males who were most similar to them and had the weakest response to males whose acoustic features suggested that they had either a higher or lower RHP.

Finally, group-level fighting ability is another aspect that can be reliably indicated in howler monkey choruses in that, at least in *A. pigra*, multiple calling males offset their roars and barks so that at least a minimum estimation of males in the group can be determined. Using playbacks to simulate invasion by strangers, Kitchen (2004) found that the relative number of males in two groups (i.e., the "numeric odds": McComb et al. 1994) influenced the responses of central males (although not the noncentral males: Kitchen et al. 2004) during playback experiments. Central males had a stronger response (defined above) the more their group outnumbered the simulated group (see also lions, *Panthera leo*: McComb et al. 1994; chimpanzees, *Pan troglodytes*: Wilson et al. 2001). Conversely, Kitchen (2000) found that numeric odds did not influence contest outcome among familiar opponents. Interestingly, only when the odds were even and thus the outcome was least clear based on group-level fighting were responses during playbacks highly correlated with specific acoustic features (as above) of individual callers. This suggests that howler monkeys might employ a system of either sequential or cumulative assessment (e.g., Payne 1998).

In a general perspective, it is likely that the structure of loud calls in howler monkeys reflect several selective forces acting upon their design. Perhaps a species like

A. guariba, which are unique among *Alouatta* in that dawn choruses are absent (reviewed in da Cunha et al. 2015) and whose roars are used almost solely during direct confrontations between groups, can shed more light on the contributions of honest signaling and long-range communication in shaping the form of howler monkey loud calls.

In this sense, perhaps the brief forms of roaring in *A. palliata* are indicative that a lower selective pressure acted in the elaboration of the roars of this species when compared to other members of the genus that evolved longer continuous roars. Sekulic and Chivers (1986) proposed that *A. palliata* living in larger, multi-male groups faced greater intragroup competition compared to *A. arctoidea*, who, they suggested, had longer calls due to the pressures of intergroup competition. However, the overall bouts (including pauses between loud calls) of *A. palliata* are very long, so this hypothesis requires more testing.

14.9.3 Deception

One last form and function nuance was pointed out by Fitch and Hauser (1995), who observed that vocal resonances could also be manipulated and therefore provide opportunities for deceptive signaling. They suggested that lip protrusion could be used as a maneuver that would lower the dominant frequency in a vocalization, simulating a longer vocal tract and, as a consequence, a larger body size (see also baboons: Fischer et al. 2004). Lip protrusion is clearly visible in roaring howler monkeys (Schön Ybarra 1986), and perhaps the wider opening of their mouths in the inhaling phase can be related to the rising modulation observed in this period (see also da Cunha et al. 2015).

Fitch and Hauser (1995) further suggested that laryngeal air sacs, which howler monkeys have, could mimic a larger body size. Additionally, the hyoid bulla in *Alouatta* is a kind of rigid laryngeal air sac (Schön Ybarra 1988, 1995), and this organ plays an important role in first formant production (see modeling studies by Riede et al. 2008; de Boer 2009), and dynamic articulations are probably responsible for the modulation of howler monkeys formants, perhaps resulting in manipulation of body size cues. This possibility is reinforced by Whitehead's (1994) study showing a lower pitch in the roars of *A. palliata* males when responding to the roars of strangers during playbacks: they could be simulating a larger body size through such a maneuver rather than a greater aggressive motivation as has been proposed. Such hypotheses are not in fact mutually exclusive given an aggressive animal could also mimic the lower dominant frequencies found in the calls of larger animals (see above), and a clear distinction between motivation versus body size simulation explanations is difficult to make. Interestingly, Schön Ybarra (1995) found that *Alouatta* was the single exception to the correlation between body size and vocal tract length among primate species, with the extension of the vocal organ in howler monkeys being close to that found in the much larger *gorillas*.

Game theory suggests that, unless it is done relatively infrequently (reviewed and tested by Hughes 2000), bluffing will not be a stable strategy because animals should periodically test and retaliate against dishonest signals. Thus, although the maneuvers of howler monkeys described here may alter the features of a call, if they are done regularly, it is possible that they would no longer be a part of the suite of features assessed by rivals. However, if some animals can exaggerate more than others based on some underlying characteristics, the signal might remain honest (but see Bee et al. 2000).

14.10 Girl Power

Although females also produce loud calls, relatively little has been published in this area, particularly with respect to the acoustic structure of these calls. In both acoustic features (da Cunha et al. 2015) and functional strategies of these vocalizations, we predict that females will have differences that are not necessarily correlated with those of conspecific males and we also anticipate strong interspecific differences.

As in males, the loud calls of females likely serve multiple functions including alarm calls and group cohesion. For example, Steinmetz (2005) suggested that female *A. guariba* produced loud calls when isolated or "lost" (a common occurrence in her study); females called alone in three such situations, during 92 days of fieldwork (see also Oliveira 2002).

Perhaps the most interesting question is why females should participate in choruses with males. In most species, it appears that females are only occasional participants. For example, Chiarello (1995) reported that *A. guariba* females participated in 31 % of howling bouts, Holzmann et al. (2012) found *A. caraya* and *A. guariba* females participated in 29 %, Whitehead (1989) reported *A. palliata* females joining in 18 %, and Kitchen (2006) reported that one or more *A. pigra* females joined in 47 % of bouts.

Because the dominant male is usually responsible for the onset of roar emissions by a group, the participation of other males and females can be seen as a form of cooperation with the alpha male (e.g., Kitchen et al. 2004; Kitchen 2006). However, the final decision about participation should be contingent on the costs and benefits of the individual in that particular situation (reviewed in Kitchen and Beehner 2007). For example, females in several species seem to be more likely to invest by joining howling bouts during close interactions with other groups rather than during interactions with distant groups or during spontaneous choruses (e.g., *A. palliata*: Baldwin and Baldwin 1976; *A. guariba*: Chiarello 1995; *A. pigra*: Kitchen 2000; *A. caraya*: da Cunha and Byrne 2006; *A. guariba* and *A. caraya*: Holzmann et al. 2012).

In multi-female groups, females may differ among themselves in their decisions to join. For example, Kitchen (2006) found that on occasions that females joined a chorus, only 60 % of the females present participated. Similar findings have been found in a group of *A. guariba*, with one of the two females in the group showing a greater degree of participation in loud calling bouts than the other (Oliveira et al.

unpubl. data). Thus, the focus should be on the conditions that vary among females that may be related to their participation in the sessions.

In the first thorough approach to the subject, Sekulic (1982b, 1983) presented a series of hypotheses that we examine here. First, she hypothesized that the roars of female *A. seniculus* incite male competition. However, this predicts that females should call first, which rarely happens in howler monkeys; in fact, it is more likely that the male's incipient roars, usually uttered at the onset of roar bouts (da Cunha et al. 2015), can act as a recruitment call, prompting other group members to join in the roar chorus (Oliveira 2002). Moreover, Sekulic's hypothesis suggests that males are in a state of constant intragroup competition, another unlikely assumption. It is also unclear why males should need female loud calls as incentive to compete.

Second, Sekulic (1982b, 1983) hypothesized that females roar to intimidate and thereby deny access of extra-group females, as a way to limit competition for food or mates. In her study, *A. arctoidea* males and females frequently roared at solitary females. In fact, Sekulic (1983) also described female-only sessions in *A. arctoidea*, apparently directed at other females. Oliveira (2002) also reported three episodes of loud calling by *A. guariba* females alone, when the central male of the group was injured and not always with the group. Additionally, Miranda et al. (2004) describe a case where an *A. guariba* female became the dominant member of a group, the most frequent caller and sometimes the only caller. However, none of the *A. guariba* incidents were apparently directed solely at females, and female-only sessions have not been recounted in other study populations. In *A. caraya*, da Cunha and Byrne (2006) reported an encounter between a group and a lone female that did not result in any howling. Thus, it remains possible that there are differences among *Alouatta* species in their tolerance to female immigration.

Future studies designed to evaluate whether female loud calls affect female emigration/immigration need to consider the following: target of the call (neighbors vs. strangers), proximity to target, sex of target, influence of calling on migration decisions, and age of the calling female (e.g., older and established females should be more resistant to migration, and, thus likely to participate in loud calling more often, particularly given hierarchy is inversely related to age in some howler monkey species: Jones 1980).

Related to the above hypothesis, Sekulic (1983) also hypothesized that, through calling, females may attempt to control access to the group's central male. In *A. arctoidea*, Sekulic documented cases of intragroup female-female competition over proximity to certain males during a howling bout. Among females within the same group, cooperation with the male could be just one more aspect of a suite of behaviors connected to status and hierarchy maintenance. If so, participation should be directly proportional to a female rank, and females may even attempt to interfere with one another, with high-ranking females preventing close access to the central male or directly interfering with the call production of lower-ranking females. The biggest obstacle to testing these predictions is determining female rank hierarchies in most *Alouatta* species.

In a third hypothesis, Sekulic (1982b, 1983) proposed that females loud calling alongside a central male could provide pair bond reinforcement and thereby encourage him to protect infants against infanticide. Under this hypothesis, female participation could reflect the risks associated with a takeover—the main source of infanticide risk. Playback experiments are an excellent way to simulate infanticidal threat because callers can be unfamiliar to subjects. To analyze female decisions to participate, aspects that could be manipulated via playback studies include the numeric odds (assuming more males in a group means better protection from invasion by potentially infanticidal males), the presence of small offspring at an age where they are still vulnerable to infanticide (see formula in Crockett and Sekulic 1984), and the number of females (maybe more females can better defend against potentially infanticidal males, even if the takeover is successful).

To test this, Kitchen (2006) measured the responses of females with and without vulnerable offspring to the sounds of unfamiliar, and therefore potentially infanticidal, males. Curiously, the presence of small, vulnerable offspring did not predict participation in a chorus (which is in contrast to the strong response of the central males in that population: see Section 14.6.3). More recently, Holzmann et al. (2012) studied the natural behavior of both *A. guariba* and *A. caraya*. The presence of infants also did not influence female decisions to join a session or not (see also da Cunha and Byrne 2006). However, Sekulic (1982d, 1983) provides several reports that, following a takeover, one cycling female *A. arctoidea* howled with the new dominant male, whereas females who were pregnant or who had a vulnerable offspring in the group continued to call with the former dominant male (the likely sire of their offspring). This strategy of choosing the usurped over the new and potentially infanticidal male leaves opens the suggestion that females may bond with males through howling as a counterstrategy to infanticide.

Finally, in a related and not mutually exclusive possibility, da Cunha and Byrne (2006) hypothesized that females could cooperate with males in the coordination of space use (be it border/resource defense, announcement of occupation, or mutual avoidance). Female decisions to participate should also be contingent on the situation, but reflecting the benefits and risks associated with guaranteeing an exclusive area and/or resources. Other aspects that could be tested in future studies of this hypothesis include numeric odds based on the entire group size, irrespective of sex, presence of relatives in the group (not just vulnerable offspring), and female status (with older, established females expected to invest more in defense than young and potentially migratory females).

Supporting a defense scenario, Whitehead (1989) found that groups of *A. palliata* were much more likely to move away from playback recordings that included the sounds of females roaring than to those with only males roaring. Using another angle to examine group-defense (although it is unclear what is being defended—space, resources, mates, or offspring), Kitchen (2006) found that females were most likely to join a howling bout if the numeric odds (resident vs. intruder males) were even. This result suggests that females join when their assistance would best improve the group's odds of winning a contest. Their responses were also different from males, who tended to join when odds were most in their favor (Kitchen 2004) or when they had a long-term relationship with other intragroup males (Kitchen et al. 2004). This provides further evidence that males and females differ in their strategies.

Future research should remain sensitive to potential sex differences in usage and function of loud calls. After all, males and females have different ecological needs (Trivers 1972) and usually differ in the strategy they use to solve problems. For example, if in a given species males migrate and females stay in the natal group, males could be more concerned with takeover attempts (especially in non-infanticidal populations), while females might be more interested in securing an area on a longer-term basis (Emlen and Oring 1977). The important message here is not to neglect the issue by considering females as merely supporting actors. Female decisions to call might reflect different pressures and reveal different functions to their calling behavior than males (e.g., Hill 1994), providing rich insights for socio-ecological theory.

14.11 Sympatric Zones and Hybrid Voices

Understanding the evolution of howler monkey loud calls would require an extensive comparative study within the *Alouatta* genus and between howler monkeys and their sister taxon, the Atelinae (Eisenberg 1976; Oliveira and Ades 2004). However, areas of sympatry and hybrid zones also represent novel scenarios for evolutionary studies (Hewitt 1988), addressing aspects of both behavioral ecology and vocal behavior.

Most of the studies of vocal behavior carried out on two (or more) sympatric primate species (none of them in howler monkeys) are related to alarm calls and mutual benefits from heterospecific associations (e.g., Fichtel 2004) or to diurnal distribution of vocal patterns (Geissman and Mutschler 2006). Another aspect of vocal behavior in sympatric primates, virtually unstudied, is related to the mutual influence from closely related species living in sympatry that could result in divergence of some vocalizations (especially long-distance calls) due to character displacement (Brown and Wilson 1956; Marler 1973) or convergence of vocal signals due to vocal learning. While character displacement (e.g., Kirschel et al. 2009) and convergence (e.g., Baker 2008) have been demonstrated in the calls of amphibians and birds, we do not know if it plays a role in the diversification of primate communication. Primates, in contrast to birds and cetaceans, have long been considered inflexible in their vocal behavior (e.g., cross-fostering between two *Macaca* spp. resulted in little vocal change: Owren et al. 1993). Although studies have demonstrated that there is learning involved in call usage (e.g., reviewed in: Seyfarth and Cheney 2010), primate repertoires and the structure of their vocalizations have been considered largely innate. However, recent studies have begun to question this assumption, demonstrating acoustic variation at different levels—regional dialects (e.g., *M. sylvanus*: Fisher et al 1998; *Pan troglodytes*: Clark Arcadi 1996; but see Mitani et al. 1999), call convergence within groups in the same population (e.g., *Cebuella pygmaea*: Elowson and Snowdon 1994; *P. troglodytes*: Crockford et al. 2004), and changes within individuals of the same population (e.g., *Nomascus concolor*: Sun et al. 2011).

Researchers (Kitchen DM, Bergman TJ, Cortes-Ortiz L, unpubl. data) are investigating the impact of sympatric species on one another in howler monkeys in a zone of contact between two species (*A. pigra* and *A. palliata*) in Tabasco, Mexico. They found that roars are at least partially genetically determined, since roars from both species living in sympatry have similar acoustic features to their allopatric conspecifics. However, these researchers also found enough notable difference between allopatric and sympatric conspecifics—with sympatric animals from the contact zone converging slightly in a few acoustic features (see Fig. 14.1)—to question whether this is the result of simple variation within a species, learned behavior, and/or extensive backcrossing of hybrids (the latter has been confirmed in this population: Kelaita and Cortés-Ortiz 2013).

Beyond acoustic features, there is also the question of how heterospecifics respond to each other within contact zones. Holzmann et al. (2012) conducted a year-long study of two groups of *A. guariba* that overlap with two groups of *A. caraya* at a site in northeastern Argentina. Both species were more likely to howl at conspecifics

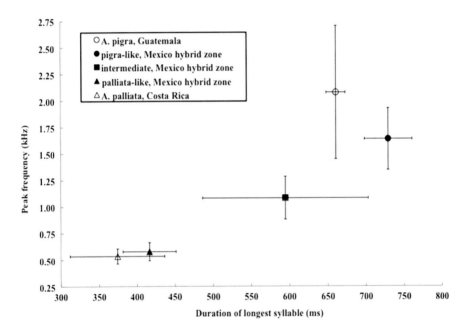

Fig. 14.1 Mean ± SD of longest sustained syllable and peak frequency of roars (see Fig. 14.2) from males recorded within and outside a hybrid zone. Legend indicates species and provenance of calls with 4–8 roars from 2–4 individuals used per point (measurements made with PRAAT 5.1.02). Data from a pilot study by Kitchen, Bergman, and Cortés-Ortiz (unpubl. data). The genetic makeup of most individuals in the hybrid zone in Tabasco, Mexico, is the result of multigenerational backcrossing, with an admixture that ranges from nearly purebred of each parental type to intermediates (Kelaita and Cortés-Ortiz 2013). Individuals were categorized as *pigra*-like, intermediates, and *palliata*-like based on morphological features (but see Kelaita and Cortés-Ortiz 2013 for problems using only morphology)

(83 % of 18 encounters) than at heterospecifics (21 % of 29 encounters) during close range intergroup encounters and never howled at distant interactions with heterospecifics (yet howled at 57 % of distant conspecifics roars). Similar findings were reported by Aguiar (2010), who studied hybrid groups (*A. caraya*×*A. guariba*) in a population where *A. guariba* predominates and found greater agonistic responses (e.g., piloerection, roaring) during intergroup encounters of conspecifics, followed by encounters with groups composed of purebreds and hybrids, and the weakest responses during interspecific encounters. Playback experiments in the contact zone in Mexico suggest similar trends in *A. pigra* and *A. palliata* (Kitchen et al. in prep).

Even more information on the evolution of vocalizations can be ascertained when species living in sympatry actually hybridize individuals is a subject relatively well studied in birds and anurans, information is almost absent for mammals and especially for primates, despite natural primate hybridization observed in many taxa (Gabow 1975; Bynum et al. 1997; Alberts and Altmann 2001; Detwiler et al. 2005) including those of howler monkeys: (Cortés-Ortiz et al. 2007). Studies carried out so far on non-primates describe three different patterns in hybrid vocalizations: (1) the hybrid can inherit one of the two parental songs; (2) the hybrid can have an intermediate song, formed by the mixture of elements from the two parental songs; and (3) the hybrid can have a unique song, different from both parental songs (anurans: Blair 1958; Gerhardt 1974; Scroggie and Littlejohn 2005; birds: Ficken and Ficken 1967; Lemaire 1977; de Kort et al. 2002). Relatively little is known about how hybridization would impact nonhuman primate communication. A study on vocalizations in *H. muelleri* and *H. lar* in captivity revealed that hybrids show both types of songs—a female hybrid produced a unique song, whereas a male hybrid produced an intermediate song (Tenaza 1985; see also *Saimiri sciureus* hybrids: Newman and Symmes 1982). Because of their distinctive calls, the few contact zones between different howler monkey species in this otherwise parapatric genus might reveal additional insights into the function and evolution of vocalizations. In some of these sympatric zones, mixed groups and hybrids have been observed (Cortés-Ortiz et al. 2007; Aguiar et al. 2007, 2008; Agostini et al 2008; Bicca-Marques et al. 2008). To the best of our knowledge, however, there is currently only one ongoing study focused on the vocal behavior of hybrids (*A. palliata*×*A. pigra* in southern Mexico: Kitchen et al. in prep). Preliminary results from this site suggest that genetically intermediate hybrids produce a roaring behavior intermediate between both parental roars (Figs. 14.1 and 14.2), further evidence for strong genetic influence on howler monkey vocal behavior.

14.12 Summary and Future Directions in Vocal Research

Both here and in da Cunha et al. (2015), variation has been our theme. It seems that howling (including roars and, in the case of Central American species, barks) has different functions within a population as well as between populations. Although calls produced at any time might "regulate space use," there is a possibility that the

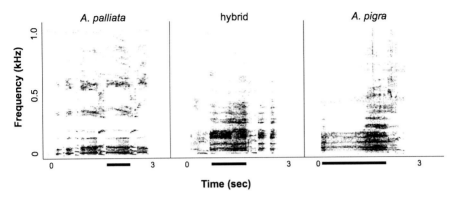

Fig. 14.2 Example spectrograms of roars from a male purebred *A. palliata* (*left*), *A. palliata* × *A. pigra* hybrid (*middle*), and purebred *A. pigra* (*right*). Dark bars under the *x*-axis indicate the duration of the longest sustained frequency in each roar (From a pilot study by Kitchen, Bergman and Cortés-Ortiz (unpubl. data))

more ritualized, spontaneous dawn choruses may function more in avoidance between groups than daytime calls. Alternatively, we argue that both dawn and daytime calls might function in sequential assessment of rivals. Playback experiments along with simultaneous documentation of movement patterns in neighboring groups (using GPS tracking technology) will be useful to test between these hypotheses.

Among species and populations, the rate and patterns of vocal battles and intergroup encounters varies, as does the likelihood that groups will escalate to physical aggression. Much of the variability probably has to do with the cost-benefit ratio of avoiding confrontations based on population density (e.g., Lichtenberg et al. 2012), the extent of home range overlap, availability of mates, habitat quality, level of feeding competition, and the threat of infanticide. However, some of this variation might have to do with the varying competitive nature among species. For example, perhaps *A. palliata* differs from other species in the likelihood that encounters will escalate because their social system is driven more by intra- rather than intergroup competition.

Although there is strong support that howling evolved at least in part under male intrasexual selective pressures, the focus of the competition is less clear. In at least one population of every species highlighted in this chapter, there is some evidence that males defend resources (either important sites, quadrants, areas, or home range boundaries). It remains unclear if such space/resource defense is driven by the mutual goals of males and females, by male defense of food/space to attract females, or if it is merely coincidental, with males acting as "hired guns" while they defend mates or infants (Wrangham and Rubenstein 1986). Despite infanticide being confirmed at several sites, only one study in *A. pigra* (Kitchen 2004, 2006) found support that howling is used to defend vulnerable infants. However, we suggest that more studies need to incorporate playback tests that include the sounds of unfamiliar, and therefore potentially infanticidal, intruders. In terms of mate defense, only a few studies have so far provided evidence (Sekulic 1982b; Holzmann et al. 2012). Endocrine research has demonstrated that central males are sensitive to threats to their reproductive success, suggesting that males compete for females (Cristóbal-Azkarate et al. 2006; Van Belle et al. 2009b; Rangel-Negrín et al. 2011), but future research should search for causal connections between reproductive access, hormone levels, and loud calling. Whereas legitimate arguments against the mate defense hypothesis have been proposed (e.g., howling happens more often than females cycle, howling should not be site specific, etc, see also Wich and Nunn 2002), most researchers seem reluctant to reject this hypothesis outright. However, the logistics of designing a mate defense study remain problematic given there are no reliable external indicators of reproductive state and both sociosexual behaviors and immigration events are observed only rarely.

After decades of speculation that males use howling bouts to assess their rivals, there is finally some evidence that howler monkey loud calling bouts may be honest indicators of RHP (e.g., Frey and Gebler 2010). Such studies are at the intersection of form and function in vocal research. However, confusion has been created in the bioacoustics literature because authors fail to define the level at which they are working (e.g., Tinbergen 1963). For example, uncovering proximate mechanisms such as evidence for a reliable relationship between acoustic features of calls and an individual's RHP (be it male age, size, condition, stamina, fighting ability, numeric odds) is not an ultimate explanation for the call. Only experimental playback tests can determine if receivers actually attend to these particular acoustic features (e.g., Kitchen 2004) and thus whether it affects contest outcome and fitness. Likewise, if studies continue to confirm that calls function in intergroup spacing in some population and thus impact fitness, this does not explain the particular strategy employed that resulted in this spatial distribution.

Because their functional significance remains unsettled, the ideal approach to future howler monkey vocal research is to simultaneously address as many alternative hypotheses within each explanatory level as possible, using a combination of systematic observational studies with carefully planned field experiments. Given questions of intergroup relationships are so central in the function of howling, the ideal observational methodology would include multiple researchers observing multiple groups simultaneously (reviewed in Kitchen and Beehner 2007). Because sites and subjects can vary in ways that are only obvious when you have visited them, the same researchers should preferably visit different populations of their species and even different species (we found this to be very enlightening!). Of course, we are aware that these approaches are not always logistically feasible or economically possible. When not cost-effective, we believe researchers would benefit from the kind of collaborative effort we have used to create this review.

Acknowledgments The coauthors are grateful to the anonymous reviewers and to the editors for constructive comments and for the opportunity to be involved in this book. RGTC is grateful to EMBRAPA, Richard Byrne, Marcelo Oliveira Maciel, Emiko Kawakami de Resende, José Anibal Comastri Filho, Gentil Cavalcanti, Marcos Tadeu, Sandra Santos, CAPES (studentship #1373/99 4), and St. Leonard's College at University of St. Andrews (for Russell Trust Award). IH wishes to thank Mario Di Bitetti, Nacho Areta, Ilaria Agostini, and Daphne Colcombet, as well as

The National Research Council from Argentina (CONICET) for funding. DMK is indebted to University of Minnesota (UMN), The Ohio State University (OSU), and NSF (grants # BCS-0962807 and 0962755) for funding. She is also grateful to UMN colleagues, advisors, and field assistants; L Cortes-Ortiz, T Bergman, PAD Dias, D Canales Espinosa, A Rangel-Negrin, C Rodriguez Muldonado, A Gomez Martinez, and A Coyohua; OSU colleagues, administrators, and students; and, most importantly, Jim, Dylan, and Nash Nicholson. DAGO wishes to thank César Ades (in memoriam), Zelinda Hirano, Júlio César de Souza Jr., Fernando Déscio, Simone Porfírio, Marcelo Marcelino, Isabel Coelho, Israel Quintani, and Esmeralda da Silva, for funding from CNPq, CAPES, FAPESP, and FINEP, as well as the following institutions: University of São Paulo, Instituto Florestal—P.E. da Cantareira, CEPESBI, FURB, Municipality of Indaial, and IBAMA/PB, RPPN Fazenda Pacatuba.

References

Agostini I, Holzmann I, Di Bitetti MS (2008) Infant hybrids in a newly formed mixed-species group of howler monkeys (*Alouatta guariba clamitans* and *Alouatta caraya*) in northeastern Argentina. Primates 49:304–307

Agostini I, Holzmann I, Di Bitetti M (2010a) Are howler monkey species ecologically equivalent? Trophic niche overlap in syntopic *Alouatta guariba clamitans* and *A. caraya*. Am J Primatol 72:173–186

Agostini I, Holzmann I, Di Bitetti M (2010b) Ranging patterns of two syntopic howler monkey species (*Alouatta guariba* and *A. caraya*) in Northeastern Argentina. Int J Primatol 31:363–381

Aguiar L (2010) Sistema social de grupos mistos de espécies de bugios (*Alouatta caraya* e *Alouatta clamitans*) e potenciais híbridos no Alto Rio Paraná, Sul do Brasil. PhD dissertation, Universidade Federal do Paraná

Aguiar LM, Mellek DM, Abreu KC, Boscarato TG, Berbardi IP, Miranda JMD, Passos FC (2007) Sympatry between *Alouatta caraya* and *Alouatta clamitans* and the rediscovery of free-ranging potential hybrids in Southern Brazil. Primates 48:245–248

Aguiar LM, Pie MR, Passos FC (2008) Wild mixed groups of howler species (*Alouatta caraya* and *Alouatta clamitans*) and new evidence for their hybridization. Primates 49:149–152

Alberts SC, Altmann J (2001) Immigration and hybridization patterns of yellow and anubis baboons in and around Amboseli, Kenya. Am J Primatol 53:139–154

Altmann SA (1959) Field observations on a howling monkey society. J Mammal 40:317–330

August PV, Anderson JGT (1987) Mammal sounds and motivation-structural rules: a test of the hypothesis. J Mammal 68:1–9

Axelrod R, Hamilton WD (1981) The evolution of cooperation. Science 211:1390–1396

Baker MC (2008) Analysis of a cultural trait across an avian hybrid zone: geographic variation in plumage morphology and vocal traits in the Australian ringneck parrot (*Platycercus zonarius*). Auk 125:651–662

Baldwin JD, Baldwin JI (1976) Vocalizations of howler monkeys (*Alouatta palliata*) in Southwestern Panama. Folia Primatol 26:81–108

Bee MA, Perrill SA, Owen PC (2000) Male green frogs lower the pitch of acoustic signals in defense of territories: a possible dishonest signal of size? Behav Ecol 11:169–177

Bernstein IS (1964) A field study of the activities of howler monkeys. Anim Behav 12:92–97

Bicca-Marques JC, Mattje Prates H, Cunha R, de Aguiar F, Jones CB (2008) Survey of *Alouatta caraya*, the black-and-gold howler monkey, and *Alouatta guariba clamitans*, the brown howler monkey, in a contact zone, State of Rio Grande do Sul, Brazil: evidence for hybridization. Primates 49:246–252

Blair FW (1958) Mating call in the speciation of anuran amphibians. Am Nat 92:27–51

Bradbury JW, Vehrencamp SL (1998) Principles of animal communication. Sinauer Associates, Sunderland

Brockett RC, Horwich RH, Jones CB (1999) Disappearance of infants following male takeovers in the Belizean black howler monkey (*Alouatta pigra*). Neotrop Primates 7:86–88

Brown WL, Wilson EO (1956) Character displacement. Syst Zool 5:49–64

Bynum EL, Bynum DZ, Suprianta J (1997) Confirmation and location of the hybrid zone between wild populations of *Macaca tonkeana* and *Macaca hecki* in Central Sulawesi, Indonesia. Am J Primatol 43:181–209

Camargo CC, Ferrari SF (2007) Interactions between tayras (*Eira barbara*) and red-handed howlers (*Alouatta belzebul*) in eastern Amazonia. Primates 48:147–150

Carpenter CR (1934) A field study of the behavior and social relations of howling monkeys *Alouatta palliata*. Comp Psychol Monogr 10:1–168

Cäsar C, Zuberbühler K (2012) Referential alarm calling behaviour in New World primates. Curr Zool 58:680–697

Charlton BD, Ellis WAH, Brumm J, Nilsson K, Fitch WT (2012) Female koalas prefer bellows in which lower formants indicate larger males. Anim Behav 84:1565–1571

Cheney DL, Seyfarth RM, Palombit RA (1996) The function and mechanisms underlying baboon 'contact' barks. Anim Behav 52:507–518

Chiarello AG (1995) Role of loud calls in brown howlers, *Alouatta fusca*. Am J Primatol 36:213–222

Chivers DJ (1969) On the daily behaviour and spacing of howling monkey groups. Folia Primatol 10:48–102

Clark Arcadi A (1996) Phrase structure of wild chimpanzee pant hoots: patterns of production and interpopulation variability. Am J Primatol 39:159–178

Clutton-Brock T (2002) Breeding together: kin selection and mutualism in cooperative vertebrates. Science 296:69–72

Clutton-Brock TH, Albon SD (1979) The roaring in red deer and the evolution of honest advertisement. Behaviour 60:145–169

Collias NE, Southwick CH (1952) A field study of population density and social organization in howling monkeys. Proc Am Philos Soc 96:144–156

Cortés-Ortiz L, Bermingham E, Rico C, Rodríguez-Luna E, Sampaio I, Ruiz-García M (2003) Molecular systematics and biogeography of the Neotropical monkey genus, Alouatta. Mol Phylogenet Evol 26:64–81

Cortés-Ortiz L, Duda TF Jr, Canales-Espinosa D, García-Orduña F, Rodríguez-Luna E, Bermingham E (2007) Hybridization in large bodied New World primates. Genetics 176:2421–2425

Cristóbal-Azkarate J, Dias PAD, Joaquim JV (2004) Causes of intraspecific aggression in *Alouatta palliata mexicana*: evidence from injuries, demography, and habitat. Int J Primatol 25:939–953

Cristóbal-Azkarate J, Chavira R, Boeck L, Rodrígez-Luna E, Veàl JJ (2006) Testosterone levels of free-ranging resident mantled howler monkey males in relation to the number and density of solitary males: a test of the challenge hypothesis. Horm Behav 49:261–267

Crockett CM (1984) Emigration by female red howler monkeys and the case for female competition. In: Small M (ed) Female primates: studies by women primatologists. Liss, New York

Crockett CM, Eisenberg JF (1987) Howlers: variations in group size and demography. In: Smuts BB, Cheney DL, Seyfarth RM, Wrangham RW, Struhsaker TT (eds) Primate societies. University of Chicago Press, Chicago

Crockett CM, Sekulic R (1984) Infanticide in red howler monkeys (*Alouatta seniculus*). In: Hausfater G, Hrdy SB (eds) Infanticide: comparative and evolutionary perspectives. Aldine, Hawthorne

Crockford C, Herbinger I, Vigilant L, Boesch C (2004) Wild chimpanzees produce group-specific calls: a case for vocal learning? Ethology 110:221–243

da Cunha RGT (2004) A functional analysis of vocalisations of black howler monkeys (*Alouatta caraya*). PhD dissertation, University of St. Andrews

da Cunha RGT, Byrne RW (2006) Roars of black howling monkeys (*Alouatta caraya*): evidence for a function in inter-group spacing. Behaviour 143:1169–1199 da Cunha RGT, Jalles-Filho E (2007) The roaring of southern brown howler monkeys (*Alouatta guariba clamitans*) as a mechanism of active defense of borders. Folia Primatol 78:259–271

da Cunha RGT, de Oliveira DAG, Holzmann I, Kitchen DM (2015) Production of loud and quiet calls in howler monkeys. In: Kowalewski M, Garber P, Cortés-Ortiz L, Urbani B, Youlatos D (eds) Howler monkeys: Adaptive radiation, systematics, and morphology. Springer, New York

de Boer B (2009) Acoustic analysis of primate air sacs and their effect on vocalization. J Acoust Soc Am 126:3329–3343

de Kort SR, den Hartog PM, ten Cate C (2002) Vocal signals, isolation and hybridization in the vinaceous dove (*Streptopelia vinacea*) and the ring-necked dove (*S. capicola*). Behav Ecol Sociobiol 51:378–385

DeGusta D, Milton K (1998) Skeletal pathologies in a population of *Alouatta palliata*: behavioral, ecological, and evolutionary implications. Int J Primatol 19:615–650

Detwiler KM, Burrell AS, Jolly CJ (2005) Conservation implications of hybridization in African Cercopithecine Monkeys. Int J Primatol 26:661–684

Di Fiore A, Campbell CJ (2007) The Atelines: variation in ecology, behavior, and social organization. In: Campbell CJ, Fuentes A, Mackinnon KC, Panger M (eds) Primates in perspective. Oxford University Press, Oxford

Dias PAD, Rangel-Negrín A, Veà JJ, Canales-Espinosa D (2010) Coalitions and male–male behavior in *Alouatta palliata*. Primates 51:91–94

Drubbel RV, Gautier JP (1993) On the occurrence of nocturnal and diurnal loud calls, differing in structure and duration, in red howlers (*Alouatta seniculus*) of French Guyana. Folia Primatol 60:195–209

Eason P (1989) Harpy eagle attempts predation on adult howler monkeys. Condor 91:469–470

Eisenberg JF (1976) Communication mechanisms and social integration in the black spider monkey, *Ateles fusciceps robustus*, and related species. Smithson Contrib Zool 213:1–108

Elowson AM, Snowdon CT (1994) Pygmy marmosets, *Cebuella pygmaea*, modify vocal structure in response to changed social environment. Anim Behav 47:1267–1277

Emlen ST, Oring LW (1977) Ecology, sexual selection, and the evolution of mating systems. Science 197:215–223

Farmer HL, Plowman AB, Leaver LA (2011) Role of vocalisations and social housing in breeding in captive howler monkeys (*Alouatta caraya*). Appl Anim Behav Sci 134:177–183

Fashing PJ (2001) Male and female strategies during inter-group encounters in guerezas (*Colobus guereza*): evidence for resource defense mediated through males and a comparison with other primates. Behav Ecol Sociobiol 50:219–230

Fialho MS, Setz EZF (2007) Extragroup copulations among brown howler monkeys in southern Brazil. Neotrop Primates 14:28–30

Fichtel C (2004) Reciprocal recognition of Sifaka (*Propithecus verreauxi verreauxi*) and red-fronted lemur (*Eulemur fulvus rufus*) alarm calls. Anim Cogn 7:45–52

Ficken MS, Ficken RW (1967) Singing behaviour of blue-winged and golden-winged warblers and their hybrid. Behaviour 28:149–181

Fischer J, Kitchen DM, Seyfarth RM, Cheney DL (2004) Baboon loud calls advertise male quality: acoustic features and their relation to rank, age and exhaustion. Behav Ecol Sociobiol 56: 140–148

Fisher J, Hammerschmidt K, Todt D (1998) Local variation in Barbary macaque shrill barks. Anim Behav 56:623–629

Fitch WT, Hauser MD (1995) Vocal production in nonhuman primates—acoustics, physiology, and functional constraints on honest advertisement. Am J Primatol 37:191–219

FitzGibbon CD, Fanshawe JH (1988) Stotting in Thomson's gazelles: an honest signal of condition. Behav Ecol Sociobiol 23:69–74

Frey R, Gebler A (2010) Mechanisms and evolution of roaring-like vocalization in mammals. In: Brudzynski SM (ed) Handbook of mammalian vocalization: an integrative neuroscience approach. Academic, London

Gabow SA (1975) Behavioral stabilization of a baboon hybrid zone. Am Nat 109:701–712

Geissman T, Mutschler T (2006) Diurnal distribution of loud calls in sympatric wild Indris (*Indri indri*) and ruffed lemurs (*Varecia variegata*): implications for call function. Primates 47:393–396

Gerhardt HC (1974) The vocalizations of some hybrid treefrogs: acoustic and behavioral analyses. Behaviour 49:130–151

Gil-da-Costa R, Palleroni A, Hauser MD, Touchton J, Kelley JP (2003) Rapid acquisition of an alarm response by a neotropical primate to a newly introduced avian predator. Proc R Soc Lond B 270:605–610

Glander KE (1978) Howling monkey feeding behavior and plant secondary compounds: a study of strategies. In: Montgomery GG (ed) The ecology of arboreal folivores. Smithsonian Institution Press, Washington, DC

Hauser MD (1993) The evolution of nonhuman primate vocalizations: effects of phylogeny, body weight, and social context. Am Nat 142:528–542

Hewitt GH (1988) Hybrid zones—natural laboratories for evolutionary studies. Trends Ecol Evol 3:158–167

Hill CM (1994) The role of female diana monkeys, *Cercopithecus diana*, in territorial defence. Anim Behav 47:425–431

Holzmann I, Agostini I, Di Bitetti M (2012) Roaring behavior of two syntopic howler species (*A. caraya* and *A. guariba clamitans*): evidence supports the mate defense hypothesis. Int J Primatol 33:338–355

Horwich RH, Gebhard K (1983) Roaring rhythms in black howler monkeys (*Alouatta pigra*) of Belize. Primates 24:290–296

Horwich RH, Lyon J (1990) A Belizean rain forest: the Community Baboon Sanctuary. Orangutan Press, Gays Mills

Horwich RH, Brockett RC, Jones CB (2000) Alternative male reproductive behaviors in the Belizean black howler monkey (*Alouatta pigra*). Neotrop Primates 8:95–98

Hughes M (2000) Deception with honest signals: signal residuals and signal function in snapping shrimp. Behav Ecol 11:614–623

Jones CB (1980) The functions of status in the mantled howler monkey, *Alouatta palliata* Gray: intraspecific competition for group membership in a folivorous Neotropical primate. Primates 21:389–405

Julliot C (1994) Predation of a young spider monkey (*Ateles paniscus*) by a crested eagle (*Morphnus guianensis*). Folia Primatol 63:75–77

Kelaita MA, Cortés-Ortiz L (2013) Morphological variation of genetically-confirmed *Alouatta pigra*×*A. palliata* hybrids from a natural hybrid zone in Tabasco, Mexico. Am J Phys Anthropol 150:223–234

Kirschel ANG, Blumstein DT, Smith TB (2009) Character displacement of song and morphology in African tinkerbirds. Proc Natl Acad Sci USA 106:8256–8261

Kitchen DM (2000) Aggression and assessment among social groups of Belizean black howler monkeys (*Alouatta pigra*). PhD dissertation, University of Minnesota

Kitchen DM (2004) Alpha male black howler monkey responses to loud calls: effect of numeric odds, male companion behaviour and reproductive investment. Anim Behav 67:125–139

Kitchen DM (2006) Experimental test of female black howler monkey (*Alouatta pigra*) responses to loud calls from potentially infanticidal males: effects of numeric odds, vulnerable offspring, and companion behavior. Am J Phys Anthropol 131:73–78

Kitchen DM, Beehner JC (2007) Factors affecting individual participation in group-level aggression in non-human primates. Behaviour 144:1551–1581

Kitchen DM, Seyfarth RM, Fischer J, Cheney DL (2003) Loud calls as indicators of dominance in male baboons (*Papio cynocephalus ursinus*). Behav Ecol Sociobiol 53:374–384

Kitchen DM, Horwich RH, James RA (2004) Subordinate male black howler monkey (*Alouatta pigra*) responses to loud calls: experimental evidence for the effects of intra-group male relationships and age. Behaviour 141:703–723

Lemaire F (1977) Mixed song, interspecific competition and hybridisation in the reed and marsh warblers (*Acrocephalus scirpaceus and palustris*). Behaviour 63:215–240

Lichtenberg SJ, Simmons J, Benefit B, McCrossin F, Kautz T, Arney I, Diaz LC, Diaz M, Diaz R, Milne R, Shendo B (2012) Differences between black howler (*Alouatta pigra*) group size and pattern of vocalizations in two ecologically different populations in northern Belize. Am J Phys Anthropol 147(suppl 54):193–194 (Abstract)

Marler P (1973) A comparison of vocalizations of red-tailed monkeys and blue monkeys, *Cercopithecus ascanius and C. mitis*, in Uganda. Z Tierpsychol 33:223–247

Maynard Smith J (1974) The theory of games and the evolution of animal conflicts. Theor Biol 47:208–221

Maynard Smith J, Parker GA (1976) The logic of asymmetric contests. Anim Behav 24:159–175

Maynard Smith J, Price GR (1973) The logic of animal conflict. Nature 246:15–18

McComb K, Packer C, Pusey A (1994) Roaring and numerical assessment in contests between groups of female lions, *Panthera leo*. Anim Behav 47:379–387

McKinney T (2009) Anthropogenic change and primate predation risk: crested caracaras (*Caracara plancus*) attempt predation on mantled howler monkeys (*Alouatta palliata*). Neotrop Primates 16:24–27

Milton K (1980) The foraging strategy of howler monkeys: a study in primates economics. Columbia University Press, New York

Miranda JMD, Bernardi IP, Moro-Rios RF, Aguiar LM, Ludwig G, Passos FC (2004) Social structure of *Alouatta guariba clamitans*: a group with a dominant female. Neotrop Primates 12:135–138

Miranda JMD, Bernardi IP, Moro-Rios RF, Passos FC (2006) Antipredator behavior of brown howlers attacked by black hawk-eagle in Southern Brazil. Int J Primatol 27:1097–1101

Mitani JC, Rodman PS (1979) Territoriality: the relation of ranging pattern and home range size to defendability, with an analysis of territoriality among primate species. Behav Ecol Sociobiol 5:241–251

Mitani JC, Hunley KL, Murdoch ME (1999) Geographic variation in the calls of wild chimpanzees: a reassessment. Am J Primatol 47:133–151

Morton ES (1977) On the occurrence and significance of motivation-structural rules in some bird and mammal sounds. Am Nat 111:855–869

Moynihan M (1967) Comparative aspects of communication in New World primates. In: Morris D (ed) Primate ethology. Aldine, Chicago

Newman JD, Symmes D (1982) Inheritance and experience in the acquisition of primate acoustic behavior. Primate communication. Cambridge University Press, New York, p 259–278

Nunn CL (2000) Collective benefits, free-riders, and male extra-group conflict. In: Kappeler P (ed) Primate males: causes and consequences of variation in group composition. Cambridge University Press, Cambridge

Nunn CL, Lewis RJ (2001) Cooperation and collective action in animal behaviour. In: Noë R, van Hooff JARAM, Hammerstein P (eds) Economics in nature: social dilemmas, mate choice and biological markets. Cambridge University Press, Cambridge

Oliveira DAG (2002) Vocalizações de longo alcance de *Alouatta fusca clamitans* e *Alouatta belzebul belzebul*: estrutura e contexto. PhD dissertation, Universidade de São Paulo

Oliveira DAG, Ades C (2004) Long-distance calls in Neotropical primates. An Acad Bras Cienc 76:393–398

Owings DH, Morton ES (1998) Animal vocal communication: a new approach. Cambridge University Press, Cambridge

Owren MJ, Linker CD (1995) Some analysis methods that may be useful to acoustic primatologists. In: Zimmermann E, Newman JD, Jürgens U (eds) Current topics in primate vocal communication. Plenum Press, New York

Owren MJ, Dieter JA, Seyfarth RM, Cheney DL (1993) Vocalizations of rhesus (*Macaca mulatta*) and Japanese (*M. fuscata*) macaques cross-fostered between species show evidence of only limited modification. Dev Psychobiol 26:389–406

Parker GA (1974) Assessment strategy and the evolution of fighting behaviour. J Theor Biol 47:223–243

Payne RJH (1998) Gradually escalating fights and displays: the cumulative assessment model. Anim Behav 56:651–662

Peres CA (1990) A harpy eagle successfully captures an adult male red howler monkey. Wilson Bull 102:560–561

Poole JH (1989) Announcing intent: the aggressive state of musth in African elephants. Anim Behav 37:140–152

Rangel-Negrín A, Dias PAD, Chavira R, Canales-Espinosa D (2011) Social modulation of testosterone levels in male black howlers (Alouatta pigra). Horm Behav 59:159–166

Riede T, Tokuda IT, Munger JB, Thomson SL (2008) Mammalian laryngeal air sacs add variability to the vocal tract impedance: physical and computational modeling. J Acoust Soc Am 124:634–647

Schön Ybarra MA (1986) Loud calls of adult male red howling monkeys (*Alouatta seniculus*). Folia Primatol 47:204–216

Schön Ybarra MA (1988) Morphological adaptations for loud phonations in the vocal organ of howling monkeys. Primate Rep 22:19–24

Schön Ybarra MA (1995) A comparative approach to the non-human primate vocal tract: implications for sound production. In: Zimmermann E, Newman JD, Jürgens U (eds) Current topics in primate vocal communication. Plenum Press, New York

Scroggie MP, LittleJohn MJ (2005) Territorial vocal behavior in hybrid smooth froglets *Geocrinia laevis* complex (Anura: Myobatrachidae). Behav Ecol Sociobiol 58:72–79

Sekulic R (1982a) Daily and seasonal patterns of roaring and spacing in four red howler *Alouatta seniculus* troops. Folia Primatol 39:22–48

Sekulic R (1982b) The function of howling in red howler monkeys (*Alouatta seniculus*). Behaviour 81:38–54

Sekulic R (1982c) Behavior and ranging patterns of a solitary female red howler (*Alouatta seniculus*). Folia Primatol 38:217–232

Sekulic R (1982d) Male relationships and infant deaths in red howler monkeys (*Alouatta seniculus*). Z Tierpsychol 61:185–202

Sekulic R (1983) The effect of female calls on male howling in red howler monkeys (*Alouatta seniculus*). Int J Primatol 4:291–305

Sekulic R, Chivers DJ (1986) The significance of call duration in howler monkeys. Int J Primatol 7:183–190

Seyfarth RM, Cheney DL (1984) Grooming, alliances and reciprocal altruism in vervet monkeys. Nature 308:541–543

Seyfarth RM, Cheney DL (2010) Production, usage, and comprehension in animal vocalizations. Brain Lang 115:92–100

Širović A, Hildebrand JA, Wiggins SM (2007) Blue and fin whale call source levels and propagation range in the Southern Ocean. J Acoust Soc Am 122:1208–1215

Snowdon CT (1989) Vocal communication in New World monkeys. J Hum Evol 18:611–633

Snowdon CT (2004) Sexual selection and communication. In: Kappeler PM, van Schaik CP (eds) Sexual selection in primates: new and comparative perspectives. Cambridge University Press, Cambridge

Southwick CH (1962) Patterns of intergroup social behavior in primates, with special reference to rhesus and howling monkeys. Ann N Y Acad Sci 102:436–454

Steenbeek R, Assink PR (1998) Individual differences long distance calls of male wild Thomas langurs (*Presbytis thomasi*). Folia Primatol 69:77–80

Steenbeek R, Piek RC, van Buul M, van Hooff JARM (1999) Vigilance in wild Thomas's langurs (Presbytis thomasi): the importance of infanticide risk. Behav Ecol Sociobiol 45:137–150

Steinmetz S (2005) Vocalizações de longo alcance como comunicação intra-grupal nos bugios (*Alouatta guariba*). Neotrop Primates 13:11–15

Stevens JR, Hauser MD (2004) Why be nice? Psychological constraints on the evolution of cooperation. Trends Cogn Sci 8:60–65

Stevens JR, Volstor J, Schooler LJ, Rieskam J (2011) Forgetting constrains the emergence of cooperative decision strategies. Front Psychol Cogn Sci 1:1–12

Sun GZ, Huang B, Guan ZH, Geissmann T, Jiang XL (2011) Individuality in male songs of wild black crested gibbons (*Nomascus concolor*). Am J Primatol 73:431–438

Tenaza R (1985) Songs of hybrid gibbons (*Hylobates lar×H. muelleri*). Am J Primatol 8:249–253

Tinbergen N (1963) On aims and methods of ethology. Z Tierpsychol 20:410–433

Trivers RL (1972) Parental investment and sexual selection. In: Campbell B (ed) Sexual selection and the descent of man. Aldine, Chicago

Van Belle S, Estrada A, Strier KB (2008) Social relationships among male *Alouatta pigra*. Int J Primatol 29:1481–1498

Van Belle S, Estrada A, Ziegler TE, Strier KB (2009a) Sexual behavior across ovarian cycles in wild black howler monkeys (*Alouatta pigra*): male mate guarding and female mate choice. Am J Primatol 71:153–164

Van Belle S, Estrada A, Ziegler TE, Strier KB (2009b) Social and hormonal mechanisms underlying male reproductive strategies in black howler monkeys (*Alouatta pigra*). Horm Behav 56:355–363

Van Belle S, Kulp AE, Thiessen-Bock R, Garcia M, Estrada A (2010) Observed infanticides following a male immigration event in black howler monkeys, *Alouatta pigra*, at Palenque National Park, Mexico. Primates 51:79–284

Waser P (1977) Individual recognition, intragroup cohesion and intergroup spacing: evidence from sound playback to forest monkeys. Behaviour 60:28–74

West SA, Griffin AS, Gardner A (2007) Social semantics: altruism, cooperation, mutualism, strong reciprocity and group selection. J Evol Biol 20:415–432

Whitehead JM (1985) Long distance vocalizations and spacing in mantled howling monkeys *Alouatta palliata*. PhD thesis, University of North Carolina, Chapel Hill

Whitehead JM (1987) Vocally mediated reciprocity between neighbouring groups of mantled howling monkeys, *Alouatta palliata palliata*. Anim Behav 35:1615–1627

Whitehead JM (1989) The effect of the location of a simulated intruder on responses to long-distance vocalizations of mantled howling monkeys, *Alouatta palliata palliata*. Behaviour 108:73–103

Whitehead JM (1994) Acoustic correlates of internal states in free-ranging primates: The example of the mantled howling monkey *Alouatta palliata*. In: Roeder JJ, Thierry B, Anderson JR, Herrenschmidt N (eds) Current primatology, vol II. Social development, learning and behaviour. Université Louis Pasteur, Strasbourg

Wich SA, Nunn CL (2002) Do male "long-distance calls" function in mate defense? A comparative study of long-distance calls in primates. Behav Ecol Sociobiol 52:47–484

Wich SA, Assink PR, Becher F, Sterck EHM (2002) Playbacks of loud calls to wild Thomas langurs (Primates; *Presbytis thomasi*): the effect of familiarity. Behaviour 139:79–87

Wiley RH, Poston J (1996) Indirect mate choice, competition for mates, and coevolution of the sexes. Evolution 50:1371–1381

Wilson ML, Hauser MD, Wrangham RW (2001) Does participation in intergroup conflict depend on numerical assessment, range location, or rank for wild chimpanzees? Anim Behav 61:1203–1206

Wrangham RW, Rubenstein DI (1986) Social evolution in birds and mammals. In: Rubenstein DI, Wrangham RW (eds) Ecological determinants of social evolution. Princeton University Press, Princeton

Ydenberg RC, Giraldeau L-A, Falls JB (1988) Neighbours, strangers, and the asymmetric war of attrition. Anim Behav 36:343–347

Zahavi A (1977) The cost of honesty. J Theor Biol 67:603–605

Zucker EL, Clarke MR (1986) Male-male interactions in a group of free-ranging howling monkeys. Am J Primatol 10:443 (Abstract)

Part V
Conclusions

Chapter 15
New Challenges in the Study of Howler Monkey Anatomy, Physiology, Sensory Ecology, and Evolution: Where We Are and Where We Need to Go?

Dionisios Youlatos, Martín M. Kowalewski, Paul A. Garber, and Liliana Cortés-Ortiz

Keywords Howlers • Fossil • Evolution • Physiology • Systematics • Challenges • Priorities

15.1 Introduction

Howler monkeys, genus *Alouatta*, are the sole survivors of a relatively long (middle Miocene to modern), geographically widespread, and ecologically and morphologically diverse clade (Rosenberger et al. 2014). The genus is currently represented by some 12 species of fruit, leaf, and flower feeding New World primates that range from southern Mexico to northern Argentina (see Cortés-Ortiz et al. 2014). In fact, *Alouatta* has the most widespread distribution of any platyrrhine genus and can exploit forest types that vary from undisturbed rainforest to severely anthropogenically impacted forest fragments adjacent to pastureland, agricultural fields, and human communities (Di Fiore et al. 2011; Estrada 2014). In many instances,

---

D. Youlatos (✉)
Department of Zoology, School of Biology, Aristotle University of Thessaloniki, Thessaloniki, Greece
e-mail: dyoul@bio.auth.gr

M.M. Kowalewski
Estación Biológica Corrientes, Museo Argentino de Ciencias Naturales, Consejo Nacional de Investigaciones Científicas y Técnicas (CONICET), Buenos Aires, Argentina

P.A. Garber
Department of Anthropology, University of Illinois at Urbana-Champaign, Urbana, IL, USA

L. Cortés-Ortiz
Museum of Zoology, Department of Ecology and Evolutionary Biology, University of Michigan, Ann Arbor, MI, USA

howlers are the only primate species found in such highly disturbed habitats (Benchimol and Peres 2014). Their extended distributional range, their survival and exploitation of highly impacted forested patches, and their energy-minimizing behavior make them relatively easy to observe and this has resulted in numerous studies of their biology and behavior (see reviews in Crockett and Eisenberg 1987; Kinzey 1997; Milton 1998; Di Fiore et al. 2011). These include short-term and long-term field studies on the ecology, behavior, and demography of individual howler species, as well as studies of morphology, genetics, and physiology aimed at evaluating the evolutionary and adaptive history of this genus. Given this relatively robust literature on *Alouatta*, we have edited two volumes on howler monkeys. The goals of these two volumes are to bring together expert scholars, many from habitat countries, to contribute to a comprehensive corpus that reviews, integrates, and evaluates current information on howler behavior, ecology, nutrition, morphology, physiology, reproduction, evolution, and conservation. Moreover, recently published studies on howler systematics, functional morphology, physiology, and nutritional ecology highlight the growing importance of the genus *Alouatta* as a comparative model for examining platyrrhine evolution and the parallel social and ecological problems faced by species of prosimians, New World monkeys, Old World monkeys, and apes (Rosenberger et al. 2009; Kowalewski and Garber 2010; Di Fiore et al. 2011; Garber and Kowalewski 2011; Benchimol and Peres 2014; Halenar and Rosenberger 2013; Van Belle and Bicca-Marques 2014; Matsushita et al. 2014).

Although there have been several recent volumes published on the morphology, systematics, evolution, ecology, and behavior of Neotropical primates including capuchins (Fragaszy et al. 2004), spider monkeys (Campbell 2008), marmosets (Ford et al. 2009), and pitheciines (Veiga et al. 2013), as well as a set of companion volumes on Mesoamerican (Estrada et al. 2006) and South American primates (Garber et al. 2009), the present volume is distinguished by its focus on integrating data on howler monkeys as a framework to understand other primate radiations. This may not be entirely new, as *Alouatta* has already served as a model for several theoretical issues in primate evolution, ecology, and behavior, such as the adaptive morphology of the locomotor system and dentition of early catarrhines and Miocene hominoids, the evolution of trichromatic vision, a dissociability model for the study of primate ontogeny, social dynamics and behavioral responses of hybrid populations, the ability to exploit and digest a low quality diet, and the production, propagation, and social function of loud and low amplitude calls (Milton 1980, 1984; Rose 1994; Ravosa and Ross 1994; Jacobs et al. 1996; Cortés-Ortiz et al. 2007; Raguet-Schofield 2010; Amato 2013; Amato et al. 2013; Zuñiga Leal and Defler 2013; Ho et al. 2014; Righini 2014). Chapters in this first volume advance our understanding of howler systematics, evolution, functional anatomy, and physiology and provide new comparative insights into the evolution of howler monkeys within the atelinae. *Alouatta* appears to be a morphologically unique branch of ateline. Increased loud call production in ancestral *Alouatta* has been proposed to relate to shifts in social and reproductive behavior, probably intergroup spacing, intense intra-male competition or attraction of females, and appears to have played a primary role in distinguishing the alouattines from the rest of the atelines.

The morpho-behavioral modifications associated with loud calling, most notably the enlarged hyoid, the elongate basicranium, the airorynchous rostrum, and the large facial skeleton, probably anteceded the derived dental (e.g., the relatively small-crowned incisors, cristodont molars with more relief, and an emphasis on relatively elevated cusps and lengthy crests) and postcranial features (e.g., the lateral position of the scapula, the relatively short forelimb long bones, the shape of the articular facets of the elbow joint emphasizing on flexion, and the reduced intracarpal mobility) that define modern howlers. It has been argued that the selective forces that shaped *Alouatta* evolution occurred outside the greater Amazon basin in response to less productive and seasonal habitats (Rosenberger et al. 2014; Youlatos et al. 2014). The earliest fossils related to howlers (*Stirtonia, Paralouatta, Protopithecus,* and perhaps *Solimoea*) indicate a tendency in alouattines towards increased howling behavior, a trend of de-encephalization, a feeding spectrum ranging from frugivory to folivory, and a positional repertoire dominated by clambering-suspensory and pronograde quadrupedalism. Chapters in this volume also provide new information on the relationships between howler monkey reproductive endocrinology and reproductive strategies, revealing non-anticipated nonaggressive intragroup male-male competition over access to females, higher sensitivity to social and ecological stress in females than in males, and more prominent reliance on high energy foods, such as ripe fruit. Additionally, chapters in this volume also test hypotheses linking gut microbiome to howler feeding ecology and bioenergetics and explore the sensorial ecology of howlers and the adaptive significance of routine trichromatic vision in *Alouatta*. Finally, chapters in this volume also describe the diversity of the howler vocal repertoire and the functional role of loud calls in both male and female communication and intergroup encounters.

15.1.1 Taxonomy, Genetics, Morphology, and Evolution

Data presented in this section of the volume fundamentally alter our view of howler evolution and systematics. *Alouatta* appears to have a very long evolutionary history, starting in the middle Miocene, as evidenced by the fossil *Stirtonia*, a close relative of *Alouatta* from the site of La Venta in Colombia. This area is reconstructed as a heterogeneous riparian, dominantly forested mosaic habitat associated with meandering rivers (Rosenberger et al. 2009). Extant *Alouatta* are medium-sized atelines, with a large airorynchous face, marked compound cranial crests, a flat reduced vertical nuchal plane, an elongate basicranium, a small posterior foramen magnum, a deep mandibular ramus hosting an enlarged hyolaryngeal apparatus related to a commitment to loud vocal communication, marked postorbital constriction, and a cylindrical reduced braincase associated with a de-encephalization when compared to other atelines (Rosenberger et al. 2014; Youlatos et al. 2014). Dentally, the reduced incisors and cristodont elongate molars denote comparatively high rates of folivory (for a platyrrhine), which varies markedly throughout the year (Di Fiore et al. 2011; Dias and Rangel-Negrín 2014; Rosenberger et al. 2014;

Youlatos et al. 2014). Moreover, the shape of the scapula; the form and articular facets of the distal humerus; the morphology of the ulna; the arrangement and shape of the carpals; the form of the proximal and distal femur and its articular facets; talar and calcaneal morphology; the curved shape of the metacarpals, metatarsals, and digits; and the comparatively short (for an ateline) prehensile tail are functionally associated with a primarily above branch form of slow quadrupedal progression complemented with climbing locomotion and tail-assisted foot hanging postural behavior (Rosenberger and Strier 1989; Jones 2008; Youlatos and Guillot 2014; Youlatos et al. 2014). It has been argued that *Alouatta* is characterized by an energy-minimizing behavioral pattern associated with small day range, small home range, and an activity budget that includes 60–80 % resting (Di Fiore et al. 2011; Barbisan-Fortes et al. 2014). These morphological, behavioral, and ecological characters distinguish *Alouatta* from other atelines. The ever-increasing fossil record of the alouattine clade (Rosenberger et al. 2014) offers insights into the morphological and associated behavioral changes that promoted the evolution of the derived morphological traits that define howlers. *Stirtonia* (Middle Miocene) is reconstructed as a medium-sized alouattine (~6 kg), with teeth barely distinguishable from *Alouatta*, suggesting a fundamentally similar diet. *Solimoea*, from the late Miocene of Rio Acre, Brazil, also weighed ~6 kg and is known from a single molar tooth, which morphologically resembles both *Stirtonia* and *Alouatta*. From the more recent fossils, *Paralouatta*, mainly known from the Pleistocene of Cuba, was a large-sized alouattine (~9.5 kg) with most of its cranial and dental features comparable to that of *Alouatta*, indicating a dietary commitment to folivory a small brain, and the presence of an enlarged hyolaryngeal apparatus. On the other hand, *Protopithecus*, a very large (17–25 kg) alouattine from the late Pleistocene of Brazil, possessed alouattine characters and differed from modern howlers in exhibiting a moderately large face, enlarged subvertical nuchal plane, enlarged incisors, but lacked the howler cristodont molar pattern implying generalized frugivory. These features suggest that *Protopithecus* is transitional from a generalized morphology towards the apomorphic nature of howlers but also retained traits present in basal members of the alouattine clade.

Despite the fact that leaf-eating adaptations are already evident in the earliest relatives of *Alouatta* in the middle Miocene, the overall primitive nature of the Cuban and Brazilian Pleistocene fossils indicates a reduced commitment to folivory and the onset of the major cranial modifications (more in *Paralouatta*, less in *Protopithecus*) that define the apomorphic skull (related to howling) of modern howler monkeys. This suggests that the basal branch of alouattines were medium-sized prehensile-tailed clambering fruit and leaf feeders characterized by high levels of male intrasexual selection associated with the production and propagation of loud calls that were imprinted on cranial morphology. These changes appear to have been followed by de-encephalization, already evident in the primitive alouattines, for which cranial evidence exists. These modifications signaled the successful differentiation of the alouattines, prior to the evolution of dental adaptations related to increased folivory and positional adaptations related to above-branch slow pronograde tail-assisted quadrupedal locomotion. This morpho-behavioral complex currently defines the apomorphic nature of howlers, but as postulated by Rosenberger et al. (2014): "We can only speculate that howling, small brains and leaf-eating are interconnected as low-energy balancing factors of potential adaptive value: long distance advertisement that requires little movement or exposure; a brain that can be metabolically maintained relatively cheaply; a food source that requires little exercise to acquire and produces energy slowly and at low dosages." Based on the biogeography and natural history of modern Alouatta, as well as the current fossil evidence, it is very likely that howlers evolved outside of Amazonia in more marginal or less productive habitats in which the exploitation of difficult-to-digest resources, such as leaves, allowed howlers to occupy niches unavailable to other platyrrhines without preventing them from expanding to more productive habitats when available.

These scenarios are presently based on the current fossil evidence. However, most of these fossils are incomplete and critical cranial and postcranial regions are missing. Furthermore, we still lack sufficient fossil evidence from time frames, such as early to middle Miocene, that are critical to understand the evolution of alouattines, as well as from areas, such as Mesoamerica or the southern cone, that could be significant for tracing the adaptive radiation and sequence of evolutionary changes in anatomy that distinguish more primitive and more derive alouattines. Such fossils are essential for understanding the succession of events that led to modern *Alouatta* by providing evidence of structural modifications associated to howling, small brain size, folivory, and pronograde quadrupedalism that shaped alouattine evolution.

The morphological, chromosomal, and genetic diversity of howler monkeys lends support for nine distinct species (*A. palliata, A. pigra, A. seniculus, A. arctoidea, A. sara, A. macconnelli, A. guariba, A. belzebul, A. caraya*) and three additional taxa that are tentatively regarded as full species (*A. nigerrima, A. ululata, A. discolor*). This diversity is further represented in five recognizable subspecies for *A. palliata* (*A. p. mexicana, A. p. palliata, A. p. coibensis, A. p. trabeata* and *A. p. aequatorialis*) and two possible subspecies for *A. pigra* (*A. p. pigra* and *A. p. luctuosa*) from Mesoamerica, three subspecies of Amazonian *A. seniculus* (*A. s. seniculus, A. s. juara,* and *A. s. puruensis*), and two subspecies of the southern peripheral *A. guariba* (*A. g. guariba* and *A. g. clamitans*). A growing number of molecular and cytogenetic studies on *Alouatta* continue to provide evidence of genetic diversity in the genus that was not been previously revealed based solely on morphological data. Furthermore, the origin and transmission of the supernumerary B chromosomes present in some howler species need to be studied in order to evaluate *Alouatta* chromosomal evolution. We continue to have a limited understanding of the levels of inter- and intraspecific genetic diversity in howler monkeys. Population genetic studies for most taxa are required and should be framed within a comparative biogeography context, which will allow us to understand mechanisms and processes that shaped the primate fauna in the Neotropics.

Another important issue that needs to be considered when analyzing genetic diversity within *Alouatta* is the potential effect of hybridization on the evolutionary history of this genus. Recent research has demonstrated a limited degree of hybridization among neighboring species, a fact well documented in Mexico between *A. palliata* and *A. pigra,* and in Argentina and Brazil between *A. guariba* and

A. caraya. Although hybridization may not be observed across extensive areas, it certainly promotes some level of genetic exchange that may have an effect (positive or negative) on the evolutionary trajectory of these sympatric howler species. The dynamic process of hybridization between well-established genetic lineages is likely a consequence of secondary contact between parapatric taxa, and may have occurred multiple times in the evolutionary history of the genus. The growing evidence of hybridization in howler monkeys should stimulate further investigation on the mechanisms that promote reproductive isolation between primate lineages and those that help maintaining species integrity despite genetic introgression.

15.1.2 Howler Physiology

Physiological studies, involving blood biochemistry, including growth, stress, and sexual hormones, are fundamental for understanding the endocrine responses of howlers to changes in their social and ecological environment. Hematological and biochemical markers are good indicators for evaluating overall health and physical condition, and both long-term and short-term responses to the social and ecological environments. Available, but limited, data indicate that parameters, such as hematocrit, white blood cell count, protein concentration, and creatinine concentration, vary considerably among species and across sexes. This reinforces the need to create a central database that maintains the hematology and blood biochemistry (e.g., white and red cell concentrations, creatine and other proteins, glucose, iron levels, chloride, sodium) of as many species as possible in the wild. This endocrine and hematological database will be extremely used for diagnosis and treatment of howler monkey individuals maintained in ex situ facilities, aiding in the development and implementation of conservation measures.

To date, data on reproductive endocrinology are available for 7 of the 12 recognized howler species (*A. palliata, A. pigra, A. arctoidea, A. seniculus, A. caraya,* and *A. belzebul*), but we continue to lack a detailed understanding of how changing social, demographic, reproductive, and ecological factors affect male and female hormonal profiles and reproductive strategies. This will provide valuable information on the mechanisms that promote age-/sex-based patterns of aggression leading to migration or tolerance and increased group size and complexity. This dynamic process also requires assessing hormonal profiles during infant development, the onset of sexual maturation, and pregnancy and establishing baseline steroid hormonal profiles with data on intra- and intergroup mating patterns, paternity assignment, and the costs and benefits of inter- and intrasexual social relationships in order to better understand the effectiveness of individual mating strategies on reproductive success (e.g., collective action, female mate choice).

Due to their herbivorous diet, howler monkeys are dependent on their gut microbiota for the breakdown of plant structural carbohydrates. The current evidence indicates that howler monkey microbial community composition differs more across habitats than across seasons, and that individual differences in the ratio or type of gut microbacteria can play a critical role in host nutrition, metabolic activity, and immune function. Adjustments to the microbiome offer the possibility for individuals consuming the same diet to differentially extract and assimilate particular nutrients. Thus, age- and sex-based shifts in microbiome diversity may enable group members to satisfy their individual nutritional requirements without major shifts in diet, activity budget, or patterns of habitat utilization (Amato and Righini 2014). Moreover, differences in host microbial communities across habitats appear to reflect differences in the nutritional profile of resources consumed or differences in gut microbial diversity in these habitats (Amato 2013). This has important implications for primate conservation and the ability of howlers to survive in altered landscapes, as the reduction or loss of natural microbial communities, or the introduction of new microbial communities in response to environmental change, pollution, or other factors, can affect host nutrition and health, and immune function. In order to understand these interactions and their adaptive significance, future research should include an analysis of environmental microbial community composition during different times of the year, in different habitats, and in different levels/substrates in the canopy, and its effect on host microbial diversity.

Martínez-Mota and collaborators (2014) offered an overview and a meta-analysis of parasites that are hosted by howler monkeys. They explored how ecological factors affect parasitic infection in this primate genus. Some factors such as human presence and annual precipitation may influence the prevalence of different parasites. In addition, the authors found that parasitic infection in howlers appears to be biased towards only a few individuals within a group. Given that infectious diseases are a serious threat to primate survival, this study provides a baseline for evaluating the dynamics of parasite-howler interactions. Three challenges researchers face in evaluating the health consequences and severity of parasite loads in howler populations exploiting habitats differentially exposed to humans, cattle, and other domesticated animals include: (a) applying recently developed molecular tools to accurately identify parasite taxa, (b) determining the life cycles of individual parasite species infecting howlers, and (c) determining exactly how habitat disturbance and forest fragmentation affect parasite survivorship. The collection of these data represents an important advancement in determining parasite pathogenic potential, and in evaluating the conditions that promote the proliferation of different parasites within their howler hosts.

15.1.3 *Ontogeny and Sensory Ecology of Howlers*

Howlers are traditionally characterized as having fast life histories compared to other atelines. However, analysis of available and new data presented in this volume Raguet-Schofield and Pavé (2014) highlights the need for a paradigm change in interpreting primate ontogeny. In the case of howlers, this new paradigm is best characterized by a dissociability framework in which, relative to other atelines, howlers possess some traits characterized by slow ontogenetic or delayed development (e.g., locomotor proficiency) whereas other traits develop quickly (e.g., reproductive maturation). Understanding the set of evolutionary, social, and ecological factors that have shaped the ontogeny of individual traits will require long-term field research that documents, in greater detail, *Alouatta* life history events (such as weaning, interbirth interval, motor development, locomotor independence, brain growth trajectories, juvenile risk) rather than considering only the adult condition. For example, although in some cases howler mothers may continue to allow their offspring to nurse or be in contact with the nipple until 12 months of age, we have no data on whether these represent true nursing events (the infant is consuming milk), how much milk the infant is consuming, and the effect of extended nipple contact on interbirth interval. A recent study by Reitsema (2012), using stable carbon and nitrogen isotope ratios present in offspring feces, offers a new methodology to document the transition from breast milk to solid food in young primates. Although this research was successfully conducted on captive Francois langurs (*Trachypithecus francoisi*), it can likely be applied in wild populations to examine a range of research questions concerning primate ontogeny, diet, and reproduction. *Alouatta* is a good model to answer these questions, because different species appear to be characterized by different ontogenetic and reproductive trajectories. Intrageneric variability in these life traits provides a critical framework for testing ecological, social, and phylogenetic effects on ontogenetic processes. Other issues that require further investigation in both the field and in museum collections are the adaptive and evolutionary significance of increased sexual dimorphism of *Alouatta* compared to other atelines (Plavcan 2001), differential growth patterns between males and females (males exhibit more rapid postnatal growth than females, yet even within the preadult time frame, male growth does not remain uniformly accelerated and instead alternates between of slower growth and growth spurts) (Leigh 1994), and evidence of age-/sex-based differences in folivory and frugivory.

Regarding sensory ecology, howling may be the most recognizable feature of the genus *Alouatta*. However, despite our knowledge of the anatomy of the vocal organs, the acoustic structure of vocalizations, the variation among different species, and functional aspects for some vocalizations in both males and females, numerous gaps in our understanding of howler monkey vocalizations continue to exist. Morphologically, we have limited understanding of anatomical correlated adaptations that make this unique sound production system possible such as lung capacity, rib cage anatomy, air flow, and hyoid vibration. In addition, few researchers have examined the function and context of low amplitude calls given by both male and female howlers, as well as loud calls given by female howlers (da Cunha et al. 2004). Although loud vocalizations are frequently performed in all howler monkey population, the understanding of the variety, frequency, and function of less-common calls requires the recording and sound spectrographic analysis of these calls. Therefore, regardless of the primary focus of a particular study, researchers should bring recording equipment to the field in an attempt to document the acoustic features and context of howler vocalizations. Similarly, although male loud calls have been the focus of several howler studies (Kitchen 2004; Van Belle et al. 2013; da Cunha et al. 2014), several questions remain. For example, it is unclear whether the acoustic features or temporal patterning of loud calls produced during predator encounters vary from calls given during intergroup contexts. Given the relative rarity of predator encounters observed by humans, future studies should focus on using acoustic and visual predator models to exam howler loud calls. Cross-species comparisons will require that researchers develop innovative and standardized methodologies to differentiate among mate defense, site/resource defense, infant defense, and spacing hypotheses and the importance of female mate choice based on potentially honest male vocal signals. The study of audiograms is a promising research tool in examining howler hearing sensitivity that can be related to behavioral studies to decipher the extent and limits of behavioral responses to a range of vocal signals given by howlers.

Another sensory element that constitutes an apomorphic character distinguishing howlers among platyrrhines is routine trichromatic color vision (in other platyrrhines there are sex-linked differences in color vision, with all males and many females dichromatic). Ever since its discovery (Jacobs et al. 1996), research on the anatomical and genetic bases of howler trichromacy has increased. However, the adaptive significance of this trait remains unclear (Hernández-Salazar et al. 2014). Extensive population-level studies of the distribution of L/M opsin genes and detailed field observations of visual behaviors of howlers are important for elucidating whether trichromacy is related to an enhanced ability to locate orange- and red-colored fruits or young leaves, or to identify changes in the intensity skin and hair pigments as social/reproductive signals (Matsushita et al. 2014; Melin et al. 2014).

15.2 Conclusions

Howlers offer an instructive model to address a broad range of research questions regarding the morphological, evolutionary, ecological, behavioral, reproductive, and social strategies of living and fossil primates. This volume has shown, however, that despite numerous field-, laboratory-, and museum-based studies, our current understanding of howler genetics, morphology, physiology, ontogeny, and sensory ecology are focused on a relatively small number of species within the genus. Clearly, there is a priority to study other howler species such as *A. sara, A. belzebul, A. macconnelli, A. discolor, A. ululata,* and *A. nigerrima*, as well as subspecies of *A. seniculus* such as *puruens*is and *juara,* and to employ new research tools and data-collecting methodologies to address the next generation of research questions.

Finally, *Alouatta* is generally considered a taxon characterized by considerable phenotypic variation. Clearly, morpho-behavioral modification or adaptability has limits, and these limits are likely to differ across phenotypes within the same species and across species. Individual traits can be described as variable or flexible relative to other traits as long as the limits of this variability are described and it is recognized that along different points across this continuum the cost/benefit ratio can increase, decrease, or remain the same. What is important to highlight is that just stating that a trait is variable provides no useful information for defining the range of variability of that trait, what factors determine this range of variability, and under what set of social and environmental conditions is the variable expression of the trait a benefit or a cost. Given that howlers can survive in highly disturbed habitats has led many to argue that they are highly adaptable or more adaptable than other primate lineages. However, howlers are often less successful than other primates in less marginal or less disturbed habitats, including habitats characterized by indigenous hunters. We anticipate that as our knowledge of the evolution, biogeography, systematics, genetics, ecology, and social behavior of understudied howler species increases, we will be able to better model the set of historical, ecological, and demographic factors that most importantly affect howler distribution. These data will allow us to develop effective management and conservation strategies to protect threatened howler populations across their range.

Acknowledgments We wish to acknowledge the following scholars for contributing ideas to this chapter: Katherine Amato, Alejandro Estrada, Tracie McKinney, Nicoletta Righini, Andrés Gómez, Marta Mudry, Rosalía Pastor-Nieto, Rodolfo Martínez-Mota, Víctor Arroyo, Dawn Kitchen, Pedro Días, Sarie Van Belle, Júlio César Bicca-Marques, Melissa Raguet-Schofield, Alfie Rosenberger, Lauren Halenar, and Marcelo Tejedor. DY thanks Alexandra and Evangelos, for their patience and support, throughout this project. MK thanks to Mariana and Bruno for their love and support. LCO was supported by the NSF (BCS-0962807) while writing this manuscript. As always PAG could not have written any of this without the love, support, and silliness of Chrissie, Sara, and Jenni.

References

Amato DR (2013) Black howler monkey (*Alouatta pigra*) nutrition: integrating the study of behavior, feeding ecology, and the gut microbial community. PhD thesis, University of Illinois, Urbana

Amato KR, Yeoman CJ, Kent A, Righini N, Carbonero F, Estrada A, Gaskins HR, Stumpf RM, Yildirim S, Torralba M, Gillis M, Wilson BA, Nelson KE, White BA, Leigh SR (2013) Habitat degradation impacts black howler monkey (*Alouatta pigra*) gastrointestinal microbiomes. ISME J 7(7):1344–1353

Amato KR, Righini N (2014) The howler monkey as a model for exploring host-gut microbiota interactions in primates. In: Kowalewski M, Garber PA, Cortés-Ortiz L, Urbani B, Youlatos D (eds) Howler monkeys: adaptive radiation, systematics, and morphology. Springer, New York

Barbisan-Fortes V, Bicca-Marques JC, Urbani B, Fernández VA, da Silva Pereira T (2014) Ranging behavior and spatial cognition of howler monkeys. In: Kowalewski M, Garber PA, Cortés-Ortiz L, Urbani B, Youlatos D (eds) Howler monkeys: behavior, ecology and conservation. Springer, New York

Benchimol M, Peres CA (2014) Predicting primate local extinctions within "real-world" forest fragments: a pan-neotropical analysis. Am J Primatol 76:289–302

Campbell CJ (2008) Spider monkeys: the biology, behavior and ecology of the genus *Ateles*. Cambridge University, New York

Cortés-Ortiz L, Duda TJ Jr, Canales-Espinosa D, García-Orduna F, Rodríguez-Luna E, Bermingham E (2007) Hybridization in large-bodied New World primates. Genetics 176:2421–2425

Cortés-Ortiz L, Rylands AB, Mittermier RA (2014) The taxonomy of howler monkeys: integrating old and new knowledge from morphological and genetic studies. In: Kowalewski M, Garber PA, Cortés-Ortiz L, Urbani B, Youlatos D (eds) Howler monkeys: adaptive radiation, systematics, and morphology. Springer, New York

Crockett CM, Eisenberg JF (1987) Howlers: variation in group size and demography. In: Smuts BB, Cheney DL, Seyfarth RM, Wrangham RW, Struhsaker TT (eds) Primate societies. University of Chicago, Chicago

da Cunha RGT (2004) A functional analysis of vocalisations of black howler monkeys (*Alouatta caraya*). PhD Dissertation, University of St. Andrews. Fife

da Cunha RGT, de Oliveira DAG, Holzmann I, Kitchen DM (2014) Production of loud and quiet calls in howler monkeys. In: Kowalewski M, Garber PA, Cortés-Ortiz L, Urbani B, Youlatos D (eds) Howler monkeys: adaptive radiation, systematics, and morphology. Springer, New York

Di Fiore A, Link A, Campbell CJ (2011) The atelines: behavior and socioecological diversity in a New World monkey radiation. In: Campbell CJ, Fuentes AF, MacKinnon KC, Bearder S, Stumpf R (eds) Primates in perspective. Oxford University, Oxford

Dias PA, Rangel-Negrin A (2014) Diets of howler monkeys. In: Kowalewski M, Garber PA, Cortés-Ortiz L, Urbani B, Youlatos D (eds) Howler monkeys: behavior, ecology and conservation. Springer, New York

Estrada A, Garber PA, Pavelka M, Luecke L (2006) New Perspectives in the Study of Mesoamerican Primates: Distribution, Ecology, Behavior and Conservation. Springer Press, New York

Estrada A (2014) Conservation of *Alouatta*: social and economic drivers of habitat loss, information vacuum and mitigating population declines. In: Kowalewski M, Garber PA, Cortés-Ortiz L, Urbani B, Youlatos D (eds) Howler monkeys: behavior, ecology and conservation. Springer, New York

Ford S, Porter LM, Davis L (2009) The smallest anthropoids: the Marmoset/Callimico radiation. Springer, New York

Fragaszy DM, Visalberghi E, Fedigan LM (2004) The complete Capuchin. Cambridge University, Cambridge

Garber PA, Estrada A, Bicca-Marques J-C, Heymann E, Strier KB (2009) South American primates: comparative perspectives in the study of behavior, ecology, and conservation. Springer, New York

Garber PA, Kowalewski MM (2011) Collective action and male affiliation in howler monkeys (*Alouatta caraya*). In: Sussman RW, Cloninger C (eds) Origins of altruism and cooperation. Springer, New York

Halenar LB, Rosenberger AL (2013) A closer look at the "Protopithecus" fossil assemblages: new genus and species from Bahia, Brazil. J Hum Evol 65:374–390

Hernández-Salazar LT, Dominy NJ, Laska M (2014) The sensory systems of *Alouatta*: evolution with an eye to ecology. In: Kowalewski M, Garber PA, Cortés-Ortiz L, Urbani B, Youlatos D (eds) Howler monkeys: adaptive radiation, systematics, and morphology. Springer, New York

Ho L, Cortés-Ortiz L, Dias PAD, Canales-Espinosa D, Kitchen DM, Bergman TJ (2014) Effect of ancestry on behavioral variation in two species of howler monkeys (*Alouatta pigra* and *A. palliata*) and their hybrids. Am J Primatol 76:855–867. doi:10.1002/ajp.22273

Jacobs GH, Neitz M, Deegan JF, Neitz J (1996) Trichromatic colour vision in New World monkeys. Nature 382(6587):156–158

Jones AL (2008) The evolution of brachiation in ateline primates, ancestral character states and history. Am J Phys Anthrop 137:123–144

Kinzey WG (1997) New world primates: ecology, evolution and behavior. De Gruyter, New York

Kitchen DM (2004) Alpha male black howler monkey responses to loud calls: effect of numeric odds, male companion behaviour and reproductive investment. Anim Behav 67(1):125–139

Kowalewski MM, Garber PA (2010) Mating promiscuity and reproductive tactics in female black and gold howler monkeys (*Alouatta caraya*) inhabiting an island on the Parana river, Argentina. Am J Primat 72(8):734–748

Leigh SR (1994) Ontogenetic correlates of diet in anthropoid primates. Am J Phys Anthropol 94(4):499–522

Martinez-Mota R, Kowalewski MM, Gillespie TR (2014) Ecological determinants of parasitism in howler monkeys. In: Kowalewski M, Garber PA, Cortés-Ortiz L, Urbani B, Youlatos D (eds) Howler monkeys: adaptive radiation, systematics, and morphology. Springer, New York

Matsushita Y, Oota H, Welker BJ, Pavelka MS, Kawamura S (2014) Color vision variation as evidenced by hybrid L/M opsin genes in wild populations of trichromatic *Alouatta* New World monkeys. Int J Primatol 35:71–87

Melin AD, Hiramatsu C, Parr NA, Matsushita Y, Kawamura S, Fedigan LM (2014) The behavioral ecology of color vision: considering fruit conspicuity, detection distance and dietary importance. Int J Primatol 35:258–287

Milton K (1980) The foraging strategy of Howler monkey: a study in primate economics. Columbia University, New York

Milton K (1984) The role of food-processing factors in primate food choice. In: Rodman PS, Cant JGH (eds) The role of food-processing factors in primate food choice. Columbia University, New York

Milton K (1998) Physiological ecology of howlers (Alouatta): energetic and digestive considerations and comparison with the Colobinae. Int J Primatol 19:513–548

Plavcan MJ (2001) Sexual dimorphism in primate evolution. Am J Phys Anthropol 116:25–53

Raguet-Schofield ML (2010) The ontogeny of feeding behavior of Nicaraguan mantled howler monkeys (*Alouatta palliata*). PhD dissertation. Urbana-Champaign, University of Illinois, Urbana

Raguet-Schofield M, Pavé RE (2014) An ontogenetic framework for *Alouatta*: infant development and evaluating models of life history. In: Kowalewski M, Garber PA, Cortés-Ortiz L, Urbani B, Youlatos D (eds) Howler monkeys: adaptive radiation, systematics, and morphology. Springer, New York

Ravosa MJ, Ross CF (1994) Craniodental allometry and heterochrony in two howler monkeys: *Alouatta seniculus* and *A. palliata*. Am J Primatol 33:277–299

Reitsema LJ (2012) Introducing fecal stable isotope analysis in primate weaning studies. Am J Primatol 74:926–939

Righini N (2014) Primate nutritional ecology: the role of food selection, energy intake, and nutrient balancing in Mexican black howler monkey (*Alouatta pigra*) foraging strategies. PhD thesis, University of Illinois, Urbana

Rose MD (1994) Quadrupedalism in some Miocene catarrhines. J Hum Evol 26:387–411

Rosenberger AL, Cooke SB, Halenar L, Tejedor M, Hartwig WC, Novo NM, Muñoz-Saba Y (2014) Fossil alouattines and the origins of *Alouatta*: craniodental diversity and interrelationships. In: Kowalewski M, Garber PA, Cortés-Ortiz L, Urbani B, Youlatos D (eds) Howler monkeys: adaptive radiation, systematics, and morphology. Springer, New York

Rosenberger AL, Strier KB (1989) Adaptive radiation of the ateline primates. J Hum Evol 18:717–750

Rosenberger AL, Tejedor MF, Cooke SB, Pekkar S (2009) Platyrrhine ecophylogenetics, past and present. In: Garber P, Estrada A, Bicca-Marques JC, Heymann EW, Strier KB (eds) South American primates: comparative perspectives in the study of behavior, ecology and conservation. Springer, New York

Van Belle S, Bicca-Marques JC (2014) Insights into reproductive strategies and sexual selection in howler monkeys. In: Kowalewski M, Garber PA, Cortés-Ortiz L, Urbani B, Youlatos D (eds) Howler monkeys: behavior, ecology and conservation. Springer, New York

Van Belle S, Estrada A, Garber PA (2013) Spatial and diurnal distribution of loud calling in black howlers (*Alouatta pigra*). Int J Primatol 34:1209–1224

Veiga LM, Barnett AA, Ferrari SF (2013) Evolutionary biology and conservation of Titis, Sakis and Uacaris. Springer, New York

Youlatos D, Couette S, Halenar L (2014) Morphology of howler monkeys. In: Kowalewski M, Garber PA, Cortés-Ortiz L, Urbani B, Youlatos D (eds) Howler monkeys: adaptive radiation, systematics and morphology. Springer, New York

Youlatos D, Guillot D (2014). Positional behavior of howler monkeys. In: Kowalewski M, Garber PA, Cortés-Ortiz L, Urbani B, Youlatos D (eds) Howler monkeys: behavior, ecology and conservation. Springer, New York

Zuñiga Leal SA, Defler TR (2013) Sympatric distribution of two species of Alouatta (*A. seniculus* and *A. palliata*: primates) in Chocó, Colombia. Neot Primatol 20(1):1–11

Subject Index

A

Adaptation, 3–5, 7, 22, 25, 33, 39, 44–50, 121, 134, 135, 140, 143, 168, 169, 197, 198, 231, 237, 301, 318, 323, 326, 329, 330, 377, 406, 410

Age, 9–11, 27, 35, 180, 195, 204, 208, 209, 217, 232, 233, 243–248, 290–302, 304–308, 358–360, 379, 383, 386, 387, 392, 408–410

Airorhynchy, 35, 135, 140

Alkaloid, 327

Allometry, 10, 33

Allomother, 296–298

Amazonia, 49, 50, 407

Anatomy, dental, 4, 10, 168

Apomorphy, 6

Argentina, 8, 56, 69, 70, 108, 110, 111, 115, 122–124, 184, 194, 196, 259–261, 294, 297–300, 389, 393, 403, 407

Atlantic forest, 4, 57, 70, 114, 122, 123

B

Backcrossing, 118, 120, 389

Banding, 79, 87, 89, 93, 94, 98, 99

Bark, 14, 296, 340–343, 346–350, 352, 355, 358–360, 362, 371, 372, 379, 381, 383, 390

Basicranial shape, 6, 34, 39–40

Basicranium, 28, 39, 45, 46, 48, 140, 142, 405

Behavior
    feeding, 297, 305, 306, 327
    positional, 5, 11, 36, 135, 309, 310, 328, 339
    postural, 169, 328, 406

Belize, 11, 60, 66, 68, 218, 221, 263, 301, 356

Biochemistry, 8, 179–198, 408

Biodiversity, 80

Biogeography, 4, 50, 329, 407, 412

Blood
    biochemistry, 8, 181, 191, 198, 408
    cells, 11, 180, 181, 186, 188, 189, 193, 197, 408

Body
    mass, 4, 5, 11, 25, 31–33, 37, 47, 197, 198, 231, 237, 241, 290–292, 301, 303, 310, 340
    size, 25–27, 31–33, 36, 44, 50, 116, 141, 143, 144, 147, 149, 151, 163, 209, 215, 243, 290, 292, 296, 303, 323, 339, 350, 356, 382–384

Bolivia, 59–61, 69, 70, 75–78, 110, 111, 146, 150

Brain, 7, 10, 22, 25, 30, 34–39, 44, 46, 47, 50, 135, 136, 140, 141, 230, 290, 291, 302, 303, 323, 405–407, 410

Braincase, 10, 30, 35, 36, 39, 44, 46, 135, 136, 140, 141, 405

Brazil, 6, 12, 24, 27, 49, 58–61, 69–79, 108, 110, 111, 114–115, 122–124, 146, 150, 184, 193, 195, 261, 262, 346, 406, 407

C

Caatinga, 69

Call
    alarm, 357, 361–362, 372, 385, 388
    female, 349–351
    loud, 14, 49, 337–363, 369–392, 404–406, 410, 411
    male, 340, 349–351, 373, 380
    quite, 337–363
    soft, 339, 357, 362

Captivity, 109, 111, 120, 183, 195, 196, 207, 209, 215, 220, 305, 318, 390
Cellulose, 143, 230, 236, 237, 249
Central America, 4, 6, 7, 67, 68, 141, 142, 147, 340, 342, 344, 346–349, 353, 354, 362, 381, 390
Cerrado, 69, 122
Cholesterol, 187, 190–192, 195, 197, 242
Chromosomal
  rearrangement, 69, 76, 87, 98–100
  syntenies, 99
Chromosome, 8, 68–70, 86–91, 93–100, 117, 120, 407
Cladistics, 25, 42, 44, 48
Colonization, 123, 233, 235, 237, 275
Communication, 5, 11, 12, 123–124, 323–325, 329, 349, 354, 356, 361, 384, 388, 390, 405
Competition, 5, 9, 12, 141, 205, 211–215, 217, 219, 221, 301, 304, 329, 339, 356, 373, 377–382, 384, 386, 391, 405
Conservation, 6, 10, 13, 80, 98, 99, 181, 183, 205, 220, 221, 237, 250, 251, 260, 404, 408, 409, 412
Cooperation, 379–380, 385, 386
Copulation
  extragroup, 5, 213
Craniodental
  diversity, 21–50, 292, 293, 302–304, 306
  morphology, 26–31, 44–47
Cranio-mandibular, 10
Cranium, 10, 25, 48, 66, 72, 135–142, 303
Creatinine, 9, 187, 190–192, 197, 198, 408
Crest, 28, 30, 34–37, 39, 41, 143, 155, 405
Crown shape, 34, 41–43
Cytogenetic, 4, 8, 66, 71, 76, 78, 85–100, 120, 122, 407

D

Deception, 356, 382, 384–385
De-encephalization, 37, 39, 47, 50, 405, 406
Defense, 231, 339, 369, 376–379, 387, 391, 392, 411
Demography, 13, 122, 125, 260, 291, 404
Dentition, 7, 11, 22, 25, 47–50, 142, 302, 303, 404
Development
  craniodental, 292, 293, 302–304, 306
  social, 222, 298–299, 309
Dichromacy, 329
Dimorphism, 11, 33, 35, 141, 149, 151, 152, 195, 301, 303, 304, 309, 311, 340, 350, 352, 410

Disease
  ecology, 260
Dissociability, 11, 291, 292, 302, 303, 305, 308–310, 404, 409
DNA, 13, 68, 70, 71, 77, 78, 89, 98, 119, 232, 249

E

Ecology
  nutritional, 5, 404
Endocrine, 5, 9, 204–207, 211, 215, 217, 220–222, 392, 408
Endocrinology, 5, 13, 203–223, 405, 408
Energy, minimizing strategy, 169
Enterotype, 234
Epidemic, 265
Estrogen, 204, 206–211, 223

F

Fast-slow continuum, 11, 12, 290–292, 308, 310, 311
Fecundity, 279
Feeding, time, 231, 238, 294–298, 301, 304, 305, 308
Fermentative process, 231
Fermenter, 231
FISH analyses, 99
Flower, 182, 243, 293–298
Foliage, 197, 319, 320
Folivory, 25, 39, 46–48, 135, 136, 140, 143, 168, 169, 301, 405–407, 410
Food, 9, 11–13, 50, 169, 181, 196, 205, 207, 217–222, 230, 231, 236–238, 241, 243, 244, 247, 260, 274, 292–294, 296, 297, 304, 307–309, 311, 318, 319, 325, 327, 328, 330, 371, 375, 377–380, 391, 407, 410
Foraging, 5, 155, 243, 293, 294, 304–306, 319, 320, 329, 357
Foramen magnum, 6, 12, 30, 34, 36–39, 46, 136, 140, 405
Forelimb, 6, 13, 134, 152–160, 169, 405
Form, 10, 12, 36, 50, 66, 71, 72, 75, 96, 102, 135, 136, 151, 203, 215, 219, 221, 305, 340, 344, 347, 353, 363, 381–384, 392, 406
Fossil, 5, 7, 21–50, 56, 405–407, 411
French Guiana, 76, 77, 182, 184, 185, 187, 189, 191, 192, 196, 262
Frugivory, 33, 36, 47, 48, 405, 406, 410
Fruit, 5, 6, 9, 33, 40, 41, 142, 143, 182, 184, 219, 221, 222, 231, 236, 238, 239, 243, 249, 250, 260, 293–298, 301, 304, 319, 320, 377, 403, 405, 406, 411

Subject Index

G
Genetics
  admixture, 109
  variation, 74, 78, 120
Genotype, 113, 121, 233, 234, 236, 250
Geometric morphometric, 5, 10, 13, 39
Glucocorticoid, 9, 204, 206, 207, 211–215, 217–223, 237, 240
Glucose, 190, 191, 197, 212, 217, 218, 230, 408
Glycoside, 327
Grasping, 11, 134, 164–166, 169
Growth, 9, 11, 94, 136, 141, 207, 212, 232, 241, 242, 248, 291, 293, 300–304, 306, 307, 310, 311, 410
Guatemala, 59, 60, 66–68, 146
Guiana, 76, 77, 182, 184, 185, 187, 189, 191, 192, 196, 197, 262
Gut microbiome, 5, 9, 10, 13, 232–235, 237, 239–241, 243–245, 247–250, 405
Guyana, 7, 8, 76, 146, 355

H
Habitat
  disturbance, 122, 123, 198, 264, 267, 276–277, 409
  fragmentation, 11, 123, 205, 206, 217, 251
Haplotype, 67, 70–72, 119, 120
Health, 6, 9, 10, 180, 181, 183, 192, 197, 230–233, 236, 237, 239–241, 248, 250, 260, 265, 274, 275, 408, 409
Hematology, 4, 8, 179–198, 408
Hemoglobin, 9, 181, 187–189, 192, 193
Heterochromatin, 89–93, 100
Hindgut, 231, 249
Hind limb, 134, 152, 160–166, 169
Hormone, 11, 13, 204–209, 222, 235, 392, 408
Humidity, 265, 267, 272, 276, 277, 363
Hurricane, 9, 221
Hybridization, 8, 98, 99, 107–126, 390, 407, 408
Hybrid zone, 8, 109, 111–115, 117–126, 388, 389
Hyoid, 4–6, 10, 33, 35, 39, 49, 69, 71, 72, 74, 75, 77–79, 135, 136, 143–152, 168, 169, 321, 339, 350–354, 384, 405, 410
Hypocone, 26, 41, 44, 143

I
Incisor, 11, 26, 29, 34, 40–41, 45, 47, 142, 143, 303, 405, 406
Infant, development, 222, 289–311, 408
Infanticide, 323, 351, 373, 379, 387, 391

Infection, 10, 184, 192, 195, 197, 219, 259–, 261, 264–268, 272, 274–277, 279, 409
Interaction, 9, 10, 12, 124–125, 209, 212, 213, 215, 218, 229–251, 274, 279, 298, 299, 309, 327–330, 353, 358–360, 374, 375, 377, 379–381, 385, 390, 409
  host-microbe, 9
Interbirth interval, 11, 232, 290–292, 410
Isolation, 8, 31, 111, 120, 122, 126, 356, 357, 408

J
Juvenile, 10, 72, 184, 187, 189, 191–196, 204, 206, 208, 209, 215, 232, 241–247, 293, 299, 301, 303, 304, 310, 325, 328, 350, 358–362, 410

K
Karyosystematics, 86
Karyotype, 74, 78, 79, 87–89, 94, 98, 99

L
Larynx, 144, 352
Leaves, 5, 11, 12, 47, 49, 136, 142, 143, 169, 182, 230, 231, 237, 239, 243, 249, 260, 275, 293–298, 301, 302, 304, 305, 319–321, 363, 377, 387, 407, 411
Leukocyte, 181, 195
Life history, 6, 11, 231, 289–311, 410
Limb, 11, 134, 152–166, 169, 310
Locomotion, 155, 157, 163–169, 296, 297, 300, 306, 310, 327, 406
Los Tuxtlas Biosphere Reserve, 213, 219, 262
Lymphocyte, 181, 187–189, 193, 195

M
Malaria, 265, 268
Mandible, 26–29, 41, 46, 136, 142, 144, 352
Mate defense, 378, 391, 392, 411
Mating, 9, 117, 124, 125, 215, 221, 373, 378, 408
Mesoamerica, 8, 66–68, 79, 86, 97, 120, 169, 404, 407
Metabolism, 181, 192, 212, 230, 232, 241, 243
Mexico, 8, 56, 59, 66–68, 108–114, 116, 117, 122, 123, 180, 182–184, 192, 196, 207, 210, 213, 214, 218, 219, 222, 237, 239, 240, 242, 243, 260, 262, 263, 275, 301, 389, 390, 407
Microchromosome, 8, 93–95

Migration pattern, 5
Molar, 7, 11, 23–29, 31, 34, 40–47, 142–144, 246, 302, 303, 405, 406
Monophyly, 135
Morpho-functional adaptation, 134
Morphometric, 5, 10, 13, 39, 72, 74, 77, 116, 117
Multiple sex chromosome systems, 87, 94, 100

N
Nicaragua, 59, 67, 293, 295, 296
Niche partitioning, 134
Nuchal plane, 6, 30, 34, 36–37, 39, 135, 136, 405, 406
Nutrition(al), 5, 9, 10, 13, 180, 196, 197, 212, 230–234, 236, 237, 240–243, 247, 248, 250, 293, 306, 307, 330, 404, 409

O
Ontogeny, 4, 7, 9, 11–12, 136, 290–292, 305, 307–309, 404, 409–411
Ovarian cycles, 11, 205–207, 209–211, 215, 222
Ovulation, 204, 209–211, 307

P
Palenque National Park, 210, 214, 218, 222, 243
Pantanal, 69, 122
Paraguay, 58, 59, 69, 70
Parasitism, 259–279
Parasitology, 5
Pathogen, 230, 235, 237, 239, 260, 264–267, 274–276, 279, 409
Pelage coloration, 66, 67, 71–74, 77, 79, 112, 114–117
Perception, 325, 327, 328, 330, 374
Phylogeny, 25, 48, 96, 233, 249, 303
Physiology, 3–13, 215, 230, 232, 234–236, 403–412
Plant, 9, 10, 13, 109, 115, 182, 204, 207, 230, 234–239, 243, 244, 249, 250, 293, 294, 296, 304, 305, 327, 330, 408
Pneumatization, 6
Position, 37, 40, 46, 49, 70, 74, 79, 98, 138, 140, 143, 150, 153, 154, 161, 162, 168, 353, 357, 363, 405
Positional
 behavior, 5, 11, 36, 135, 306, 309, 310, 328, 339
 repertoire, 11, 169, 405
Postcranial anatomy, 10, 168
Postcranium, 31, 152–168

Postorbital constriction, 30, 34, 35, 405
Precipitation, 10, 265, 267–269, 271–274, 409
Predator, 6, 12, 219, 323, 329, 339, 362, 372, 411
Pressure
 environmental, 134
 selective, 12, 233–236, 330, 355, 371, 384, 391
Prisoner's dilemma, 375
Prognathism, 33, 35, 140
Protein, 9, 97, 190–192, 194, 196–198, 234, 241, 243, 244, 305, 307, 408
Puberty, 204, 205, 207–209, 301

Q
Quadrivalent, 8, 69, 94, 96, 97, 120
Quadrupedalism, 134, 155, 405, 407

R
Rainfall, 222, 261–263, 265, 267, 272, 274, 277, 294
Red blood cell (RBC), 181, 188, 189, 193, 195, 197
Reproduction, 5, 11, 186, 204, 205, 212, 232, 241, 242, 248, 290, 291, 307, 310, 311, 318, 325, 328, 373, 404, 410
Reproductive isolation, 8, 120, 122, 126, 356, 408
Resource, holding potential, 375, 382–384
r-/K-selection theory, 290
Roar(s), 12, 321, 340–350, 352–356, 359, 371, 372, 379, 381–386, 389–391

S
Selection, 6, 12, 33, 37, 40, 46, 47, 50, 121, 122, 290, 318, 320, 323, 325, 327, 328, 351, 356, 372, 373, 406
Sensory
 ecology, 3–13, 403–412
 system, 11, 317–330
Sex, heterogametic, 8, 117
Sexual dichromatism, 69, 75
Shearing crest, 143
Signal(s), 6, 99, 115–117, 120, 135, 322, 329, 354, 358, 359, 372–373, 381–385, 388, 406, 411
Skeleton, 22, 27–29, 31, 33, 152, 168, 169, 376, 405
Skull, 5, 6, 10, 28–32, 36, 37, 39, 46, 49, 66–68, 70, 72, 76, 135, 137, 140, 144, 145, 150, 168, 169, 321, 352, 406
 airorhynchous, 6

Smell, 12, 318, 323, 325, 329
Social dynamic, 124–125, 404
Socioendocrinology, 204, 205, 211–217
Sound
    production, 6, 11, 135, 136, 140, 168, 169, 410
    propagation, 354–352, 406
South America, 4, 6, 8, 12, 62, 67, 69–79, 97, 99, 100, 141, 147, 169, 182, 268, 275, 276, 340, 342, 344, 346–349, 351, 353, 362, 381, 404
Spermatogenesis, 207, 211
Steroid, 206, 207, 209, 211, 215, 221, 408
Stress, 5, 9, 13, 162, 163, 168, 180, 181, 195, 197, 205–207, 212, 213, 217–220, 234, 240, 261, 274, 318, 340, 355, 358–360, 362, 371, 405, 408
Subfossil, 26, 136
Sympatry, 111, 112, 114, 388, 389
System
    auditory, 322, 323
    immune, 212, 230, 235
    organ, 291
    vocal, 5
Systematics, 3, 4, 6, 13, 87, 109, 125, 143, 353, 363, 392, 404, 405, 412

T

Taste, 12, 325–327, 329, 330
Teeth, 24–29, 31, 33, 35, 36, 41, 44, 45, 47–49, 68, 135, 144, 168, 302, 303, 311, 406
Temperature, 180, 217, 218, 222, 265, 266, 277, 294, 327, 328, 355

Terra firme, 4
Testicular growth, 207
Tissue, 4, 10, 230, 233, 291, 352, 353
Touch, 14, 296, 318, 327–330
Translocation, 9, 87, 94, 95, 99, 181, 182, 197, 205, 207, 220, 221, 233, 275
Trichromacy, 325, 329, 411

V

Vector, 265
Vertebral column, 36, 152, 166–168
Vision
    dichromatic, 6
    trichromatic, 12, 319–321, 329, 404, 405
Vocal
    apparatus, 339, 351–353
    repertoire, 6, 12, 329, 349, 405
    system, 5
    tract, 145, 147, 352, 353, 382, 384
Vocalization, 12, 33, 49, 112, 114, 123, 124, 140, 144, 321–324, 329, 339, 340, 344, 346, 348–352, 355, 356, 360–363, 372, 373, 381, 384, 385, 388, 390, 410

W

Weaning, 11, 233, 290–294, 297, 302, 303, 306–308, 410
White blood cells (WBC), 9, 181, 186–189, 193, 195, 198

Z

Zygomatic arches, 39, 140, 141

Taxonomic Index

A

African elephant, 339
Alouatta
   A. aequatorialis, 57, 63, 79, 146
   A. amazonica, 61
   A. arctoidea, 4–7, 9, 56, 58, 61, 64, 69, 74–76, 78, 79, 91, 99, 141, 204, 205, 208, 209, 213, 214, 261, 342, 344–350, 354–356, 371–374, 377, 380, 384, 386, 387, 407, 408
   A. auratus, 77
   A. barbatus, 58
   A. belzebul, 5, 7–10, 56, 58, 59, 62, 64, 69–73, 78, 79, 85, 88, 91, 95, 96, 137, 141, 145–147, 149–151, 204, 205, 261, 267, 268, 274, 292, 303, 341–346, 348, 349, 354, 372, 407, 408, 411
   A. beniensis, 61
   A. bicolor, 59
   A. bogotensis, 60
   A. caquetensis, 60
   A. caraya, 4–11, 56–59, 63, 64, 66, 68–70, 73, 76, 87–90, 92–99, 108, 110, 111, 113–117, 120, 122–124, 134, 137, 141, 145–147, 149–151, 181, 183, 184, 192–198, 204, 205, 208, 209, 267, 268, 274, 291, 292, 294–305, 307–311, 319, 321, 325, 327, 341–348, 350, 351, 353, 356, 357, 360–361, 372–380, 382, 385–390, 389, 390, 407, 408, 260, 261
   A. caucensis, 59
   A. chrysurus, 58
   A. clamitans, 65
   A. coibensis, 65, 67
   A. discolor, 8, 56, 58, 61, 64, 71–73, 78, 79, 81, 110, 111, 407, 411
   A. fusca, 57, 65, 71, 292, 342
   A. guariba, 4, 5, 7, 8, 10, 58, 59, 61, 63, 65, 70, 71, 75, 79, 88, 92, 96–100, 108, 110, 111, 113–117, 120, 122–124, 137, 141, 147, 149–151, 261, 267, 268, 274, 303, 304, 341–348, 350–352, 351–357, 361, 362, 371–373, 376–379, 381–383, 389, 390, 407, 209, 260
   A. iheringi, 61
   A. inclamax, 60
   A. inconsonans, 60
   A. insulanus, 57, 62
   A. juara, 57, 60–62, 64
   A. juruana, 61
   A. laniger, 59
   A. luctuosa, 60
   A. macconnelli, 7, 8, 10, 56–59, 62, 64, 69, 70, 74–77, 79, 88, 91, 93, 94, 110, 111, 146, 180–187, 189, 191, 192, 195–197, 262, 267, 268, 342–344, 347–349, 355, 372, 377, 378, 381, 407, 411
   A. mauroi, 24, 27
   A. metagalpa, 59
   A. mexianae, 59
   A. mexicana, 59
   A. niger, 58
   A. nigerrima, 8, 56, 60, 61, 64, 66, 68, 71, 74, 75, 78–79, 88, 90. 93, 100, 110, 111, 137, 141, 142, 407, 411
   A. nigra, 59

Alouatta (cont.)
  A. puruensis, 61, 64
  A. quichua, 60
  A. rubicunda, 59
  A. rufimanus, 58
  A. sara, 9, 10, 12, 56, 57, 60, 61, 63, 64, 69, 70, 74, 76–79, 91, 93, 94, 96, 97, 99, 100, 110, 111, 263, 267, 268, 407
  A. stramineus, 58, 88, 94
  A. tapajozensis, 61
  A. trabeata, 60
  A. ululata, 8, 56, 57, 60, 62, 64, 72–73, 79, 407, 411
  A. ursina, 57, 63
  A. villosus, 57, 63
  palliata
    A. p. aequatorialis, 8, 56, 59, 60, 63, 64, 66, 67, 108, 146
    A. p. coibensis, 8, 56, 59, 63, 65–67, 100, 407
    A. p. mexicana, 8, 56, 59, 63, 64, 66, 67, 79, 196, 407
    A. p. palliata, 8, 56, 59, 63, 64, 66, 67, 146, 407
    A. p. trabeata, 8, 56, 60, 63, 65–67, 407
  pigra
    A. p. luctuosa, 8, 56, 63, 65, 66, 68, 407
    A. p. pigra, 8, 56, 63, 66, 67, 407
  seniculus
    A. s. juara, 8, 56, 62, 64, 73–75, 407
    A. s. puruensis, 8, 56, 62, 64, 73–75, 110, 111, 209, 407
    A. s. seniculus, 8, 56, 62–64, 73, 74, 78, 146, 407
Alouattinae, 22, 24, 26, 47, 56
Alouattine, 7, 21–50, 169, 404–407
Alouattini, 56
Amazonian black howler, 78
Amoebae parasite, 268, 271, 274, 276
Anaerobe, 232, 233
Anopheles, 265
Antillothrix bernensis, 30
Aotus azarae, 94
Ape, 248, 249, 353
Ascaris lumbricoides, 265
Ateles
  A. fusciceps, 325
  A. geoffroyi, 11, 28, 29, 94, 99, 217, 301, 321, 323, 325, 330
  A. paniscus, 94
Atelinae, 26, 56, 388, 404

Ateline, 4, 5, 7, 10–12, 23, 24, 26, 29–31, 34–43, 45–49, 135, 136, 139, 140, 142, 143, 152, 159–162, 164–169, 231, 232, 291, 292, 300–302, 360, 308–311, 329, 404–406, 409

B
Baboon, 22, 33, 75, 263, 301
Bacteria, 230, 232–237, 239, 240, 242, 245, 248–250, 264
Bacteroides, 234, 242, 245, 250
Bacteroidete, 234, 242, 244, 245, 248, 250
Bellbird, 339
Bifidobacterium, 235, 242, 245
Black and gold howler, 4, 7, 69, 193, 194, 205, 209–211, 215, 267, 294–300, 327
Black-and-white colobus, 275
Black-hawk eagle, 372
Bolivian red howler, 7, 77
Brachyteles hypoxanthus, 212
Brown howler, 7, 57, 70, 71, 267

C
Cacajao, 45, 94, 100
Caipora, 7, 26, 28–39, 41, 43, 56
Callicebus moloch, 350
Callimico, 94, 100
Callithrix jacchus, 196, 325, 330
Canis lupus, 339
Capybara, 372
Caracara cheriway, 372
Caribbean primate, 24, 50
Cartelles, 56
Catarrhine, 26, 31, 100, 160, 248, 319, 321, 322, 325, 404
Cebine, 4, 291
Cebuella pygmaea, 388
Cebus
  C. a. paraguayanus, 94
  C. capucinus, 195, 305, 310
  C. libidinosus, 94, 97, 100
Central American black howler, 7, 68
Cercocebus galeritus galeritus, 264
Cercopithecine, 109, 116, 231, 301
Cercopithecus
  C. ascanius, 249, 264, 275
  C. neglectus, 249
Cervus elaphus, 373
Cestode, 10, 268–270, 272, 273, 276

Taxonomic Index

Chacma baboon, 212, 217, 266
Chimpanzee, 212, 248, 265, 383
Clostridia, 239, 244, 249
Clostridium, 236, 248
Colobine, 231, 249
Colobus guereza, 249, 275
Colombian red howler, 73, 78
Crested eagle, 372

D
Dog, 36

E
Eira barbara, 372
Endolimax nana, 274
Entamoeba
    E. coli, 242, 274
    E. histolytica, 274
Enterobacteriaceae, 242, 245
Enterobius vermicularis, 265
Erythrocyte, 181, 211
Eubacterium oxidoreducens, 249
Eulemur fulvus, 305

F
Ficus, 239, 247, 275, 377
Fig, 59
Firmicute(s), 234, 242, 244, 245, 248

G
Giardia, 10, 268, 271–274, 276
Gibbon(s), 265, 345
Golden howler, 76
Gorilla
    G. beringei, 212, 249
    G. gorilla, 249
Gorilla, mountain, 212
Greater Antillean primate, 30
Guapinol tree, 377
Guianan red howler, 76, 182
Guyanan red howler, 7

H
Helminth, 195, 265, 266, 268–270, 272, 277–279
Homo sapiens, 373
Hookworm, 265
Hydrochoerus hydrochaeris, 372

Hylobates
    H. lar, 390
    H. muelleri, 390
Hylobatid, 350
Hymenaea courbaril, 377

I
Indriid, 231, 249, 301

J
Jaguar, 339
Juruá red howler monkey, 74

K
Koala, 136, 373

L
Lagothrix
    lagotricha
        L. l. lagotricha, 146
        L. l. lugens, 146
Lemur catta, 212
Lemurid, 301
Leopardus pardalis, 372
Lepilemur, 231
Lion, 383
Long-tailed macaque, 212
Loxodonta africana, 339

M
Macaca
    M. fascicularis, 212
    M. fuscata, 221
    M. mulatta, 97, 234, 307, 326
    M. nemestrina, 326
    M. sylvanus, 388
Mandrill, 212
Mandrillus sphinx, 212
Mangabey, 217, 264
Mantled howler, 7, 67, 188, 190, 205, 207–209, 213, 218–221, 248, 267, 277, 293–294, 305, 320, 321, 324–329
Maranhão red-and-black howler, 72
Microbe, 5, 9, 230–234, 236, 237, 242, 248, 250, 251
Microflora, 235
Morphnus guianensis, 372

N

Necator americanus, 265
Nematode, 10, 268, 269, 272, 276
Neotropical primate, 6, 7, 80, 97, 109, 181, 192, 195, 357, 404
New World monkey (NWM), 23, 28, 31, 33, 36, 37, 39, 41, 43, 48, 87, 116, 196, 209, 231, 301, 318, 322, 404
Nomascus concolor, 388
Northern muriqui, 212
Nycticebus pygmaeus, 248

O

Ocelot, 372
Oesophagostomum, 264
Old World monkey, 12, 116, 144, 231, 329, 338, 404
Orangutan, 265

P

Pan
  P. paniscus, 221, 249
  troglodytes
    P. t. schweinfurthii, 248
Panthera
  P. leo, 339, 383
  P. onca, 339
Papio cynocephalus, 305
Paraguayan howler, 69
Paralouatta
  P. marianae, 24, 27, 29, 49
  P. varonai, 24, 25, 27–30, 49
Parasite, protozoan, 266, 268, 274, 275
Phascolarctos cinereus, 373
Pinworm, 265
Pitheciid, 24, 33
Pitheciidae, 56
Plasmodium, 10, 192, 197, 265, 268, 271, 273, 274, 276
Platyrrhine, 4, 5, 10, 12, 23, 25, 26, 28, 32–37, 39–41, 43–47, 49, 56, 87, 89, 94, 96, 99, 100, 135, 136, 155, 157, 160, 161, 164, 166, 168, 209, 303, 319, 329, 330, 371, 403–405, 407, 411
Platyrrhini, 89, 94, 99
Plica sublingualis, 325, 326
Presbytis cristata, 100

Prevotella, 234, 239
Procnias, 339
Propithecus
  P. edwardsi, 264
  P. verreauxi, 212
Prosimian(s), 231, 326, 404
Proteobacteria, 244, 248, 250
Protopithecus brasiliensis, 24, 27, 28, 49
Pygmy loris, 248, 250

R

Red-and-black howler, 71
Red colobus monkey, 217, 249, 277
Red deer, 307, 373, 383
Red-handed howler, 7, 71, 205, 217, 267
Red howler, 4, 7, 70, 73–78, 87, 182, 189, 191, 197, 205, 207, 219, 220, 251, 267
Red langur, 265
Redtail guenons, 264, 275
Ring-tailed lemur, 212, 217
Roundworm, 265
Ruminococcus, 236, 237, 239, 248
  R. flavefaciens, 249

S

Saguinus mystax, 212
Saimiri
  S. boliviensis, 94
  S. sciureus, 305, 325, 330, 390
Sapajus nigritus, 212
Sifaka, 212, 264
Solimoea acrensis, 24, 27, 31, 42, 44
Spix's red-handed howler monkey, 72
Spizaetus tyrannus, 372
Stirtonia
  S. tatacoensis, 23, 27–29, 41, 49
  S. victoriae, 23, 27, 49

T

Tamarin, moustached, 212
Tayra, 372
Titi monkey, 349–350
Trematode, 10, 268, 269, 272–274, 276
Trichuris trichiura, 265
Trypanoxyuris, 268–271, 276
Trypanoxyuris, 268–271, 276
Tufted capuchin monkey, 212, 221

Taxonomic Index

U
Ursine red howler, 75

V
Venezuelan red howler, 75
Verreaux's sifaka, 212

W
Whipworm, 265
Wolves, 339

X
Xenothrix mcgregori, 30